Biofuels and Biorefineries

Volume 10

Aims and Scope of the series

The Biofuels and Biorefineries Series aims at being a comprehensive and integrated reference for biomass, bioenergy, biofuels, and bioproducts. The Series provides leading global research advances and critical evaluations of methods for converting biomass into biofuels and chemicals. Scientific and engineering challenges in biomass production and conversion are covered that show technological advances and approaches for creating new bio-economies in a format that is suitable for both industrialists and environmental policy decision-makers

The Biofuels and Biorefineries Series provides readers with clear and concisely-written chapters that are peer-reviewed on significant topics in biomass production, biofuels, bio-products, chemicals, catalysts, energy policy, economics, thermochemical and processing technologies. The text covers major fields of plant science, green chemistry, economics and economy, biotechnology, microbiology, chemical engineering, mechanical engineering and energy.

Series description

Annual global biomass production is about 220 billion dry tons or 4,500 EJ, equivalent to 8.3 times the world's energy consumption in 2014 (543 EJ). On the other hand, world-proven oil reserves at the end of 2011 reached 1652.6 billion barrels, which can only meet just over 50 years of global production. Therefore, alternative resources are needed to both supplement and replace fossil oils as the raw material for transportation fuels, chemicals and materials in petroleum-based industries. Renewable biomass is a likely candidate, because it is prevalent over the Earth and is readily converted to other products. Compared with coal, some of the advantages of biomass are: (i) its carbon-neutral and sustainable nature when properly managed; (ii) its reactivity in biological conversion processes; (iii) its potential to produce bio-oil (ca. yields of 75%) by fast pyrolysis because of its high oxygen content; (iv) its low sulphur and lack of undesirable contaminants (e.g. metals, nitrogen content) (v) its wide geographical distribution and (vi) its potential for creating jobs and industries in energy crop productions and conversion plants. Many researchers, governments, research institutions and industries are developing projects for converting biomass including forest woody and herbaceous biomass into chemicals, biofuels and materials and the race is on for creating new "biorefinery" processes needed for future economies. The development of biorefineries will create remarkable opportunities for the forestry sector, biotechnology, materials, chemical processing industry, and stimulate advances in agriculture. It will help to create a sustainable society and industries that use renewable and carbon-neutral resources.

More information about this series at http://www.springer.com/series/11687

Zhen Fang • Richard L. Smith Jr • Lujiang Xu
Editors

Production of Biofuels and Chemicals with Pyrolysis

Springer

Editors
Zhen Fang
Biomass Group, College of Engineering
Nanjing Agricultural University
Nanjing, Jiangsu, China

Richard L. Smith Jr
Graduate School of Environmental Studies
Tohoku University
Aoba-ku, Sendai, Japan

Lujiang Xu
Biomass Group, College of Engineering
Nanjing Agricultural University
Nanjing, Jiangsu, China

ISSN 2214-1537 ISSN 2214-1545 (electronic)
Biofuels and Biorefineries
ISBN 978-981-15-2731-9 ISBN 978-981-15-2732-6 (eBook)
https://doi.org/10.1007/978-981-15-2732-6

This Springer imprint is published by the registered company Springer Nature Singapore Pte Ltd.
The registered company address is: 152 Beach Road, #21-01/04 Gateway East, Singapore 189721, Singapore

Preface

Lignocellulosic biomass consists of biopolymers (cellulose, hemicellulose and lignin) that form a matrix with structural similarities but with uniqueness among its many forms. As one of the most abundant renewable resources, lignocellulosic biomass can be transformed into materials, chemicals and energy with sustainable chemistry and engineering. The substitution of traditional fossil resources by the three major biopolymers as sustainable feedstocks is being extensively investigated for the manufacture of high value-added products including biofuels, commodity chemicals, bio-based functional materials, and heterogeneous catalysts that can be directly applied to promoting the development of sustainable manufacturing processes. Aimed at improving the awareness of effective conversion protocols and for developing innovative biomass conversion processes, this text was conceived as a collection of studies on state-of-the-art techniques and know-how for producing biofuels and chemicals from biomass by pyrolysis. Discussion on related topics in terms of recent advances and their assessment and the promise and prospects of new methods or new technological strategies are provided to readers in a concise and informative format. Each individual chapter was contributed by globally-selected experts or professionals and was peer-reviewed and edited for content and consistency in terminology.

This book is the tenth book of the series entitled, "Biofuels and Biorefineries", and contains 13 chapters contributed by leading experts in the field. The text is arranged into five key areas:

- **Part I. Fundamentals of pyrolysis** (Chapters 1-4)
- **Part II. Production of liquid biofuels by pyrolysis and catalytic pyrolysis** (Chapters 5 and 6)
- **Part III. Production of Liquid biofuels with microwave pyrolysis** (Chapters 7 and 8)
- **Part IV. Production of bio-chemicals by pyrolysis** (Chapters 9-11)
- **Part V. Design of pyrolysis units and models** (Chapters 12 and 13)

Chapter 1 introduces thermo-chemical conversion methods that produce biochar, bio-oil and bio-gases via slow pyrolysis, torrefaction, intermediate pyrolysis, fast pyrolysis, flash pyrolysis, and catalytic pyrolysis. Pyrolysis chemistry of biomass, especially for biomass-related compounds such as cellulose, hemicellulose and lignin is summarized. **Chapter 2** introduces catalytic effects of ash on pyrolysis products, secondary or successive gas phase reactions of pyrolysis products and covers kinetic models that allow study of optimal conditions for bio-oil production, mathematical modeling of the thermochemical processes and coupled transport and kinetic processes on the scale of both the particle and the reactor. **Chapter 3** focuses on experimental and theoretical studies regarding free radical and concerted reactions of lignin model compounds for the production of phenolic and aromatic compounds, and provides the essence of fast pyrolysis chemistry of lignin and its model compounds, and associated reaction kinetics. **Chapter 4** focuses on individual reaction mechanisms for cellulose, hemicellulose, and lignin, and their interaction as well as the effect of inorganic species, based on the formation mechanism of bio-oil, char and gaseous products. **Chapter 5** introduces upgrading of bio-oils by use of several typically highly-active solid catalysts with metal modification, and gives a state-of-the-art overview of the effects of pore size and acidic-basic properties of the solid catalysts on their activity, selectivity, stability and deactivation. **Chapter 6** presents a comprehensive review of the development of biomass catalytic fast pyrolysis (CFP) and bio-oil upgrading routes via choice of catalysts, feedstocks, reaction methods and reactors. **Chapter 7** discusses mechanisms of microwave-assisted pyrolysis (MAP) of lignocellulosic biomass and developments in MAP in use of microwave absorbers, modeling and simulation of MAP of biomass, reaction kinetics and mass transfer, and challenges in the scaling up of MAP from laboratory to industrial scales. **Chapter 8** introduces MAP of waste biomass together with an exhaustive coverage of interactions of microwave with materials that shows the main mechanisms involved. **Chapter 9** shows approaches for integrating pyrolysis and microbial processes and summarizes opportunities and challenges involving microbial conversion of pyrolysis products. **Chapter 10** presents a comprehensive strategy for lignocellulose saccharification and investigates pretreatment of components and structures of lignocellulose and the influence of alterations on levoglucosan production. **Chapter 11** deals with production of renewable phenols, especially phenol, from lignocellulosic biomass pyrolysis, and discusses types of lignocellulosic biomass used in pyrolysis processes, and the effect of reaction conditions on phenol production along with applications for phenolic-rich bio-oils. **Chapter 12** analyzes syngas production through pyrolysis and gasification, its compression and its use in gas turbines, and discusses important points related to syngas ignition, syngas explosion limits at high-temperatures and high-pressures and syngas combustion kinetics. **Chapter 13** introduces three main biomass conversion models denoted as the molecular model, the single particle model and the reactor model and provides applications and how they can be used in analysis from a practical point of view.

The text should be of interest to professionals in academia and industry who are working in the fields of natural renewable materials, biorefinery of lignocellulose, biofuels and environmental engineering. It can also be used as comprehensive references for university students with backgrounds of chemical engineering, material science and environmental engineering.

Nanjing, China — Zhen Fang
Sendai, Japan — Richard L. Smith Jr
Nanjing, China — Lujiang Xu

Acknowledgements

First and foremost, we would like to cordially thank all the contributing authors for their great efforts in writing and revising the chapters and insuring the reliability of the information given in their chapters. Their contributions have really made this project realizable.

Apart from the efforts of authors, we would also like to acknowledge the individuals listed below for carefully reading the book chapters and giving many constructive comments that significantly improved the quality of the chapters:

- Prof. Abhijeet P. Borole, University of Tennessee, USA;
- Dr. Stephen Dooley, Trinity College Dublin, the University of Dublin, Ireland;
- Dr. LiLu Tian Funkenbusch, University of Florida, USA;
- Dr. Lucía García, Universidad de Zaragoza, Spain;
- Prof. Kevin van Geem, Gent University, Belgium;
- Dr. Paola Giudicianni, Istituto di Ricerche sulla Combustione IRC-CNR, Italy;
- Mr. Tong Han, KTH Royal Institute of Technology, Sweden;
- Prof. Hero Jan Heeres, University of Groningen, the Netherlands;
- Dr. Carlos Herce, ENEA - Italian National Agency for New Technologies, Energy and Sustainable Economic Development, Italy;
- Prof. Anker Degn Jensen, Technical University of Denmark, Denmark;
- Dr. Liqun Jiang, Guangzhou Institute of Energy Conversion, Chinese Academy of Sciences, China;
- Dr. Konstantinos G. Kalogiannis, Chemical Process & Energy Resources Institute (CPERI), Greece;
- Prof. Song-Charng Kong, Iowa State University, USA;
- Dr. Shogo Kumagai, Tohoku University, Japan;
- Prof. Mª Ángeles Martín Lara, Universidad de Granada, Spain;
- Dr. Heather Mayes, University of Michigan, Ann Arbor, USA;
- Prof. Bernhard Peters, Université du Luxembourg, Luxembourg;
- Dr. Armando T. Quitain, Kumamoto University, Japan;
- Prof. Eliseo Ranzi, Politecnico di Milano, Italy;
- Dr. M Soledad Callén Romero, Instituto de Carboquímica (ICB-CSIC), Spain;

- Prof. Roger Ruan, University of Minnesota, USA;
- Dr. M. Pilar Ruiz, University of Twente, The Netherlands;
- Dr. Agustin Valera-Medina, Cardiff University, United Kingdom;
- Dr. Francesco Vizza, Istituto di Chimica dei Composti OrganoMetallici (ICCOM), Consiglio Nazionale delle Ricerche (CNR), Italy;
  Prof. Vinicyus Wiggers, Universidade de Blumenau, Brazil;
- Dr. Hsi-Wu Wong, University of Massachusetts, USA;
- Prof. Shu Zhang, Nanjing Forestry University, China;
- Dr. Xuesong Zhang, University of Illinois, Urbana-Champaign, USA;
- Dr. Yifeng Zhu, Pacific Northwest National Laboratory, USA

We are also grateful to Dr. Peng Zhang (senior editor), Ms. Becky Zhao (senior editor) and Ms. Abbey Huang (editorial assistant) for their encouragement, assistance and guidance during preparation of the book.

Finally, we would like to express our deepest gratitude towards our families for their love, understanding and encouragement, which help us in completion of this project.

Contents

Editors and Contributors

About the Editors

Zhen Fang is a Professor and Leader of the Biomass Group at Nanjing Agricultural University. He is the inventor of the "fast hydrolysis" process, and was listed in the "Most Cited Chinese Researchers" in energy for 2014-2019 (Elsevier-Scopus). Professor Fang specializes in thermal/biochemical conversion of biomass, nanocatalyst synthesis and their applications, pretreatment of biomass for biorefineries, and supercritical fluid processes. He holds Ph.D.s from China Agricultural University and McGill University. Professor Fang is an associate editor of *Biotechnology for Biofuels* and the *Journal of Supercritical Fluids*. He has more than 20 years of research experience in the field of renewable energy and green technologies at top universities and institutes around the globe, including 1 year in Spain (University of Zaragoza), 3 years in Japan (Biomass Technology Research Center, AIST; Tohoku University), and more than 8 years in Canada (McGill). He has worked for 7 years as engineer in energy, bioresource utilization and engine design in industry before moving to academia.

Richard L. Smith Jr is a Professor of Chemical Engineering at the Graduate School of Environmental Studies, Research Center of Supercritical Fluid Technology, Tohoku University, Japan. He has a strong background in physical properties and separations and holds a Ph.D. in Chemical Engineering from the Georgia Institute of Technology (USA). His research focuses on developing green chemical processes, especially those that use water and carbon dioxide as the solvents in their supercritical state. He is an expert on physical property measurements and separation techniques with ionic liquids, and has published more than 270 scientific papers, patents and reports in the field of chemical engineering. Professor Smith is the Asia regional editor for the *Journal of Supercritical Fluids* and has served on the editorial boards of major international journals.

Lujiang Xu is a Lecturer in the Biomass Group at Nanjing Agricultural University, College of Engineering. He holds a Ph.D. in Renewable and Clean Energy from the Department of Chemistry of University of Science and Technology of China (USTC) and a B.S. in Light Chemical Engineering from Nanjing Forestry University. His research focuses on the selective thermo-chemical conversion of biomass and derived compounds into value-added chemicals and liquid fuels.

Contributors

Abuliti Abudula Graduate School of Science and Technology, Hirosaki University, Aomori, Japan

Andrés Anca-Couce Graz University of Technology, Graz, Austria

Pietro Bartocci Department of Engineering, University of Perugia, Perugia, Italy

Gianni Bidini Department of Engineering, University of Perugia, Perugia, Italy

Nichaboon Chaihad Graduate School of Science and Technology, Hirosaki University, Aomori, Japan

He Chao School of Chemical and Biomedical Engineering, Nanyang Technological University, Singapore, Singapore

Wei Chen State Key Laboratory of Coal Combustion, School of Power and Energy Engineering, Huazhong University of Science and Technology, Wuhan, China

Yingquan Chen State Key Laboratory of Coal Combustion, School of Power and Energy Engineering, Huazhong University of Science and Technology, Wuhan, China

Hanping Chen State Key Laboratory of Coal Combustion, School of Power and Energy Engineering, Huazhong University of Science and Technology, Wuhan, China

P. Debiagi Simulation of Reactive Thermo-Fluid Systems, TU Darmstadt, Darmstadt, Germany

Tharaka Rama Krishna C. Doddapaneni Chair of Biosystems Engineering, Institute of Technology, Estonian University of Life Sciences, Tartu, Estonia

Zhen Fang Biomass Group, College of Engineering, Nanjing Agricultural University, Nanjing, Jiangsu, China

Francesco Fantozzi Department of Engineering, University of Perugia, Perugia, Italy

T. Faravelli Department of Chemistry, Materials and Chemical Engineering "Giulio Natta", Politecnico di Milano, Milan, Italy

Guoqing Guan Graduate School of Science and Technology, Hirosaki University, Aomori, Japan
Laborary of Energy Conversion Engineering, Institute of Regional Innovation (IRI), Hirosaki University, Aomori, Japan

Eid Gul China-EU Institute for Clean and Renewable Energy, Huazhong University of Science and Technology, Wuhan, Hubei, China

C. Hasse Simulation of Reactive Thermo-Fluid Systems, TU Darmstadt, Darmstadt, Germany

Liqun Jiang CAS Key Laboratory of Renewable Energy, Guangdong Provincial Key Laboratory of New and Renewable Energy Research and Development, Guangzhou Institute of Energy Conversion, Chinese Academy of Sciences, Guangzhou, China

Surachai Karnjanakom Department of Chemistry, Faculty of Science, Rangsit University, Pathumthani, Thailand

Mohsin Ali Khan University of Haripur Pakistan, Haripur, KPK, Pakistan

Timo Kikas Chair of Biosystems Engineering, Institute of Technology, Estonian University of Life Sciences, Tartu, Estonia

Joo-Sik Kim Graduate School of Energy and Environmental System Engineering, University of Seoul, Seoul, Republic of Korea

Suwadee Kongparakul Department of Chemistry, Faculty of Science and Technology, Thammasat University, Pathumthani, Thailand

Jiby Kurian Department of Bioresource Engineering, McGill University, St. Anne-de-Bellevue, QC, Canada

Hai-bin Li CAS Key Laboratory of Renewable Energy, Guangdong Provincial Key Laboratory of New and Renewable Energy Research and Development, Guangzhou Institute of Energy Conversion, Chinese Academy of Sciences, Guangzhou, China

Przemyslaw Maziarka Department Green Chemistry and Technology, Ghent University, Ghent, Belgium

Upendra Natarajan Department of Chemical Engineering, Indian Institute of Technology Madras, Chennai, India

Ki-Bum Park Graduate School of Energy and Environmental System Engineering, University of Seoul, Seoul, Republic of Korea

G. S. Vijaya Raghavan Department of Bioresource Engineering, McGill University, St. Anne-de-Bellevue, QC, Canada

E. Ranzi Department of Chemistry, Materials and Chemical Engineering "Giulio Natta", Politecnico di Milano, Milan, Italy

Frederik Ronsse Department Green Chemistry and Technology, Ghent University, Ghent, Belgium

Chanatip Samart Department of Chemistry, Faculty of Science and Technology, Thammasat University, Pathumthani, Thailand

Oyvind Skreiberg SINTEF Energy Research, Trondheim, Norway

Katarzyna Slopiecka Department of Engineering, University of Perugia, Perugia, Italy

Richard L. Smith Jr Graduate School of Environmental Studies, Tohoku University, Aoba-ku, Sendai, Japan

R. Vinu Department of Chemical Engineering, Indian Institute of Technology Madras, Chennai, India
National Centre for Combustion Research and Development, Indian Institute of Technology Madras, Chennai, India

Xiao-bo Wang CAS Key Laboratory of Renewable Energy, Guangdong Provincial Key Laboratory of New and Renewable Energy Research and Development, Guangzhou Institute of Energy Conversion, Chinese Academy of Sciences, Guangzhou, China

Liang Wang SINTEF Energy Research, Trondheim, Norway

Kai Wu Key Laboratory of Energy Thermal Conversion and Control of Ministry of Education, School of Energy and Environment, Southeast University, Nanjing, China

Lujiang Xu Biomass Group, College of Engineering, Nanjing Agricultural University, Nanjing, Jiangsu, China

Minjiao Yang China-EU Institute for Clean and Renewable Energy, Huazhong University of Science and Technology, Wuhan, Hubei, China

Haiping Yang China-EU Institute for Clean and Renewable Energy, Huazhong University of Science and Technology, Wuhan, Hubei, China
State Key Laboratory of Coal Combustion, Huazhong University of Science and Technology, Wuhan, Hubei, China

Qing Yang China-EU Institute for Clean and Renewable Energy, Huazhong University of Science and Technology, Wuhan, Hubei, China

Haiping Yang State Key Laboratory of Coal Combustion, School of Power and Energy Engineering, Huazhong University of Science and Technology, Wuhan, China

Attada Yerrayya Department of Chemical Engineering, Indian Institute of Technology Madras, Chennai, India

Huan Zhang Biomass Group, College of Engineering, Nanjing Agricultural University, Nanjing, Jiangsu, China

Bo Zhang Key Laboratory of Energy Thermal Conversion and Control of Ministry of Education, School of Energy and Environment, Southeast University, Nanjing, China

Jing Zhang Key Laboratory of Energy Thermal Conversion and Control of Ministry of Education, School of Energy and Environment, Southeast University, Nanjing, China

Huiyan Zhang Key Laboratory of Energy Thermal Conversion and Control of Ministry of Education, School of Energy and Environment, Southeast University, Nanjing, China

Zeng-li Zhao CAS Key Laboratory of Renewable Energy, Guangdong Provincial Key Laboratory of New and Renewable Energy Research and Development, Guangzhou Institute of Energy Conversion, Chinese Academy of Sciences, Guangzhou, China

Haibo Zhao State Key Laboratory of Coal Combustion, Huazhong University of Science and Technology, Wuhan, Hubei, China

Siying Zhong Key Laboratory of Energy Thermal Conversion and Control of Ministry of Education, School of Energy and Environment, Southeast University, Nanjing, China

Hewen Zhou China-EU Institute for Clean and Renewable Energy, Huazhong University of Science and Technology, Wuhan, Hubei, China

Part I
Fundamentals of Pyrolysis

Chapter 1
Introduction to Pyrolysis as a Thermo-Chemical Conversion Technology

Lujiang Xu, Liqun Jiang, Huan Zhang, Zhen Fang, and Richard L. Smith Jr

Abstract Pyrolysis is the thermal decomposition of materials under an inert atmosphere to produce biofuels or chemicals. This chapter introduces the following thermo-chemical conversion methods that produce biochar, bio-oil and bio-gases: slow pyrolysis, torrefaction, intermediate pyrolysis, fast pyrolysis, flash pyrolysis, and catalytic pyrolysis. Pyrolysis chemistry of biomass, especially for biomass-related compounds such as cellulose, hemicellulose and lignin is summarized. Pyrolysis is compared with other common technologies to define its scope as a method for biomass conversion.

Keywords Pyrolysis · Classification · Products · Chemistry · Parameters

1.1 Introduction

Energy is an important part of social development and provides the basis for health, welfare and economic security of a country or region [1, 2]. Present energy consumption (2018) in units of tonne of oil equivalent (toe) is 13,978 Mtoe that is derived from roughly 32% oil, 22% gas, 27% coal, and 10% biomass [3]. The Energy consumption for 2030 is forecast to increase at a rate of 2% per year based on 2015–2018 trends, which is driven by changing lifestyles of the present population. The United Nations Sustainable Development Goals (UN SDGs) were initiated to

L. Xu · H. Zhang · Z. Fang (✉)
Biomass Group, College of Engineering, Nanjing Agricultural University, Nanjing, Jiangsu, China
e-mail: zhenfang@njau.edu.cn

L. Jiang
CAS Key Laboratory of Renewable Energy, Guangdong Provincial Key Laboratory of New and Renewable Energy Research and Development, Guangzhou Institute of Energy Conversion, Chinese Academy of Sciences, Guangzhou, China

R. L. Smith Jr
Graduate School of Environmental Studies, Tohoku University, Aoba-ku, Sendai, Japan

Z. Fang et al. (eds.), *Production of Biofuels and Chemicals with Pyrolysis*, Biofuels and Biorefineries 10, https://doi.org/10.1007/978-981-15-2732-6_1

promote efficient use of resources for human and world development including the environment. As the world's population increases from its present value of 7.7 billion (10^9) people to 8.6 billion people in 2030, it can be expected that many environmental issues will emerge related to energy consumption. For example, the world carbon dioxide emissions reached 37.1×10^9 tonnes (37.1 Gt) in 2018 and are rising at a rate of more than 2% per year [4]. The consumption of oil, gas and coal produces not only a large number of greenhouse gases, but also many environmental pollutants such as nitrogen oxides and sulfur oxides, which greatly affects the environment and the health of living beings [5]. Biomass, on the other hand, if converted to energy and chemicals with efficient and sustainable methods, offers a resource that can be used beneficially for human welfare.

Biomass (wood, grass, agricultural waste, palm oil processing residues, lipids and animal wastes) is potentially a carbon neutral feedstock [6]. Biomass is a mixture of organic compounds and usually contains minor amounts of inorganic compounds, and includes carbon, oxygen, hydrogen elements, but also can contain nitrogen, sulphur, or chlorine elements depending on its source and its contamination. Presently, biomass is the third most important resource used to generate heat and electricity in the world. Biomass is used for the production of renewable bio-fuels via appropriate technologies, such as physical, bio-chemical and thermo-chemical processes as shown in Fig. 1.1 [7, 8]. Physical processes (Fig. 1.1) use crushing, heat or pressure for converting biomass into solid fuels [9]. Bio-chemical processes (Fig. 1.1) use enzymes and micro-organisms for converting biomass into desired chemical products such as ethanol or biogas [10]. Thermo-chemical processes (Fig. 1.1) use heat for converting biomass into energy products [11] or combustion in air to directly produce heat or gas [12], whereas gasification of biomass uses the presence of air and steam to produce syngas and fuel gas [13]. Pyrolysis, on the other hand, is a thermo-chemical method that uses heat in an inert atmosphere for converting biomass into gaseous, liquid or solid products [14].

Compared with physical methods, thermo-chemical methods have realized a large number of practical bio-products [15], whereas compared with bio-chemical methods, the thermo-chemical methods tend to be more efficient (short conversion time) and more robust (variable feedstocks, insensitive to presence of impurities). Among the thermo-chemical conversion methods (Fig. 1.1), pyrolysis is one of the most simple and economic methods to convert biomass into liquid fuels, and has therefore attracted a great deal of research over the past two decades [16]. Techno-economic analysis of three conversion methods (pyrolysis, gasification, and bio-chemical) for near-term biomass-to-liquid fuels technology scenarios shows that pyrolysis has the lowest capital and operating costs [17]. Thus, the method has achieved commercialization in many aspects of biomass processing.

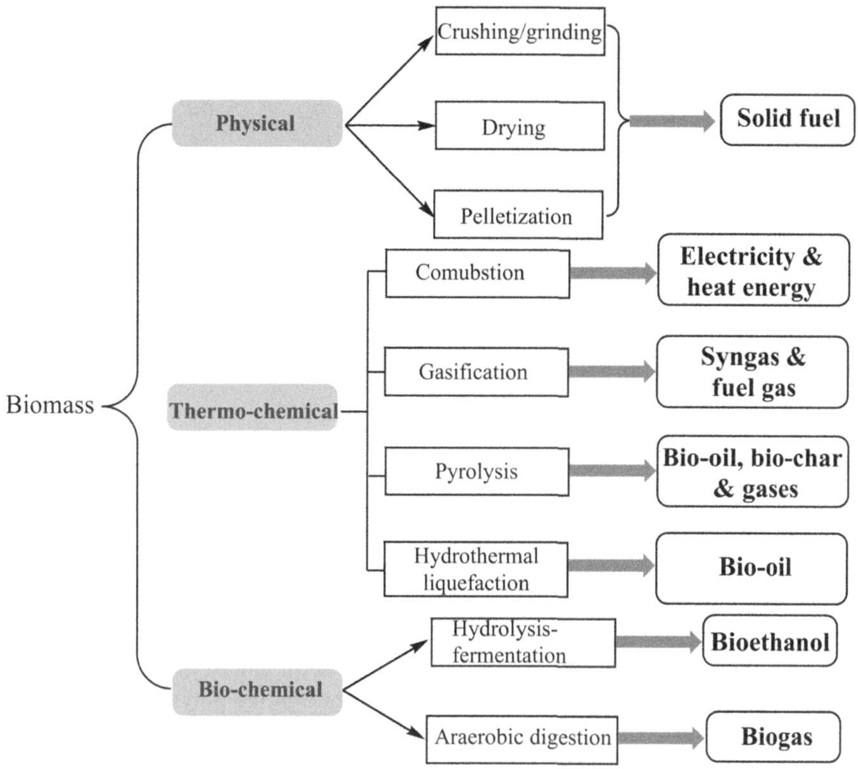

Fig. 1.1 Primary methods and pathways for biomass conversion into energy and chemical products

1.2 Composition of Biomass

Biomass is a diverse resource derived from plant or animal materials such as wood, bark, agricultural waste, lipids, algae, microalgae, animal residues or municipal solid wastes [18]. Table 1.1 summarizes the main components of several major biomass resources [19–23]. Lignocellulosic biomass is mainly composed of cellulose, hemicellulose and lignin, while the main component in chlorella biomass is extractives. For lignocellulosic biomass, the content of cellulose, hemicellulose and lignin is in the range of 35–55%, 15–31%, and 10–30%, respectively. Besides cellulose, hemicellulose and lignin, a certain amount of ash and extractives exist in biomass. Certain amounts of cellulose, hemicellulose and lignin are found in the animal residues of manure.

Table 1.1 Component analyses (%) of selected types of biomass [19–23]

Biomass	Cellulose	Hemicellulose	Lignin	Ash	Extractives[a]
Bamboo	39.8	19.5	20.8	6.7	1.2
Birch wood	53.1	20.4	17.2	0.4	8.9
Cattle manure	25	22	13	–	–
Chlorella	–	–	–	–	67.2
Miscanthus	34.4	25.4	22.8	1.2	6.8
Pine bark	24.9	31.1	29.6	10.7	4.6
Pine wood	44.5	22.9	27.7	0.3	5.1
Spruce wood	45.6	20.0	28.2	0.3	5.9
Rice straw	37	16.5	13.6	13.1	19.8

[a]Extractives in chlorella are mainly composed of proteins and lipids

1.2.1 Lignocellulose

Lignocellulose (Fig. 1.2) is composed of C, H and O elements with the total content of these three elements being more than 95%. Cellulose is the major component of the plant cell walls of lignocellulosic biomass and is composed of D-glucose through polymerization of β(1·4) glycoside bonds that form of a linear polymer [24]. Hemicellulose is typically the second most abundant component of lignocellulosic biomass (Table 1.1), and is present along with cellulose in almost all plant cell walls. Hemicellulose is composed of sugar monomers through the polymerization linked by glycoside bonds, however, it is composed of C5 and C6 sugar monomers including glucose, galactose, mannose, xylose, and arabinose [25]. Lignin is a complex three-dimensional polymer of propyl-phenol groups bound together by C-O (β-O-4, α-O-4, 4-O-5 linkage) and C-C (β-5, 5-5, β-1, β–β linkage) bonds [26]. Lignin forms a "ligno-cellulosic" structure to provide a natural shield against rapid microbial or fungal destruction of cellulosic fibers. The three basic phenol-containing components of lignin are *p*-coumaryl/*p*-hydroxylphenyl, coniferyl/guaiacyl, and sinapyl/syringyl alcohol units [27].

1.2.2 Microalgae

Microalgae is a photosynthetic microorganism, and its three major chemical components are lipids, proteins and carbohydrates (Fig. 1.2b), making it very different from lignocellulosic biomass. The primary elemental constituents of microalgae are C, H, O and N [28]. The lipid content of microalgae is around 7~23%, thus it is considered to a promising energy or food source [29]. The content of proteins in microalgae is around 6–52% and they are mainly composed of amino acids, which are the source of nitrogen in microalgae [30]. The content of carbohydrates in microalgae is much less than that of lignocellulosic biomass. The carbohydrates in microalgae are homopolymers consisting of D-glucopyranose units linked via

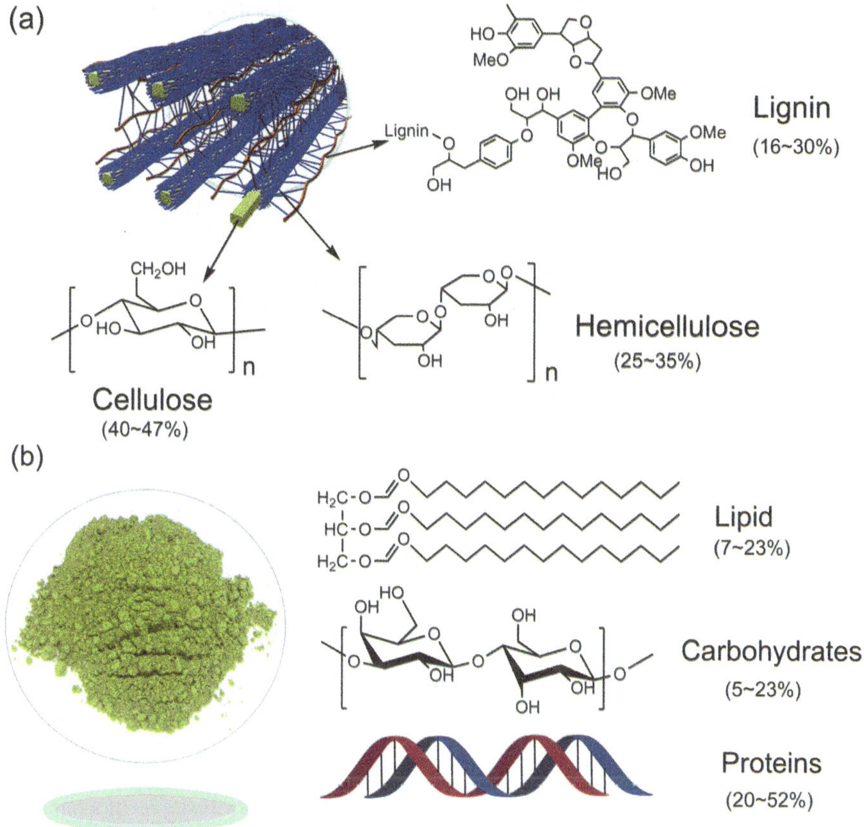

Fig. 1.2 Composition of lignocellulose (**a**) and microalgae (**b**)

β-glycosidic bonds or α-glycosidic bonds, and they are the main components of the cell walls [31]. Microalgae contain more functional groups than lignocellulosic biomass giving it high potential for producing many different kinds of value-added chemicals.

1.2.3 Lipids

Lipids, namely triglycerides, are usually defined as organic components of biomass rather than water-soluble components, and they are the main constituents of vegetable oils, microalgae and animal fats [32]. Lipids are composed of a glycerol molecule attached to three fatty acid molecules (Fig. 1.2b). The length of the carbon chains and number of the double bonds in the fatty acids vary depending on the source. Presently, lipids can be readily converted into liquid bio-fuels compared with lignocellulosic biomass because of their properties and their relatively uniform chemical structure that contains the glycerol backbone [33].

1.3 Classification of Pyrolysis Technology

Since the production of charcoal thousands of years ago, biomass pyrolysis has been used to manufacture many chemical products, such as levoglucosan, levoglucosenone, furfural, pyrrole, phenols and aromatics. Pyrolysis yields tar (mixtures of aromatic liquid fuel components), acetic acid, various gaseous species, and products that are of interest in recent years such as levoglucosan, furan, furfural, guaiacols, and 5-hydroxymethylfurfural. Pyrolysis operating conditions (temperature, heating rate, reaction time, catalyst, carrier gas) affect the yield and composition of the obtained products so that names are given to each process as shown in Table 1.2: slow pyrolysis, intermediate pyrolysis, fast pyrolysis (catalytic fast pyrolysis), flash pyrolysis, torrefaction and gasification based on the different parameters [34]. Different pyrolysis processes are described below to help distinguish between the conditions and application.

1.3.1 Slow Pyrolysis

Slow pyrolysis (Table 1.2) uses low heating rate, has longer reaction time and low reaction temperature [35]. The main product of slow pyrolysis is biochar and biogases, which results from the longer reaction time and secondary reactions of

Table 1.2 Overview of pyrolysis process name and key parameters

	Feedstock characteristics		Pyrolysis parameters			
Type	Feed mass scale	Moisture	Reaction temperature (°C)	Pressure (bar)	Heating rate (°C/s)	Reaction time/min
Slow pyrolysis	M-L	low	<500	1	<1	10–2000
Intermediate pyrolysis	S-L	low	400–500	1	1–1000	1–10
Fast pyrolysis	S	low	450–650	1	~1000	0.5–5
Flash pyrolysis	S	low	500–850	0.1–1	~1000	<0.5
Torrefaction	S-L	low	<250	1	–	60–2000
Gasification	S-L	low-high	700–1450	1–100	–	–

S small, *M*: moderate, *L* large

pyrolytic products during the pyrolysis process to give yield of biochar and biogases, and low yields of pyrolytic bio-oil [14]. Brown et al. found the higher yield of biochar than the yield of bio-oil could be obtained during the slow pyrolysis of corn stover process, the biochar yield could up to around 40 wt% at round 500 °C [36]. Slow pyrolysis produces lower bio-oil yields and higher biochar and biogases yields than other pyrolysis processes.

1.3.2 Intermediate Pyrolysis

Intermediate pyrolysis of biomass (Table 1.2) is carried out at process conditions between slow pyrolysis and fast pyrolysis [37]. Compared with slow pyrolysis, the intermediate pyrolysis process uses a faster heating rate and shorter reaction time [38]. Bio-oil obtained from intermediate pyrolysis is stable and can be used directly as a fuel in engines and boilers. The intermediate pyrolysis process is suitable for converting agricultural waste, woody materials, grass, and sewage sludge into bio-oils and biochar [39]. Ahmed et al. studied conversion of *Acacia cincinnata* and *Acacia holosericea* species into bio-oil and biochar through the intermediate pyrolysis process (the pyrolysis temperature at 500 °C and at heating rates at 25 °C/min, and the highest yield of bio-oil reached 53% from the trunk of *A. cincinnata* species [40].

1.3.3 Fast Pyrolysis

1.3.3.1 Direct Fast Pyrolysis

Fast pyrolysis proccss has thc following charactcristics: (1) high heating rate and high heat transfer rate (10–100) °C/s; (2) pyrolytic temperature strictly (450–600) °C; (3) short reaction time (0.5–5) s; and (4) rapid cooling of pyrolytic vapors that form water-soluble/-insoluble components in the bio-oil [41]. The short reaction time of the fast pyrolysis process prevents secondary reactions and reduces the biochar formation, thus leading to high pyrolytic oil yields.

1.3.3.2 Catalytic Fast Pyrolysis (CFP)

Although high yields of bio-oil can be produced by fast pyrolysis, bio-oil has the disadvantages of having relatively high oxygen content, low calorific value, complex composition, high corrosiveness and low stability, and it is difficult to be used directly as a fuel [42]. To solve the above issues, researchers have proposed to improve bio-oil quality by introducing a catalyst during the fast pyrolysis process, and the process is known as catalytic fast pyrolysis (CFP). The advantages of CFP

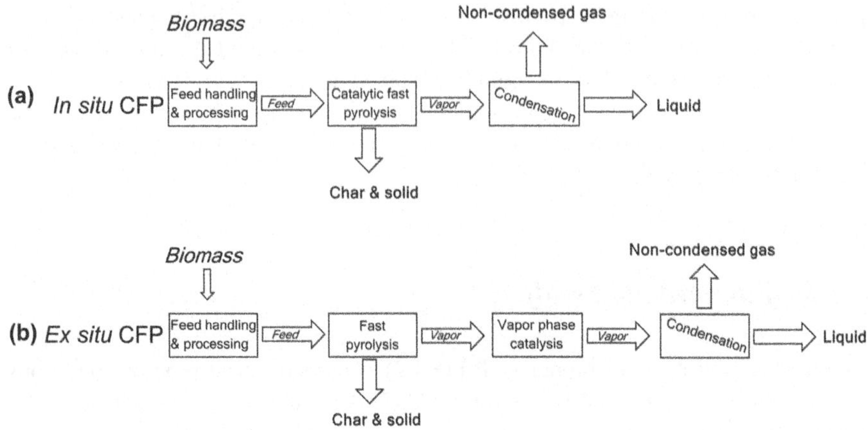

Fig. 1.3 The schematic of in situ CFP (**a**) and ex situ CFP (**b**) (Reproduced with permission from [44]. Copyright © 2013 Royal Society of Chemistry)

process are the simplification of the thermo-chemical conversion process, since it avoids condensation and re-evaporation of the bio-oil [43]. Currently, CFP process is one of the most attractive processes to remove the oxygen and improve the bio-oil quality. CFP process configurations can be divided into two different types (in situ CFP and ex situ CFP) according to the use of catalysts internal to the pyrolyzer or external to the pyrolyzer [44]. Figure 1.3 shows block flow diagrams of the in situ CFP and ex situ CFP processes. The process that the catalyst is packed/fed together with the feedstock in the pyrolysis reactor is referred to as in situ catalytic fast pyrolysis (in situ CFP, Fig. 1.3a). In the in situ CFP process, catalysts intimately mix with biomass and intervene in the pyrolysis and cracking reactions, which enhances the decomposition of large fragments and reduces the secondary char formation. In the ex situ CFP process, catalysts are placed in a reactor separated from the pyrolyzer and only contacted with the pyrolysis vapor is referred to ex situ catalytic fast pyrolysis (ex situ CFP, Fig. 1.3b). During the ex situ CFP process, the catalyst is separated from the feed. The deoxygenating and upgrading operation can flexibly run under the environment independent of pyrolysis, and thus enable the optimum catalyst performance to obtain the desired products under the optimized conditions.

1.3.4 Flash Pyrolysis

Flash pyrolysis process is characterized by rapid heating rates (>1000 °C/s) and high reaction temperatures (500–850) °C and affords high yields of bio-oil and low water content with conversion efficiencies being high as 70% (based on mass) [45]. The reaction time of flash pyrolysis process is typically less than 0.5 s, which is shorter than that of fast pyrolysis. To obtain such high heating and heat transfer rates, the particle size of the biomass feedstock must be as small as possible [46].

1.3.5 Torrefaction

Although torrefaction is not strictly pyrolysis, it may be regarded as a low temperature variant of pyrolysis, which is typically carried out at (200–300) °C under atmospheric pressure with low particle heating rates (<50 °C/min) and in the absence of oxygen [47]. The torrefaction of biomass is generally used as a pre-treatment method to ensure that biomass materials are roasting as this makes biomass raw materials become less tough and more brittle, so that they can be fed into a pyrolysis reactor [48]. During torrefaction, biomass partially decomposes and releases CO_2 and CO, but retains most of its hydrogen to improve its pyrolytic characteristics (*e.g.* density and calorific value) [49]. Therefore, torrefaction is a promising thermal pretreatment technology that improves the properties of biomass and allows a bio-oil of high quality to be obtained by pyrolysis.

1.3.6 Gasification

Gasification has been well studied more than 100 years and is a high temperature variant of pyrolysis of biomass with catalyst under a partial oxidizing atmosphere, in which the reaction temperature is usually (700–1450) °C [50]. The partial oxidation gasifying agents are mixtures of air, H_2O, N_2 or Ar/He [51]. Gasification is divided into four steps: drying, pyrolysis, oxidation, and reduction [52]. During gasification, thermal degradation of biomass at high temperatures takes place and results in intermediates (bio-oil) and final products (syngas). The components of syngas are CO, CO_2, H_2, and CH_4, which can be used for the production of energy, chemicals, and bio-fuels. Because of the advantages of the obtained product being easily accessible and requiring less post-recovery work, the gasification of biomass is practiced on commercial scale [53]. Currently, the global capacity of biomass gasification installation is more than 2600 MW [54].

1.4 Pyrolytic Products

Three different products, solid products (biochar), liquid products (bio-oil) and gas products (bio-gases) can be produced from biomass pyrolysis process. All products have commercial merit, whereas the chemical composition and the properties of the products is regulated by changing pyrolysis conditions, reactor types and feedstock components and quality. When fine particles of feedstock are pyrolyzed under

Table 1.3 Pyrolysis mechanisms and products of cellulose pyrolysis at different temperatures

Temperature (°C)	Pyrolytic mechanism	Pyrolytic products
<350	• Free radical formation • Water elimination • Depolymerization	• Carbonyl and carboxyl containing species • CO and CO_2 • Charred residue
350–450	• Split of glycosidic connections • Substitution	• Levoglucosan • Anhydrides • Oligosaccharides as a tar segment
450–500	• Dehydration • Rearrangement • Fission of sugar units	• Carbonyl compounds
500–700	• Combination of above	• Combination of above
>700	• Combination of above • Thermal cracking	• Combination of above products • Mainly gas products

conditions of high heating rates, short reaction time and medium pyrolytic temperatures, high yields of bio-oil can be produced. In contrast, when large particles are pyrolyzed under conditions of low temperature, low heating rate and long reaction time, the main pyrolytic product is biochar. Yields of gaseous can be maximized by changing pyrolytic conditions to high temperatures, low heating rates and long reaction time [55].

Table 1.3 shows pyrolysis mechanisms and the main products for cellulose pyrolysis at different temperatures. Herein, the chemical and physical properties of three main pyrolytic products are introduced [56].

1.4.1 Biochar

Biochar is the solid residue remaining at the end of the pyrolysis process, and it is a stable carbon-rich solid. Pyrolytic conditions that use low pyrolytic temperatures, low heating rates and long reaction times promote formation of biochar. The chemical and physical properties of biochar vary with pyrolysis conditions [57]. Low pyrolytic temperature produce biochars with more functional groups, and with higher oxygen content than that obtained at higher temperatures. However, biochar of high pyrolytic temperature is more stable than that produced at low temperatures. The content of ash of biochar prepared at high temperatures and with long reaction times is higher. In addition, if the biochar is prepared under an atmosphere of CO_2 or steam, the specific surface area of obtained biochar will be much larger than that obtained under inert atmospheres (N_2, Ar) [58].

Biochar is made by varying ratios of a highly carbonaceous material (sometimes referred to as charcoal) and ash, which comprises various inorganic residues with the carbon content being usually more than 50% [59]. Therefore, biochar is a rich stable carbon, that has a microporous structure, and functional groups of C=C, –OH, CH_2- and C=O, and rich N P, K, Ca, Mg [60]. Biochar has been used in industrial applications that include: (1) as alternative solid fuel in industrial boilers [61]; (2) as raw material for production of activated carbon [62]; (3) as reagent for sustainable production of syngas via the further gasification processes [63]; (4) as a soil improver or as the basis for fertilizer [64]; (5) as a carbon raw material for synthesis of carbon nanotubes/supercapacitor [65]; and (6) as a renewable catalyst/catalyst support for catalytic reactions [66].

1.4.2 Gases

Gases are another product of biomass pyrolysis process. In gasification processes, light gases are the main products. The yields, components and contents of the gas products are linked to the pyrolytic conditions and type, including pyrolytic temperature, heating rate, reaction time, feedstock size, and feedstock type [52]. In the slow pyrolysis process, the yield of gases is similar to char and reaches around 10–35%, while a higher yield of gases can be obtained with gasification process, in which the yield of gases can reach around 80%. He et al. investigated gas production from biomass with a bench-scale downstream fixed bed reactor using flash pyrolysis of municipal solid wastes, and the gases yield reached 79% [67]. The gas yield is highly affected by pyrolytic temperature, and in general it is possible to increase gas yields by increasing pyrolytic temperature. With the increasing pyrolytic temperature, heating and mass transfer during the inner biomass become fast, and thermal degradation and devolatilization of biomass increase, too. Simultaneously, tar is found during the pyrolysis process, which easy undergoes a series of secondary reactions, such as decarboxylation, dehydrogenation, deoxygenation, and cracking, to form gas products [55]. Therefore, higher pyrolytic temperature promotes tar decomposition and thermal cracking of tar that increases the proportion of gas products.

The main components of the gas products are hydrogen (H_2), methane (CH_4), carbon oxide (CO), carbon dioxide (CO_2), ethane (C_2H_6), ethane (C_2H_4), propane (C_3H_8), propylene (C_3H_6), and butane (C_4H_{10}). In addition, the gas products may also contain small amounts of larger carbon-containing molecules, such as hexane (C_6H_{14}) or benzene (C_6H_6). The gas products from biomass pyrolysis have many applications that include use as: (1) an alternative fuel to generate electricity or heat [68]; (2) a gas fuel for internal combustion engines in vehicles [69]; or (3) raw materials for synthesis of gasoline and diesel via Fischer-Tropsch synthesis [70].

1.4.3 Bio-Oil

Liquid product formed in biomass pyrolysis is referred to as pyrolytic oil, and it is also called bio-oil. Other names for bio-oil are crude bio-oil, pyrolytic tar, pyrolytic liquor, wood liquid, wood oil, smoke condensate, or distillate [71]. Typically, the collected bio-oil is dark-colored and free-flowing, but it is a viscous fluid. Bio-oil is the main product of most biomass pyrolysis processes. As with biochar and gases, the yields and quality of bio-oil are affected by the pyrolytic conditions and feedstock types. High heating rates, short reaction times and medium pyrolytic temperatures prefer production of bio-oil. Bio-oil has many applications that include use as: (1) a transport fuel after the further upgrading process [72]; (2) a source for producing chemicals via the subsequent separation and purification [73]; or (3) an alternative fuel for turbines and electric power generation engines [74].

Bio-oil is a complex mixture of water, oxygenated compounds, hydrocarbons, and lignin-derived oligomers [75]. Therefore, the quality of bio-oil obtained by direct pyrolysis is typically very poor and the bio-oil is difficult be used directly. To use the bio-oil, understanding of its chemical composition and physical properties is necessary. Details of bio-oil properties, namely water content, composition, oxygen content, corrosiveness and stability are introduced as next.

Water Content The water content of bio-oil is very high, and is around 15–30 wt% [76]. Water in the bio-oil is both directly from moisture present in the biomass feed and also that generated from dehydration reactions of pyrolytic intermediates during the pyrolysis process. High content of water of the bio-oil gives it a low heating value and flame temperature, and bigger ignition delay, and a lower combustion rate.

Oxygen Content Generally, the oxygen content in bio-oil can also reach 35–40% [77]. Oxygen mainly exists in the water of bio-oil and in the oxygenated compounds of the bio-oil. High oxygen content gives a bio-oil a low calorific value and low H/C ratio.

Composition The composition of the bio-oil is mainly determined from the characteristics of the starting biomass raw material. The composition of a typical bio-oil is very complex and contains more than 400 organic compounds (Table 1.4) [78]. The complex composition of bio-oil makes it corrosive and unstable.

Corrosiveness Based on component analysis of bio-oil, many kinds of acidic organic compounds are present in bio-oil that generally has a pH in the range of 2.5–3.0, and so that it is incompatible with common metal materials such as carbon steel and aluminum [79].

Stability The high acidity (low pH about 2.5) and complex chemical composition of bio-oil causes its stability to be low. Large amounts of aldehydes, ketones and phenols in the bio-oil easily undergoes reaction, such as dehydration or polymerization to generate macromolecular compounds that leads to deterioration and aging of the bio-oil and affects its direct use [80].

Table 1.4 Categories of organic compounds in bio-oil via direct pyrolysis of pine wood at 500 °C

Category	Typical organic compounds
Acids	Formic acid, acetic acid, propionic acid
Alcohols	Methanol, ethanol, ethylene glycol, 2-propene-1-ol
Aldehydes and ketones	Formaldehyde, acetaldehyde, 2-butenal, pentanal, glyoxal, acetone, 2-butanone, 2-pentanone, 2-cyclopentanone
Esters	Methyl formate, methyl propionate, butyrolactone
Furans	Furan, 2-methyl furan, 2-furanone, furfural, 5-hydroxylmethyllfurfural
Hydrocarbons	2-methyl propene, dimethylcyclopentene, alpha-pinene, benzene, toluene, xylenes, naphthalenes
Oxygenates	Hydroxyacetaldehyde, hydroxyacetone, dimethyl acetal, acetal
Nitrogen compounds	Pyrrole, methylamine, pyridine, methylpyridine
Oligomers	Lignin dimers, trimers
Phenols	Phenol, methyl phenols, guaiacol, 4-methyl guaiacol, ethyl guaiacol, eugenol, methyl syringol, 4-ethyl syringol, propyl syringol
Sugars	Levoglucosan, glucose, fructose, d-xylose, d-arabinose

Due to the above characteristics, bio-oils cannot be used directly but require further refining and upgrading.

1.5 Biomass Pyrolysis Chemistry

Pyrolytic products of biomass tend to be complex so that mechanistic studies on biomass pyrolysis are important to achieve production of value-added chemicals and high quality bio-oil [81]. Currently, studies on biomass pyrolysis mechanisms focus on two aspects: (1) reaction kinetics in the biomass pyrolysis process; (2) chemical reactions in the biomass pyrolysis process, especially formation mechanisms [22, 82–84]. Biomass pyrolysis mechanisms are readily studied by thermogravimetric differential thermogravimetry (TG-DTG), Fourier transform infrared spectroscopy (FT-IR), pyrolysis-chromatography-mass spectrometry (Py-GCMS) and quantum chemical theoretical calculation [9, 22, 85–87]. Three possible biomass pyrolysis mechanisms have been proposed: free radical mechanisms, concerted mechanisms, and ionic mechanisms [88–90]. Herein, the pyrolysis chemistry of lignocellulosic biomass components (cellulose, hemicelluloses and lignin) is briefly introduced.

1.5.1 Pyrolysis Chemistry of Cellulose

Many studies on the pyrolysis of cellulose and its mechanism have been reported [22]. During the pyrolysis process, cellulose first forms a liquid and then decomposes to pyrolytic products via two pathways [91]. The first pathway in cellulose pyrolysis is that it directly undergoes reactions to form small molecular products (acetaldehyde, hydroxyl acetone, furans, and anhydrosugars); the second pathway is that cellulose pyrolyzes to form oligomers and then further pyrolyzed to form the liquid pyrolysis products. Figure 1.4 shows the pyrolysis chemistry of cellulose [92–94]. Cellulose is depolymerized into oligosaccharides, and then the glucosidic bonds of the oligosaccharides are broken to produce D-glucopyranose, which undergoes an intramolecular rearrangement to form anhydrosugars (*e.g.* levoglucosan). Levoglucosan can be converted to form levoglucosone (LGO) via further dehydration reactions. The pyran ring-based glucose can also break the C-O bond to form a glucose chain compound, and then undergo isomerization reaction to form intermediates having furan structures to form 5-hydroxylmethylfurfural (5-HMF). The obtained 5-HMF can undergo further cracking reactions to form furfural (FF). Furfural can also be obtained via five-carbon intermediates formed by glucose fracture. Besides dehydration compounds and furans, some small molecule compounds (*e.g.* formaldehyde, acetaldehyde, hydroxyl acetone) can also be produced during the pyrolysis process. The small molecule compounds can be produced via the three pathways: (1) is directly from glucose via the cracking reaction; (2) via glucose that undergoes cracking to form formaldehyde and five-carbon intermediate; and (3) five-carbon intermediates undergo cracking reactions to from glycolic aldehyde and glyceraldehydes.

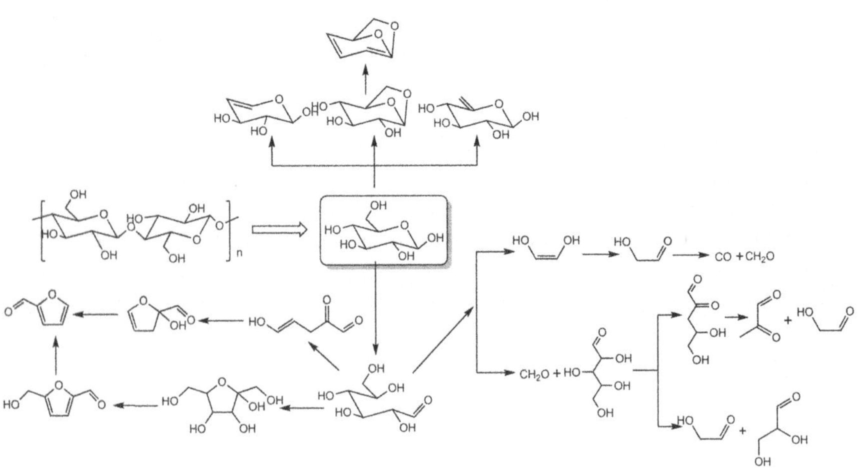

Fig. 1.4 Pathways of cellulose pyrolysis

1.5.2 Pyrolysis Chemistry of Hemicellulose

Actual hemicelluloses are difficult to separate from biomass completely so that xylan is often used as a model compound to investigate pyrolysis chemistry of hemicelluloses [95]. Because the structure of hemicelluloses is similar to that of cellulose, the pyrolysis chemistry of hemicelluloses is also similar to that of cellulose. The chemistry of hemicellulose pyrolysis is mainly through a free radical mechanism with some hydrosugars, furans and small molecule compounds being formed during hemicellulose pyrolysis [96]. Figure 1.5 shows the pathways of xylan pyrolysis.

1.5.3 Pyrolysis Chemistry of Lignin

The structure of lignin is more complex and disordered than that of cellulose and hemicelluloses, the pyrolysis chemistry of lignin is more complex than those of cellulose and hemicelluloses [26, 27]. The lignin pyrolysis process occurs over a very broad temperature range and consists of three stages from low pyrolysis temperature to high pyrolysis temperature: drying, fast degradation, and slow degradation stages [97]. Lignin pyrolysis starts with the breaking of weaker bonds (*e.g.* hydrogen binding and C–OH binding) at low temperature and proceeds through cleavage of stronger bonds (*e.g.* β-O-4, 5-5) with increasing temperature [98]. The phenolic compounds (*e.g.* phenol type, guaiacol type, syringol type, and catechol type) are the main products of the lignin pyrolysis. Similar to cellulose and hemicellulose, lignin pyrolysis is mainly through free radical chemistry [99]. Figure 1.6 shows the main lignin pyrolysis reaction pathways. In the pyrolysis oflignin, free radicals are generated in the breaking of the β-O-4 linkage of the lignin molecules,

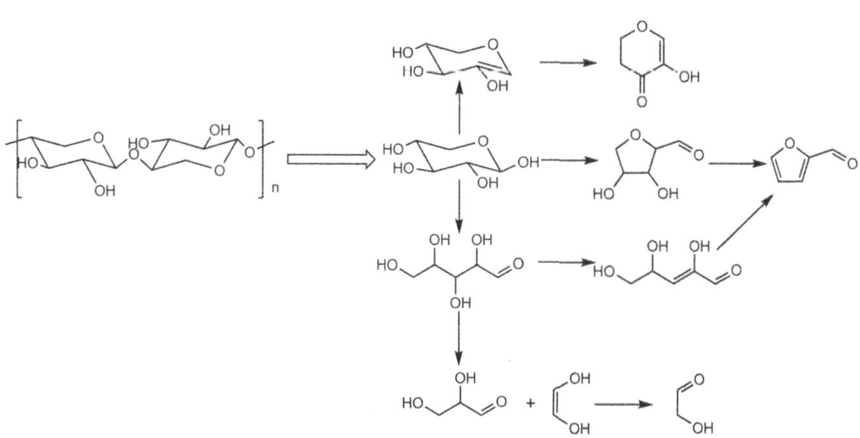

Fig. 1.5 Pathways of xylan pyrolysis

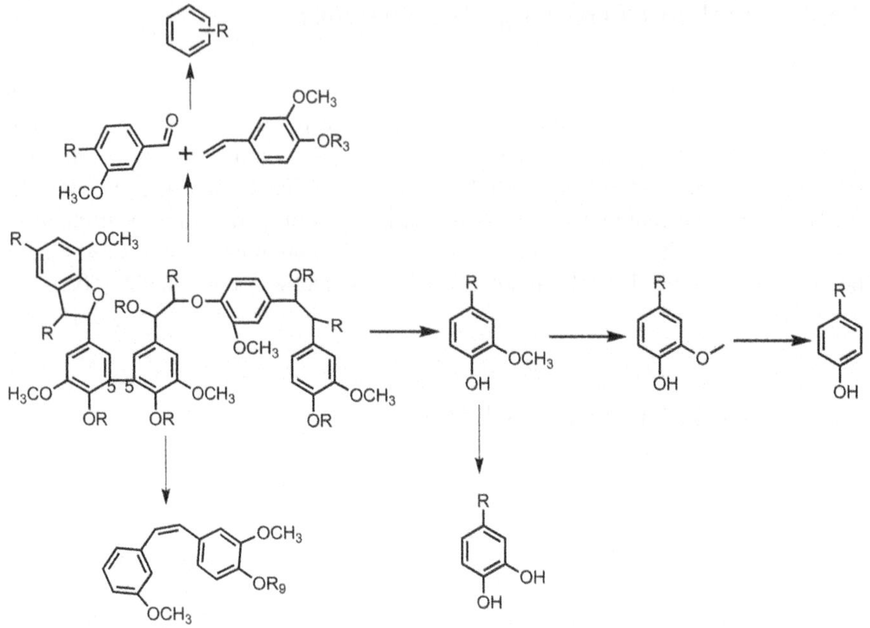

Fig. 1.6 Pathway of lignin pyrolysis

which is the initial step for the reaction of free radical chain. More pyrolytic biochar is produced in lignin pyrolysis than that when cellulose and hemicelluloses is used.

1.6 Pyrolysis Process Parameters

The biomass conversion, yield and quality of the products are affected by the feedstock type and operating conditions. Herein, some important parameters (*e.g.* feedstock type, feedstock particle size, heating rate and pyrolytic temperature, reaction time and reactor type) that affect biomass pyrolysis are discussed.

1.6.1 Feedstock Type

Lignocellulosic biomass is composed of cellulose, hemicellulose, lignin, extractives and ash; microalgae are mainly composed of proteins, lipids, carbohydrates and ash. The different components ratios of biomass will affect the composition and quality of the pyrolytic products [100]. During pyrolysis, the contribution of biomass components to product quality and yield can be summarized as follows. Cellulose and hemicellulose in the lignocellulosic biomass mainly contribute to produce the bio-oil and gas products. Lignin mainly contributes to form the biochar and gases [101]. Proteins, lipids and carbohydrates in the microalgae are prone to produce bio-oil and gases [102]. Sugars, furans, acids and ketones in the bio-oil are mainly derived from the cellulose and hemicellulose; the phenols and oligomers in the bio-oil are derived from the lignin; long chain hydrocarbons and long-chain fatty acids are derived from the lipids; and nitrogen compounds are derived from proteins. Extractives in biomass have similar contributions as cellulose and hemicelluloses and are prone to produce bio-oil and gases via simple volatilization or decomposition reactions [103]. Ash in the biomass usually remains in the biochar after pyrolysis, and has an effective catalytic effect on the formation of biochar and gas products, and has a negative effect on the yield of bio-oil [104].

1.6.2 Heating Rate and Pyrolytic Temperature

In addition to direct pyrolysis to form pyrolysis vapors, the secondary reactions readily occur during biomass pyrolysis that can reduce bio-oil yield and have a negative impact on bio-oil quality [105]. To obtain maximum bio-oil yield, fast heating rates and cooling of primary vapors are required [41]. Slow pyrolysis is prone to produce biochar. Fast pyrolysis and flash pyrolysis can produce more bio-oil rather than biochar. Safdari et al. found the average tar yield was 58 wt% for fast pyrolysis compared with 49 wt% for slow pyrolysis [106].

Pyrolytic temperature also has a significant effect on pyrolytic product yields. Hemicellulose usually decomposes from 200 to 300 °C, cellulose decomposes from 300 to 380 °C, and lignin decomposes from 200 to 500 °C [107]. Higher temperatures promote to gases formation and less biochar during pyrolysis that may be due to by higher pyrolytic temperatures causing more volatiles to be released from biochar [108]. Qin et al. studied the effect of temperature on physicochemical characteristics of pine nut shell pyrolytic products for the range of 300–700 °C and found that biochar yields gradually decreased from 52% (300 °C) to 27% (700 °C), and gas yields increased from 17% (300 °C) to 42% (700 °C) [109]. The optimal pyrolytic temperatures for obtaining the maximum yields of bio-oil depend on the biomass source [110]. Generally, maximum bio-oil yields are obtained with pyrolysis of biomass in the range of 450–600 °C. For cellulose, Gao et al. found that maximum bio-oil yields were obtained at around 450 °C from cellulose in a fixed bed reactor [111]. For pine nut shell, maximum bio-oil yields were obtained at 500 °C. For wheat straw, rice straw, rape stalk and cotton stalk, the highest bio-oil yield were obtained around 500 °C, while corn stover required a lower temperature of 450 °C [111].

The content of volatile matter in biochar decreases with increasing pyrolytic temperature, while the contents of fixed carbon and ash in biochar increase [112]. For gas products, CO_2 concentration decreases with increasing temperature whereas that of CO increases [113]. For components of bio-oil, more light molecules are formed with increasing pyrolytic temperature [114]. Therefore, pyrolytic temperature is an important parameter for biomass pyrolysis.

1.6.3 Residence Time

The residence time is the contact time between pyrolytic vapors and biochar during pyrolysis [115]. Biomass firstly undergoes thermal cracking to form primary pyrolytic vapors during biomass pyrolysis. The primary pyrolytic vapors readily undergo secondary reactions of thermal cracking, re-polymerization, and re-condensation with biochar that can lead to a decrease in bio-oil yield and also affect volatile product properties. Thus, rapid removal of pyrolytic vapors from the reaction zone (short residence time) is necessary to minimize secondary reactions and to improve bio-oil yield [34]. The residence time can be varied by changing the flow rate of the carrier gas. A higher gas flow rate in the process leads to a shorter vapor residence time in the hot pyrolysis zone. Asadullah et al. observed an increase of 14% in bio-oil yield and reduction of char and gas by increasing N_2 flow rate from 1 to 2 L/min in the pyrolysis of palm kernel shell with a fluidized bed reactor [116]. Putun et al. studied the effect of vapor residence time in different biomass pyrolysis processes [117–119]. For *Euphorbia rigida*, higher liquid yields were achieved at 400 mL/min, while for sunflower pressed residual and hazelnut shell flow rates were 100 and 200 mL/min, respectively, which indicate means that biomass composition also influences the process. Akhtar and Amin found that high temperatures and longer residence times are suitable for the production of oxygen-free bio-oil [105]. Therefore, the optimization of the vapor residence times, considering the other variables can help to obtain bio-oil of desired quality better quality and yields.

1.6.4 Feedstock Particle Size

The mass and heat transfer in the feedstock is affects production of pyrolytic products can be controlled by changing the feedstock particle size. Larger feedstock particle size causes larger thermal gradients and increases reaction time that leads to secondary reactions [120]. Smaller particles allow faster and possibly more uniform heating. Therefore, reducing feedstock particle size for improving heat and mass transfer is essential to maximizing bio-oil yield. Shen et al. found an increase in bio-oil yield of 12–14 wt% by reducing particle size from 1.5 to 0.3 mm with a fluidized bed reactor [121]. Kang et al. found the similar results for bio-oil yield increases with particle size reduction for Radiana pine pyrolysis with a fluidized bed reactor [122]. However, the smaller particle sizes do not necessarily gives higher bio-oil yields. Abnisa et al. found that for palm shell pyrolysis, increasing particle size from 0.5 to 2 mm caused an increase in liquid production to 70 wt% [123]. Onay et al. found particle sizes of (0.6–1.8) mm are suitable for obtaining high liquid yields for fast pyrolysis of rapeseed with a fixed bed reactor [124]. Conflicting information about feedstock particle size found makes it difficult to generalize this parameter for a pyrolysis system. Guedes et al. concluded that the suitable particle size for obtaining the maximum bio-oil yield may varies depending on the type of biomass and other conditions (*e.g.* pyrolysis type, pyrolysis reactor type) [125]. A small particle size is required for total decomposition of biomass with fast pyrolysis method.

1.6.5 Pyrolytic Reactor Types

Pyrolytic reactor types also influence the overall yield and energy requirements of biomass pyrolysis processes. Pyrolytic reactor types commonly used are shown in Table 1.5. Differences in pyrolysis reactors given in Table 1.5 are mainly due to heating transfer, solid removal, liquid collection and scale-up aspects [126]. Currently, fluidized bed reactors are one of the most studied and applied pyrolysis reactors, and these can classified into three basic types, namely bubbling and circulating [127]. The advantages of the fluidized bed reactors are that they have high heat and mass transfer coefficients, a simpler design, easy scale-up and proficient control over reaction time, and they can produce high yield of bio-oil from different feedstock types [128].

Ablative pyrolysis reactor is another kind of pyrolysis reactor, in which biomass pyrolysis takes place by pressing feed against a hot reactor wall. The advantages of the ablative pyrolysis reactor are: (1) it can deal with biomass feedstock having large particle sizes; (2) it allows good mechanical abrasion of char; and (3) it has great biomass handling capacity (up to 2 t/h) [129]. Ablative pyrolysis reactors have disadvantages such as low heat transfer co-efficient, coke formation, clogging and incomplete biomass conversion [130].

Table 1.5 Commercial and development pyrolysis reactors [126–129]

Reactor type	Operational complexity	Particle size	Biomass variability	Inert gas flow rate	Capacity (t/h)	Organization
Fluidized bed	M	S	M	H	0.3	BEST Energy, Australia, Ensyn, USA
Fixed bed	M	L	H	S	2	Bio-alternative, USA
Ablative	H	L	M	S	0.2	PYTEC, Germany
Rotating cone	M	M-L	S	S	0.2	BTG, The Netherlands
Screw/ auger	L	M	S	S	0.02	University of Saskatchewan, Canada

S small, *M* medium, *L* large, *H* high

Rotating cone reactors feed biomass feedstock on the bottom of a rotating cone and biomass spirals along the wall surface of the cone under centrifugal force. The advantages of rotating cone reactors are those of rapid heating, high heat transfer and no carrier gas being required [131].

The entrained flow reactor is similar to a fluidized bed reactor, and it also has the advantage of simplicity. The entrained flow reactor can handle large gas flow rate, but has high capital cost, scale-up limitations and gives low bio-oil yield [45]. Thus, it has not been applied on a large scale yet, and is only studied on the laboratory scale.

Screw/auger reactor is a tubular, continuous reactor, in which solid biomass is transported through a rotating screw [132]. Different from other pyrolysis reactor types, the heat required for pyrolysis in screw/auger reactor is transported along the tubular wall of the reactor. The advantages of screw/auger reactor is that it can be built very compact as a mobile pyrolysis unit and it can save operation cost for biomass pyrolysis, while its disadvantages are longer reaction time; low heat transfer and lower liquid yield. Therefore, the important factors for selecting the suitable pyrolysis reactor are the medium of heat transfer to biomass particles inside the reactor and the bio-oil yield during the pyrolysis process.

1.7 Conclusions and Future Outlook

Pyrolysis is a promising method for converting biomass into fuels and chemicals. Pyrolysis products are greatly affected by biomass source, feedstock composition and reactor type such that feedstock characterization, choice of proper conditions and reactor configuration are important in obtaining the desired products. Areas of research interest are:

1. To further investigate the biomass pyrolysis mechanism using biomass components that have been obtained with little damage as feedstocks. For example, mild extractions and oligomers with specific linkages and functional groups can be obtained with many methods so that the materials can be used as feedstock. Advanced and in-situ technologies, such two-dimensional perturbation correlation infrared spectroscopy (2D-PCIS), TG-MS, Py-GC-MS should be employed to study pyrolysis mechanisms.
2. Although pyrolytic bio-oils are composed of hundreds of compounds, the main compounds are oxygenated compounds so that suitable reaction methods should be developed to obtained useful fractions. According to the properties and composition of bio-oil, pathways can be designed to convert bio-oil into higher value-added chemicals, such as producing aromatics from phenolic-rich bio-oil via selective HDO reactions, producing value-added chemicals via catalytic pyrolysis.
3. Biomass pyrolysis process is greatly affected by many parameters, such as particle size, pyrolysis temperature, heating rate, reaction time and reactor type.

For future work, it is recommended to establish relationships between pyrolysis operating parameters and product yields and properties through optimization studies, and to identify key parameters needed to obtain desired products. The understanding of factors to scale-up laboratory experiments should be investigated to further promote industrial application.

4. Pyrolytic bio-oils cannot be used directly but require further refining and upgrading. The properties of acidity/basicity, porosity, hydrothermal stability, and resistance to deactivation are keys to understanding the reaction chemistry and bio-oil reactivity. Therefore, technologies for upgrading the bio-oil should be developed, such as catalytic pyrolysis, catalytic hydrogenation/hydrodeoxygenation. Furthermore, the catalysts for the bio-oil upgrading should be studied for improving the bio-oil quality. Further research is needed to explore fundamental reaction mechanisms, inactivation mechanisms of reactants on active sites and the desorption of biomass or bio-oils from the catalysts, and to synthesize stable and active catalysts.

Acknowledgments The authors are grateful to the financial supports from the Natural Science Foundation of Jiangsu Province (No. BK20180548), Natural Science Foundation of China (No. 51906112), China Postdoctoral Science Foundation (2019M651852), Open Funding of Guangdong Provincial Key Laboratory of New and Renewable Energy Research and Development, Guangzhou Institute of Energy Conversion, Chinese Academy of Sciences (Y909ks1001), Nanjing Agricultural University (77J-0603), "Innovation and Entrepreneurship Talents" Introduction Plan of Jiangsu Province.

References

1. Midilli A, Dincer I, Ay M. Green energy strategies for sustainable development. Energy Policy. 2006;34:3623–33. https://doi.org/10.1016/j.enpol.2005.08.003.
2. Asif M, Muneer T. Energy supply, its demand and security issues for developed and emerging economies. Renew Sust Energ Rev. 2007;111:1388–413. https://doi.org/10.1016/j.rser.2005.12.004.
3. Enerdata. (2019). Global Energy statistics year book 2019. https://yearbook.enerdata.net/total-energy/world-consumption-statistics.html. Accessed 20 Jan 2020
4. World Resources Institute. (2019). Greenhouse gas emissions over 165 years, 2019. https://www.wri.org/tags/ghg-emissions. Accessed 20 Jan 2020
5. Atilgan B, Azapagic A. Life cycle environmental impacts of electricity from fossil fuels in Turkey. J Clean Prod. 2015;106:555–64. https://doi.org/10.1016/j.jclepro.2014.07.046.
6. Tilman D, Hill J, Lehman C. Carbon-negative biofuels from low-input high-diversity grassland biomass. Science. 2006;314:1598–600. https://doi.org/10.1126/science.1133306.
7. Luo S, Liu C, Xiao B, Xiao L. A novel biomass pulverization technology. Renew Energy. 2011;36:578–82. https://doi.org/10.1016/j.renene.2010.08.003.
8. Sharma A, Pareek V, Zhang D. Biomass pyrolysis-A review of modelling, process parameters and catalytic studies. Renew Sust Energ Rev. 2015;50:1081–96. https://doi.org/10.1016/j.rser.2015.04.193.
9. Blasi CD. Modeling chemical and physical processes of wood and biomass pyrolysis. Prog Energy Combust Sci. 2008;34:47–90. https://doi.org/10.1016/j.pecs.2006.12.001.

10. Simone B, Michael S. Biochemical conversion processes of lignocellulosic biomass to fuels and chemicals – a review. CHIMIA Int J Chem. 2015;69:572–81. https://doi.org/10.2533/chimia.2015.572.
11. Demirbas A, Arin G. An overview of biomass pyrolysis. Energy Sour. 2002;24:471–82. https://doi.org/10.1080/00908310252889979.
12. Demirbas A. Combustion characteristics of different biomass fuels. Prog Energy Combust Sci. 2004;30:219–30. https://doi.org/10.1016/j.pecs.2003.10.004.
13. Ruiz JA, Juárez MC, Morales MP, Muñoz P, Mendívil MA. Biomass gasification for electricity generation: review of current technology barriers. Renewa Sustain Energy Rev. 2013;18:174–83. https://doi.org/10.1016/j.rser.2012.10.021.
14. Yaman S. Pyrolysis of biomass to produce fuels and chemical feedstocks. Energy Convers Manag. 2004;45:651–71. https://doi.org/10.1016/S0196-8904(03)00177-8.
15. Zhang LT, Xu CB, Champagne P. Overview of recent advances in thermo-chemical conversion of biomass. Energy Convers Manag. 2010;51:969–82. https://doi.org/10.1016/j.enconman.2009.11.038.
16. Liu H, Wang H, Karim AM, Sun J, Wang Y. Catalytic fast pyrolysis of lignocellulosic biomass. Chem Soc Rev. 2014;43:7594–623. https://doi.org/10.1039/C3CS60.
17. Ringer M, Putsche V, Scahill J. Large-scale pyrolysis oil production: a technology assessment and economic analysis. Golden, CO: National Renewable Energy Laboratory; 2006.
18. Paudel SR, Banjara SP, Choi OK, Park KY, Kim YM, Lee JW. Pretreatment of agricultural biomass for anaerobic digestion: current state and challenge. Bioresour Technol. 2017;245:1194–205. https://doi.org/10.1016/j.biortech.2017.08.182.
19. Wang S, Luo ZY. Pyrolysis of biomass components. Beijing, CN: Beijing Science Press; 2013. p. 4.
20. Chen Y, Wu Y, Hua D, Li C, Harold MP, Wang J, Yang M. Thermochemical conversion of low-lipid microalgae for the production of liquid fuels: challenges and opportunities. RSC Adv. 2015;5:18673–701. https://doi.org/10.1039/C4RA13.
21. Paudel SR, Banjara SP, Choi OK, Park KY, Kim YM, Lee JW. Pretreatment of agricultural biomass for anaerobic digestion: current state and challenges. Bioresour Technol. 2017;245:1194–205. https://doi.org/10.1016/j.biortech.2017.08.182.
22. Wang S, Dai G, Yang H, Luo Z. Lignocellulosic biomass pyrolysis mechanism: a state-of-the-art review. Prog Energy Combust Sci. 2017;62:33–86. https://doi.org/10.1016/j.pecs.2017.05.004.
23. Rasi S, Kilpeläinen P, Rasa K, Korpinen R, Raitanen J-E, Vainio M, Kitunen V, Pulkkinen H, Jyske T. Cascade processing of softwood bark with hot water extraction, pyrolysis and anaerobic digestion. Bioresour Technol. 2019;292:121893–9. https://doi.org/10.1016/j.biortech.2019.121893.
24. Huang YB, Fu Y. Hydrolysis of cellulose to glucose by solid acid catalysts. Green Chem. 2013;15:1095–111. https://doi.org/10.1039/C3GC40136G.
25. Peng F, Peng P, Xu F, Sun RC. Fractional purification and bioconversion of hemicelluloses. Biotechnol Adv. 2012;30:879–903. https://doi.org/10.1016/j.biotechadv.2012.01.018.
26. Li C, Zhao X, Wang A, Huber GW, Zhang T. Catalytic transformation of lignin for the production of chemicals and fuels. Chem Rev. 2015;115:11559–624. https://doi.org/10.1021/acs.chemrev.5b00155.
27. Zakzeski J, Bruijnincx PC, Jongerius AL, Weckhuysen BM. The catalytic valorization of lignin for the production of renewable chemicals. Chem Rev. 2010;110:3552–9. https://doi.org/10.1021/cr900354u.
28. Babich IV, Hulst MV, Lefferts L, Moulijn JA, Connor PO, Seshan K. Catalytic pyrolysis of microalgae to high-quality liquid bio-fuels. Biomass Bioenergy. 2011;35:3199–207. https://doi.org/10.1016/j.biombioe.2011.04.043.
29. Silva LS, González DL, Minguillan AMG, Valverde JL. Pyrolysis, combustion and gasification characteristics of Nannochloropsis gaditana microalgae. Bioresour Technol. 2013;130:321–31. https://doi.org/10.1016/j.biortech.2012.12.002.

30. Ho SH, Chen CY, Lee DJ, Chang JS. Perspectives on microalgal CO_2-emission mitigation systems—a review. Biotechnol Adv. 2011;29:189–98. https://doi.org/10.1016/j.biotechadv.2010.11.001.
31. Chen CY, Zhao XQ, Yen HW, Ho SH, Cheng CL, Lee DJ, Bai FW, Chang JS. Microalgae-based carbohydrates for biofuel production. Biochem Eng J. 2013;78:1–10. https://doi.org/10.1016/j.bej.2013.03.006.
32. Maher KD, Bressler DC. Pyrolysis of triglyceride materials for the production of renewable fuels and chemicals. Bioresour Technol. 2007;98:2351–68. https://doi.org/10.1016/j.biortech.2006.10.025.
33. Demirbas A. Competitive liquid biofuels from biomass. Appl Energy. 2011;88:17–28. https://doi.org/10.1016/j.apenergy.2010.07.016.
34. Bridgwater AV. Review of fast pyrolysis of biomass and product upgrading. Biomass Bioenergy. 2012;28:68–94. https://doi.org/10.1016/j.biombioe.2011.01.048.
35. Parthasarathy P, Sheeba KN. Combined slow pyrolysis and steam gasification of biomass for hydrogen generation - a review. Int J Energy Res. 2015;29:147–64. https://doi.org/10.1002/er.3218.
36. Brown TR, Wright MM, Robert C, Brown RC. Estimating profi tability of two biochar production scenarios: slow pyrolysis vs fast pyrolysis. Biofuel Bioprod Biorefin. 2010;5:54–68. https://doi.org/10.1002/bbb.254.
37. Funke A, Morgano MT, Dahmen N, Leibold H. Experimental comparison of two bench scale units for fast and intermediate pyrolysis. J Anal Appl Pyrolysis. 2017;124:504–14. https://doi.org/10.1016/j.jaap.2016.12.033.
38. Yang Y, Wang J, Chong K, Bridgwater AV. A techno-economic analysis of energy recovery from organic fraction of municipal solid waste (MSW) by an integrated intermediate pyrolysis and combined heat and power (CHP) plant. Energy Convers Manag. 2018;174:406–16. https://doi.org/10.1016/j.enconman.2018.08.033.
39. Yang Y, Brammer JG, Wright DG, Scott JA, Serrano C, Bridgwater AV. Combined heat and power from the intermediate pyrolysis of biomass materials: performance, economics and environmental impact. Appl Energy. 2017;191:639–52. https://doi.org/10.1016/j.apenergy.2017.02.004.
40. Ahmed A, Bakar MSA, Azad AK, Sukri RS, Phusunti N. Intermediate pyrolysis of Acacia cincinnata and Acacia holosericea species for bio-oil and biochar production. Energy Convers Manag. 2018;176:393–408. https://doi.org/10.1016/j.enconman.2018.09.041.
41. Perkins G, Perkins T, Perkins M. Process development status of fast pyrolysis technologies for the manufacture of renewable transport fuels from biomass. Renew Sust Energ Rev. 2018;90:282–315. https://doi.org/10.1016/j.rser.2018.03.048.
42. Dabros TMH, Stummann MZ, Høj M, Jensen PA, Grunwaldt JD, Gabrielsen J, Mortensen PM, Jensen AD. Transportation fuels from biomass fast pyrolysis, catalytic hydrodeoxygenation, and catalytic fast hydropyrolysis. Prog Energy Combust. Sci. 2018;68:268–309. https://doi.org/10.1016/j.pecs.2018.05.002.
43. Dickerson T, Soria J. Catalytic fast pyrolysis: a review. Energies. 2013;6:514–38. https://doi.org/10.3390/en6010514.
44. Ruddy DA, Schaidle JA, Ferrell JR III, Wang J, Moens L, Hensley JE. Recent advances in heterogeneous catalysts for bio-oil upgrading via "ex situ catalytic fast pyrolysis": catalyst development through the study of model compounds. Green Chem. 2014;16:454–90. https://doi.org/10.1039/C3GC41354C.
45. Maliutina K, Tahmasebi A, Yu J, Saltykov SN. Comparative study on flash pyrolysis characteristics of microalgal and lignocellulosic biomass in entrained-flow reactor. Energy Convers Manag. 2017;151:426–38. https://doi.org/10.1016/j.enconman.2017.09.013.
46. Nzihou A, Stanmore B, Lyczko N. The catalytic effect of inherent and adsorbed metals on the fast/flash pyrolysis of biomass: a review. Energy. 2019;170:326–37. https://doi.org/10.1016/j.energy.2018.12.174.

47. Ribeiro JMC, Godina R, Matias JCO, Nunes LJR. Future perspectives of biomass torrefaction: review of the current state-of-the-art and research development. Sustainability. 2018;10:2323–40. https://doi.org/10.3390/su10072323.
48. Dai L, Wang Y, Liu Y, Ruan R, He C, Yu Z, Jiang L, Zeng Z, Tian X. Integrated process of lignocellulosic biomass torrefaction and pyrolysis for upgrading bio-oil production: a state-of-the-art review. Renew Sustain Energy Rev. 2019;107:20–36. https://doi.org/10.1016/j.rser.2019.02.015.
49. Barskov S, Zappi M, Buchireddy P, Dufreche S, Guillory J, Gang D, Hernandez R, Bajpai R, Baudier J, Cooper R, Sharp R. Torrefaction of biomass: a review of production methods for biocoal from cultured and waste lignocellulosic feedstocks. Renew Energy. 2019;142:624–42. https://doi.org/10.1016/j.renene.2019.04.068.
50. Sikarwar VS, Zhao M, Fennell PS, Shah N, Anthony EJ. Progress in biofuel production from gasification. Progr Energy Combust Sci. 2017;61:189–248. https://doi.org/10.1016/j.pecs.2017.04.001.
51. Asadullah M. Biomass gasification gas cleaning for downstream applications: a comparative critical review. Renew Sustain Energy Rev. 2014;40:118–32. https://doi.org/10.1016/j.rser.2014.07.132.
52. Sansaniwal SK, Pal K, Rosen MA, Tyagi SK. Recent advances in the development of biomass gasification technology: a comprehensive review. Renew Sust Energ Rev. 2017;72:363–84. https://doi.org/10.1016/j.rser.2017.01.038.
53. Widjaya ER, Chen G, Bowtell L, Hills C. Gasification of non-woody biomass: a literature review. Renew Sustain Energy Rev. 2018;89:184–93. https://doi.org/10.1016/j.rser.2018.03.023.
54. Molino A, Larocca V, Chianese S, Musmarra D. Biofuels production by biomass gasification: a review. Energies. 2018;11:811–42. https://doi.org/10.3390/en11040811.
55. Kan T, Strezov V, Evans TJ. Lignocellulosic biomass pyrolysis: a review of product properties and effects of pyrolysis parameters. Renew Sust Energ Rev. 2016;57:1126–40. https://doi.org/10.1016/j.rser.2015.12.185.
56. Uddin MN, Techato K, Taweekun J, Rahman MM, Rasul MG, Mahlia TMI, Ashrafur SM. An overview of recent developments in biomass pyrolysis technologies. Energies. 2018;11:3115. https://doi.org/10.3390/en11113115.
57. Manyà JJ. Pyrolysis for biochar purposes: a review to establish current knowledge gaps and research needs. Environ Sci Technol. 2012;46:7939–54. https://doi.org/10.1021/es301029g.
58. Lee J, Yang X, Cho SH, Kim JK, Lee SS, Tsang DCW, Ok YS, Kwon EE. Pyrolysis process of agricultural waste using CO_2 for waste management, energy recovery, and biochar fabrication. Appl Energy. 2017;185:214–22. https://doi.org/10.1016/j.apenergy.2016.10.092.
59. Kambo HS, Dutta A. A comparative review of biochar and hydrochar in terms of production, physico-chemical properties and applications. Renew Sust Energ Rev. 2015;45:359–78. https://doi.org/10.1016/j.rser.2015.01.050.
60. Qian K, Kumar A, Zhang H, Bellmer D. Recent advances in utilization of biochar. Renew Sust Energ Rev. 2015;42:1055–64. https://doi.org/10.1016/j.rser.2014.10.074.
61. Liu Z, Quek A, Hoekman SK, Balasubramanian R. Production of solid biochar fuel from waste biomass by hydrothermal carbonization. Fuel. 2013;103:943–9. https://doi.org/10.1016/j.fuel.2012.07.069.
62. Gratuito MKB, Panyathanmaporn T, Chumnanklang PA, Sirinuntawittaya N, Dutta A. Production of activated carbon from coconut shell: optimization using response surface methodology. Bioresour Technol. 2008;99:4887–95. https://doi.org/10.1016/j.biortech.2007.09.042.
63. Yao Z, You S, Ge T, Wang C. Biomass gasification for syngas and biochar co-production: energy application and economic evaluation. Appl Energy. 2018;209:43–55. https://doi.org/10.1016/j.apenergy.2017.10.077.
64. Jha P, Biswas A, Lakaria B, Rao AS. Biochar in agriculture—prospects and related implications. Curr Sci. 2010;99:1218–25.. https://www.jstor.org/stable/24068517

65. Titirici MM, White RJ, Falco C, Sevilla M. Black perspectives for a green future: hydrothermal carbons for environment protection and energy storage. Energy Environ Sci. 2012;5:6796–822. https://doi.org/10.1039/C2EE21166A.
66. Mani S, Kastner JR, Juneja A. Catalytic decomposition of toluene using a biomass derived catalyst. Fuel Process Technol. 2013;114:118–25. https://doi.org/10.1016/j.fuproc.2013.03.015.
67. He N, Bi X, Liu S, Hu Z, Guo X, Luo S, Yang F. Syngas production from pyrolysis of municipal solid waste (MSW) with dolomite as downstream catalysts. J Anal Appl Pyrolysis. 2010;87:181–7. https://doi.org/10.1016/j.jaap.2009.11.005.
68. Baratieri M, Baggio P, Bosio B, Grigiante M, Longo GA. The use of biomass syngas in IC engines and CCGT plants: a comparative analysis. Appl Thermal Eng. 2009;29:3309–18. https://doi.org/10.1016/j.applthermaleng.2009.05.003.
69. Hagos FY, Aziz RA, Sulaiman SA. Trends of syngas as a fuel in internal combustion engines. Adv Mechan Eng. 2014;2014:401587. https://doi.org/10.1155/2014/401587.
70. Siedlecki M, Jong WD, Verkooijen AHM. Fluidized bed gasification as a mature and reliable technology for the production of bio-syngas and applied in the production of liquid transportation fuels-A review. Energies. 2011;4:389–434. https://doi.org/10.3390/en4030389.
71. Abnisa F, Daud WMAW, Husin WNW, Sahu JN. Utilization possibilities of palm shell as a source of biomass energy in Malaysia by producing bio-oil in pyrolysis process. Biomass Bioenergy. 2011;35:1863–72. https://doi.org/10.1016/j.biombioe.2011.01.033.
72. Xiu S, Shahbazi A. Bio-oil production and upgrading research: a review. Renew Sust Energ Rev. 2012;16:4406–14. https://doi.org/10.1016/j.rser.2012.04.028.
73. Wang Z, Lu Q, Zhu X, Zhang Y. Catalytic fast pyrolysis of cellulose to prepare levoglucosenone using sulfated zirconia. ChemSusChem. 2011;4:79–84.
74. Solantausta Y, Nylund NO, Westerholm M, Koljonen T, Oasmaa A. Wood pyrolysis oil as fuel in a diesel power plant. Bioresour Technol. 1993;46:177–88. https://doi.org/10.1002/cssc.201000210.
75. Lu Q, Li WZ, Zhu XF. Overview of fuel properties of biomass fast pyrolysis oils. Energy Convers Manag. 2009;50:1376–83. https://doi.org/10.1016/0960-8524(93)90071-I.
76. Zhang Q, Chang J, Wang T, Xu Y. Review of biomass pyrolysis oil properties and upgrading research. Energy Convers Manag. 2007;48:87–92. https://doi.org/10.1016/j.enconman.2006.05.010.
77. Oasmaa A, Czernik S. Fuel oil quality of biomass pyrolysis oils-state of the art for the end-users. Energy Fuel. 1999;13:914–21. https://doi.org/10.1021/ef980272b.
78. Guo Y, Wang Y, Wei F. Research progress in biomass flash pyrolysis technology for liquids production. Chem Ind Eng Progr. 2001;8:13–7.
79. Sipilaè K, Kuoppala E, Fagernaès L. Characterization of biomass-based flash pyrolysis oils. Biomass Bioenergy. 1998;14:103–13. https://doi.org/10.1016/S0961-9534(97)10024-1.
80. He R, Ye XP, Harte F, English B. Effects of high-pressure homogenization on physicochemical properties and storage stability of switchgrass bio-oil. Fuel Process Technol. 2009;90:415–21. https://doi.org/10.1016/j.fuproc.2008.11.003.
81. Mettler MS, Vlachos DG, Dauenhauer PJ. Top ten fundamental challenges of biomass pyrolysis for biofuels. Energy Environ Sci. 2012;5:7797–809. https://doi.org/10.1039/C2EE21679E.
82. White JE, Catallo WJ, Legendre BL. Biomass pyrolysis kinetics: a comparative critical review with relevant agricultural residue case studies. J Anal Appl Pyrolysis. 2011;91:1–33. https://doi.org/10.1016/j.jaap.2011.01.004.
83. Collard FX, Blin J. A review on pyrolysis of biomass constituents: mechanisms and composition of the products obtained from the conversion of cellulose, hemicelluloses and lignin. Renew Sust Energ Rev. 2014;38:594–608. https://doi.org/10.1016/j.rser.2014.06.013.
84. Mahinpey N, Gomez A. Review of gasification fundamentals and new findings: reactors, feedstock, and kinetic studies. Kinetic study on pyrolysis of biomass components: a critical review. Chem Eng Sci. 2016;148:14–3. https://doi.org/10.1016/j.ces.2016.03.037.

85. Liu Q, Wang S, Zheng Y, Luo Z, Cen K. Mechanism study of wood lignin pyrolysis by using TG–FTIR analysis. J Anal Appl Pyrol. 2008;82:170–7. https://doi.org/10.1016/j.jaap.2008.03.007.
86. Lv G, Wu S. Analytical pyrolysis studies of corn stalk and its three main components by TG-MS and Py-GC/MS. J Anal Appl Pyrolysis. 2012;97:11–8. https://doi.org/10.1016/j.jaap.2012.04.010.
87. Zhang H, Deng S. Density functional theory investigation of gasification mechanism of a lignin dimer with β-5 linkage. Renew Energy. 2018;115:937–45. https://doi.org/10.1016/j.renene.2017.08.095.
88. Choi YS, Singh R, Zhang J, Balasubramanian G, Sturgeon MR, Katahira K, Chupka G, Beckham GT, Shanks BH. Pyrolysis reaction networks for lignin model compounds: unraveling thermal deconstruction of β-O-4 and α-O-4 compounds. Green Chem. 2016;18:1762–73. https://doi.org/10.1039/C5GC02268A.
89. Van de Velden M, Baeyens J, Brems A, Janssens B, Dewil R. Fundamentals, kinetics and endothermicity of the biomass pyrolysis reaction. Renew Energy. 2010;35:232–42. https://doi.org/10.1016/j.renene.2009.04.019.
90. Mettler MS, Mushrif SH, Paulsen AD, Javadekar AD, Vlachos DG, Dauenhauer PJ. Revealing pyrolysis chemistry for biofuels production: conversion of cellulose to furans and small oxygenates. Energy Environ Sci. 2012;5:5414–24. https://doi.org/10.1039/C1EE02743C.
91. Shen D, Gu S. The mechanism for thermal decomposition of cellulose and its main products. Bioresour Technol. 2009;100:6496–504. https://doi.org/10.1016/j.biortech.2009.06.095.
92. Lu Q, Zhang Y, Dong C, Yang Y, Yu H. The mechanism for the formation of levoglucosenone during pyrolysis of β-d-glucopyranose and cellobiose: a density functional theory study. J Anal Appl Pyrolysis. 2014;110:34–43. https://doi.org/10.1016/j.jaap.2014.08.002.
93. Assary RS, Redfern PC, Greeley J, Curtiss LA. Mechanistic insights into the decomposition of fructose to hydroxy methyl furfural in neutral and acidic environments using high-level quantum chemical methods. J Phys Chem B. 2011;115:4341–9. https://doi.org/10.1021/jp1104278.
94. Lu Q, Tian H, Hu B, Jiang X, Dong C, Yang Y. Pyrolysis mechanism of holocellulose-based monosaccharides: the formation of hydroxyacetaldehyde. J Anal Appl Pyrolysis. 2016;120:15–26. https://doi.org/10.1016/j.jaap.2016.04.003.
95. Wang S, Ru B, Lin H, Luo Z. Degradation mechanism of monosaccharides and xylan under pyrolytic conditions with theoretic modeling on the energy profiles. Bioresour Technol. 2013;143:378–83. https://doi.org/10.1016/j.biortech.2013.06.026.
96. Patwardhan PR, Brown RC, Shanks BH. Product distribution from the fast pyrolysis of hemicelluloses. ChemSusChem. 2011;4:636–43. https://doi.org/10.1002/cssc.201000425.
97. Janković B. The comparative kinetic analysis of Acetocell and Lignoboost® lignin pyrolysis: the estimation of the distributed reactivity models. Bioresour Technol. 2011;102:9763–71. https://doi.org/10.1016/j.biortech.2011.07.080.
98. Chen L, Ye X, Luo F, Shao J, Lu Q, Fang Y, Dong C, Yang Y. Pyrolysis mechanism of β O 4 type lignin model dimer. J Anal Appl Pyrol. 2015;115:103–11. https://doi.org/10.1016/j.jaap.2015.07.009.
99. Faravelli T, Frassoldati A, Migliavacca G, Ranzi E. Detailed kinetic modeling of the thermal degradation of lignins. Biomass Bioenergy. 2010;34:290–301. https://doi.org/10.1016/j.biombioe.2009.10.018.
100. Tröger N, Richter D, Stahl R. Effect of feedstock composition on product yields and energy recovery rates of fast pyrolysis products from different straw types. J Anal Appl Pyrolysis. 2013;100:158–65. https://doi.org/10.1016/j.jaap.2012.12.012.
101. Di Blasi C, Branca C. Kinetics of primary product formation from wood pyrolysis. Ind Eng Chem Res. 2001;40:47–56. https://doi.org/10.1021/ie000997e.
102. Wang K, Brown RC, Homsy S, Martinez L, Sidhu SS. Fast pyrolysis of microalgae remnants in a fluidized bed reactor for bio-oil and biochar production. Bioresour Technol. 2013;127:494–9. https://doi.org/10.1016/j.biortech.2012.08.016.

103. Guo X, Wang S, Wang K, Liu Q, Luo Z. Influence of extractives on mechanism of biomass pyrolysis. J Fuel Chem Technol. 2010;38:42–6. https://doi.org/10.1016/S1872-5813(10)60019-9.
104. Nik-Azar M, Hajaligol MR, Sohrabi M, Dabir B. Mineral matter effects in rapid pyrolysis of beech wood. Fuel Process Technol. 1997;51:7–17. https://doi.org/10.1016/S0378-3820(96)01074-0.
105. Akhtar J, Amin NS. A review on operating parameters for optimum liquid oil yield in biomass pyrolysis. Renew Sustain Energy Rev. 2012;16:5101–9. https://doi.org/10.1016/j.rser.2012.05.033.
106. Safdari MS, Amini E, Weise DR, Fletcher TH. Heating rate and temperature effects on pyrolysis products from live wild land fuels. Fuel. 2019;242:295–304. https://doi.org/10.1016/j.fuel.2019.01.040.
107. Bridgwater AV. Principles and practice of biomass fast pyrolysis processes for liquids. J Anal Appl Pyrolysis. 1999;51:3–22. https://doi.org/10.1016/S0165-2370(99)00005-4.
108. Demirbas A. Effects of temperature and particle size on biochar yield from pyrolysis of agricultural residues. J Anal Appl Pyrolysis. 2004;72:243–8. https://doi.org/10.1016/j.jaap.2004.07.003.
109. Qin L, Shao Y, Hou Z, Jiang E. Effect of temperature on the physicochemical characteristics of pine nut shell pyrolysis products in a screw reactor. Energy Sourc A. https://doi.org/10.1080/15567036.2019.1618993.
110. Park HJ, Dong JI, Jeon JK, Park YK, Yoo KS, Kim SS, Kim J, Kim S. Effects of the operating parameters on the production of bio-oil in the fast pyrolysis of Japanese larch. Chem Eng J. 2008;143:124–32. https://doi.org/10.1016/j.cej.2007.12.031.
111. Gao Z, Li N, Yin S, Yi W. Pyrolysis behavior of cellulose in a fixed bed reactor: residue evolution and effects of parameters on products distribution and bio-oil composition. Energy. 2019;175:1067–74. https://doi.org/10.1016/j.energy.2019.03.094.
112. Angın D. Effect of pyrolysis temperature and heating rate on biochar obtained from pyrolysis of safflower seed press cake. Bioresour Technol. 2013;128:593–7. https://doi.org/10.1016/j.biortech.2012.10.150.
113. Encinar JM, Beltrán FJ, Bernalte A, Ramiro A, González JF. Pyrolysis of two agricultural residues: olive and grape bagasse. Influence of particle size and temperature. Biomass Bioenergy. 1996;11:397–409. https://doi.org/10.1016/S0961-9534(96)00029-3.
114. Horne PA, Williams PT. Influence of temperature on the products from the flash pyrolysis of biomass. Fuel. 1996;75:1051–9. https://doi.org/10.1016/0016-2361(96)00081-6.
115. Scott DS, Majerski P, Piskorz J, Radlein D. A second look at fast pyrolysis of biomass - the RTI process. J Anal Appl Pyrolysis. 1999;51:23–37. https://doi.org/10.1016/S0165-2370(99)00006-6.
116. Asadullah M, Rahman MA, Ali MM, Motin MA, Sultan MB, Alam MR, Rahman MS. Jute stick pyrolysis for bio-oil production in fluidized bed reactor. Bioresour Technol. 2008;99:44–50. https://doi.org/10.1016/j.biortech.2006.12.002.
117. Putun AE, Ozcan A, Putun E. Pyrolysis of hazelnut shells in a fixed-bed tubular reactor: yields and structural analysis of bio-oil. J Anal Appl Pyrol. 1999;52:33–49. https://doi.org/10.1016/S0165-2370(99)00044-3.
118. Putun AE, Apaydin E, Putun E. Bio-oil production from pyrolysis and steam pyrolysis of soybean-cake: product yields and composition. Energy. 2002;27:703–13. https://doi.org/10.1016/S0360-5442(02)00015-4.
119. Putun E, Ates F, Putun AE. Catalytic pyrolysis of biomass in inert and steam atmospheres. Fuel. 2008;87:815–24. https://doi.org/10.1016/j.fuel.2007.05.042.
120. Chan WCR, Kelbon M, Krieger BB. Modelling and experimental verification of physical and chemical processes during pyrolysis of a large biomass particle. Fuel. 1985;64:1505–13. https://doi.org/10.1016/0016-2361(85)90364-3.

121. Shen J, Wang XS, Garcia-Perez M, Mourant D, Rhodes MJ, Li CZ. Effects of particle size on the fast pyrolysis of oil mallee woody biomass. Fuel. 2009;88:1810–7. https://doi.org/10.1016/j.fuel.2009.05.001.
122. Kang BS, Lee KH, Park HJ, Park YK, Kim JS. Fast pyrolysis of radiata pine in a bench scale plant with a fluidized bed: in fluence of a char separation system and reaction conditions on the production of bio-oil. J Anal Appl Pyrolysis. 2006;76:32–7. https://doi.org/10.1016/j.jaap.2005.06.012.
123. Abnisa F, Daud WMAW, Wan WMA, Husin WNW, Sahu JN. Utilization possibilities of palm shell as a source of biomass energy in Malaysia by producing bio-oil in pyrolysis process. Biomass Bioenergy. 2011;35:1863–72. https://doi.org/10.1016/j.biombioe.2011.01.033.
124. Onay O, Beis SH, Koçkar OM. Fast pyrolysis of rape seed in a well-swept fi xed-bed reactor. J Anal Appl Pyrolysis. 2001;58-59:995–1007. https://doi.org/10.1016/S0165-2370(00)00133-9.
125. Guedes RE, Luna AS, Torres AR. Operating parameters for bio-oil production in biomass pyrolysis: a review. J Anal Appl Pyrolysis. 2018;129:134–49. https://doi.org/10.1016/j.jaap.2017.11.019.
126. Garcia-Nunez JA, Pelaez-Samaniego MR, Garcia-Perez ME, Fonts I, Abrego J, Westerhof RJM, Garcia-Perez M. Historical developments of pyrolysis reactors: a review. Energy Fuel. 2017;31:5751–75. https://doi.org/10.1021/acs.energyfuels.7b00641.
127. Venderbosch RH, Prins W. Fast Pyrolysis technology development. Biofuels Bioprod Biorefin. 2010;4:178–208. https://doi.org/10.1002/bbb.205.
128. Warnecke R. Gasification of biomass: comparison of fixed bed and fluidized bed gasifier. Biomass Bioenergy. 2000;18:489–97. https://doi.org/10.1016/S0961-9534(00)00009-X.
129. Lédé J. Comparison of contact and radiant ablative pyrolysis of biomass. J Anal Appl Pyrolysis. 2003;70:601–18. https://doi.org/10.1016/S0165-2370(03)00043-3.
130. Luo G, Chandler DS, Anjos LCA, Eng RJ, Jia P, Resende FLP. Pyrolysis of whole wood chips and rods in a novel ablative reactor. Fuel. 2017;194:229–38. https://doi.org/10.1016/j.fuel.2017.01.010.
131. Wagenaar BM, Prins W, von Swaaij WPM. Pyrolysis of biomass in the rotating cone reactor: modelling and experimental justification. Chem Eng Sci. 1994;49:5109–26. https://doi.org/10.1016/0009-2509(94)00392-0.
132. Funke A, Henrich E, Dahmen N, Sauer J. Dimensional analysis of auger-type fast pyrolysis reactors. Energ Technol. 2017;5:119–29. https://doi.org/10.1002/ente.201600095.

Chapter 2
Kinetic Modeling of Solid, Liquid and Gas Biofuel Formation from Biomass Pyrolysis

P. Debiagi, T. Faravelli, C. Hasse, and E. Ranzi

Abstract Modeling of biomass pyrolysis can be understood as several critical multicomponent, multiphase and multiscale processes. The characterization of the biomass and selection of the reference species of cellulose, hemicellulose, lignins and extractives have a major effect on the results. Intrinsic differences exist between hardwood, softwood and grass/cereals and must be taken into account. Thermo-chemical processes such as pyrolysis, gasification and combustion involve several kinetic mechanisms, first in the solid phase for the devolatilization of the biomass, then in the gas phase for the secondary reactions of released products, and finally for the heterogeneous reactions of the char residue. These mechanisms involve a large number of chemical species and reactions and make modeling computationally intensive. For reactor-scale simulations, mechanistic equations need to be simplified, while maintaining their descriptive capability. For example, lumping procedures can allow detailed compositions of oil, gas and char residue to be obtained. In this chapter, the catalytic effect of ash on pyrolysis products is discussed. Secondary or successive gas phase reactions of pyrolysis products complete the kinetic model and allow optimal conditions for bio-oil production to be determined. On the scale of both the particle and the reactor, mathematical modeling of the thermochemical process requires descriptions of coupled transport and kinetic processes. Examples and comparisons with experimental data are used to show the validation and the reliability of a general model. Additional examples for the application of models are taken from the large-scale German project *Oxyflame*, which works on combustion of solid fuels in oxy-fuel atmospheres.

Keywords Biomass pyrolysis · Bio-oil from fast pyrolysis of biomass · Detailed chemical kinetics · Multi-scale modeling of biomass pyrolysis

---

P. Debiagi (✉) · C. Hasse
Simulation of Reactive Thermo-Fluid Systems, TU Darmstadt, Darmstadt, Germany
e-mail: debiagi@stfs.tu-darmstadt.de

T. Faravelli · E. Ranzi
Department of Chemistry, Materials and Chemical Engineering "Giulio Natta", Politecnico di Milano, Milan, Italy

Z. Fang et al. (eds.), *Production of Biofuels and Chemicals with Pyrolysis*, Biofuels and Biorefineries 10, https://doi.org/10.1007/978-981-15-2732-6_2

Nomenclature

Bi	Biot number
$\widehat{C}$	specific heat
Da	Darcy tensor
$\mathcal{D}$	diffusion coefficient
g	gravitational acceleration
h	heat exchange coefficient
$\widehat{h}$	specific mass enthalpy
I	Identity matrix
j	gas diffusive flux
k_c	convective mass exchange coefficient
k_R	rate constant
$\dot{m}$	mass flow rate
n	outward pointing unit normal
NC	number of species
N_p	number of particles
p	pressure
Py	pyrolysis number
q	conductive heat flux
q_{rad}	radiative heat flux
$\dot{Q}_R$	reaction heat
r	radius
S	surface
T	temperature
t	time
Th	Thiele number
u	velocity
$\mathbf{u}^*$	relative velocity
v	diffusion velocity of gas species
V	volume

Greek symbols

ε	solid porosity
λ	thermal conductivity
μ	dynamic viscosity
ξ	emissivity
ρ	density
ω	mass fraction
$\dot{\Omega}_k$	net formation rate

∇ nabla – vector differential operator
$\otimes$ vertex position

Superscripts

bulk region outside the particle
G gas phase
(I) interface
S solid phase

Subscripts

eff *effective*
J species solid
k species gas
p particle

2.1 Introduction

Pyrolysis is the thermal treatment of biomass in the absence of oxygen producing a liquid fuel and/or a gas stream consisting mainly of CO_2, CO and CH_4, together with minor amounts of C_2 hydrocarbons and H_2 [1, 2]. Syngas can be used as a raw material for the synthesis of methanol (CH_3OH) and liquid fuels [3] or directly as a fuel for electricity generation. Gasification is the partial oxidation of the solid fuel with steam and air and has several potential advantages compared to traditional combustion, mainly related to the possibility of combining the temperature and equivalence ratio to obtain an appropriate syngas [4, 5]. BTL (biomass-to-liquids) and IGCC (integrated gasification combined cycle) are emerging technologies based on biomass gasification [6, 7].

Chemical kinetics influence the pyrolysis of biomass particles at three different levels: pyrolysis or devolatilization of the biomass, heterogeneous reactions of the residual char and successive or secondary gas-phase reactions of released volatile products.

Pyrolysis is also often the initial step in gasification and combustion processes and represents the primary release of volatile products. Gases, condensable hydrocarbons and oxygenated species (tars), and char residues are always produced by biomass pyrolysis, but their nature and quantity vary significantly depending on the nature of the biomass and the process conditions.

The mathematical modeling of biomass pyrolysis is a challenging problem, because it is complex on several levels [8, 9]:

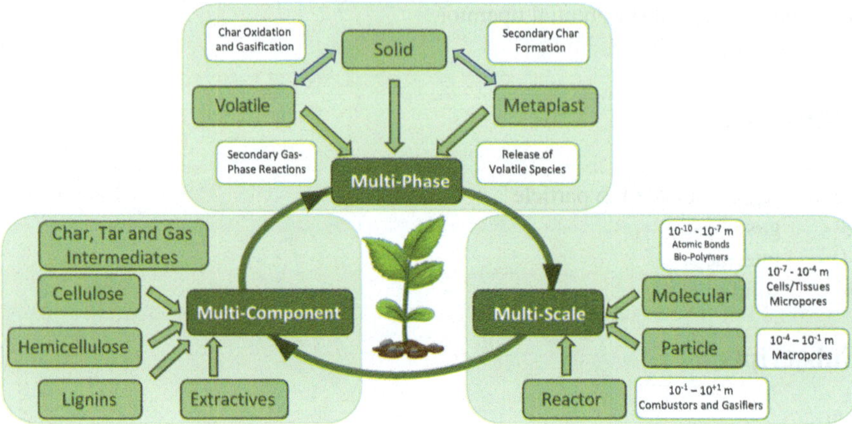

Fig. 2.1 Multicomponent, multiphase and multiscale nature of thermochemical conversion of biomass

- *Multicomponent.* Biomass is a complex material and requires detailed characterization.
- *Multiphase.* Biomass reacts in a condensed phase forming a solid, a liquid and a gas phase. Heterogeneous gas-solid reactions involve the residual char, whereas gas and tar species react in the gas phase.
- *Multiscale.* Kinetic and transport processes need to be coupled on both the particle and the reactor scale.

Figure 2.1 offers a schematic representation of the complexity of this problem. The development of a comprehensive model for biomass conversion is challenging because of the complex composition of biomass, the feed material, and the complex physicochemical interactions that take place in multiple phases and across the wide range of time and length scales [10].

Modeling the thermochemical conversion processes which produce solid fuels clearly requires these multiple complexities to be taken into account. The multiscale nature of this problem is evident when the size of the molecules (order of angstroms) and size of pyrolysis reactors (order of meters) are taken into account. Timescales also vary from the very short lifetimes of the radicals in the reacting system, passing through the minutes required to heat and devolatilize large biomass particles, to the hours of residence time of fuels in the reactor. Thus, the multi-scale mathematical modelling of thermochemical units of solid fuels requires complex chemical mechanisms to be combined with transport phenomena, on both the particle and the reactor scale. A further complexity of the problem derives from the non-ideal, anisotropic nature of the biomass particles, with possible fractures and comminution during the decomposition process. Moreover, for this reason, the thermal and transport properties of the solid residue vary continuously as the conversion progresses. This complexity demands considerable simplification and lumping procedures in the solid and gas-phase kinetic mechanisms and an appropriate level of description in the mass, momentum and energy balance equations. As usual, well-balanced efforts are required to develop a mathematical model of pyrolysis, gasification and combustion units, on both the particle and the reactor scale [11].

This chapter updates and summarizes the research activities carried out at Politecnico di Milano in the field of the mathematical modeling of biomass pyrolysis, gasification and oxidation. The multistep kinetic mechanism of biomass pyrolysis discussed here is an extension of that originally presented by [12] and progressively extended and upgraded [13] to account for new available experimental data and theoretical findings. One of the peculiarities of this model lies in its ability to provide detailed information on the composition of released volatiles and solid residues. The chemistry in the model also involves secondary gas phase reactions of volatiles released during the biomass pyrolysis. This very large number of kinetic mechanisms of pyrolysis and combustion of hydrocarbon and oxygenated species takes advantage of well-consolidated experience, both in pyrolysis [14] and in combustion processes [15].

After this general introduction, the chapter is structured as follows. Section 2.2 describes the characterization and the kinetic mechanisms of biomass pyrolysis. Namely, biomass is first characterized by means of a limited number of reference components. Then, biomass pyrolysis products are obtained simply from a linear combination of char, tar and gas products released by the individual reference components. Attention is also devoted to the catalytic effect of ash and to the characterization of algae, as third-generation biofuels. Section 2.3 discusses the secondary gas-phase reactions of volatile species released from biomass pyrolysis, paying special attention to phenolic species. Section 2.4 presents mass and energy balances on the particle and the reactor scales, emphasizing the effect of reaction kinetics being coupled with mass and heat transfer resistances. The fast pyrolysis of biomass and the crucial role of the residence time of the gas and the biomass to maximize bio-oil production is then discussed. Finally, some conclusions are drawn in Sect. 2.5.

2.2 Biomass Characterization and Multi-Step Pyrolysis Model

The characterization of biomass is briefly discussed in this section, along with corresponding devolatilization models. Choosing several reference components, together with their multistep pyrolysis models, allows the biomass decomposition reactions to be described [8].

2.2.1 Lignocellulosic Biomass Characterization and Reference Species

Much of the literature contains one of the three typical compositional analyses of biomass samples (proximate, biochemical and ultimate), but few papers contain all three analyses on the same sample. A large collection of data from the literature, in the form of a database, was reported by Debiagi et al. [16]. Table 2.1 reports the composition of several typical biomass samples for which all the three analysis were performed.

It is well known that cellulose (25–60) wt%, hemicellulose (15–40) wt%, and lignin (15–45) wt% are the building blocks of lignocellulosic biomass [21]. The abundance of each component is connected to the type of biomass considered, which can be roughly grouped as softwood, hardwood and grass/cereals because of their similarities. The softwood category groups all gymnosperms—coniferous trees—of which the most common species are pine, fir, larch and spruce. Hardwood encompasses all woody flowering angiosperms, including beech, birch, oak and olive. The grass/cereals category includes all non-woody biomass, some of which is grown for energy purposes, such as Miscanthus and short rotation coppice (SRC), while others are a residue from agriculture, such as wheat straw, rice husks and corn stalks.

Lignocellulosic biomass has a porous structure, with cellulose micro-fibrils being the most important element, surrounded by hemicellulose and pectin that act as ligands and embed lignin materials in the cell walls, resulting in the biomass's macroscopic structure [22]. Present in minor amounts, extractives are bio-active, non-structural compounds which play many important roles in plant metabolism. The inorganic matter is usually measured and accounted for as ash, which is mostly metal oxides, formed during the devolatilization and oxidation of the original inorganic compounds in the biomass [16]. Moisture is found as hygroscopic water, capillary water in the lumens, and water vapor in the gas phase [23].

Cellulose is a long chain polymer built by monomeric units of a six-carbon sugar (glucose), bonded through β-1,4 glycosidic bonds. The chains are kept together by hydrogen bonds, which endow the polymer with a crystalline structure whose elementary micro-fibrils contain 36 chains. Cellulose is the most abundant structural component in biomass (25–60%, dry mass basis). Hemicellulose is a second structural compound, consisting in a polymeric chain of hexoses (six-carbon sugars, mainly glucose and mannose) and pentoses (five-carbon sugars, mainly galactose and arabinose). It forms a microfibril network which is closely connected to the rigid cellulose structure. Compared with cellulose, hemicellulose has shorter chains and more amorphous structures because of the branches present on the chain and its less regular composition. The amount of hemicellulose usually present in biomass ranges from 15% to 40%, rarely in quantities greater than cellulose. Hardwood plants have average hemicellulose contents of 10–15%, lower than that of softwood and herbaceous materials (20–30%) [24]. Table 2.2 shows the average composition of monosaccharides in hemicellulose and the significant variability among the different types of biomass.

Table 2.1 Typical proximate, biochemical and ultimate analyses of biomass samples

	Proximate analysis				Biochemical analysis (DAF wt%)				Ultimate analysis (DAF wt%)				
Biomass	Moist.	Ash	Volatile	F.C.[a]	Cellulose	Hemicell.	Lignin	Extractives	C	H	O	Ref	ID
Beech	7.4	0.4	74.0	24.6	44.8	31.1	21.4	2.7	51.1	6.4	42.5	[17, 18]	983
Tobacco Stalks	8.9	2.4	79.6	18.0	38.4	26.1	24.4	11.1	50.8	5.7	43.6	[17, 18]	985
Wheat Straw	8.5	13.5	63.0	23.5	30.6	41.6	19.8	8.0	53.7	6.0	40.3	[17, 18]	977
Hazelnut Shell	9.0	1.4	70.0	28.6	25.2	29.1	41.4	4.2	52.3	5.5	42.2	[17, 18]	975
Walnut Shell	1.8	0.5	76.7	20.9	33.9	39.2	15.3	10.2	52.3	5.8	41.9	[19]	–
Wood Bark	8.8	1.6	66.6	31.8	25.2	30.3	44.5	n.a.[b]	53.2	6.1	40.7	[17, 18]	978
Olive Husk	9.2	3.6	70.3	26.1	23.0	22.6	46.3	8.1	50.8	6.3	42.9	[17, 18]	974
Tea Waste	6.5	1.4	85.0	13.6	29.6	19.5	39.2	11.7	51.9	5.8	42.2	[17, 18]	980
Corncob	12.1	1.0	86.5	12.5	51.8	31.8	15.0	1.4	49.5	5.5	45.1	[17, 18]	979
Ailanthus Wood	8.1	1.7	73.5	24.8	45.7	26.0	25.6	2.6	51.2	6.4	42.4	[17, 18]	986
Spruce Wood	7.6	1.5	70.2	28.3	49.5	20.6	26.7	3.1	52.5	6.2	41.4	[17, 18]	982
Olive Branches	n.a.[b]	3.5	77.9	18.6	33.2	23.0	36.2	7.6	51.1	6.4	42.5	[20]	–
Kiwi Branches	n.a.[b]	2.2	77.5	20.3	28.8	29.9	26.7	14.6	50.5	5.8	43.7	[20]	–
Rice Husks	n.a.[b]	11.2	70.7	18.1	40.9	16.5	30.4	12.1	49.6	6.2	44.3	[20]	–
Wheat Straw	n.a.[b]	10.9	72.1	17.0	25.7	23.3	28.9	22.1	48.3	6.1	45.6	[20]	–
Pine Bark	n.a.[b]	1.1	69.9	29.0	28.6	12.6	45	13.8	51.9	5.3	42.8	[20]	–

[a]Fixed carbon
[b]Not available
ID refers to the sample number in the Phyllis2 Database [17]

Table 2.2 Average monomeric composition of hemicellulose in different biomass types [24]

Distribution of monomers (mol %)						
Biomass type	Mannose	Glucose	Galactose	Arabinose	Xylose	Uronic Acids
Hardwood	6.3	3.9	5.9	2.8	65.2	15.9
Softwood	34.7	19.4	12.3	7.2	13.3	13.2
Grass/Cereals	3.3	24.2	3.6	9.8	52.1	6.9

Holocellulose is commonly referred to as a combination of cellulose and hemicellulose. Lignin is typically present in biomass in amounts ranging from 15% to 45%. Lignin is a racemic polymer made up of monomeric units of aromatic alcohols (coniferyl, sinapyl and p-coumaryl), whose composition differs widely in hardwood, softwood and grass/cereal biomass.

Biomass offers important advantages as a solid fuel due to the high volatility and high reactivity of the fuel and the residual char. In comparison with coal, biomass has a lower density and a lower heating value, because of the higher oxygen and moisture content. Biomass is primarily composed of C, H and O elements, with smaller amounts of N, S and Cl. Several correlations between the heating value and ultimate or elemental analysis have been proposed in the literature [18]. The more complete structural analysis of biomass samples provides significant information on the relative content of carbohydrates (glucose, xylose, galactose, arabinose and mannose), lignin, extractable materials, protein and ash. Compared with elemental analysis, the analytical techniques and methods required are more complex and involve thermal, chemical and/or enzymatic separations which could also modify the original biomass structure. Despite several research efforts in this direction [25], data reporting both elementary and biochemical composition are not easily available in the open-access literature. This lack of information creates some difficulties when characterizing biomass for modeling purposes.

Several years ago, a method to characterize the biomass feedstock was proposed based only on elemental analysis [12]. As already mentioned, cellulose, hemicellulose and lignin, together with extractives, constitute the largest portion of the biomass, and these are the main reference species. Biomass pyrolysis products are then assumed to be a linear combination of the pyrolysis products of these reference compounds. When direct information on biochemical composition is unavailable, the cellulose, hemicellulose, lignin and extractive content are derived from the elemental biomass composition in terms of H/C/O [8, 21]. As well as cellulose and hemicellulose, three different types of lignin which are rich in carbon, hydrogen and oxygen, respectively, are used as reference species [26]. Two more lumped reference species account for the hydrophobic and the hydrophilic extractives. Figure 2.2 reports the structure and formula of seven typical reference species. To reduce the total number of degrees of freedom, the characterization procedure is based on different empirical parameters, which reflect the nature of the biomass. Therefore, the ratio between hemicellulose and cellulose, and the internal composition of hemicellulose, lignins and extractives, are defined on an empirical basis according to the different types of biomass considered. Three reference mixtures

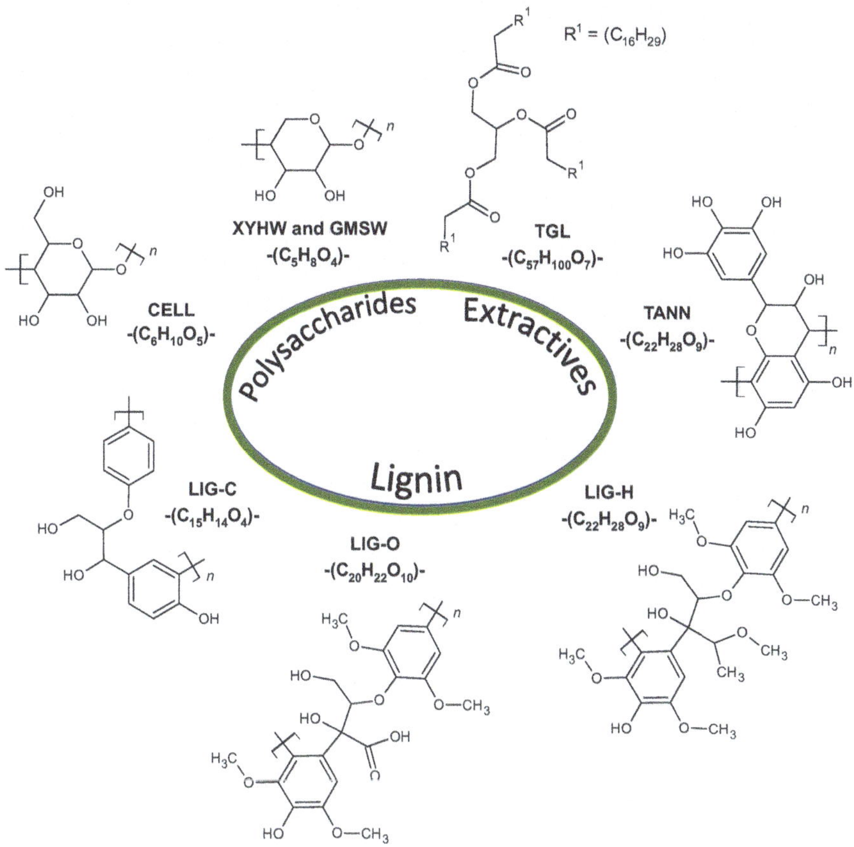

Fig. 2.2 Reference species for lignocellulosic biomass characterization

(linear combinations of the seven reference components) can then describe a wide range of lignocellulosic biomasses. Full details of biomass characterization methods are reported in previous papers [16].

The chemical percolation devolatilization model (bio-CPD) uses a very similar approach, assuming that biomass pyrolysis occurs as a weighted average of the individual cellulose, hemicellulose and lignin components. The light gas and tar yields of a particular biomass are then calculated, together with the residual char, as a weighted average of the pyrolysis products of the reference components [27].

2.2.2 *Kinetic Model of Biomass Pyrolysis*

As summarized by Dhahak et al. [28], two simplified approaches are generally proposed to describe the thermal conversion of biomass.

In the first approach, one or more global chemical reactions may be sufficient to describe how biomass devolatilizes into volatile compounds and a solid residue. This type of mechanism is generally used to model coupled chemical kinetics and physical phenomena, for example using a CFD (Computational Fluid Dynamics) approach [29–33]. Unlike the global one-step mechanism, independent reactions can competitively form tar, gas and coal [34]. Usually, these models are coupled with secondary reactions of the volatile products [35–38].

In the second approach, the contribution of each biomass constituent is taken into account [12, 39]. The pyrolytic behavior of the overall biomass is derived from that of its major constituents through independent parallel reactions, representing cellulose, hemicellulose and lignin decomposition, respectively. These types of mechanisms can be applied to a variety of biomasses, since the content of cellulose, hemicellulose and lignin differs from one type to another. This allows a better prediction of product yields. The mechanisms are an extension to the lignocellulosic material used in Broido-Shafizadeh's scheme to model cellulose pyrolysis [40]. Biomass is first converted into intermediate 'metaplastic and active species', which account for the depolymerization of long chains, forming smaller molecules that then evaporate or decompose through competitive reactions into gaseous, liquid and solid products. A pioneering semi-detailed, multi-step mechanism is that described by Ranzi et al. [12], who first extended Broido-Shafizadeh's approach to include lignocellulosic biomass pyrolysis.

More advanced approaches involve micro-kinetic mechanistic models based on theoretical calculations as proposed by the team around Broadbelt, who extensively studied the fast pyrolysis of cellulose and hemicellulose [41–45], from both theoretical and experimental points of view. They developed a detailed mechanistic model, which involved 100 species and ~300 reactions and described the decomposition of cellulosic polymer chains, reactions of intermediates, and the formation of species of low molecular weight. Seshadri and Westmoreland [46] highlight the implications of concerted molecular reactions for cellulose and hemicellulose kinetics, and Horton et al. [47] present a biomass pyrolysis and gasification model at the molecular level. Yanez et al. [48] published a detailed microkinetic model of lignin pyrolysis based on more than 1500 species and 4000 reactions.

Modeling the pyrolysis of large molecules at the molecular level is still very computationally expensive and impractical. Even at today's most powerful clusters, it would take decades to simulate the pyrolysis of one cellulose chain composed of 100 monomeric units, and thousands of years of CPU time to simulate the pyrolysis of one cellulose micro-fibril via Car-Parrinello molecular dynamics [10]. Thus, model compounds which are smaller in size, such as α-cyclodextrin and glucose, are first studied as surrogates to extrapolate the fundamental chemistry of biomass pyrolysis [24]. The main possibilities and drawbacks of the molecular modeling of wood as a composite of cellulose, hemicellulose and lignin have been very recently recapped by Westmoreland [49]. Research efforts have obtained an increasingly clear picture for cellulose pyrolysis. Modeling hemicellulose decomposition and char formation remains very challenging, while the modeling of lignin seems generally good, despite its structural complexity [49].

The Polimi lumped and multistep kinetic mechanism of biomass decomposition is discussed and analyzed here in terms of its main characteristics. The differences in the composition of biomass and in the operating conditions of the thermal treatment significantly change the resulting product distribution, but similar products are always formed (on a qualitative basis): light permanent gases, water and sugars together with small quantities of oxygenated species (alcohols, ketones, aldehydes, phenols, etc.) and a char residue. The species formed and their distribution can be tracked back to the initial reference components of the biomass. Table 2.3 reports the multicomponent and multistep kinetic mechanism of primary biomass pyrolysis for the seven reference components given in Fig. 2.2.

Each reference component independently decomposes through a multi-step, branched mechanism of first-order reactions. These lumped reactions describe the formation of char, solid and chemisorbed intermediate species, tars and permanent gases. The apparent and global reactions, both in terms of rates and stoichiometries, have been derived from experimental findings [12] and are progressively extended and updated based on the latest experimental data and the range of experimental conditions. Experimental data on temperature profiles in large particles with overshooting of the center temperature have allowed the endothermic devolatilization of tars and the exothermic carbonization process to be validated [50]. More recent works are also dedicated to describing the thermophysical properties of the species involved in the mechanism [51].

One peculiarity of the kinetic model is the detailed characterization of pyrolysis products, including not only water vapor and permanent gases (H_2, CO, CO_2, CH_4 and C_2H_4), different alcohols and carbonyl compounds, but also different carbohydrates together with phenolic and heterocyclic components. At high temperatures, different chemisorbed species contribute to describing the successive charification steps through the progressive release of H_2, CO and CO_2 [16].

Both cellulose and hemicellulose are characterized by polymeric sugar chains, which are released together with tar components, permanent gases and several oxygenated species [52]. The pyrolysis mechanism of cellulose [53–55] is characterized by the production of active cellulose through an initial depolymerization step, with an apparent activation energy of 47 kcal/mol and with no significant release of volatile species. Active cellulose then decomposes with two competitive reactions: a main reaction that releases levoglucosan and a slow reaction that forms char and permanent gases. At high temperatures, the decomposition reaction prevails over the release of tar (levoglucosan). Similarly, the multistep kinetic mechanism of hemicellulose pyrolysis produces two different active intermediates, with a successive release of tar and gas components. The ratio of the two intermediates depends on the nature of the biomass and is different for hardwood, softwood and grass/cereals [16, 56].

The multistep kinetic scheme of lignin pyrolysis in Table 2.3 is a considerable simplification of the detailed mechanism described by Faravelli et al. [57]. The pyrolysis reactions of the three lignins are active across a wide temperature range and explain the formation of phenolic components. Phenol, anisole and a few selected lumped species are representatives of these compounds. This kinetic

Table 2.3 Multi-step kinetic scheme of biomass pyrolysis [16]

Reactions			Kinetic parameters A (s^{-1}), Eact (kcal/mol)	ΔH Reaction (700 K) (kcal/mol)
Cellulose				
1	CELL →	CELLA	1.5×10^{14} exp(−47.0/RT)	0
2	CELLA →	0.05 CH_2OHCH_2CHO + 0.4 CH_2OHCHO + 0.03 CHOCHO + 0.17 CH_3CHO + 0.25 $C_6H_6O_3$ + 0.35 C_2H_5CHO + 0.2 CH_3OH + 0.15 CH_2O + 0.49 CO + 0.43 CO_2 + 0.13 H_2 + 0.93 H_2O + 0.02 HCOOH + 0.05 CH_4 + 0.66 CHAR + 0.05 G{CO} + 0.05 G$\{COH_2\}_{Loose}$ + 0.1 G$\{H_2\}$	2.5×10^{6} exp(−19.1/RT)	21.09
3	CELLA →	$C_6H_{10}O_5$	3.3 × T × exp(−10.0/RT)	24.11
4	CELL →	0.125 H_2 + 4.45 H_2O + 5.45 CHAR + 0.12 G$\{COH_2\}_{Stiff}$ + 0.25 G {CO} + 0.18 G$\{COH_2\}_{Loose}$ + 0.125 G$\{H_2\}$	9.0×10^{7} exp(−31.0/RT)	−49.69
Hemicelluloses				
5	GMSW →	0.70 HCE1 + 0.30 HCE2	1.0×10^{10} exp(−31.0/RT)	0.72
6	XYHW →	0.35 HCE1 + 0.65 HCE2	1.25×10^{11} exp(−31.4.0/RT)	0.72
7	XYGR →	0.12 HCE1 + 0.88 HCE2	1.25×10^{11} exp(−30.0/RT)	0.72
8	HCE1 →	0.06 CH_2OHCH_2CHO + 0.16 FURFURAL + 0.1 CHOCHO + 0.13 $C_6H_6O_3$ + 0.09 CO_2 + 0.02 H_2 + 0.54 H_2O + 0.25 $C_6H_{10}O_5$ + 0.1 CH_4 + 0.25 $C_5H_8O_4$ + 0.1 CHAR	16.0 x T x exp(−12.9/RT)	20.67
9	HCE1 →	0.4 CH_2O + 0.49 CO + 0.39 CO_2 + 0.1 H_2 + 0.4 H_2O + 0.05 HCOOH + 0.1 C_2H_4+ 0.3 CH_4 + 0.975 CHAR + 0.37 G$\{COH_2\}_{Stiff}$ + 0.51 G $\{CO_2\}$ + 0.01 G{CO} + 0.325 G$\{CH_4\}$ + 0.075 G$\{C_2H_4\}$ + 0.43 G $\{COH_2\}_{Loose}$ + 0.05 G$\{H_2\}$ + 0.2 G$\{C_2H_6\}$	$3.0\ 10^{-3}$ × T × exp(−3.6/RT)	−9.11
10	HCE2 →	0.145 FURFURAL + 0.105 CH_3CO_2H + 0.035 CH_2OHCHO + 0.3 CO + 0.5125 CO_2+ 0.5505 H_2 + 0.056 H_2O + 0.0175 HCOOH + 0.049 C_2H_5OH + 0.1895 CH_4 + 0.7125 CHAR + 0.78 G$\{COH_2\}_{Stiff}$ + 0.45 G $\{CO_2\}$ + 0.105 G$\{CH_3OH\}$ + 0.05 G$\{CH_4\}$ + 0.1 G$\{C_2H_4\}$ + 0.18 G $\{COH_2\}_{Loose}$ + 0.21 G$\{H_2\}$ + 0.2 G$\{C_2H_6\}$	7.0×10^{9} exp(−30.5/RT)	3.06

Lignins				
11	LIGH →	0.2 CH_2OHCHO + 0.5 C_2H_5CHO + 0.1 CO + 0.4 C_2H_4+ 0.1 C_2H_6 + LIGOH	$6.7 \times 10^{12} \exp(-37.5/RT)$	30.66
12	LIGO →	CO_2 + LIGOH	$3.3 \times 10^{8} \exp(-25.5/RT)$	26.289
13	LIGC →	0.1 $C_6H_5OCH_3$ + 0.22 CH_2O + 0.21 CO + 0.1 CO_2 + H_2O + 0.27 C_2H_4+ 0.1 VANILLIN + 0.35 LIGCC + 5.85 CHAR + 0.4 G $\{COH_2\}_{Stiff}$ + 0.36 G$\{CH_4\}$ + 0.17 G$\{COH_2\}_{Loose}$ + 0.1 G$\{H_2\}$ + 0.2 G$\{C_2H_6\}$	$1.0 \times 10^{11} \exp(-37.2/RT)$	−7.12
14	LIGCC →	0.15 $C_6H_5OCH_3$ + 0.35 CH_2OHCHO + 1.15 CO + 0.7 H_2 + 0.7 H_2O + 0.3 C_2H_4+ 0.45 CH_4 + 0.25 VANILLIN + 0.15 CRESOL + 0.4 C_2H_6 + 6.8 CHAR + 0.4 G{CO}	$1.0 \times 10^{4} \exp(-24.8/RT)$	−27.39
15	LIGOH →	0.025 $C_{24}H_{28}O_4$ + 0.1 C_2H_3CHO + 0.6 CH_3OH + 0.65 CO + 0.05 CO_2 + H_2O + 0.05 HCOOH + 0.1 CH_4 + 0.9 LIG + 4.25 CHAR + 0.4 G $\{COH_2\}_{Stiff}$ + 0.6 G{CO} + 0.3 G$\{CH_3OH\}$ + 0.25 G$\{CH_4\}$ + 0.1 G$\{C_2H_4\}$ + 0.45 G$\{COH_2\}_{Loose}$ + 0.15 G$\{C_2H_6\}$	$1.5 \times 10^{8} \exp(-30.0/RT)$	−23.85
16	LIG →	0.1 $C_6H_5OCH_3$ + 0.3 CH_3CHO + 0.6 CO + 0.5 C_2H_4+ VANILLIN + 0.1 CHAR	$4.0 \times T \times \exp(-12.0/RT)$	43.21
17	LIG →	0.4 CH_2O + 0.3 CO + 0.1 CO_2 + 0.6 H_2O + 0.2 CH_4 + 6.1 CHAR + 0.65 G$\{COH_2\}_{Stiff}$ + 0.2 G{CO} + 0.4 G$\{CH_3OH\}$ + 0.4 G$\{CH_4\}$ + 0.5 G$\{C_2H_4\}$ + 1.25 G$\{COH_2\}_{Loose}$ + 0.1 G$\{H_2\}$	$8.3 \times 10^{-2} \times T \times \exp(-8/RT)$	−39.02
18	LIG →	0.4 CH_3OH + 0.4 CH_2O + 2.6 CO + 0.6 H_2O + 0.75 C_2H_4+ 0.6 CH_4 + 0.5 C_2H_6 + 4.5 CHAR	$1.5 \times 10^{9} \exp(-31.5/RT)$	11.32
Extractives				
19	TGL →	C_2H_3CHO + 0.5 U2ME12 + 2.5 MLINO	$7.0 \times 10^{12} \exp(-45.7/RT)$	23.90
20	TANN →	H_2O + 0.85 C_6H_5OH + ITANN + G{CO} + 0.15 G$\{C_6H_5OH\}$	$2.0 \times 10^{1} \exp(-10.0/RT)$	11.95
21	ITANN →	2 CO + H_2O + 5 CHAR + 0.45 G$\{COH_2\}_{Stiff}$ + 0.55 G$\{COH_2\}_{Loose}$	$1.0 \times 10^{3} \exp(-25.0/RT)$	−9.25
Metaplastic species release				
22	G$\{CO_2\}$ →	CO_2	$1.0 \times 10^{6} \exp(-24.5/RT)$	−23.90
23	G{CO} →	CO	$5.0 \times 10^{12} \exp(-52.5/RT)$	−13.43

(continued)

Table 2.3 (continued)

Reactions			Kinetic parameters A (s^{-1}), Eact (kcal/mol)	ΔH Reaction (700 K) (kcal/mol)
Cellulose				
24	G{CH_3OH} →	CH_3OH	2.0×10^{12} exp(−50.0/RT)	0
25	G {COH_2}$_{Loose}$ →	0.2 CO + 0.2 H_2 + 0.8 H_2O + 0.8 CHAR	6.0×10^{10} exp(−50.0/RT)	−6.74
26	G{C_2H_6} →	C_2H_6	1.0×10^{11} exp(−52.0/RT)	0
27	G{CH_4} →	CH_4	1.0×10^{11} exp(−53.0/RT)	0
28	G{C_2H_4} →	C_2H_4	1.0×10^{11} exp(−54.0/RT)	0
29	G{C_6H_5OH} →	C_6H_5OH	1.5×10^{12} exp(−55.0/RT)	0
30	G{COH_2}$_{Stiff}$ →	0.8 CO + 0.8 H_2 + 0.2 H_2O + 0.2 CHAR	1.0×10^{9} exp(−59.0/RT)	48.69
31	G{H_2} →	H_2	1.0×10^{8} exp(−70.0/RT)	0
H_2O evaporation				
32	Moisture →	H_2O	1 × T × exp(−8.0/RT)	9.73

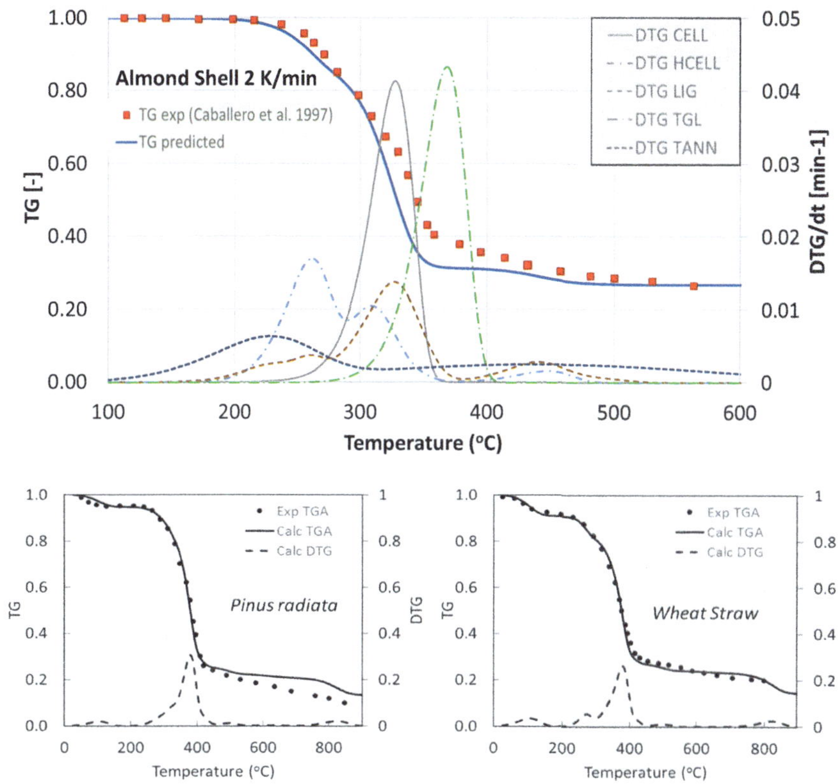

Fig. 2.3 TG Pyrolysis of: almond shell at 2 K/min (top); *Pinus radiata (bottom-left)* and wheat straw (bottom-right) at 80 K/min. Comparisons between experimental data (points) and model predictions (lines) [62–64]. DTG curves of individual reference components are also shown for the almond shell. Reprinted with permission from [8]. Copyright © 2017 American Chemical Society

mechanism is in agreement with the most recent mechanism discussed by Zhou et al. [42]. As already shown in previous papers [8, 16], the model of cellulose, hemicellulose and lignin pyrolysis agree well with the experimental data at different heating rates [58–61]. Similar considerations can be applied to the pyrolysis of tannins and triglyceride species, representing hydrophilic and hydrophobic extractives. The tannin mainly forms phenolic species, while the triglycerides easily decompose to produce a lumped species representative of free fatty acids (FFAs) [21].

As discussed further in the next section, the kinetic model and related reaction heats have also been validated by comparing model predictions with different experimental pyrolysis data on large biomass particles [50].

Figure 2.3 shows a comparison between predicted and experimental TG pyrolysis curves for three different kinds of biomass: almond shell [62], *Pinus radiata* [63] and wheat straw [64]. The biomass pyrolysis is treated as a linear combination of the pyrolysis of the seven reference components, derived from the ultimate or elemental composition. For instance, the almond shell composition (C/H/O = 0.509/0.061/

Table 2.4 Analysis and characterization of walnut shell ("Oxyflame project", DFG TRR 129)

	Proximate analysis			Analysis of Components (DAF wt%)					Ultimate Analysis (DAF wt%)			
	Moist.	Ash	Volatile	Fixed C	Cellulose	Hemicellulose	Lignins	Extractives	C	H	O	N
Analysis of fibers	2	0.9	–		60.8	4.2	32.4	n.a.				
Biochemical analysis	–	0.42	81	18.5	33.9	39.2	15.3	11.6				
Ultimate analysis									52.16	5.77	41.79	0.28

0.430) [21] corresponds to the distribution of reference species as follows: cellulose = 0.446, hemicellulose = 0.203, lignin = 0.283. Along with these main species, a small number of triglycerides and tannins are also considered. The composition and corresponding characterization of the other samples is reported elsewhere [65]. In the almond shell example, the differential contributions (DTG—Derivative ThermoGravimetry) of the individual reference components to the overall TG curve is also shown. While cellulose shows a single visible peak, both hemicellulose and lignins exhibit a three-step decomposition path over a wide temperature range. For wheat straw and *Pinus radiata*, the overall predicted DTG of the samples are reported, and they clearly show the peak of cellulose devolatilization at about 400 °C. *Pinus radiata* presents a shoulder before the cellulose peak, which correspond to GMSW hemicellulose decomposition. On the other hand, wheat straw has a smaller peak before 300 °C, corresponding to the thermal behavior of XYGR hemicellulose pyrolysis.

To further explore the subject, a more recent example of walnut shell pyrolysis is presented. Walnut shell is one of the standard biomass samples being investigated in the "OxyFlame" project (DFG – TRR 129). The composition of this sample was analyzed employing several analytical methods, which are summarized in Table 2.4. Proximate analysis was obtained with the standard procedure ASTM D5142 [19]. The components were analyzed in two ways: (1) using the detergent method (analysis of fibers) [66] and (2) by a combination of procedures, isolating and quantifying the sample's extractives [67], holocellulose [68] and hemicellulose [69], while cellulose and lignin contents were obtained by difference. Lastly, an ultimate analysis was carried out in accordance to the EN ISO 16948 standard and as reported previously in Senneca et al. [70].

Characterization in terms of reference components (DAF wt%)							
	Cellulose	Hemicellulose	Lignin_C	Lignin_H	Lignin_O	Tannins	Triglycerides
From ultimate analysis	32.8	17.8	5.3	3.4	31.5	7.1	2.1
From fiber analysis	62.1	4.3	11.2	11.2	11.2	0	0

(continued)

Characterization in terms of reference components (DAF wt%)							
	Cellulose	Hemicellulose	Lignin_C	Lignin_H	Lignin_O	Tannins	Triglycerides
From biochemical analysis	33.9	39.2	5.1	5.1	5.1	10	1.6

The walnut shell example provides an indication of the sensitivity of the model, in terms of decomposition profiles and product distribution, to the different methods and procedures used to characterize the biomass. Here, the simulations of the TGA of walnut biomass are analyzed during the characterization step based on three different assumptions. The first assumption uses the triangulation procedure based on the ultimate analysis [16]. The second assumption directly correlates the analysis of fibers with the reference species. Finally, the third assumption correlates the component or biochemical analysis to the reference species.

The left panel of Fig. 2.4 reports on the TGA of this biomass sample in an inert atmosphere at 10 [70] and 20 K/min [19]. The experimental procedures adopted are described in detail elsewhere [70, 71]. Comparison of the experimental results is useful to verify the reproducibility of the data in different facilities and for varying heating rates.

The center panel of Fig. 2.4 shows TGA under inert gas at a heating rate of 20 K/min and in comparison with model predictions. The differences between the TGA simulations and the experimental data are clear but not very extensive. It should be noted that the fiber method is usually applied for the nutritional value of animal feed and does not seem sufficiently precise for the characterization of biomass as a fuel. The uncertainty of the analysis is high; in fact, the amount of hemicellulose seems excessively low and far from the typical composition of almond and walnut shells [17]. However, not explicitly reported in Fig. 2.4, the distribution of volatiles is significantly different, since each reference component will release different pyrolysis products. It can be concluded that each analytical method has its flaws and uncertainties, and the best way to effectively characterize a biomass sample is to combine the results of the available analyses and cross-compare them with data from the literature, thus reducing invalid assumptions and the possibility of large errors. Finally, the right panel of Fig. 2.4 compares TGA in a CO_2 atmosphere at a heating rate of 10 K/min [70] with model predictions. The characterization applied is that used in biochemical analysis and confirms that the pyrolysis step is not clearly influenced by the CO_2 atmosphere, which is only reactive at higher temperatures, when the char is consumed by gasification reactions. The kinetic parameters for char gasification reactions were obtained from Tufano et al. [72].

The lumped mechanism in Table 2.3 is oversimplified, but effective for use not only on the particle scale, but also on the reactor scale. In fact, the computational time limits are very serious when simulating thermochemical conversions of biomass on the reactor scale [73, 74]. However, it is evident that this multistep kinetic mechanism has been and can be further improved and extended in terms of new reactions, kinetic parameters, and details of pyrolysis products, based on continuous research in biomass pyrolysis [56, 75–77].

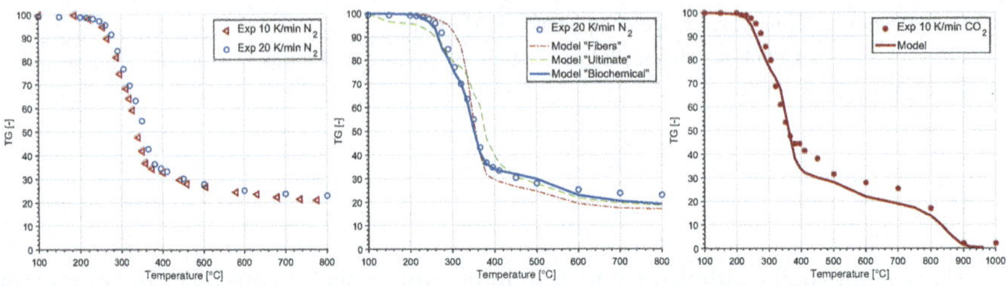

Fig. 2.4 Left: Experimental pyrolysis of walnut shell at (10 and 20) K/min in TGA. Center: Comparison between experimental TGA at 20 K/min (marks) and model predictions with different characterization assumptions (lines). Right: Experimental and model prediction of TGA at 10 K/min in CO_2 atmosphere

It is worth emphasizing that the interactions among reference species are not addressed in the present model [78]. Cellulose-lignin interaction can lead to a decrease in the levoglucosan yield [79] and an increase in light (C_1-C_3) compounds, especially glycolaldehyde and furans. This effect on pyrolysis products is most pronounced in grasses and softwood, possibly due to the increased prevalence of covalent bonds between cellulose and lignin in the cell wall. Hemicellulose–lignin interactions, and especially xylan–lignin interactions, may increase coniferyl alcohol yields. Further compositional features that may impact product distributions are the moisture content and the degree of acetylation of hemicellulose [80].

2.2.3 *Catalytic Effect of Ash*

Together with C/H/O, elemental analysis reveals the nitrogen and sulfur content by measuring NO_X and SO_X formation. The ash content is typically less than 1 wt% in wood, while it can reach more than 10% and 25% in grass and rice husks, respectively. The major elements in ash include Al, Ca, Fe, Mg, P, K, Si, Na and Ti and are usually expressed in terms of their oxides. Grass materials are rich in K and Na, and have a higher Si and lower Ca content than wood. Minor elements include As, Cd, Co, Cr, Cu, Hg, Mn, Mo, Ni, Pb, Sb, V and Zn.

It is well known that ash catalyzes and significantly modifies the overall biomass pyrolysis process. Lv et al. [81] investigated and observed a significant interaction between alkali and alkaline earth metals (AAEMs) and biomass components in terms of a decrease in initial gasification temperature with an increase in peak gasification value. The DFT study by Arora et al. [82] provides a detailed mechanistic insight into the catalytic effect of AAEMs on the first step of biomass pyrolysis (i.e. glycosidic bond cleavage). Zhou et al. [83] and Zhu et al. [84] investigated the catalytic effects of Na and Ca on the pyrolysis of carbohydrates from experimental and theoretical viewpoints. The sharp reduction in levoglucosan is mainly caused by the catalytic effect of Na on dehydration reactions, while Ca is more active than Mg in promoting the progressive decomposition of levoglucosan into char and permanent gases. With respect to levoglucosan reduction, the following reactivity trend was observed [85]:

$$K > Na > Ca > Mg \tag{2.1}$$

Accordingly, the ash catalytic effect is further confirmed by a decrease in bio-oil yields from fast biomass pyrolysis in the presence of a high ash content [2]. In several other investigations, the effect of minerals in the decomposition of single components and in biomass was investigated [71, 85–93]. These works show that similar effects in the decomposition of cellulose and hemicellulose are observed in the presence minerals. However, no clear understanding of the effect on lignin decomposition was achieved, and the authors mention that the extraction method employed lead to different effects of the minerals. To take into account the ash's effect on pyrolysis products, some simple modifications of the pyrolysis mechanism of cellulose and hemicellulose shown in Table 2.3 have been proposed [8, 94]. Because of the inconclusive effects on lignin decomposition, the corresponding changes were not employed in the lignin mechanism.

2.2.4 Characterization and Kinetic Model of Algae Pyrolysis

Macro- and microalgae-derived fuels are known as third-generation biofuels and have a promising role as alternative energy sources due to several advantages compared with lignocellulosic biomass: they do not compete with agriculture and produce better yields than oleaginous crops, their growth is fast, they are more capable of fixing CO_2, they have a high oil content and high heating value, and are often rich in proteins. Algae harvesting is important in terms of achieving high energy levels. The biochemical composition of algae is very complex and varies widely among species and cultivation conditions. Both micro- and macroalgae contain lipids, proteins, carbohydrates and ash. Ross et al. [95] discussed the differences in the pyrolysis behavior of algal and cellulosic biomass samples. Figure 2.5 shows a comparison between the TGA of almond shell and Macrocystis sp. (a macroalgae species). It is clear that the onset of pyrolytic decomposition occurs at a lower temperature for algae compared with lignocellulosic biomass. In the region of 250 °C, the main weight loss of macroalgae is consistent with high carbohydrate content, even if the catalytic effect of ash cannot be neglected. Then, a progressive loss of mass is observed from 300 °C to 800 °C, accounting for about 30% of the total mass. These behaviors reveal the different reactivity of algae carbohydrates compared with cellulose, and the presence of large amounts of complex non-crystalline compounds.

Algae have constituents of proteins, carbohydrates and lipids, which are present in various amounts depending on the taxonomy and growing conditions, but generally contain higher quantities of these constituents than lignocellulosic biomass. Debiagi et al. [96] developed a large database listing the main features of algae and

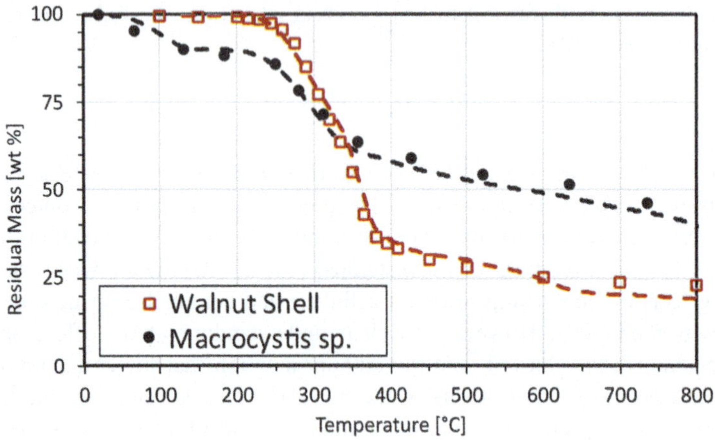

Fig. 2.5 TGA of fuels in N_2 showing mass loss at 25 K/min for algae (Macrocystis sp. [95] and 20 K/min for almond shell ("OxyFlame" project). Comparisons between experimental data (symbols) and model predictions (lines)

characterization methods together with a multistep pyrolysis mechanism for algae fuels. Starting out from the ultimate analysis and ash content, the biochemical composition of algal species is defined in terms of proteins, carbohydrates and lipids. A limited number of reference species are first defined, based on the atomic mass balances, i.e. on the elemental analysis. Three reference proteins were considered, which were rich in carbon, hydrogen and oxygen, respectively. Then, a multistep semi-detailed kinetic mechanism of algae pyrolysis was developed for the reference components, following the same approach applied for lignocellulosic biomass. Moreover, the further release of ammonium, nitrates and carbonate groups is taken into account and related to the ash content. A complete description of the pyrolysis model is found in Debiagi et al. [96].

Figure 2.6 refers to the pyrolysis of Spirulina (Arthrospira platensis), which is among the richest sources of proteins. This figure compares the model prediction and experimental data obtained with a Pyroprobe 5150 pyrolyzer [98] and in a thermogravimetric unit at a heating rate of 10 K/min [97].

2.3 Secondary Gas-Phase Reactions of Released Products

During biomass pyrolysis, primary volatile products can be exposed to high temperatures, with successive gas phase reactions playing a significant role [10, 28, 99, 100]. These secondary reactions of gas and tar products need to be taken into account using detailed kinetic mechanisms of pyrolysis and the combustion of hydrocarbon and oxygenated fuels [101].

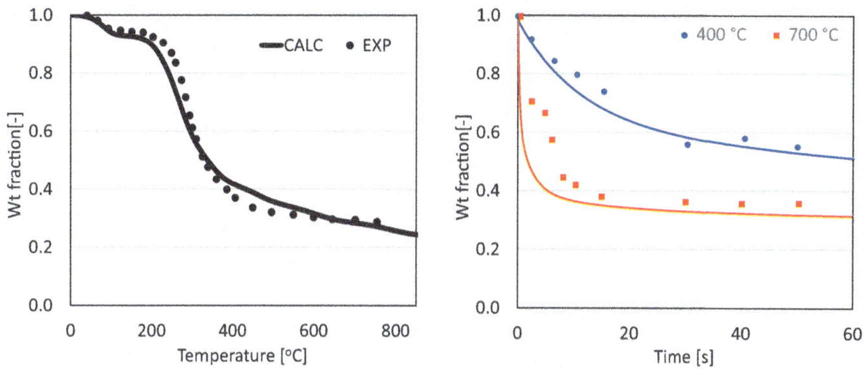

Fig. 2.6 Pyrolysis of Arthrospira platensis (Spirulina) in a TG unit at 10 K/min [97] (left) and in a pyroprobe [98] (right). Comparisons between experimental data (symbols) and model predictions (lines). Reprinted with permission from [96]. Copyright © 2017 Elsevier B.V

Biomass pyrolysis products are typically carbohydrates, phenols, alcohols and aldehydes, together with species containing two or more oxygenated groups. To limit the species and reactions in the kinetic scheme to a reasonable number, tars and heavy species are grouped into equivalent or lumped components representative of species and/or isomers with similar reactivity. Table 2.5 provides a list of major oxygenated species considered in the CRECK kinetic mechanism. Since alcohols and aldehydes have been discussed previously [57, 102–105], the discussion here is limited to aromatic species, which are particularly relevant as potential precursors of PAHs and soot particles. Thus, substituted aromatic compounds will be discussed in greater detail below [106–108]. Due to the large dimension of the kinetic mechanism, it is not feasible to perform ab-initio high-level calculations of the rate constants for all reactions of the biomass pyrolysis products. It is more suitable to systematically derive the kinetic law for a reaction class from first-principle calculations, based on a series of small and simple reactants. Then, the rate estimation rules can be extrapolated to all members of the same reaction class [99, 109].

Because of the hierarchical and modular structure of the kinetic scheme, its extension to include new species released by the pyrolysis of biomass simply requires the inclusion of the primary reactions of these species. Usually, the reaction classes to be included are initiation, H-abstraction and addition reactions, together with successive propagation reactions until products form which were already considered in the mechanism. At high temperatures, the decomposition reactions of volatile products from biomass pyrolysis are mainly responsible for decreases in bio-oil yields and increases in gas products. Therefore, it is possible to find the conditions which produce the highest yield of bio-oil and the corresponding optimal temperature and operating conditions [1, 9, 110, 111].

Mainly for oxygenated species, molecular reactions constitute a further important class. As a simple example, the two successive dehydration reactions in the pyrolysis of glycerol, through four center molecular reactions, are very important to explain

Table 2.5 Formation enthalpy ΔH_f,298 and entropy ΔS_f,298 of relevant oxygenated species released from biomass pyrolysis and involved in secondary gas-phase reactions

Chemical name	Formula	ΔH_f [kcal/mol]	ΔS_f [cal/(mol K)]
Glyoxal	$C_2H_2O_2$	−50.6	65.4
Acetaldehyde	C_2H_4O	−39.5	63
Acetic acid	$C_2H_4O_2$	−103.9	67.4
Hydroxy-acetaldehyde	$C_2H_4O_2$	−73.5	73.6
Ethylene-glycol	$C_2H_6O_2$	−92.0	76.3
Acrolein	C_3H_4O	−20.3	67.4
Propanedial	$C_3H_4O_2$	−62.4	73.7
3-Hydroxy-2-oxo-propanal	$C_3H_4O_3$	−102.7	88.4
Propanal	C_3H_6O	−45.3	72.8
1-Propanol	C_3H_8O	−60.9	76.4
2-Propanol	C_3H_8O	−65.5	74.5
Acetol	$C_3H_6O_2$	−87.4	80.6
3-Hydroxypropanal	$C_3H_6O_2$	−80.3	83.3
1,3-Propanediol	$C_3H_8O_2$	−45.5	86.0
Glycerol	$C_3H_8O_3$	−137.1	95.8
Furan	C_4H_4O	−10.2	60.2
Butanedione	$C_4H_6O_2$	−78.4	84.2
C4 O-heterocycles	C_4H_8O	−27.7	73.6
Furfural	$C_5H_4O_2$	−36.1	77.8
Xylosan	$C_5H_8O_4$	−151.6	104.8
Phenol	C_6H_6O	−23.0	75.3
Catechol	$C_6H_6O_2$	−65.9	86.3
Hydroxymethyl-furfural	$C_6H_6O_3$	−79.8	98.2
Levoglucosan	$C_6H_{10}O_5$	−200.9	113.5
Anisole	C_7H_8O	−17.1	84.0
Guaiacol	$C_7H_8O_2$	−60.0	94.9
Syringol	$C_8H_{10}O_3$	−95.3	111.0
Vanillin	$C_8H_8O_3$	−88.7	116.4
Coumaryl alcohol	$C_9H_{10}O_2$	−49.2	109.0
Heavy Molecular Weight Lignin (HMWL)	$C_{24}H_{28}O_4$	40.0	186.7

the initial formation of the reactive acetol and 3-hydroxypropanal intermediates, which rapidly decompose to form the most stable species: acetaldehyde and acrolein [112, 113].

The complete kinetic scheme in CHEMKIN format is available on the website: www.creckmodeling.chem.polimi.it, together with the thermodynamic and transport properties of all species involved.

2.3.1 Reference Kinetic Parameters and Rate Rules for H-Abstraction Reactions

The H, OH and CH_3 are the dominant reactive radicals in pyrolysis and oxidation conditions. The rate rules of H-abstraction or metathesis reactions of hydrocarbons have been well defined for several years [15, 114]. H-abstraction reactions can be written in the generic form:

$$R\cdot + R'H \leftrightarrow RH + R'\cdot \tag{2.2}$$

where R● is the H-abstracting radical and R'H the hydrocarbon. The rate constant of this reaction can be decomposed into the product of two terms:

$$k_f = k^0_{\mathrm{ref},R} \cdot C_{R'H} \tag{2.3}$$

where $\boldsymbol{k^0_{\mathrm{ref},R}}$ is the reference kinetic parameter of the R radical to abstract an H atom from a methyl group and $\boldsymbol{C_{R'H}}$ is the reactivity of the specific H atom with respect to the primary one. Figure 2.7 shows the rate constants of H-abstraction reactions of H, OH and CH_3 radicals from primary, secondary, tertiary and vinyl positions. These rate constants are strongly correlated with the corresponding C-H dissociative bond dissociation energies (BDEs). The rate constants of H abstraction reactions from aromatics, forming phenyl-like radicals, are similar to those in the case of H-abstraction from a vinyl H atom, while the rate constants attributed to benzyl radicals formation are more similar to the ones required to form allyl radicals [115, 116]. Very similar rate rules have recently been summed up in a review by Wang et al. [109]. Generic rate rules can be formulated for abstraction reactions involving different H sites not only in hydrocarbons but also in oxygenated species. The difference in BDEs explains the relative selectivities of H-abstraction reactions of the different H sites [104].

Removal of the acyl H atom is highly favored in aldehydes due to the low BDE of the C-H bond in the carbonyl group [117], as only short-range forces (i.e. in the order of magnitude of the bond length) can affect the reaction rates [118]. Therefore, the influence of the acyl or hydroxyl group on the reactivity of C-H bonds practically

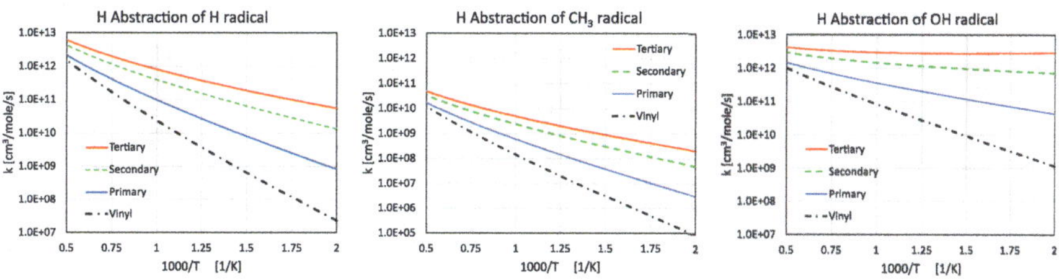

Fig. 2.7 H-abstraction reactions. Rate constants (per H atom) for primary, secondary, tertiary and vinyl H atoms

vanishes after the β position, which is in agreement with the conclusions drawn by Carstensen et al. [99].

All of these rate rules for H-abstraction reactions are useful to produce an initial set of rate parameters for secondary gas phase reactions. Successive rate and sensitivity analyses are then able to highlight sensitive reactions which require more detailed evaluation.

2.3.2 *Phenols and Substituted Aromatic Species*

Together with the kinetics of benzene and toluene, phenol reactions are important, firstly for their presence as tar components released by lignins and secondly for their role as precursors of dibenzofurans, dibenzodioxins and PAHs. Kinetic studies on phenol and cresol chemistry highlight the importance of CO elimination from unsubstituted and substituted phenoxy radicals [99]; successive reactions of cyclopentadienyl radicals are responsible for the formation of naphthalene and heavier PAHs [119]. While phenol and cresol have been investigated to define their role in combustion systems, anisole ($C_6H_5OCH_3$) has been studied as a simple surrogate of tar from lignin pyrolysis [107, 120, 121]. Chain initiation reactions of aromatic species containing one or more methoxy groups (-OCH_3) involve the breaking of the weak O-CH_3 bond. The presence of different functional groups on the aromatic ring greatly affects the bond dissociation energies of vicinal bonds. Figure 2.8 compares the BDEs of a series of mono- and poli-substituted aromatic components calculated at the CCSD(T)/aug-cc-pVTZ level, corrected for basis-size effects on M06-2X/6-311+G(d,p) geometries (T = 298 K) [108]. These BDEs provide the basis for defining rate rules to describe the pyrolysis and oxidation of the aromatic species.

Catechol has the lowest BDE of O-H bond (78 kcal/mol), while salicylaldehyde has the highest value (94.1 kcal/mol). Similarly, there is a difference of 8 kcal/mol for the phenyl-OH bond energy between salicylaldehyde (118.1 kcal/mol) and

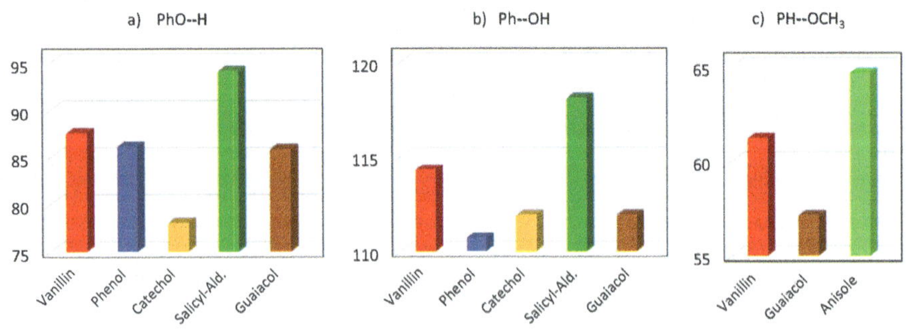

Fig. 2.8 Comparison of bond dissociation energies (BDEs) [kcal/mol] in different aromatic compounds. Panel (**a**) PhO--H bond. Panel (**b**) Ph--OH bond. Panel (**c**) Ph--OCH3 bond

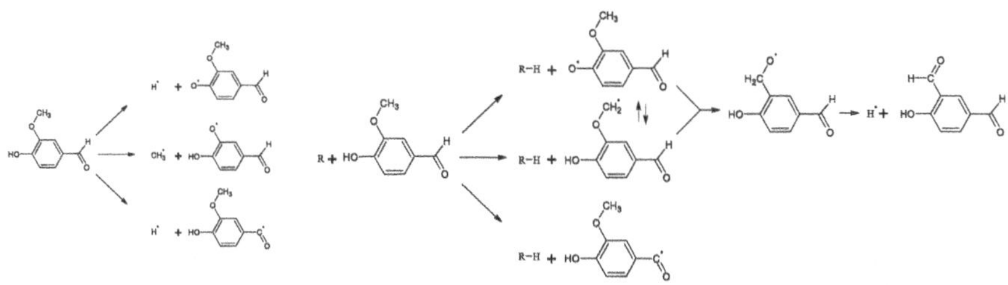

Fig. 2.9 Chain initiation and H abstraction reactions of vanillin

phenol (110.7 kcal/mol). Finally, the bond energy of the methoxy group is lowest in guaiacol (57.1 kcal/mol) and highest in anisole (64.7 kcal/mol). These large differences in BDEs not only allow different selectivities to be defined in H-abstraction reactions, but also involve great variability in the initial decomposition reactions.

As an example, Fig. 2.9 shows favored chain initiation and H-abstraction reactions of vanillin pyrolysis based on BDEs. The phenoxy radical produced by the most favored vanillin initiation reaction easily forms 3,4-dihydroxybenzaldehyde, an important intermediate whose subsequent extraction reactions form the guaiacol radical and 4-hydroxy-isophthalaldehyde.

While further details on the chemistry of these aromatic species are reported in Pelucchi et al. [108], Fig. 2.10 shows a very detailed comparison of pyrolysis products from catechol decomposition. Experiments were conducted at temperatures from 600 °C to 1000 °C and with a residence time of 0.3 s [122]. It is clear from the data that catechol decomposition is very significant after 700 °C, reaching total conversion in this very short residence time at temperatures above 800 °C. Despite the fact that little decomposition is observed below 700 °C, it is important to highlight that the residence time is very short (0.3 s). In industrial scale reactors, such a short residence time of the vapor phase is difficult to attain, typically resulting in a few seconds of contact time before being swept out of the reacting zone. Therefore, decomposition of lignin-derived tars cannot be completely neglected below 700 °C when the operating conditions allow longer exposition time for the released volatiles. Satisfactory agreement is shown for both light hydrocarbons and aromatic species. There is a notably high yield of butadiene, formed mainly through a molecular reaction and a subsequent high quantity of PAH species.

Norinaga et al. [123, 124] and Yang et al. [125] developed a two-stage tubular reactor for evaluating the rapid biomass pyrolysis and then the successive decomposition of biomass pyrolysis products, while minimizing the interactions amongst char and volatile species. These data are very useful to validate the secondary gas-phase reactions of biomass pyrolysis products and the time evolution of the most abundant light products (CO, H_2O, CH_4, H_2 and methane) as well as the formation of benzene, PAHs and soot [100]. Saggese et al. [126, 127] revised and discussed the high temperature reactions of benzene and aromatics, paying particular

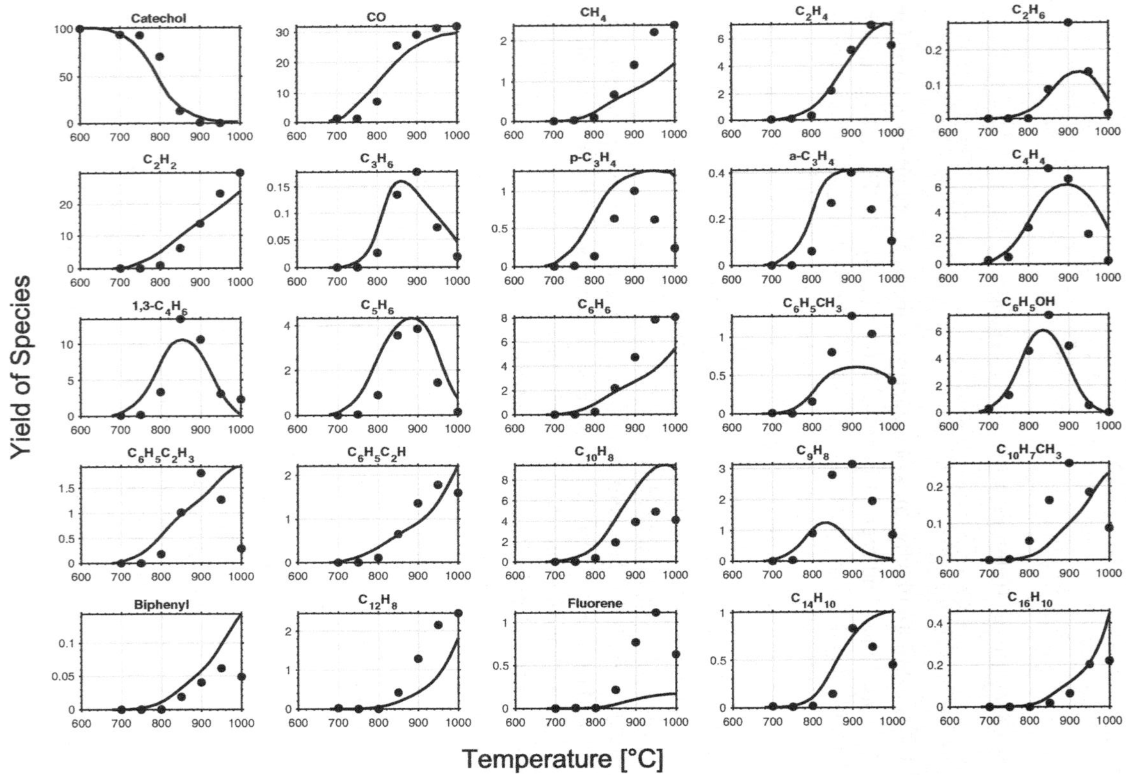

Fig. 2.10 Catechol pyrolysis at a pressure of 1 atm and $\tau = 0.3$ s. Species yields versus temperature [°C]. Comparison between model predictions (lines) and experimental data (symbols) [122]

attention to the successive reactions forming PAHs and soot particles. Pejpichestakul et al. [128] analyzed and discussed the soot kinetic model in premixed laminar flames under fuel-rich conditions.

2.4 Biomass Pyrolysis and Balance Equations on the Particle and Reactor Scales

The type of biomass and process operating conditions greatly influence the gas, tar and residual char products from biomass pyrolysis. Torrefaction, pyrolysis and gasification are three modes of thermal treatment for biomass, depending on the heating rate, temperature and residence times. Table 2.6 shows the operating temperatures, timescales and product yields of different pyrolysis and gasification processes [1, 110].

Torrefaction is a mild, gentle heat treatment useful for improving energy density and biomass grindability. Slow pyrolysis maximizes biochar production at low temperatures (300–500 °C) and when released volatiles have long residence times. Tar species favor the charification process through cross-reticulation and condensation reactions. By contrast, fast pyrolysis operates at high heating rates and short vapor residence times, and optimizes bio-oil yields. Small biomass particles in fluidized-bed reactors are optimal for the fast heating process, whereas large particles (3–6 cm), pellets, or biomass briquettes in packed-bed reactors are the usual biomass feed for the slow pyrolysis process. Safarian et al. [129] published an interesting review on the modeling of biomass gasification.

The fast pyrolysis process centers on the reactor, and research efforts are mostly devoted to developing new reactor configurations, with particular attention paid to the optimal way of providing process heat. Bridgwater [1] recapped and discussed the major features of fast pyrolysis reactors.

- Bubbling and circulating fluid beds are simple, proven technologies, which provide efficient heat transfer and good temperature control. Char products can

Table 2.6 Biomass pyrolysis, torrefaction and gasification processes [1]

Mode	Temperature (°C)	Residence time		Product yields (wt%)		
		Vapor	Solid	Liquid	Solid	Gas
Torrefaction	280		(10–60) min	0	80	20
Carbonization (Slow Pyrolysis)	400	days	hours	30	35	35
Intermediate Pyrolysis	500	(5–30) s		50	25	25
Fast Pyrolysis	500	(1–2) s		75	12	13
Gasification	750–900	(1–5) s		3	1	95

be used to provide the pyrolysis heat. Because of the high velocities, char attrition in circulating beds can become a negative issue.

- The rotating cone reactor, with rapid heating and a short solid residence time, produces the flash biomass pyrolysis with negligible char formation.
- In ablative pyrolysis, heat is transferred from the hot reactor wall to the wood surface. The pyrolysis front moves through the biomass particle, and the reaction rate strongly depends on the pressure applied to the wood on the heated surface.
- The auger pyrolysis reactor is characterized by a double screw, where chopped biomass particles are mixed with hot sand and decomposed into vapors and char.
- Hydropyrolysis combines pyrolysis and hydrocracking, adding hydrogen to reduce the oxygen content of the bio-oil product.
- Heating and pyrolysis in microwave reactors are largely different from previous techniques as biomass particles are rapidly heated from the inside. The reduction of thermal gradients allows us to study the fundamentals of fast pyrolysis kinetics.

Intra- and inter-phase heat and mass transfer phenomena need to be considered and coupled with kinetics when modeling reactors treating large particles. According to previous works [73], a convenient way to present the mass and energy balance equations is to distinguish between the particle and the reactor scale. The particle model should be able to predict temperature profiles and product distribution as a function of time. This means that the model requires not only reaction kinetics, but also reliable rules for estimating effective transport properties to account for morphological changes during the pyrolysis process. Biomass particles shrink by as much as 60–70% in different directions during the conversion process. Heat transfer should account for variable transport properties of the reacting biomass and the char residue [130, 131].

2.4.1 Balance Equations on the Particle Scale

The mathematical model for the evolution of the biomass particle is based on fundamental governing equations of conservation of total mass, momentum and energy, for both the fluid and solid phases. The particle is considered as a porous medium, i.e. the solid volume and the fluid contained inside its pores. Heat transfer occurs by means of conduction, convection and radiation. There is assumed to be a local thermal equilibrium between the solid and the gas phase, with the Péclet number for heat transfer being sufficiently large. Rigorous governing equations are discussed in depth in Gentile et al. [132].

The simplified condition of isotropic spherical particles is considered here. These equations include the gas and solid phase conservation equations. In particular, in the solid phase it is possible to refer to Eqs. (2.4) and (2.5):

Continuity equation:

$$\frac{\partial}{\partial t}\left[\rho^S(1-\varepsilon)\right]+\nabla\cdot\left(\rho^S\mathbf{u}_r^S\right)=\sum_{j=1}^{NC^S}\dot{\boldsymbol{\Omega}}_j^S \tag{2.4}$$

Species equations:

$$\frac{\partial}{\partial t}\left[\rho^S(1-\varepsilon)\omega_j^S\right]+\nabla\cdot\left(\rho^S\mathbf{u}_r^S\omega_j^S\right)=\dot{\boldsymbol{\Omega}}_j^S\left[j=1,\ldots,NC^S\right] \tag{2.5}$$

whereas the gas phase conservation equations (Eqs. (2.6), (2.7), (2.8), and (2.9)) are:

Continuity equation:

$$\frac{\partial}{\partial t}\left(\rho^G\varepsilon\right)+\nabla\cdot\left(\rho^G\mathbf{u}_r^G\right)=\sum_{k=1}^{NC^G}\dot{\boldsymbol{\Omega}}_k^G \tag{2.6}$$

Species equations:

$$\frac{\partial}{\partial t}\left(\rho^G\varepsilon\omega_k^G\right)+\nabla\cdot\left(\rho^G\mathbf{u}_r^G\omega_k^G\right)=-\nabla\cdot\left(\rho^G\omega_k^G\mathbf{v}_\mathbf{k}^\mathbf{C}\right)+\dot{\boldsymbol{\Omega}}_k^G\left[k=1,\ldots,NC^G\right] \tag{2.7}$$

Momentum equation:

$$\begin{aligned}\frac{\partial}{\partial t}\left(\rho^G\varepsilon\mathbf{u}\right)+\nabla\cdot\left(\rho^G\mathbf{u}_r^G\otimes\mathbf{u}\right)=&-\nabla p+\nabla\\&\cdot\left[\mu\left(\nabla\mathbf{u}+\nabla\mathbf{u}^T\right)-\frac{2}{3}\mu(\nabla\cdot\mathbf{u})\mathbf{I}\right]+\rho^G\mathbf{g}\\&+(\mu\mathbf{Da})\mathbf{u}\end{aligned} \tag{2.8}$$

Energy equation:

$$\begin{aligned}&\widehat{C}_p^G\frac{\partial\left(\rho^G\varepsilon T\right)}{\partial t}+\widehat{C}_p^G\nabla\cdot\left(\rho^G\mathbf{u}_r^GT\right)+\widehat{C}^S\frac{\partial\left(\rho^S(1-\varepsilon)T\right)}{\partial t}+\widehat{C}^S\nabla\left(\rho^S\mathbf{u}_r^ST\right)\\&=\nabla\cdot\left(\boldsymbol{\lambda}_{\mathbf{eff}}\nabla T\right)+\dot{Q}_R-\rho^G\sum_{k=1}^{NC^G}\widehat{C}_{p,k}^G\omega_k^G\mathbf{v}_\mathbf{k}\end{aligned} \tag{2.9}$$

The conservation equations above require the proper definition of boundary conditions at the particle surface for pressure, velocity, the species mass fractions (for both the gas and the solid phase) and temperature, as summarized below (Eqs. (2.10), (2.11), (2.12), (2.13), and (2.14)):

Pressure:

$$p = p_{\text{ext}} \tag{2.10}$$

Velocity:

$$\nabla \mathbf{u} = 0 \tag{2.11}$$

Solid phase species mass fraction:

$$\nabla \omega_j^S = 0 \left[j = 1, \ldots, NC^S \right] \tag{2.12}$$

Gas phase species mass fraction:

$$-\mathbf{n}\left(-\mathcal{D}_{\text{eff},k}\rho^G \nabla \omega_k^G\right) = k_{\text{ext}}\rho^G \left(\omega_{k,\text{ext}}^G - \omega_k^G\right) \left[k = 1, \ldots, NC^G\right] \tag{2.13}$$

Temperature:

$$-\mathbf{n}(-\lambda_{\text{eff}} \nabla T) = h_{\text{ext}}(T_e - T) + \xi\sigma\left(T_e^4 - T^4\right) \tag{2.14}$$

This system can be conveniently simplified in the case of isotropic spherical particles, discretizing the particles with an onion-like structure of concentric iso-volumetric shells [11].

The specific heats are evaluated neglecting mixing effects, and the very low-pressure work induced by gas expansion is not considered.

A couple of examples of biomass pyrolysis are investigated on the particle scale, including the effect of secondary gas-phase reactions. The model is first applied to analyze the temperature profiles during the pyrolysis of large biomass particles. A second example discusses fast pyrolysis of biomass and the optimal operating temperature for the highest bio-oil yield.

2.4.2 Pyrolysis of Large Biomass Particles

Large biomass particles are often used when charcoal is the desired product or when rapid heating rates are not required. Slow pyrolysis of wood chips and centimeter-scale wood particles is useful to optimize the production of biochar or charcoal for soil amendment [133, 134]. From a modeling point of view, the pyrolysis of large particles provides a sensitive and useful test for kinetic models of biomass pyrolysis, mainly with respect to thermochemical properties.

Park et al. [135] studied the pyrolysis of large spherical particles in the temperature range of 638–879 K and measured global mass loss and temperature evolution at the surface and center of the particle. The center temperature profile exhibits an initial increase, then a plateau followed by a sharp peak which can exceed the surface temperature. Corbetta et al. [50] discuss how the competition between the char

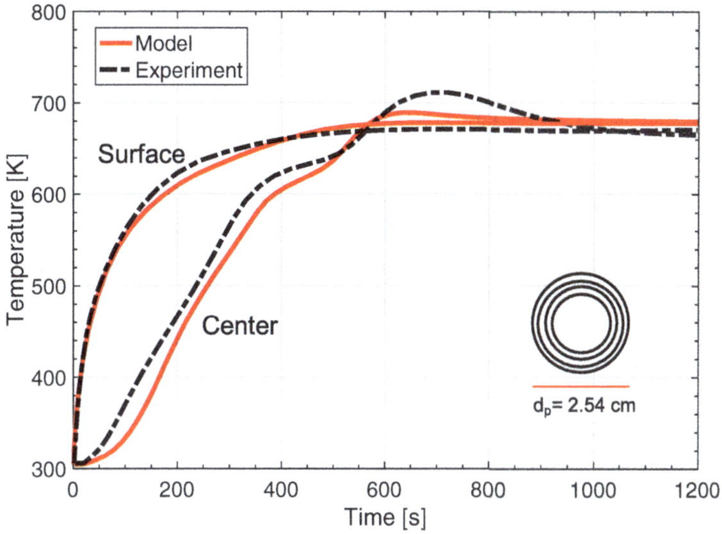

Fig. 2.11 Surface and center temperature profiles in a wood sphere at 688 K. Comparison between experimental data (dashed lines) and model predictions (solid lines) [135]

formation exothermic reactions and the heat transport resistance in the wood particle can explain the possible overshooting of the internal temperature.

Figure 2.11 shows the experimental and predicted temperature profiles of center and surface temperatures from the pyrolysis experiment on a wood sphere which is 2.54 cm in diameter at 688 K [135]. The center temperature initially increases until reaching an inflection point at about 600–650 K, where there is a plateau due to the latent heat required for the devolatilization of tar products. Thereafter, the temperature increases, exceeding the nominal surface temperature.

Following the approach proposed by Paulsen et al. [136], it is convenient to use the pyrolysis number and the Biot number to compare the timescales of heat transfer and pyrolysis reactions for the fuel particle. The Biot number (Eq. (2.15)), which is the ratio of the conduction and convection timescales, provides the relative importance of external and internal heat transfer:

$$\mathrm{Bi} = \frac{h \cdot R_p}{k_p} \tag{2.15}$$

where h is the external heat-transfer coefficient, k_p is the thermal conductivity of the particle and R_p is the particle radius. The high external heating rates and low thermal conductivity of large particles correspond to a large Biot number, which causes large temperature gradients within the particle. Biomass particles larger than (100–200) μm usually have Biot numbers greater than 1.

Pyrolysis numbers (Py_1 and Py_2) (Eqs. (2.16) and (2.17)) are the ratios of the reaction timescale and the conduction or the convection timescales:

$$Py_1 = \frac{k_p}{\rho_p \widehat{C} k_R R_p^2} = \frac{1}{\mathrm{Th}^2} \tag{2.16}$$

$$Py_2 = \frac{h}{\rho_p \widehat{C} k_R R_p} = Py_1 \cdot \mathrm{Bi} \tag{2.17}$$

where k_R is the rate constant of the biomass devolatilization reaction and Py_1 is equivalent to the reverse of the squared thermal Thiele modulus (Th).

By using Pyrolysis and Biot numbers and comparing the timescale of pyrolysis reactions and conductive and convective heat transfer, it is possible to distinguish between thermally thin and thermally thick particles, highlighting at least two typical regimes. At $\mathrm{Bi} < 1$ and $Py > 1$, there is an isothermally and kinetically limited region where the thermally thin particle has a uniform temperature. At $\mathrm{Bi} > 1$ and $Py < 1$, there is a region of limited conduction where there are significant temperature gradients within the thermally thick particle [50, 136].

At temperatures higher than 750 K, secondary reactions in the gas surrounding the particle can play an important role. A large fraction of tar components can therefore decompose with a corresponding increase in the gas fraction. The effect of secondary gas phase reactions will be further discussed in Sect. 2.4.3, where biomass fast pyrolysis and bio-oil formation are analyzed.

2.4.3 Fast Biomass Pyrolysis and Bio-Oil Formation

The fast pyrolysis process typically involves high biomass particle heating rates and released products with short residence times. Bio-oil yields can be as high as 50–70% on a weight basis, while the flash pyrolysis process can produce even higher bio-oil yields [137]. Small biomass particles in fluidized-bed reactors, with or without recirculation, are common practice for fast biomass pyrolysis, where the contact times of bio-oil products at high temperatures are minimized. Namely, fine biomass particles are rapidly heated to the optimum temperature, with minimal exposure to low temperatures which favor secondary char formation. There are two reasons for the lower bio-oil yield at low temperatures: incomplete devolatilization of the solid particles and a favored charification process due to the condensation reactions of tar species. Moreover, the gas-phase decomposition reactions of tar components at high temperatures lower yields of bio-oil. Thus, it is clear that both chemical processes and heat and mass transfer play a fundamental role in identifying optimal operating conditions to maximize the bio-oil yields from the fast pyrolysis process.

Currently, bubbling and circulating fluidized-bed processes produce bio-oil on a commercial scale, using wood or wood waste [1]. Circulating fluidized-bed reactors are suitable for larger throughputs than bubbling reactors even though their hydrodynamics are more complex.

Table 2.7 Optimal conditions for fast biomass pyrolysis

Biomass composition	Pine spruce sawdust		Sesame stalk		Pine wood	
Cellulose	48.7		26.1		35.0	
Hemicellulose	21.4		21.3		29.0	
Lignin	21.9		43.9		28.0	
Moisture	8.0		8.7		8.0	
Reactor type	Spouted bed		Fixed bed		Fluidized bed	
Temperature (K)	720		770		750	
Weight fractions	Model	Experiment	Model	Experiment	Model	Experiment
Solid residue	0.14	0.12	0.23	0.22	0.27	0.17
Gases	0.17	0.17	0.23	0.26	0.18	0.23
Total liquids	0.69	0.70	0.54	0.52	0.55	0.58
$-H_2O$	0.12	0.09	0.14	0.16	0.13	0.12
−organic liquids	0.57	0.61	0.40	0.36	0.42	0.45

Comparison of experimental data and model predictions [141–143]

Schematically, in a fast pyrolysis process, small particles of dried and grinded biomass are fed into a fluidized-bed reactor. The residence time of the product gases must be short. After the tar products have condensed, a dark red-brown liquid bio-oil with a density of ~1200 kg/m^3 is obtained. The combustion of the char product provides the heat required by the endothermic pyrolysis reactions. Although large particles produce slightly lower oil yields, grinding to these sizes is less expensive than to finer particles.

Torrefaction, which is a light thermal pre-treatment of the biomass in anoxic conditions, is very useful to improve the quality of the feed in terms of energy density and grindability properties. Through the initial decomposition of hemicelluloses, coupled with the partial depolymerization of cellulose and the thermal softening of the lignin, the cell wall in the biomass sample is considerably weakened. For pyrolysis and combustion processes, torrefaction also guarantees that the biomass fuel is more homogeneous [7, 138–140].

As discussed by Calonaci et al. [111], Table 2.7 compares experimental data and model predictions for the fast pyrolysis of three different biomass samples:

- pine spruce sawdust in a conical spouted-bed reactor [141];
- sesame stalk in a fixed-bed reactor [142];
- pine wood in a fluidized-bed reactor [143].

In line with the experimental data, the model predicts the highest bio-oil yields to be between 50 and 70% in the temperature range of 720–770 K.

The model predictions in Table 2.7 have been obtained using a comprehensive mathematical model of biomass pyrolysis on the particle scale, including the surrounding gas phase. Therefore, bio-oil yields are evaluated considering the coupling of chemical and transport processes within the biomass particles, as well as fluid dynamics inside the reactor, which play a crucial role in defining the residence times of volatile products and therefore the evolution of secondary pyrolysis reactions of tar components in the gas phase.

For small particles, the model predicts flat bio-oil yields and gas formations in a temperature range of ±50 K around the maximum, where biomass devolatilization goes to completion. Carbon oxides and water, together with small quantities of CH_4 and C_2 hydrocarbons, are the main gas species from the primary devolatilization. The H_2 yield is very limited and only occurs at high temperatures, where the residual char is nearly constant. The chemical compositions of product liquids predicted by the model agree fairly well with other experimental data available in the literature, as reported in more details in [8, 16]

Fast pyrolysis bio-oil is non-flammable, non-distillable and has only limited volatility. It contains both an oil and an aqueous fraction (15–30% water), and for this reason it is immiscible with traditional liquid hydrocarbon fuels. The aqueous fraction contains low-molecular-weight oxygenated compounds, whereas the tar fraction is constituted by high-molecular-weight, water-insoluble lignin fragments (pyrolytic lignin). Besides water, bio-oils are composed of complex mixtures of hundreds of organic compounds such as phenolic components, acids, aldehydes, ketones, alcohols, esters, anhydro-sugars, furans and nitrogen-containing compounds, as well as large anhydro-oligosaccharides and lignin-derived oligomers. Oxygenated compounds make up 50–60 wt% of the products [108].

Because of its complexity, the kinetic modeling of bio-oil combustion benefits from the definition of a limited number of reference species accounted for in surrogate fuel formulations. Surrogate mixtures for bio-oils typically include, alongside phenol, a relevant amount of more complex phenolic components, such as guaiacol, catechol and vanillin. Table 2.8 reports on a surrogate mixture of pyrolysis bio-oil as proposed by Pelucchi et al. [108]. Vanillin (4-hydroxy-3-methoxybenzaldehyde) is one of the most interesting representatives of the phenolic fraction derived from lignin pyrolysis.

To highlight the variability in the bio-oil yield from the pyrolysis of different biomass samples, Fig. 2.12 illustrates model predictions from the pyrolysis of pure cellulose, beech and pine wood, switchgrass and rice husks. This figure shows that the highest bio-oil yields vary between 45% and 75%, and also shows that in these pyrolysis conditions, the primary biomass pyrolysis ends before 650–700 K, while the decomposition reactions of tar species are particularly relevant at temperatures above 800 K. The biomass with the highest cellulose content produces the highest

Table 2.8 Surrogate mixture of pyrolysis bio-oil

Component	wt%	Component	wt%
Water	22	Acetic acid	3.9
Ethylene glycol	5.5	Glycol aldehyde	5.5
Vanillin	17.9	Lignin, alkali	7.8
Levoglucosan	29.6	2,5-Dimethylfuran	5.5
Oleic acid	2.34		

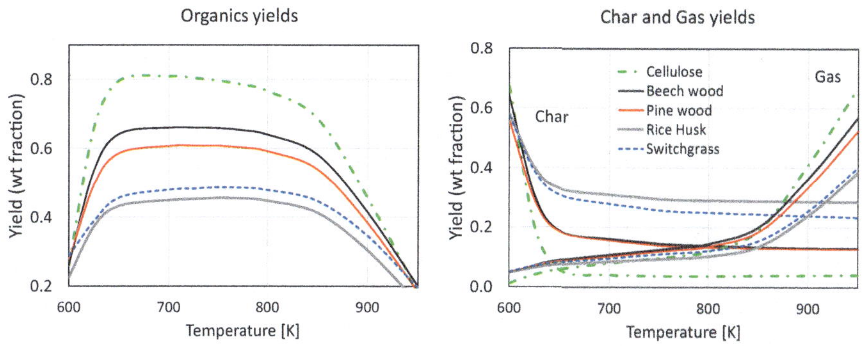

Fig. 2.12 Predicted yields of organics, char and gas from the fast pyrolysis of typical biomass samples. The residence time of biomass particles was set to 5 min, and that of released volatiles was set to 2 s

bio-oil yields, and yields continuously increase with the temperature. At temperatures below 700 K, an appropriate residence time (mainly based on the particle size) is essential to complete the pyrolysis process and maximize the bio-oil yield. The bio-oil yields shown in Fig. 2.12 refer to average ash conditions, while a higher or lower content of inorganics can modify the bio-oil production by up to about 10% due to the catalytic effect of alkali and alkali-earth metals [9].

2.4.4 *Biochar Formation, Yield and Composition*

As discussed in depth in Debiagi et al. [16], the pyrolysis model also accounts for the formation of biochar, which consists of a solid carbonaceous structure containing significant amounts of oxygen and hydrogen, a minor amount of nitrogen and sulfur, together with metal oxides (i.e. ash). The carbon content of biochar usually ranges between 65% and 95% depending on the initial biomass composition and pyrolysis operating conditions. The hydrogen and oxygen content of biochar progressively drop when the pyrolysis temperature increases [144], and their content directly influences the rate of biochar conversion during oxidation and gasification processes. In particular, H and O sites disturb the organization of the crystalline carbon matrix, with the formation of amorphous areas and weakly bonded functional groups in the biochar. At high temperatures ($T > 1000$ °C) biochar undergoes progressive graphitization or annealing [145]. Heterogeneous secondary reactions during slow pyrolysis can strongly affect biochar reactivity, mainly in the case of large particles. The mechanism reported in Table 2.3 is able to predict the biochar formation process, as clearly shown by the parity diagram in Fig. 2.13.

Being able to reliably predict the yield and composition of the residual solid is essential when seamlessly describing the heterogeneous reactions taking place during the biochar conversion step of biomass combustion.

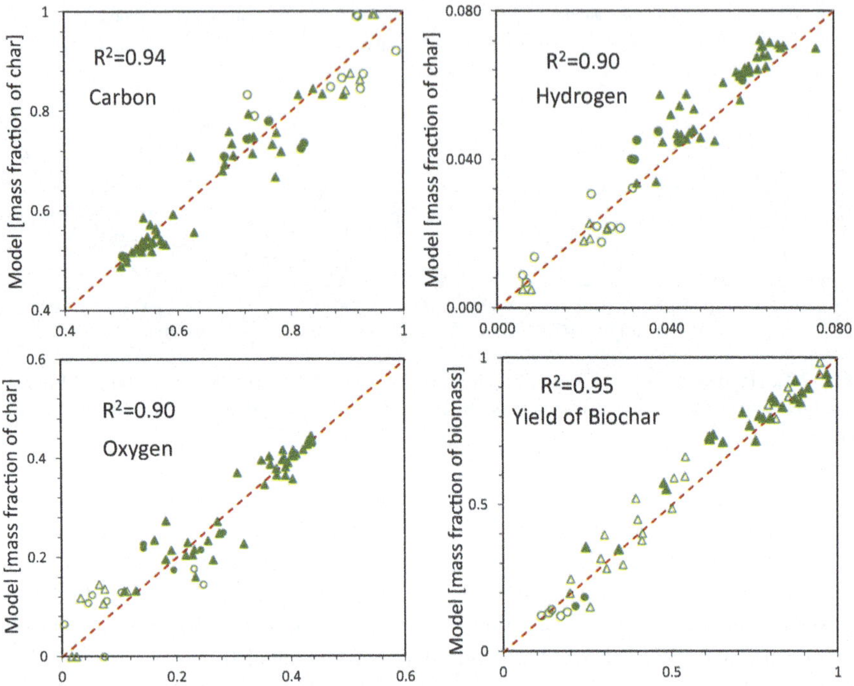

Fig. 2.13 Parity diagrams of model predictions and experimental values of the biochar yield and composition (carbon, hydrogen and oxygen) [16]

2.4.5 Balance Equations on the Reactor Scale

While the description of entrained and fluidized-bed reactors can simply refer to the previous particle model, the modeling of packed or fixed-bed reactors can greatly benefit from the definition of an elemental reactor layer that characterizes gas-solid interactions. The packed bed of biomass particles can be described as a series of *NR* elemental layers [11, 73] with the height of each layer being in the order of the dimension of the biomass particle to account for vertical dispersion. Both the gas and solid phases are assumed to be perfectly mixed inside the layer. In fact, mixing of the gaseous phase within the layer is further improved by jets of volatile species released during biomass pyrolysis [146].

The mass balance equations (Eqs. (2.18) and (2.19)) for the gas phase of each elemental reactor are:

$$\frac{dm_k}{dt} = \dot{m}_{k,\text{in}} - \dot{m}_{k,\text{out}} + J_k S N_p + V_R \dot{\Omega}_k^G \tag{2.18}$$

where m_k is the mass of the kth gas species within the reactor volume V_R, $\dot{m}_{k,\text{in}}$ and $\dot{m}_{k,\text{out}}$ are the inlet and outlet flow rates, $V_R \dot{\Omega}_k^G$ is the net formation from gas-phase reactions, the term J_k is the gas-solid mass exchange multiplied by the particle surface $S$and N_p is the number of particles inside the layer. The energy balance equation for the gas phase of each elemental reactor is:

$$\frac{d\sum_{k=1}^{NC^G} m_k \widehat{h}_k^G}{dt} = \sum_{k=1}^{NC^G} \dot{m}_{k,\text{in}} \widehat{h}_{k,\text{in}}^G - \sum_{k=1}^{NC^G} \dot{m}_{k,\text{out}} \widehat{h}_{k,\text{out}}^G + \sum_{k=1}^{NC^G} J_k \widehat{h}_k^G SN_p + h\left(T - T^{\text{bulk}}\right) SN_p + V_R \dot{Q}_R^G \tag{2.19}$$

whereT^{bulk} is the gas-phase temperature, the terms $\sum_{k=1}^{NC^G} \dot{m}_{k,\text{in}} \widehat{h}_{k,\text{in}}^G - \sum_{k=1}^{NC^G} \dot{m}_{k,\text{out}} \widehat{h}_{k,\text{out}}^G$ are the species enthalpies of the inlet and outlet flow rates and $J_k \widehat{h}_k^G$ is the flux of enthalpy relating to the mass transfer of each component of a single particle. Finally, $\dot{Q}_R^G$ is the overall heat of gas-phase reactions. The size of the global problem easily becomes prohibitive in terms of calculation time due to the discretization of both the particle and the reactor, and due to the large number of species involved, both in the solid and in the gas phase. Especially considering the large dimension of the problem, and the relative numerical difficulties in solving the corresponding balance equations, it has been and is still necessary to adopt key simplifications and lumping procedures when describing the chemistry of the process.

A couple of application examples relating to a traveling grate combustor [73] and to a biomass gasifier [147] have been discussed elsewhere, and are beyond the main scope of this chapter. Nevertheless, it seems relevant to observe that the model for biomass pyrolysis, gasification and combustion described in this chapter is able to provide a wide range of useful predictions in a feasible way.

2.5 Conclusions and Future Outlook

This chapter presents the CRECK modeling group's approach in the field of the detailed kinetics of biomass pyrolysis and combustion; a complex multicomponent, multiscale and multiphase process. The first step is the characterization of the biomass, which is discussed in detail for lignocellulosic biomass. Suitable reference species are introduced for cellulose, hemicellulose, lignins and extractives, requiring an appropriate determination of the biochemical composition. For each of these species, a multi-step kinetic model for pyrolysis is formulated. The combination of these parallel reactions yields a comprehensive mechanism which also considers the catalytic effects of the different ash components. An extension to include algae is presented and the differences to lignocellulosic biomass are discussed in detail. To understand the final spectrum of products resulting from biomass pyrolysis, it is crucial to include the secondary reactions of the released small gas-phase and tar species. This is achieved by coupling with the modular gas phase CRECK mechanism.

Finally, to understand and predict the pyrolysis in technical systems, the kinetic description must be coupled to particle-scale models and eventually to reactor-scale models. This coupling is especially necessary for larger biomass particles for which intra-particle transport processes, characterized the pyrolysis and Biot numbers, cannot be neglected. A model with full chemistry-transport coupling is presented and the results are discussed. Corresponding equations are formulated for the reactor scale.

This kinetic model offers a favorable compromise between predictiveness and efficiency. As computational power is ever-increasing, it is now suitable for coupling with 3D CFD approaches which will allow the interplay to be investigated between heat, mass and momentum transport with gas- and solid-phase chemistry, even locally, with high temporal resolution. Such comprehensive approaches will provide new insights and will have a substantial impact on future reactor and process designs.

Acknowledgments Our sincere acknowledgments to Paola Giudicianni, Corinna Grottola and Raffaele Ragucci from the IRC/CNR, for their support and for providing additional data presented in this work. Funded by the Deutsche Forschungsgemeinschaft (DFG, German Research Foundation) – projektnummer 215035359 – TRR 129.

References

1. Bridgwater AV. Review of fast pyrolysis of biomass and product upgrading. Biomass Bioenergy. 2012;38:68–94. https://doi.org/10.1016/j.biombioe.2011.01.048.
2. Oasmaa A, Sundqvist T, Kuoppala E, Garcia-Perez M, Solantausta Y, Lindfors C, Paasikallio V. Controlling the phase stability of biomass fast pyrolysis bio-oils. Energy Fuels. 2015;29:4373–81. https://doi.org/10.1021/acs.energyfuels.5b00607.
3. Olah GA, Goeppert A, Prakash GKS. Beyond oil and gas: the methanol economy. Weinheim: Wiley; 2009.
4. Ardolino F, Arena U. Biowaste-to-biomethane: an LCA study on biogas and syngas roads. Waste Manag. 2019;87:441–53. https://doi.org/10.1016/j.wasman.2019.02.030.
5. Arena U. Process and technological aspects of municipal solid waste gasification. A review. Waste Manag. 2012;32:625–39. https://doi.org/10.1016/j.wasman.2011.09.025.
6. Leckner B. Process aspects in combustion and gasification waste-to-energy (WtE) units. Waste Manag. 2015;37:13–25. https://doi.org/10.1016/j.wasman.2014.04.019.
7. Basu P. Biomass gasification, pyrolysis and torrefaction: practical design and theory. 2nd ed. Boston: Academic Press; 2013.
8. Ranzi E, Debiagi PEA, Frassoldati A. Mathematical modeling of fast biomass pyrolysis and bio-oil formation. Note I: kinetic mechanism of biomass pyrolysis. ACS Sustain Chem Eng. 2017;5:2867–81. https://doi.org/10.1021/acssuschemeng.6b03096.
9. Ranzi E, Debiagi PEA, Frassoldati A. Mathematical modeling of fast biomass pyrolysis and bio-oil formation. Note II: secondary gas-phase reactions and bio-oil formation. ACS Sustain Chem Eng. 2017;5:2882–96. https://doi.org/10.1021/acssuschemeng.6b03098.

10. Mettler MS, Vlachos DG, Dauenhauer PJ. Top ten fundamental challenges of biomass pyrolysis for biofuels. Energy Environ Sci. 2012;5:7797–809. https://doi.org/10.1039/C2EE21679E.
11. Ranzi E, Corbetta M, Manenti F, Pierucci S. Kinetic modeling of the thermal degradation and combustion of biomass. Chem Eng Sci. 2014;110:2–12. https://doi.org/10.1016/j.ces.2013.08.014.
12. Ranzi E, Cuoci A, Faravelli T, Frassoldati A, Migliavacca G, Pierucci S, Sommariva S. Chemical kinetics of biomass pyrolysis. Energy Fuels. 2008;22:4292–300. https://doi.org/10.1021/ef800551t.
13. Debiagi PEA, Gentile G, Cuoci A, Frassoldati A, Ranzi E, Faravelli T. Yield, composition and active surface area of char from biomass pyrolysis. Chem Eng Trans. 2018;65:97–102. https://doi.org/10.3303/CET1865017.
14. Dente M, Ranzi E, Goossens AG. Detailed prediction of olefin yields from hydrocarbon pyrolysis through a fundamental simulation model (SPYRO). Comput Chem Eng. 1979;3:61–75. https://doi.org/10.1016/0098-1354(79)80013-7.
15. Ranzi E, Sogaro A, Gaffuri P, Pennati G, Faravelli T. A wide range modeling study of methane oxidation. Combust Sci Technol. 1994;96:279–325. https://doi.org/10.1080/00102209408935359.
16. Debiagi P, Gentile G, Cuoci A, Frassoldati A, Ranzi E, Faravelli T. A predictive model of biochar formation and characterization. J Anal Appl Pyrolysis. 2018;134:326–35. https://doi.org/10.1016/j.jaap.2018.06.022.
17. ECN (2014) Phyllis2: database for biomass and waste - Energy research Centre of the Netherlands
18. Demirbas A. Combustion characteristics of different biomass fuels. Prog Energy Combust Sci. 2004;30:219–30. https://doi.org/10.1016/j.pecs.2003.10.004.
19. Grottola CM, Giudicianni P, Ragucci R (2019) Compositional and thermogravimetric analysis of walnut shells. Biomass samples from the DFG TRR-129 "Oxyflame" project.
20. Ferreiro AI, Giudicianni P, Grottola CM, Rabaçal M, Costa M, Ragucci R. Unresolved issues on the kinetic modeling of pyrolysis of woody and nonwoody biomass fuels. Energy Fuels. 2017;31:4035–44. https://doi.org/10.1021/acs.energyfuels.6b03445.
21. Debiagi PEA, Pecchi C, Gentile G, Frassoldati A, Cuoci A, Faravelli T, Ranzi E. Extractives extend the applicability of multistep kinetic scheme of biomass pyrolysis. Energy Fuels. 2015;29:6544–55. https://doi.org/10.1021/acs.energyfuels.5b01753.
22. Himmel ME, Ding SY, Johnson DK, Adney WS, Nimlos MR, Brady JW, Foust TD. Biomass recalcitrance: engineering plants and enzymes for biofuels production. Science. 2007;315(80):804–7. https://doi.org/10.1126/science.1137016.
23. Grønli MG. A theoretical and experimental study of the thermal degradation of biomass. Trondheim: Norges teknisk-naturvitenskapelige universitet trondheim; 1996.
24. Wang S, Dai G, Yang H, Luo Z. Lignocellulosic biomass pyrolysis mechanism: a state-of-the-art review. Prog Energy Combust Sci. 2017;62:33–86. https://doi.org/10.1016/j.pecs.2017.05.004.
25. Sluiter JB, Ruiz RO, Scarlata CJ, Sluiter AD, Templeton DW. Compositional analysis of lignocellulosic feedstocks. 1. Review and description of methods. J Agric Food Chem. 2010;58:9043–53. https://doi.org/10.1021/jf1008023.
26. Faravelli T, Frassoldati A, Migliavacca G, Ranzi E. Detailed kinetic modeling of the thermal degradation of lignins. Biomass Bioenergy. 2010;34:290–301. https://doi.org/10.1016/j.biombioe.2009.10.018.
27. Lewis AD, Fletcher TH. Prediction of sawdust pyrolysis yields from a flat-flame burner using the CPD model. Energy Fuels. 2013;27:942–53. https://doi.org/10.1021/ef3018783.
28. Dhahak A, Bounaceur R, Le Dreff-Lorimier C, Schmidt G, Trouve G, Battin-Leclerc F. Development of a detailed kinetic model for the combustion of biomass. Fuel. 2019;242:756–74. https://doi.org/10.1016/j.fuel.2019.01.093.

29. Papari S, Hawboldt K. A review on the pyrolysis of woody biomass to bio-oil: focus on kinetic models. Renew Sustain Energy Rev. 2015;52:1580–95. https://doi.org/10.1016/j.rser.2015.07.191.
30. Galgano A, Di Blasi C, Horvat A, Sinai Y. Experimental validation of a coupled solid- and gas-phase model for combustion and gasification of wood logs. Energy Fuels. 2006;20:2223–32. https://doi.org/10.1021/ef060042u.
31. Galgano A, Di Blasi C, Ritondale S, Todisco A. Numerical simulation of the glowing combustion of moist wood by means of a front-based model. Fire Mater. 2014;38:639–58. https://doi.org/10.1002/fam.2203.
32. Yuen RKK, Yeoh GH, de Vahl DG, Leonardi E. Modelling the pyrolysis of wet wood - II. Three-dimensional cone calorimeter simulation. Int J Heat Mass Transf. 2007;50:4387–99. https://doi.org/10.1016/j.ijheatmasstransfer.2007.01.018.
33. Ding Y, Wang C, Lu S. Modeling the pyrolysis of wet wood using fire FOAM. Energy Convers Manag. 2015;98:500–6. https://doi.org/10.1016/j.enconman.2015.03.106.
34. Haberle I, Skreiberg Ø, Łazar J, Haugen NEL. Numerical models for thermochemical degradation of thermally thick woody biomass, and their application in domestic wood heating appliances and grate furnaces. Prog Energy Combust Sci. 2017;63:204–52. https://doi.org/10.1016/j.pecs.2017.07.004.
35. Gronli MG, Melaaen MC. Mathematical model for wood pyrolysis-comparison of experimental measurements with model predictions. Energy Fuels. 2000;14:791–800. https://doi.org/10.1021/ef990176q.
36. Sand U, Sandberg J, Larfeldt J, Bel Fdhila R. Numerical prediction of the transport and pyrolysis in the interior and surrounding of dry and wet wood log. Appl Energy. 2008;85:1208–24. https://doi.org/10.1016/j.apenergy.2008.03.001.
37. Sadhukhan AK, Gupta P, Goyal T, Saha RK. Modelling of pyrolysis of coal-biomass blends using thermogravimetric analysis. Bioresour Technol. 2008;99:8022–6. https://doi.org/10.1016/j.biortech.2008.03.047.
38. Kwiatkowski K, Bajer K, Celińska A, Dudyński M, Korotko J, Sosnowska M. Pyrolysis and gasification of a thermally thick wood particle - effect of fragmentation. Fuel. 2014;132:125–34. https://doi.org/10.1016/j.fuel.2014.04.057.
39. Miller RS, Bellan J. A generalized biomass pyrolysis model based on superimposed cellulose, hemicellulose and lignin kinetics. Combust Sci Technol. 1997;126:97–137. https://doi.org/10.1080/00102209708935670.
40. Bradbury AGW, Sakai Y, Shafizadeh F. A kinetic model for pyrolysis of cellulose. J Appl Polym Sci. 1979;23:3271–80. https://doi.org/10.1002/app.1979.070231112.
41. Vinu R, Broadbelt LJ. A mechanistic model of fast pyrolysis of glucose-based carbohydrates to predict bio-oil composition. Energy Environ Sci. 2012;5:9808–26. https://doi.org/10.1039/c2ee22784c.
42. Zhou X, Nolte MW, Mayes HB, Shanks BH, Broadbelt LJ. Experimental and mechanistic modeling of fast pyrolysis of neat glucose-based carbohydrates. 1. Experiments and development of a detailed mechanistic model. Ind Eng Chem Res. 2014;53:13274–89. https://doi.org/10.1021/ie502259w.
43. Mayes HB, Nolte MW, Beckham GT, Shanks BH, Broadbelt LJ. The alpha-bet(a) of glucose pyrolysis: computational and experimental investigations of 5-hydroxymethylfurfural and levoglucosan formation reveal implications for cellulose pyrolysis. ACS Sustain Chem Eng. 2014;2:1461–73. https://doi.org/10.1021/sc500113m.
44. Zhou X, Li W, Mabon R, Broadbelt LJ. A mechanistic model of fast pyrolysis of hemicellulose. Energy Environ Sci. 2018;11:1240–60. https://doi.org/10.1039/c7ee03208k.
45. Mayes HB, Broadbelt LJ. Unraveling the reactions that unravel cellulose. J Phys Chem A. 2012;116:7098–106. https://doi.org/10.1021/jp300405x.
46. Seshadri V, Westmoreland PR. Concerted reactions and mechanism of glucose pyrolysis and implications for cellulose kinetics. J Phys Chem A. 2012;116:11997–2013. https://doi.org/10.1021/jp3085099.

47. Horton SR, Mohr RJ, Zhang Y, Petrocelli FP, Klein MT. Molecular-level kinetic modeling of biomass gasification. Energy Fuels. 2016;30:1647–61. https://doi.org/10.1021/acs.energyfuels.5b01988.
48. Yanez AJ, Natarajan P, Li W, Mabon R, Broadbelt LJ. Coupled structural and kinetic model of lignin fast pyrolysis. Energy Fuels. 2018;32:1822–30. https://doi.org/10.1021/acs.energyfuels.7b03311.
49. Westmoreland PR. Pyrolysis kinetics for lignocellulosic biomass-to-oil from molecular modeling. Curr Opin Chem Eng. 2019;23:123–9. https://doi.org/10.1016/j.coche.2019.03.011.
50. Corbetta M, Frassoldati A, Bennadji H, Smith K, Serapiglia MJ, Gauthier G, Melkior T, Ranzi E, Fisher EM. Pyrolysis of centimeter-scale woody biomass particles: kinetic modeling and experimental validation. Energy and Fuels. 2014;28:3884–98. https://doi.org/10.1021/ef500525v.
51. Gorensek MB, Shukre R, Chen CC. Development of a thermophysical properties model for flowsheet simulation of biomass pyrolysis processes. ACS Sustain Chem Eng. 2019;7:9017–27. https://doi.org/10.1021/acssuschemeng.9b01278.
52. Shen D, Jin W, Hu J, Xiao R, Luo K. An overview on fast pyrolysis of the main constituents in lignocellulosic biomass to valued-added chemicals: structures, pathways and interactions. Renew Sustain Energy Rev. 2015;51:761–74. https://doi.org/10.1016/j.rser.2015.06.054.
53. Antal MJ, Varhegyi G. Cellulose pyrolysis kinetic: the current state of knowledge. Ind Eng Chem Res. 1995;34:703–17. https://doi.org/10.1021/ie00042a001.
54. Lédé J. Cellulose pyrolysis kinetics: an historical review on the existence and role of intermediate active cellulose. J Anal Appl Pyrolysis. 2012;94:17–32. https://doi.org/10.1016/j.jaap.2011.12.019.
55. Burnham AK, Zhou X, Broadbelt LJ. Critical review of the global chemical kinetics of cellulose thermal decomposition. Energy Fuels. 2015;29:2906–18. https://doi.org/10.1021/acs.energyfuels.5b00350.
56. Dussan K, Dooley S, Monaghan R. Integrating compositional features in model compounds for a kinetic mechanism of hemicellulose pyrolysis. Chem Eng J. 2017;328:943–61. https://doi.org/10.1016/j.cej.2017.07.089.
57. Frassoldati A, Cuoci A, Faravelli T, Niemann U, Ranzi E, Seiser R, Seshadri K. An experimental and kinetic modeling study of n-propanol and iso-propanol combustion. Combust Flame. 2010;157:2–16. https://doi.org/10.1016/j.combustflame.2009.09.002.
58. Jakab E, Faix O, Till F, Székely T. Thermogravimetry/mass spectrometry study of six lignins within the scope of an international round robin test. J Anal Appl Pyrolysis. 1995;35:167–79. https://doi.org/10.1016/0165-2370(95)00907-7.
59. Antal MJ, Várhegyi G, Jakab E. Cellulose pyrolysis kinetics: revisited. Ind Eng Chem Res. 1998;37:1267–75. https://doi.org/10.1021/ie970144v.
60. Williams PT, Besler S. The influence of temperature and heating rate on the slow pyrolysis of biomass. Renew Energy. 1996;7:233–50. https://doi.org/10.1016/0960-1481(96)00006-7.
61. Milosavljevic I, Suuberg EM. Cellulose thermal decomposition kinetics: global mass loss kinetics. Ind Eng Chem Res. 1995;34:1081–91. https://doi.org/10.1021/ie00043a009.
62. Caballero JA, Conesa JA, Font R, Marcilla A. Pyrolysis kinetics of almond shells and olive stones considering their organic fractions. J Anal Appl Pyrolysis. 1997;42:159–75. https://doi.org/10.1016/S0165-2370(97)00015-6.
63. Moore AM The effect of biomass characteristics on bio-oil produced via fast pyrolysis. M.Sc. thesis, North Carolina State University. 2015.
64. Pasangulapati V (2012) Devolatilization characteristics of cellulose, hemicellulose, lignin and the selected biomass during thermochemical gasification: experiment and modeling studies. Oklahoma State University
65. Debiagi PEA A kinetic model of thermochemical conversion of biomass. Ph.D. thesis, Politecnico di Milano. (2018).

66. Wolfrum EJ, Lorenz AJ, DeLeon N. Correlating detergent fiber analysis and dietary fiber analysis data for corn stover collected by NIRS. Cellulose. 2009;16:577–85. https://doi.org/10.1007/s10570-009-9318-9.
67. Sluiter A, Ruiz R, Scarlata C, Sluiter J, Templeton D (2008) Determination of extractives in biomass: laboratory analytical procedure (LAP); Issue Date 7/17/2005.9. https://doi.org/NREL/TP-510-42621
68. Rowell RM. Handbook of wood chemistry and wood composites. Boca Raton: CRC Press; 2012.
69. Yang H, Yan R, Chen H, Zheng C, Lee DH, Liang DT. In-depth investigation of biomass pyrolysis based on three major components: hemicellulose, cellulose and lignin. Energy Fuels. 2006;20:388–93. https://doi.org/10.1021/ef0580117.
70. Senneca O, Cerciello F, Heuer S, Ammendola P. Slow pyrolysis of walnut shells in nitrogen and carbon dioxide. Fuel. 2018;225:419–25. https://doi.org/10.1016/j.fuel.2018.03.094.
71. Giudicianni P, Gargiulo V, Grottola CM, Alfè M, Ragucci R. Effect of alkali metal ions presence on the products of xylan steam assisted slow pyrolysis. Fuel. 2018;216:36–43. https://doi.org/10.1016/j.fuel.2017.11.150.
72. Tufano GL, Stein OT, Kronenburg A, Gentile G, Stagni A, Frassoldati A, Faravelli T, Kempf AM, Vascellari M, Hasse C. Fully-resolved simulations of coal particle combustion using a detailed multi-step approach for heterogeneous kinetics. Fuel. 2019;240:75–83. https://doi.org/10.1016/j.fuel.2018.11.139.
73. Ranzi E, Pierucci S, Aliprandi PC, Stringa S. Comprehensive and detailed kinetic model of a traveling grate combustor of biomass. Energy Fuels. 2011;25:4195–205. https://doi.org/10.1021/ef200902v.
74. Stark AK, Bates RB, Zhao Z, Ghoniem AF. Prediction and validation of major gas and tar species from a reactor network model of air-blown fluidized bed biomass gasification. Energy Fuels. 2015;29:2437–52. https://doi.org/10.1021/ef5027955.
75. Dussan K, Dooley S, Monaghan RFD. A model of the chemical composition and pyrolysis kinetics of lignin. Proc Combust Inst. 2019;37:2697–704. https://doi.org/10.1016/j.proci.2018.05.149.
76. Anca-Couce A. Reaction mechanisms and multi-scale modelling of lignocellulosic biomass pyrolysis. Prog Energy Combust Sci. 2016;53:41–79. https://doi.org/10.1016/j.pecs.2015.10.002.
77. Anca-Couce A, Obernberger I. Application of a detailed biomass pyrolysis kinetic scheme to hardwood and softwood torrefaction. Fuel. 2016;167:158–67. https://doi.org/10.1016/j.fuel.2015.11.062.
78. Yu J, Paterson N, Blamey J, Millan M. Cellulose, xylan and lignin interactions during pyrolysis of lignocellulosic biomass. Fuel. 2017;191:140–9. https://doi.org/10.1016/j.fuel.2016.11.057.
79. Giudicianni P, Cardone G, Sorrentino G, Ragucci R. Hemicellulose, cellulose and lignin interactions on Arundo donax steam assisted pyrolysis. J Anal Appl Pyrolysis. 2014;110:138–46. https://doi.org/10.1016/j.jaap.2014.08.014.
80. Lin F, Waters CL, Mallinson RG, Lobban LL, Bartley LE. Relationships between biomass composition and liquid products formed via pyrolysis. Front Energy Res. 2015;3:45. https://doi.org/10.3389/fenrg.2015.00045.
81. Lv D, Xu M, Liu X, Zhan Z, Li Z, Yao H. Effect of cellulose, lignin, alkali and alkaline earth metallic species on biomass pyrolysis and gasification. Fuel Process Technol. 2010;91:903–9. https://doi.org/10.1016/j.fuproc.2009.09.014.
82. Arora JS, Chew JW, Mushrif SH. Influence of alkali and alkaline-earth metals on the cleavage of glycosidic bond in biomass pyrolysis: a DFT study using cellobiose as a model compound. J Phys Chem A. 2018;122:7646–58. https://doi.org/10.1021/acs.jpca.8b06083.
83. Zhou X, Mayes HB, Broadbelt LJ, Nolte MW, Shanks BH. Fast pyrolysis of glucose-based carbohydrates with added NaCl part 2: validation and evaluation of the mechanistic model. AIChE J. 2016;62:778–91. https://doi.org/10.1002/aic.15107.

84. Zhu C, Maduskar S, Paulsen AD, Dauenhauer PJ. Alkaline-earth-metal-catalyzed thin-film pyrolysis of cellulose. ChemCatChem. 2016;8:818–29. https://doi.org/10.1002/cctc.201501235.
85. Patwardhan PR, Satrio JA, Brown RC, Shanks BH. Influence of inorganic salts on the primary pyrolysis products of cellulose. Bioresour Technol. 2010;101:4646–55. https://doi.org/10.1016/j.biortech.2010.01.112.
86. Collard F-X, Blin J, Bensakhria A, Valette J. Influence of impregnated metal on the pyrolysis conversion of biomass constituents. J Anal Appl Pyrolysis. 2012;95:213–26. https://doi.org/10.1016/j.jaap.2012.02.009.
87. Rutkowski P. Pyrolysis of cellulose, xylan and lignin with the K2CO3 and ZnCl2 addition for bio-oil production. Fuel Process Technol. 2011;92:517–22. https://doi.org/10.1016/j.fuproc.2010.11.006.
88. Giudicianni P, Gargiulo V, Alfè M, Ragucci R, Ferreiro AI, Rabaçal M, Costa M. Slow pyrolysis of xylan as pentose model compound for hardwood hemicellulose: a study of the catalytic effect of Na ions. J Anal Appl Pyrolysis. 2019;137:266–75. https://doi.org/10.1016/j.jaap.2018.12.004.
89. Trubetskaya A, Timko MT, Umeki K. Prediction of fast pyrolysis products yields using lignocellulosic compounds and ash contents. Appl Energy. 2020;257:113897. https://doi.org/10.1016/j.apenergy.2019.113897.
90. Nzihou A, Stanmore B, Lyczko N, Minh DP. The catalytic effect of inherent and adsorbed metals on the fast/flash pyrolysis of biomass: a review. Energy. 2019;170:326–37. https://doi.org/10.1016/j.energy.2018.12.174.
91. Guo D, Wu S, Liu B, Yin X, Yang Q. Catalytic effects of NaOH and Na2CO3 additives on alkali lignin pyrolysis and gasification. Appl Energy. 2012;95:22–30. https://doi.org/10.1016/j.apenergy.2012.01.042.
92. Wang S, Ru B, Lin H, Sun W, Luo Z. Pyrolysis behaviors of four lignin polymers isolated from the same pine wood. Bioresour Technol. 2015;182:120–7. https://doi.org/10.1016/j.biortech.2015.01.127.
93. Jakab E, Faix O, Till F. Thermal decomposition of milled wood lignins studied by thermogravimetry/mass spectrometry. J Anal Appl Pyrolysis. 1997;40:171–86. https://doi.org/10.1016/S0165-2370(97)00046-6.
94. Trendewicz A, Evans R, Dutta A, Sykes R, Carpenter D, Braun R. Evaluating the effect of potassium on cellulose pyrolysis reaction kinetics. Biomass Bioenergy. 2015;74:15–25. https://doi.org/10.1016/j.biombioe.2015.01.001.
95. Ross AB, Jones JM, Kubacki ML, Bridgeman T. Classification of macroalgae as fuel and its thermochemical behaviour. Bioresour Technol. 2008;99:6494–504. https://doi.org/10.1016/j.biortech.2007.11.036.
96. Debiagi PEA, Trinchera M, Frassoldati A, Faravelli T, Vinu R, Ranzi E. Algae characterization and multistep pyrolysis mechanism. J Anal Appl Pyrolysis. 2017;128:423–36. https://doi.org/10.1016/j.jaap.2017.08.007.
97. Gai C, Zhang Y, Chen WT, Zhang P, Dong Y. Thermogravimetric and kinetic analysis of thermal decomposition characteristics of low-lipid microalgae. Bioresour Technol. 2013;150:139–48. https://doi.org/10.1016/j.biortech.2013.09.137.
98. Ojha DK, Viju D, Vinu R. Fast pyrolysis kinetics of alkali lignin: Evaluation of apparent rate parameters and product time evolution. Bioresour Technol. 2017;241:142–51. https://doi.org/10.1016/j.biortech.2017.05.084.
99. Carstensen H-H, Dean AM. Development of detailed kinetic models for the thermal conversion of biomass via first principle methods and rate estimation rules. In: Computational modeling in lignocellulosic biofuel production, ACS Symposium Series. Washington: ACS Publications; 2010. p. 201–43. https://doi.org/10.1021/bk-2010-1052.ch010.
100. Debiagi PEA, Gentile G, Pelucchi M, Frassoldati A, Cuoci A, Faravelli T, Ranzi E. Detailed kinetic mechanism of gas-phase reactions of volatiles released from biomass pyrolysis. Biomass Bioenergy. 2016;93:60–71. https://doi.org/10.1016/j.biombioe.2016.06.015.

101. Ranzi E, Frassoldati A, Grana R, Cuoci A, Faravelli T, Kelley AP, Law CK. Hierarchical and comparative kinetic modeling of laminar flame speeds of hydrocarbon and oxygenated fuels. Prog Energy Combust Sci. 2012;38:468–501. https://doi.org/10.1016/j.pecs.2012.03.004.
102. Frassoldati A, Grana R, Faravelli T, Ranzi E, Oßwald P, Kohse-Höinghaus K. Detailed kinetic modeling of the combustion of the four butanol isomers in premixed low-pressure flames. Combust Flame. 2012;159:2295–311. https://doi.org/10.1016/j.combustflame.2012.03.002.
103. Grana R, Frassoldati A, Faravelli T, Niemann U, Ranzi E, Seiser R, Cattolica R, Seshadri K. An experimental and kinetic modeling study of combustion of isomers of butanol. Combust Flame. 2010;157:2137–54. https://doi.org/10.1016/j.combustflame.2010.05.009.
104. Pelucchi M, Cavallotti C, Ranzi E, Frassoldati A, Faravelli T. Relative reactivity of oxygenated fuels: alcohols, aldehydes, ketones, and methyl esters. Energy Fuels. 2016;30:8665–79. https://doi.org/10.1021/acs.energyfuels.6b01171.
105. Pelucchi M, Ranzi E, Frassoldati A, Faravelli T. Alkyl radicals rule the low temperature oxidation of long chain aldehydes. Proc Combust Inst. 2017;36:393–401. https://doi.org/10.1016/j.proci.2016.05.051.
106. Ince A, Carstensen HH, Reyniers MF, Marin GB. First-principles based group additivity values for thermochemical properties of substituted aromatic compounds. AIChE J. 2015;61:3858–70. https://doi.org/10.1002/aic.15008.
107. Wagnon SW, Thion S, Nilsson EJK, Mehl M, Serinyel Z, Zhang K, Dagaut P, Konnov AA, Dayma G, Pitz WJ. Experimental and modeling studies of a biofuel surrogate compound: laminar burning velocities and jet-stirred reactor measurements of anisole. Combust Flame. 2018;189:325–36. https://doi.org/10.1016/j.combustflame.2017.10.020.
108. Pelucchi M, Cavallotti C, Cuoci A, Faravelli T, Frassoldati A, Ranzi E. Detailed kinetics of substituted phenolic species in pyrolysis bio-oils. React Chem Eng. 2019;4:490–506. https://doi.org/10.1039/c8re00198g.
109. Wang K, Dean AM. Rate rules and reaction classes, Comput. Aided Chem. Eng. Amsterdam: Elsevier; 2019. p. 203–57. https://doi.org/10.1016/B978-0-444-64087-1.00004-8.
110. Bridgwater AV. Renewable fuels and chemicals by thermal processing of biomass. Chem Eng J. 2003;91:87–102. https://doi.org/10.1016/S1385-8947(02)00142-0.
111. Calonaci M, Grana R, Barker Hemings E, Bozzano G, Dente M, Ranzi E. Comprehensive kinetic modeling study of bio-oil formation from fast pyrolysis of biomass. Energy Fuels. 2010;24:5727–34. https://doi.org/10.1021/ef1008902.
112. Fantozzi F, Frassoldati A, Bartocci P, Cinti G, Quagliarini F, Bidini G, Ranzi EM. An experimental and kinetic modeling study of glycerol pyrolysis. Appl Energy. 2016;184:68–76. https://doi.org/10.1016/j.apenergy.2016.10.018.
113. Hemings EB, Cavallotti C, Cuoci A, Faravelli T, Ranzi E. A detailed kinetic study of pyrolysis and oxidation of glycerol (Propane-1,2,3-triol). Combust Sci Technol. 2012;184:1164–78. https://doi.org/10.1080/00102202.2012.664006.
114. Dente M, Pierucci S, Ranzi E, Bussani G. New improvements in modeling kinetic schemes for hydrocarbons pyrolysis reactors. Chem Eng Sci. 1992;47:2629–34. https://doi.org/10.1016/0009-2509(92)87104-X.
115. Violi A, Truong TN, Sarofim AF. Kinetics of hydrogen abstraction reactions from polycyclic aromatic hydrocarbons by H atoms. J Phys Chem A. 2004;108:4846–52. https://doi.org/10.1021/jp026557d.
116. Pelucchi M, Cavallotti C, Faravelli T, Klippenstein SJ. H-Abstraction reactions by OH, HO2, O, O2 and benzyl radical addition to O2 and their implications for kinetic modelling of toluene oxidation. Phys Chem Chem Phys. 2018;20:10607–27. https://doi.org/10.1039/C7CP07779C.
117. Pelucchi M, Somers KP, Yasunaga K, Burke U, Frassoldati A, Ranzi E, Curran HJ, Faravelli T. An experimental and kinetic modeling study of the pyrolysis and oxidation of n-C3C5 aldehydes in shock tubes. Combust Flame. 2015;162:265–86. https://doi.org/10.1016/j.combustflame.2014.07.027.

118. Ackermann T. S. W. Benson: Thermochemical kinetics. Methods for the estimation of thermochemical data and rate parameters. John Wiley & Sons, Inc., New York, 1968. XII und 223 Seiten, 4 Abbildungen, Preis: 94 s. Berichte der Bunsengesellschaft für Phys Chemie. 1969;73:119–244. https://doi.org/10.1002/bbpc.19690730226.
119. Djokic MR, Van Geem KM, Cavallotti C, Frassoldati A, Ranzi E, Marin GB. An experimental and kinetic modeling study of cyclopentadiene pyrolysis: first growth of polycyclic aromatic hydrocarbons. Combust Flame. 2014;161:2739–51. https://doi.org/10.1016/j.combustflame.2014.04.013.
120. Hemings EB, Bozzano G, Dente M, Ranzi E. Detailed kinetics of the pyrolysis and oxidation of anisole. Chem Eng Trans. 2011;24:61–6. https://doi.org/10.3303/CET1124011.
121. Nowakowska M, Herbinet O, Dufour A, Glaude PA. Detailed kinetic study of anisole pyrolysis and oxidation to understand tar formation during biomass combustion and gasification. Combust Flame. 2014;161:1474–88. https://doi.org/10.1016/j.combustflame.2013.11.024.
122. Thomas S, Ledesma EB, Wornat MJ. The effects of oxygen on the yields of the thermal decomposition products of catechol under pyrolysis and fuel-rich oxidation conditions. Fuel. 2007;86:2581–95. https://doi.org/10.1016/j.fuel.2007.02.003.
123. Norinaga K, Shoji T, Kudo S, Hayashi J. Detailed chemical kinetic modelling of vapour-phase cracking of multi-component molecular mixtures derived from the fast pyrolysis of cellulose. Fuel. 2013;103:141–50. https://doi.org/10.1016/j.fuel.2011.07.045.
124. Norinaga K, Yang H, Tanaka R, Appari S, Iwanaga K, Takashima Y, Kudo S, Shoji T, Ichiro HJ. A mechanistic study on the reaction pathways leading to benzene and naphthalene in cellulose vapor phase cracking. Biomass Bioenergy. 2014;69:144–54. https://doi.org/10.1016/j.biombioe.2014.07.008.
125. Yang HM, Appari S, Kudo S, Hayashi JI, Norinaga K. Detailed chemical kinetic modeling of vapor-phase reactions of volatiles derived from fast pyrolysis of lignin. Ind Eng Chem Res. 2015;54:6855–64. https://doi.org/10.1021/acs.iecr.5b01289.
126. Saggese C, Frassoldati A, Cuoci A, Faravelli T, Ranzi E. A wide range kinetic modeling study of pyrolysis and oxidation of benzene. Combust Flame. 2013;160:1168–90. https://doi.org/10.1016/j.combustflame.2013.02.013.
127. Saggese C, Ferrario S, Camacho J, Cuoci A, Frassoldati A, Ranzi E, Wang H, Faravelli T. Kinetic modeling of particle size distribution of soot in a premixed burner-stabilized stagnation ethylene flame. Combust Flame. 2015;162:3356–69. https://doi.org/10.1016/j.combustflame.2015.06.002.
128. Pejpichestakul W, Ranzi E, Pelucchi M, Frassoldati A, Cuoci A, Parente A, Faravelli T. Examination of a soot model in premixed laminar flames at fuel-rich conditions. Proc Combust Inst. 2019;37:1013–21. https://doi.org/10.1016/j.proci.2018.06.104.
129. Safarian S, Unnþórsson R, Richter C. A review of biomass gasification modelling. Renew Sustain Energy Rev. 2019;110:378–91. https://doi.org/10.1016/j.rser.2019.05.003.
130. Di Blasi C. Modeling and simulation of combustion processes of charring and non-charring solid fuels. Prog Energy Combust Sci. 1993;19:71–104. https://doi.org/10.1016/0360-1285(93)90022-7.
131. Di Blasi C. Modeling chemical and physical processes of wood and biomass pyrolysis. Prog Energy Combust Sci. 2008;34:47–90. https://doi.org/10.1016/j.pecs.2006.12.001.
132. Gentile G, Debiagi P, Cuoci A, Frassoldati A, Ranzi E, Faravelli T. A computational framework for the pyrolysis of anisotropic biomass particles. Chem Eng J. 2017;321:458–73. https://doi.org/10.1016/j.cej.2017.03.113.
133. Bennadji H, Smith K, Shabangu S, Fisher EM. Low-temperature pyrolysis of woody biomass in the thermally thick regime. Energy Fuels. 2013;27:1453–9. https://doi.org/10.1021/ef400079a.
134. Wang L, Skreiberg Ø, Gronli M, Specht GP, Antal MJ. Is elevated pressure required to achieve a high fixed-carbon yield of charcoal from biomass? Part 2: The importance of particle size. Energy Fuels. 2013;27:2146–56. https://doi.org/10.1021/ef400041h.

135. Park WC, Atreya A, Baum HR. Experimental and theoretical investigation of heat and mass transfer processes during wood pyrolysis. Combust Flame. 2010;157:481–94. https://doi.org/10.1016/j.combustflame.2009.10.006.
136. Paulsen AD, Mettler MS, Dauenhauer PJ. The role of sample dimension and temperature in cellulose pyrolysis. Energy Fuels. 2013;27:2126–34. https://doi.org/10.1021/ef302117j.
137. Kan T, Strezov V, Evans TJ. Lignocellulosic biomass pyrolysis: a review of product properties and effects of pyrolysis parameters. Renew Sustain Energy Rev. 2016;57:1126–40. https://doi.org/10.1016/j.rser.2015.12.185.
138. van der Stelt MJC, Gerhauser H, Kiel JHA, Ptasinski KJ. Biomass upgrading by torrefaction for the production of biofuels: A review. Biomass Bioenergy. 2011;35:3748–62. https://doi.org/10.1016/j.biombioe.2011.06.023.
139. Shankar Tumuluru J, Sokhansanj S, Hess JR, Wright CT, Boardman RD. A review on biomass torrefaction process and product properties for energy applications. Ind Biotechnol. 2011;7:384–401. https://doi.org/10.1089/ind.2011.7.384.
140. Ribeiro J, Godina R, Matias J, Nunes L. Future perspectives of biomass torrefaction: review of the current state-of-the-art and research development. Sustainability. 2018;10:2323. https://doi.org/10.3390/su10072323.
141. Aguado R, Olazar M, Barona A, Bilbao J. Char-formation kinetics in the pyrolysis of sawdust in a conical spouted bed reactor. J Chem Technol Biotechnol. 2000;75:583–8. https://doi.org/10.1002/1097-4660(200007)75:7<583::AID-JCTB261>3.0.CO;2-B.
142. Ateş F, Pütün E, Pütün AE. Fast pyrolysis of sesame stalk: yields and structural analysis of bio-oil. J Anal Appl Pyrolysis. 2004;71:779–90. https://doi.org/10.1016/j.jaap.2003.11.001.
143. Westerhof RJM, Wim BDWF, Van Swaaij WPM, Kersten SRA. Effect of temperature in fluidized bed fast pyrolysis of biomass: oil quality assessment in test units. Ind Eng Chem Res. 2010;49:1160–8. https://doi.org/10.1021/ie900885c.
144. Neves D, Thunman H, Matos A, Tarelho L, Gómez-Barea A. Characterization and prediction of biomass pyrolysis products. Prog Energy Combust Sci. 2011;37:611–30. https://doi.org/10.1016/j.pecs.2011.01.001.
145. Senneca O, Salatino P. A semi-detailed kinetic model of char combustion with consideration of thermal annealing. Proc Combust Inst. 2011;33:1763–70. https://doi.org/10.1016/j.proci.2010.08.011.
146. Frigerio S, Thunman H, Leckner B, Hermansson S. Estimation of gas phase mixing in packed beds. Combust Flame. 2008;153:137–48. https://doi.org/10.1016/j.combustflame.2007.05.006.
147. Ravaghi-Ardebili Z, Manenti F, Corbetta M, Pirola C, Ranzi E. Biomass gasification using low-temperature solar-driven steam supply. Renew Energy. 2015;74:671–80. https://doi.org/10.1016/j.renene.2014.07.021.

Chapter 3
Production of Valuable Chemicals and Fuel Molecules from Lignin Via Fast Pyrolysis: Experimental and Theoretical Studies Using Model Compounds

Attada Yerrayya, Upendra Natarajan, and R. Vinu

Abstract Lignin is the second-most abundant compound in lignocellulosic biomass (up to 30% dry weight) and a major by-product of the pulp and paper industries. Even though it is projected as a primary source of renewable phenolic compounds, its complex and highly-condensed structure with phenyl propane monomers, viz., p-hydroxyphenyl, guaiacyl, and syringyl sub-units, makes lignin conversion challenging. The use of lignin as a source of phenolic compounds is also exacerbated by its wide molecular weight distribution and branching. Pyrolysis and catalytic fast pyrolysis have emerged as promising thermochemical conversion technologies to convert lignin into phenols and aromatic hydrocarbons. Pyrolysis of lignin model compounds is valuable to unravel the mechanism of formation of phenols through the cleavage of specific linkages in lignin, and their secondary gas phase decomposition reactions. This chapter focuses on experimental and theoretical studies of free radical and concerted reactions of lignin model compounds for the production of phenolic and other aromatic compounds. The purpose of this chapter is to provide an essence of fast pyrolysis chemistry of lignin and its model compounds, and the associated reaction kinetics. Challenges in obtaining a mechanistic understanding of lignin pyrolysis are highlighted, and the need for a synergistic combination of experimental and computational studies is emphasized.

Keywords Lignin · Fast pyrolysis · Catalytic pyrolysis · Density functional theory · Kinetic model · Phenolics

A. Yerrayya · U. Natarajan
Department of Chemical Engineering, Indian Institute of Technology Madras, Chennai, India

R. Vinu (✉)
Department of Chemical Engineering, Indian Institute of Technology Madras, Chennai, India

National Centre for Combustion Research and Development, Indian Institute of Technology Madras, Chennai, India
e-mail: vinu@iitm.ac.in

Z. Fang et al. (eds.), *Production of Biofuels and Chemicals with Pyrolysis*, Biofuels and Biorefineries 10, https://doi.org/10.1007/978-981-15-2732-6_3

3.1 Introduction

Lignin is a major by-product from paper and pulp-making industries, and second generation biorefineries that produce ethanol from agricultural residues. An estimated 70 million tons out of the 20 billion tons of lignin that are synthesized in nature are generated as a by-product of the Kraft pulping process. Sulfate pulp or Kraft pulp constitutes nearly 58.2% of the total pulp generated in the world, followed by 21.2% of mechanical pulp and only 9.2% of sulfite pulp [1]. Nevertheless, nearly 99% of lignin produced via Kraft pulping is used in paper and pulp industry for energy recovery. Present day paper and pulp industries are highly integrated facilities wherein the black liquor is initially concentrated in multiple effect evaporators to remove water, and then combusted in a recovery boiler. The high-pressure steam generated in the boiler is used to run a turbine, which in turn generates electricity [2, 3]. Simultaneously, the oxidized sulfur compounds are reduced to sulfides, and thus, the initially added reagents are recovered. The low-pressure steam exhaust from the turbine is used for process heat applications in the paper and pulp mill. Therefore, lignin that is available for other applications is only about 2% of the total processed quantity [4]. A major portion of nearly one million tons of the marketed lignin is available in the form of sulfonated lignin or lignosulfonates from the sulfite process, followed by 60,000 tons of non-sulfonated or hydrophobic lignin from Kraft pulping process, and 10,000 tons from soda pulping process.

Energy production is presently the high volume, yet a low value utilization of lignin. Available lignosulfonates from sulfite process and hydrophobic lignin from Kraft and soda pulping processes are being used for high value, low volume applications in the production of additives, binders, adhesives, composites, phenolic resin substitutes, oxidized products like vanillin and its derivatives, and syngas products [4–7]. Nearly two thirds of the lignosulfonates are used for dispersant applications while the remaining one third is used for making concrete admixtures, binders, adhesives and emulsifiers [4–7]. Sulfonated or sulfomethylated Kraft lignin also finds similar applications, while unmodified Kraft lignin is used as a stabilizer in asphalt formulations, as an antioxidant, as carbon black and in rubber reinforcements [4, 5]. Importantly, Kraft lignin contains aliphatic thiol moieties in minor amounts that are beneficial for the manufacture of aliphatic sulfur compounds like dimethyl sulfide and methyl mercaptan in small volumes, which find application in solvent and agrochemical industries. Nevertheless, the presence of sulfur causes an irritating odor during thermal processing [8]. Recently, sulfur-free, water insoluble lignins obtained via (1) biomass conversion techniques, (2) solvent/organosolv pulping and (3) soda pulping processes are becoming common owing to their interesting properties like structural similarity to native lignin and low glass transition temperature [8, 9]. Even for power generation, organosolv lignin is superior to Kraft lignin owing to its high fluidity index and low ash content. Hence, organosolv lignin can be easily fed into combustion chambers and it results in cleaner combustion. Therefore, sulfur-free lignins are promising candidates as substitutes for phenolic resins, polyurethane foams, epoxy resins and biodispersants. Table 3.1 [10–13] depicts the

Table 3.1 Elemental composition and ash content in different varieties of lignin

Wt %	Indulin AT [10]	Acetocell [10]	Lignoboost [10]	ALM [11]	ETEK [11]	Kraft [12]	Organosolv [13]
C	64.5	66.5	67.3	61	54.3	61.8	64.6
H	5.4	5.1	5.6	7	6.1	5.6	6.3
N	1	<0.5	<0.5	0.9	0	0.8	0.2
O	24.8	26.5	22.3	27	39.2	28.3	28.1
S	1.9	<0.05	2.8	<0.03	0.2	1.5	0
Ash	2.4	1.8	1.4	<4	0.2	2.0	0.8

elemental composition of different types of lignin. It is evident that the methodology adopted to extract lignin from biomass significantly affects the sulfur, oxygen and ash content in lignin.

In recent years, there is a significant focus on altering the reactivity of lignin to achieve breakthroughs in the field of lignin valorization. Efficient utilization of lignin through thermochemical conversion technologies can produce fine chemicals, bio-oil and carbon materials [14–19]. The objectives of this chapter are three-fold. Firstly, to describe the basic structure of lignin and elucidate the key transformations involved in both non-catalytic and catalytic pyrolysis of lignin to different chemicals and fuel molecules. Secondly, to provide a holistic overview of pyrolysis mechanism of different model compounds of lignin via quantum chemical modeling. Thirdly, to review the recent advancements and to emphasize the need of composition models of lignin, and mechanistic kinetic models of lignin pyrolysis.

3.1.1 *Lignin Structure*

After cellulose, lignin is the most abundant source of carbon on earth. It is an amorphous, three-dimensional polyphenolic material with complex structure [20]. Lignin fills up the space in the cell wall between the cellulosic fibers and hemicellulosic bundles. It is covalently linked to hemicellulose and cellulose via lignin-carbohydrate complexes, and acts as a binder of fibrous carbohydrate structures. The structure of lignin is complex, and is yet to be fully understood. In raw lignocellulosic biomass, lignin is a crosslinked and branched macromolecule that adds strength and rigidity to the cell walls. It is widely accepted that lignin composition and its content in biomass vary with the type of plant species. As shown in the Fig. 3.1, lignin is made up of three major phenolic sub-units, viz. p-coumaryl (4-hydroxycinnamyl), coniferyl (3-methoxy, 4-hydroxycinnamyl), and sinapyl (3,5-dimethoxy, 4-hydroxycinnamyl), linked by etheric C-O-C and C-C linkages [21]. The three monolignol compounds are also known as p-hydroxyphenyl (H), guaiacyl (G) and syringyl (S) sub-units [22]. The typical molecular structure of softwood lignin is shown in Fig. 3.2 [23]. These monolignols are connected by a number of C-O-C and C-C linkages.

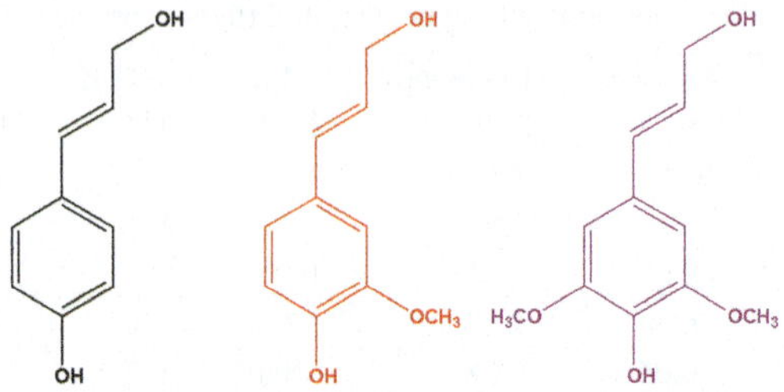

p-Coumaryl alcohol Coniferyl alcohol Sinapyl alcohol

Fig. 3.1 Structure of monomeric phenols (p-coumaryl, coniferyl, sinapyl alcohols), the building blocks of lignin

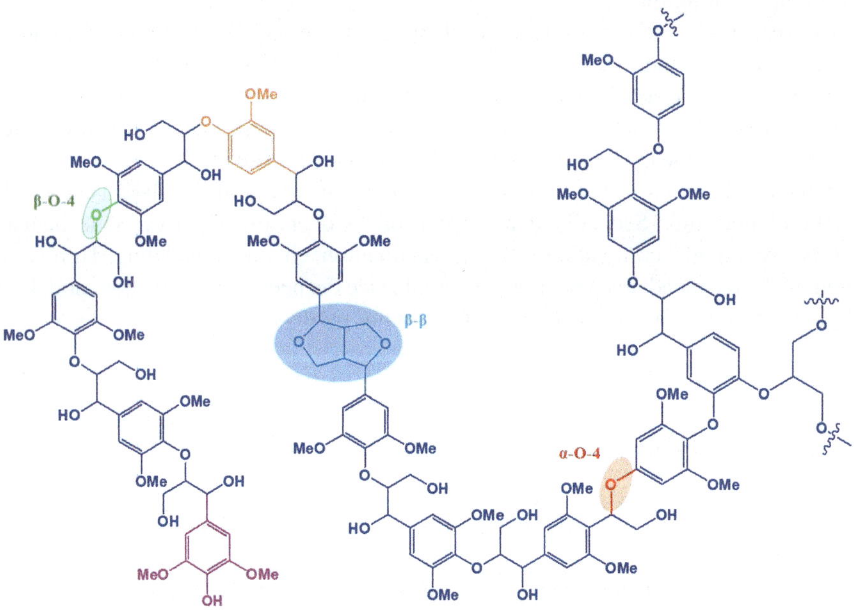

Fig. 3.2 Structure of softwood lignin (Redrawn with permission from [23]. Copyright © 2013 Elsevier)

3.1.2 *Major Linkages Present in Lignin*

The dominant linkages in lignin include β-O-4, α-O-4, 4-O-5, 5-5, β-β, β-1, and β-5 as shown in Fig. 3.3 [24]. The proportion of these linkages in softwood and hardwood lignins are depicted in Table 3.2. The dominant linkage in lignin is β-O-4, which accounts for nearly 46–60% of the total linkages depending on the source of lignin. Around 6–8% of the total linkages are constituted by α-O-4 type bonds [25, 26]. The plant lignins can be classified into softwood, hardwood, and grass lignins. Softwood lignin contains around 90–95% coniferyl alcohol-type monomers,

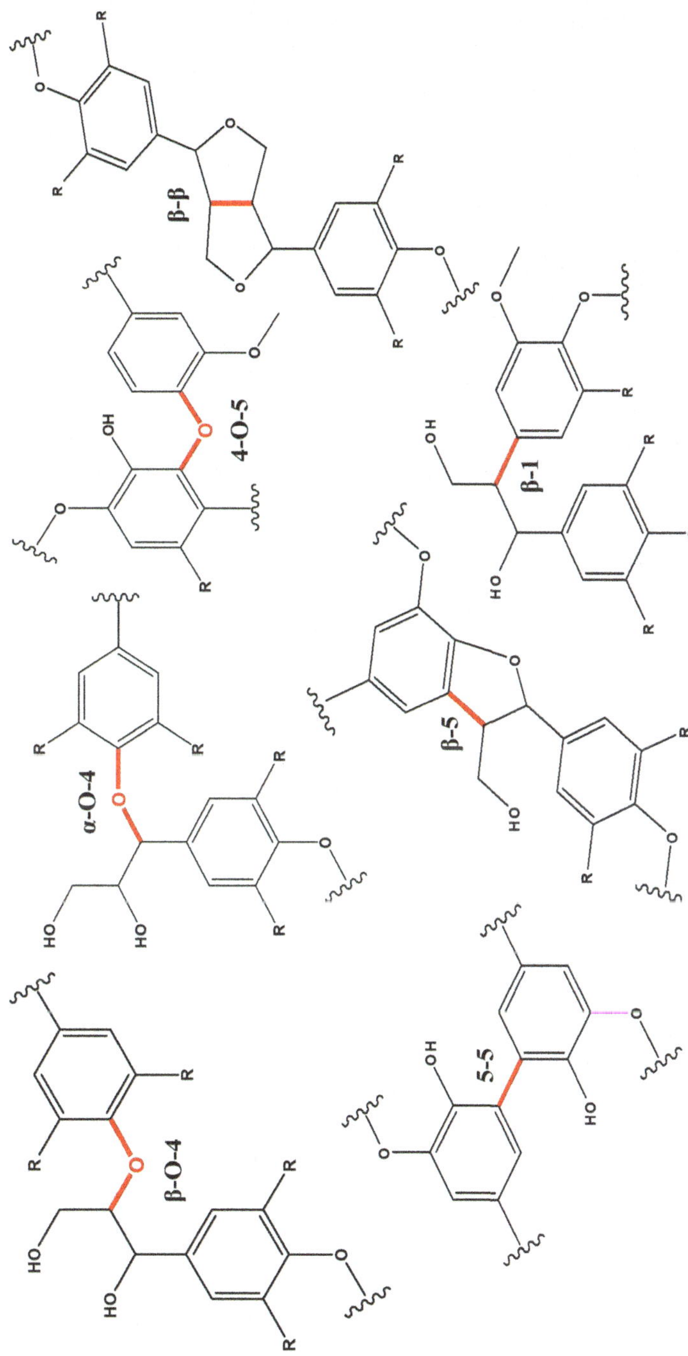

Fig. 3.3 Major links present in native lignin (Redrawn with permission from [26]. Copyright © 2010 American Chemical Society)

Table 3.2 Major links and their proportions in softwood and hardwood lignin

SI. no.	Linkage type	Softwood lignin (%)	Hardwood lignin (%)
1	β-O-4	45–50	60
2	α-O-4	2–8	7
3	4-O-5	4–8	7–9
4	β-β	2–6	3–12
5	5-5	10–27	3–9
6	β-5	9–12	6
7	β-1	7–10	1–7

Data taken from [6, 23, 30]

while hardwood lignin contains both coniferyl and sinapyl alcohol-types, whose composition vary in the range of 25–50% and 45–75%, respectively. Grass lignin contains significant amount of p-coumaryl-type monomers, whose composition varies in the range 5–35% [20, 27, 28]. The fraction of lignin in biomass varies for different plant types. The highest fraction of lignin is reported in softwoods (25–32) wt% as compared to hardwoods (18–25) wt% [29].

3.1.3 Thermochemical Conversion Platforms

Thermochemical techniques such as combustion, gasification, hydrothermal liquefaction, and pyrolysis are used to convert biomass into chemicals, liquid fuels, gases and bio-char. Combustion is a commercially developed process, wherein lignocellulosic biomass is directly burnt in presence of air to convert the chemical energy stored in biomass to heat and electricity at higher temperatures (1000–1200) °C. This process can handle any type of biomass with up to 50% of moisture [31, 32]. Despite combustion being an established thermochemical technique, with application in major industries for power generation, these is a need to control the emissions of SO_x, NO_x and other particulate matter. The major problem in biomass combustion is the presence of ash, which reduces the energy efficiency, and leads to higher maintenance cost due to slagging, fouling and corrosion in boilers [33, 34].

Gasification is a process of converting carbonaceous material such as biomass and coal into mainly gases (synthesis gas or syngas), with minor quantities of char and liquid tar products at higher temperature (700–1000) °C in the presence of sub-stoichiometric amount of air or oxygen, steam or CO_2. Syngas comprises of carbon monoxide (CO) and hydrogen (H_2) as the major constituents with smaller amount of carbon dioxide (CO_2), methane (CH_4) and light hydrocarbons. Syngas can be used to produce heat via combustion, hydrogen through water gas shift reaction [35], and hydrocarbons and other organics like acids, alcohols and esters via Fischer-Tropsch synthesis [36] followed by reforming reactions [37]. Depending on the gasifying agent (air, oxygen, steam or CO_2) the composition of product gases varies. Lignin gasification results in the formation of a mixture of non-condensable gases such as H_2, CO, CO_2, and CH_4 in different molar ratios depending on the gasification temperature, pressure, presence of steam/oxygen/CO_2, heating rate, catalysts and the type of lignin [38–40]. Yu et al. [38] studied lignin gasification with two different catalysts, viz., dolomite and Na_2CO_3, and found that, with the addition of dolomite, the production of CO and CO_2 increased slightly, while that of H_2 and CH_4 was unchanged. With the addition of Na_2CO_3, the composition of H_2, CH_4 and CO_2 decreased, and the composition of CO increased significantly. Recently, the combined approach of gasification with solid oxide fuel cells is gaining attention as it is an efficient and environment-friendly method for power generation [41, 42].

Hydrothermal liquefaction is a processing technique that can handle wet biomass feedstocks with as high as 50 wt% moisture to produce high quality bio-crude. This process is generally carried out at moderate temperatures of (200–400) °C, and high pressures of (50–300) bar in water or in a solvent mixture containing water and a co-solvent such as methanol, ethanol, propanol or glycerol [43, 44]. The process partly mimics the geological formation of crude oil, but in a controlled reactor environment. Liquefaction temperature, reaction time and heating rate are the main factors that affect the yield and product distribution from lignin [45]. Increasing the reaction time leads to effective depolymerization of high molecular weight components to smaller oligomeric phenols, while high reaction temperature results in increase and decrease in production of alkyl-substituted phenols and methoxy-substituted aromatic compounds, respectively [46, 47]. The major chemicals produced from hydrothermal liquefaction of lignin in near-critical and supercritical water include phenolic compounds such as phenol, catechol, and cresols.

Pyrolysis refers to thermochemical degradation of organic matter in the absence of oxygen at elevated temperatures, (450–650) °C [48]. Pyrolysis can be broadly classified into three types based on the residence time and heating rate. These include slow, fast, and flash pyrolysis. Slow pyrolysis involves the decomposition of biomass at slow heating rates (1–100) °C min^{-1} with long residence time of the order of minutes to hours. The char yield is usually high and liquid yield is low from this process. Fast pyrolysis involves high heating rate, (~100–500) °C/s of the sample with shorter residence time to produce high yield of liquids [20, 49, 50]. Flash pyrolysis occurs at high temperatures (700–1000) °C, high heating rate (>1000 °C/s) and very short residence time of the order of few seconds to produce high yield of gaseous products. Fast pyrolysis technology has received immense attention as a viable method to convert various biomass feedstocks into useful liquid fuels and chemicals that compete with and replace conventional fossil-derived fuels such as gasoline and diesel, and petroleum-derived products [51].

3.2 Conversion of Lignin into Valuable Fuel Intermediates and Chemicals

3.2.1 *Pyrolysis of Lignin*

Figure 3.4 shows a typical thermogram of lignin collected in a Thermogravimetric Analyzer (TGA) [52]. Due to the complex and heterogeneous structure, the pyrolytic conversion of lignin occurs in a broad temperature range (150–800) °C with highest degradation rate observed between (360 and 400) °C [53]. The major transformations occurring at different temperatures are also illustrated in the Fig. 3.4. The decomposition of bio-oil to biogas is rapid when the final temperature is over 500 °C. The typical mass loss and differential mass loss profiles of different lignins obtained from TGA are presented in Fig. 3.5 [54]. It is evident that these four lignins exhibit different thermal degradation behavior. The major decomposition of lignin occurs in the temperature range of (150–500) °C. Sulfonated wheat straw lignin exhibited higher moss loss compared to other lignins at lower temperatures owing to the high sulfur and carboxylic functional groups in it. These groups are expected to decompose and form gases such as SO_2 and CO_2 at low temperatures.

Nair and Vinu [55] performed fast pyrolysis of alkali lignin, also known as Kraft lignin, in an analytical micropyrolyzer coupled with gas chromatograph/mass spectrometry (GC/MS) at 500 °C. In this study, the pyrolysates were classified into guaiacols, simple phenols, and non-phenolic compounds. The yield of guaiacol and its derivatives in the bio-oil fraction was high as compared to simple phenols. The primary products of fast pyrolysis of alkali lignin include monomeric guaiacols formed by the cleavage of β-O-4, α-O-4, aryl-alkyl and alkyl-alkyl linkages [56–58]. The formation of guaiacol involves the scission of β-O-4 ether linkage followed by hydrogen abstraction reactions [58]. Guaiacol can also be formed by demethoxylation of the sinapyl unit of lignin. The yields of simple phenols, guaiacols and syringols from lignin vary based on the type of lignin, and the

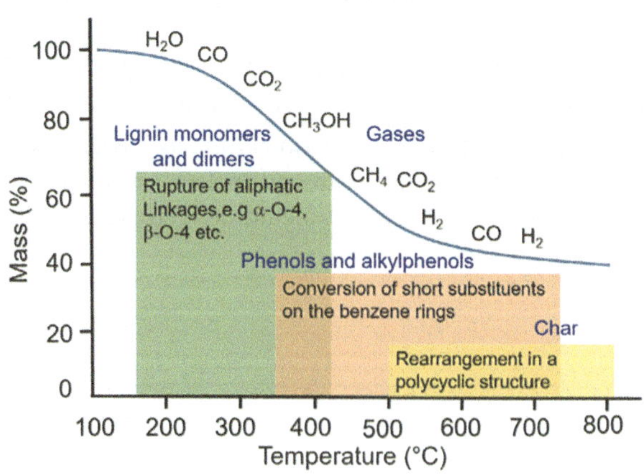

Fig. 3.4 Typical mass loss profile of lignin depicting the salient transformation regimes (Adapted with permission from [52]. Copyright © 2014 Elsevier). The temperature regimes where the major bond cleavages would occur and the gaseous products would evolve via different transformations are indicated by the shaded region

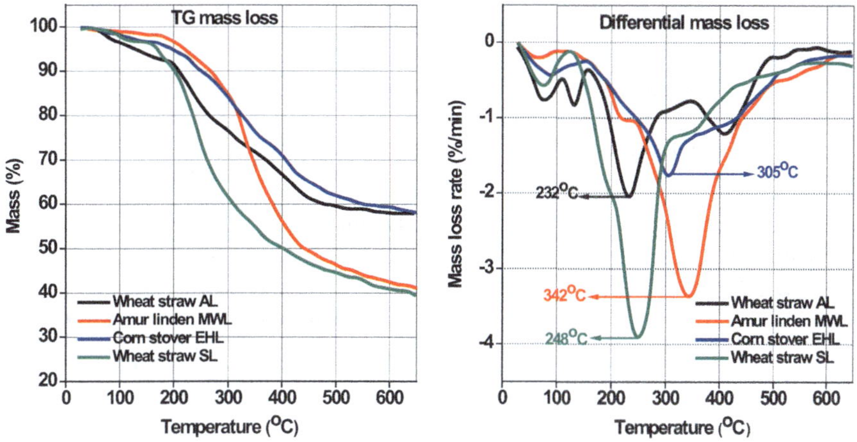

Fig. 3.5 TGA and differential thermogravimetric plots for four different lignins at 10 °C min^{-1} heating rate. *MWL* milled wood lignin, *EHL* enzymatic hydrolysis lignin, *AL* alkali lignin, *SL* sulfonate lignin (Redrawn with permission from [54]. Copyright © 2015 MDPI)

operating conditions like heating rate, temperature and gas ambience [59]. For example, the total yield of guaiacols and simple phenols obtained from cornstover lignin is shown to be 16.7 wt% [60]. The major pyrolysates from lignin are now well established in various studies to be phenol, cresol, ethyl phenol, 4-vinyl phenol, guaiacol, 4-methyl guaiacol, 4-ethyl guaiacol, 4-vinylguaiacol, isoeugenol and eugenol [11, 59–63].

Several studies have been conducted on the pyrolysis of lignin to elucidate the pyrolysis mechanism by using analytical techniques such as TGA coupled with Fourier transform infrared spectroscopy (TGA-FTIR), and pyrolysis coupled with GC/MS (Py-GC/MS) [64–67]. The aromatic methoxy groups are stable during the primary pyrolysis stage, and become very reactive in the temperature range of (400–450) °C. Hence, the aromatic compounds produced during the primary pyrolysis stage are predominantly 2-methoxy phenols from G-type lignins, and 2,6-dimethoxyphenols from S-type lignins. A majority of side chains are unsaturated alkyl groups with a smaller amount of saturated alkyl groups. The major volatile products from G-type lignins include isoeugenol, vanillin, acetovanillone, coniferyl aldehyde and dihydro-coniferyl alcohol. When the pyrolysis temperature is increased, secondary pyrolysis reactions take place, and G/S-type lignins rapidly transform to catechols, pyrogallols, o-cresols, xylenols and phenols, as shown in Fig. 3.6 [68]. In this temperature range, cracking of C-C side-chain occurs, which increases the yield of monomers. The product distribution is altered from unsaturated to saturated alkyl side chains such as methyl, ethyl, propyl, and 3-hydroxypropyl groups. Around 550 °C, catechols and pyrogallols disappear, and the production of non-condensable gases increases significantly. At temperatures greater than 700 °C, the formation of polyaromatic hydrocarbons (PAHs) and coke is enhanced. Phenols

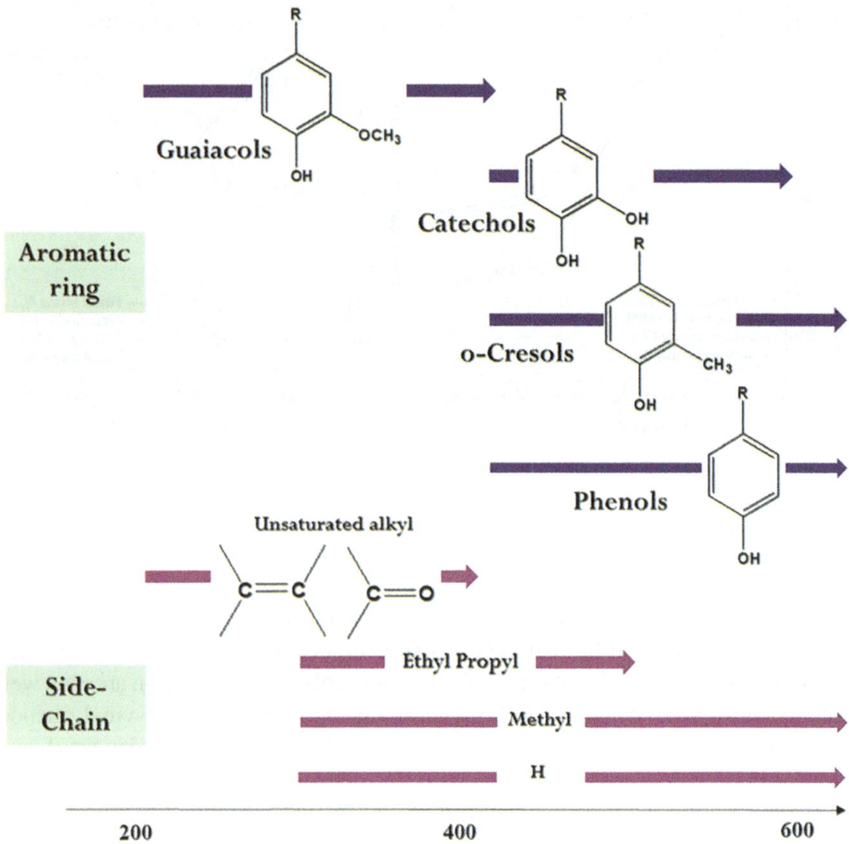

Fig. 3.6 Effect of pyrolysis temperature on aromatic substitution pattern and side chain structure of the products from G-type lignin (Redrawn from [68]. Copyright © 2017 Springer)

and o-cresols are relatively stable at such high temperatures, and hence, these are also observed at high pyrolysis temperatures along with PAHs.

Lou et al. [69] conducted fast pyrolysis experiments at (400–800) °C in a tubular reactor using lignin extracted from bamboo via enzymatic/mild hydrolysis method. The reaction temperature is shown to play a vital role in determining the yields of pyrolysis products and product distribution. The yield of bio-oil increased to a maximum 57.1 wt% at 500 °C, and then decreased to 53.3 wt% at 800 °C. The yield of gaseous products increased from 6.2 wt% to 20.3 wt%, and the char yield decreased from 42 wt% to 26.6 wt% with increase in temperature. The total phenolics, including guaiacols, syringols, vanillin and simple phenols, was maximum (79.3%) in bio-oil at 600 °C. At temperatures lower than 500 °C, pyrolysis of lignin was incomplete, while at temperatures greater than 700 °C, secondary reactions such as decarbonylation, decarboxylation, dehydration, demethoxylation, demethylation and radical rearrangement occur, and generate non-condensable gases like CO, CO_2, CH_4 and H_2.

Patwardhan et al. [60] studied pyrolysis of cornstover lignin at different temperatures. The char yield decreased from 60 wt% to 20 wt% with increasing temperature from 300 °C to 700 °C. The formation of low molecular weight compounds and gaseous products was higher at 700 °C, whereas phenolic products were higher at 600 °C. The major phenolic compounds include phenol, o-cresol, p-cresol, xylenols and ethyl phenol. Figure 3.7 depicts the major pathways involved in lignin pyrolysis to form different aromatic oxygenates. Jiang et al. [70] investigated pyrolysis of two different lignins, Asian and Alcell lignin, over a temperature range of (400–800) °C. The maximum yield of phenolic compounds at 600 °C was 17.2% and 15.5% for Alcell and Asian lignins, respectively. The major product from Alcell lignin was 5-hydroxyvanillin, whereas 2-methoxy-4-vinylphenol was the major product from Asian lignin. The yield of bio-oil and its composition mainly depend on particle size, temperature, heating rate and catalyst. The heating rate has a significant effect on bio-oil yield. Low heating rate leads to greater char formation whereas high heating rates are beneficial for the production of bio-oil.

3.2.2 *Catalytic Pyrolysis*

Catalysts play a vital role in the conversion of lignin to valuable fuels and chemicals. The use of a catalyst leads to improvement in selectivity to a particular aromatic compound from lignin. Various catalysts have been tested for fast pyrolysis of different feedstocks including lignin and its model compounds. Various types of catalysts such as inorganic metal, transition metal, noble metal, alumina and zeolites have been applied for pyrolysis of lignin and its model compounds, as shown in Table 3.3 [64, 71–76]. It is important to note that these are either in situ or ex situ catalytic fast pyrolysis studies, without the use of any reactive gas like hydrogen. Hence, the typical transformations only include deoxygenation, while hydrodeoxygenation or hydrogenation reactions do not occur. Among these catalysts, zeolites are effective deoxygenation catalysts to produce bio-oil with reduced oxygen content and improved aromatic hydrocarbons. Zhang et al. [71] conducted catalytic fast pyrolysis of aspen lignin in the presence of H-ZSM5 and HY catalysts using Py-GC/MS. They found that HZSM-5 catalyst was more effective than HY catalyst for the conversion of phenolics into aromatic hydrocarbons. The production of aromatic hydrocarbons was found to be maximum at a catalyst-to-lignin mass ratio of 3:1. Toluene and p-xylene were the two abundant compounds formed in the presence of catalysts.

Lee et al. [72] performed catalytic pyrolysis of lignin using Py-GC/MS at 500 °C in presence of mesoporous Y zeolite. They reported that, in the presence of catalyst, the major products were simple phenols, monoaromatics and polyaromatic hydrocarbons. The typical pathways involved in the formation of different aromatic compounds from catalytic fast pyrolysis of lignin are depicted in Fig. 3.8. The

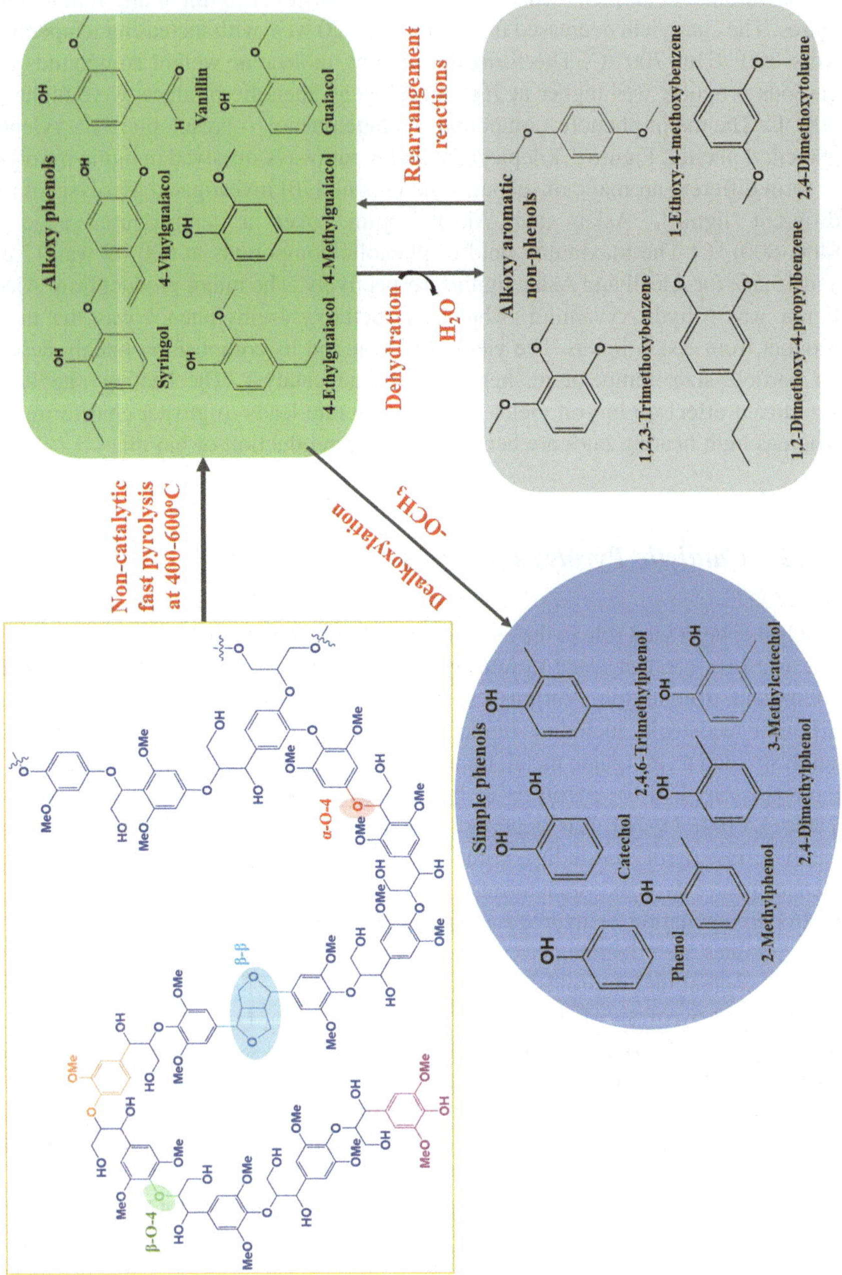

Fig. 3.7 Reaction scheme of non-catalytic fast pyrolysis of lignin

Table 3.3 Catalytic fast pyrolysis of different lignins conducted in various pyrolysis reactors

SI. no.	Feedstock	Operating conditions	Catalyst	Catalyst: Lignin (wt/wt)	Yield of aromatics	Major aromatic products in bio-oil	References
1	Aspen lignin	Reactor: Pyroprobe® 5000, 0.5 mg sample, 600 °C	HZSM-5 HY	3:1	23.0 wt%	Toluene and p-xylene	[71]
2	Kraft lignin	Reactor: vertical furnace type pyrolyzer, 1.5 mg sample, 500 °C	Meso-Y	3:1	18.0 wt%	Toluene, xylene, ethyl benzene, and benzene	[72]
3	Alkali lignin	Reactor: curie-point pyrolyzer, 1 mg sample, 650 °C	HZSM-5	20:1	68.3 wt%	Benzene, toluene, and xylene	[73]
4	Milled wood lignin	Reactor: micro furnace pyrolyzer, 0.5 mg sample, 700 °C	HZSM-5	20:1	8.5 wt%	Benzene, toluene, xylene, indanes, indenes, alkyl benzenes, and naphthalenes	[74]
5	Alkaline lignin	Reactor: platinum coil pyrolyzer 5150, 1.5 mg sample, 650 °C	H-USY	---	42.0 wt%	Benzene, toluene, and xylene	[64]
6	Alkaline lignin	Reactor: platinum coil pyrolyzer 5150, 1.5 mg sample, 650 °C	Ni/HZSM-5	4:1	71.0 area %	Benzene, toluene, xylene, and naphthalene	[75]
7	Alkaline lignin	Reactor: platinum coil pyrolyzer 5150, 1.5 mg sample, 650 °C	H-beta	4:1	75.0 area %	Benzene, toluene, xylene, and naphthalene	[75]
8	Kraft lignin	Reactor: coil type Pyroprobe 5200, 650 °C	ZSM-5	10:1	90.0 area %	o-xylene, toluene, 1,2,3-trimethylbenzene, and naphthalene	[76]

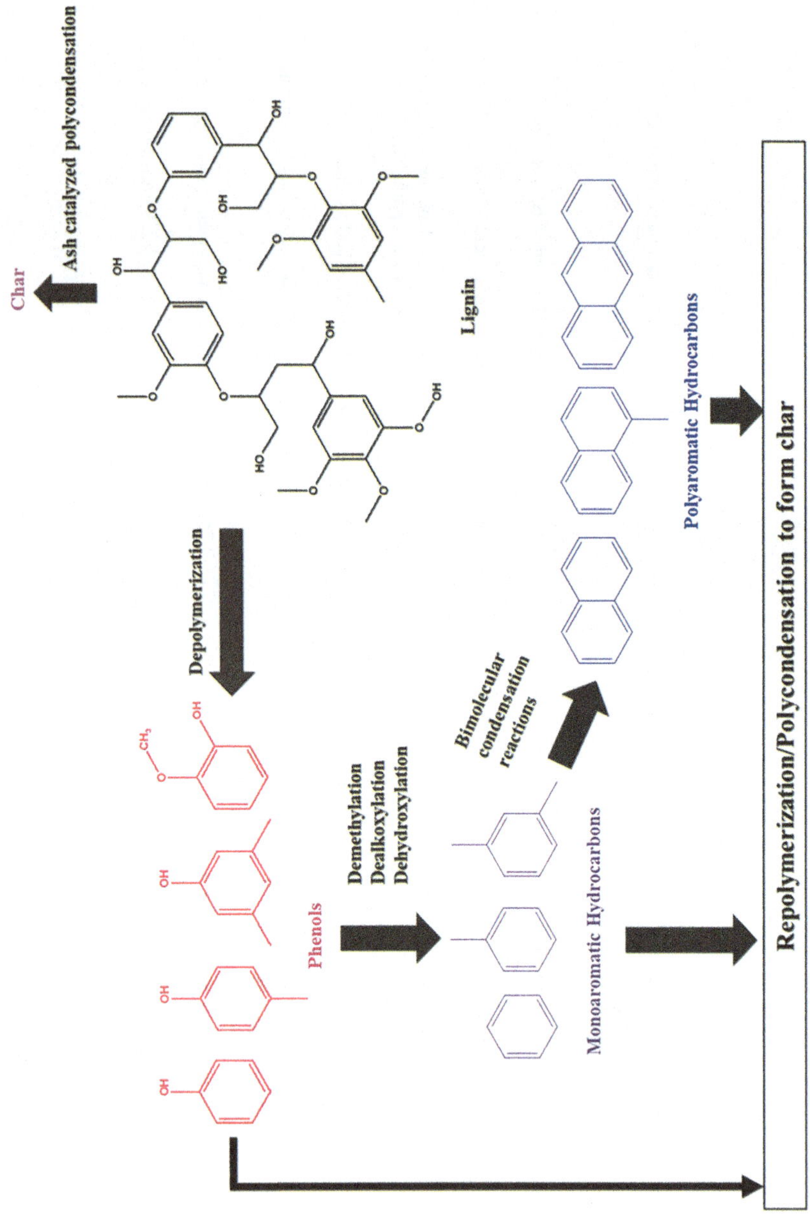

Fig. 3.8 General scheme of catalytic fast pyrolysis of lignin showing the salient transformations

yield of total phenolics decreased, and simple phenols, monoaromatic hydrocarbons and PAHs increased with increasing mesoporous Y-to-lignin mass ratio.

Jackson et al. [77] performed catalytic pyrolysis of lignin over various catalysts like HZSM-5, KZSM-5, Al-MCM-41, solid phosphoric acid and Co/Mo/Al_2O_3. In the presence of HZSM-5, the production of single ring aromatics like benzene, toluene and xylene (46.7%) and naphthalene-type aromatics (46.2%) were higher, whereas the proportion of oxygenated aromatics was higher in the presence of Al-MCM-41. Li et al. [73] conducted fast pyrolysis of two Kraft lignins using Curie point pyrolyzer in the presence of HZSM-5. They found that, by decreasing the Si/Al ratio from 200 to 25, and by increasing the catalyst-to-lignin mass ratio from 1 to 20, the production of aromatic hydrocarbons increased substantially with a concomitant reduction in lignin-derived oxygenates. Increase in pyrolysis temperature from 500 °C to 650 °C led to increase of yield of aromatics and phenols.

Kim et al. [78] studied catalytic pyrolysis of lignin for the production of aromatic hydrocarbons by varying the catalyst-to-lignin mass ratio, and acidity (Si/Al ratio of 30–280) of the HZSM-5 catalyst. They reported that the yield of aromatic hydrocarbons increased from (0.76 to 2.62) wt% with increase in catalyst acidity and decrease in Si/Al ratio. The optimal temperature and catalyst: lignin mass ratio were 600 °C and 2:1, respectively, for the production of aromatic hydrocarbons. Lignin pyrolysis products can be converted to aromatic hydrocarbons only at the external surface of zeolite catalyst, as almost all lignin-derived phenols are larger in size to pass through the pores of the catalyst. The yield of char typically increased with increasing catalyst amount, owing to the coking reactions.

Catalytic fast pyrolysis of alkaline lignin was investigated using different zeolite catalysts of different pore sizes and acidity by Ma et al. [63]. They reported that highest yield of liquid (75 wt%) was obtained at 650 °C in the presence of H-USY catalyst, which was highly acidic (1.2 mmol g^{-1} total acidity) with Si:Al ratio of 7 and large pore size of 7.4 Å among the different catalysts (H-ZSM5, Na-ZSM5, and H-beta). Ma et al. [75] performed catalytic fast pyrolysis of lignin in Pyroprobe® reactor using different types of catalysts such as non-acidic alumina silicates, transition metal oxides (Co, Mo, Ni, Fe, Mn and Cu) and zeolite-supported transition metals. The major products from non-catalytic fast pyrolysis of lignin were alkoxyphenols. Alkoxyphenols were still present in the pyrolysate with the addition of transition metal oxides, although its yield was low as compared to the case without the catalyst. The production of aromatic hydrocarbons increased with the addition of Co and Ni-supported zeolites as compared to that of only the zeolite without transition metals. The major reactions involved were shown to be dehydration, decarboxylation, dealkylation, cracking and isomerization.

Zhang et al. [71] investigated the effect of the catalyst loading and temperature on the production of aromatic hydrocarbons. The yield of aromatic hydrocarbons was 12.8 wt% at 650 °C and ZSM-5: lignin mass ratio of 5:1. The abundant aromatic hydrocarbons were toluene and p-xylene. Generally, with more amount of catalyst, the pyrolysis vapors are trapped in the catalyst pores, and are exposed to higher temperatures for longer time periods. This can potentially lead to their conversion to coke via bimolecular condensation reactions.

The use of catalysts such as $AlCl_3$ and $ZnCl_2$ is shown to increase the yields of bio-oil and gases, at the expense of lower bio-char yields [79]. The bio-oil from pyrolysis of lignin using these catalysts mainly contains substituted phenols, monocarboxylic acids and cyclic esters [79]. In the presence of alkaline catalysts (NaOH, KOH, Na_2CO_3, and K_2CO_3), the formation of alkylphenols and methoxyphenols is enhanced. In the presence of hydroxy alkali like NaOH and KOH, more alkylphenols are produced, whereas in the presence of carbonate alkali like Na_2CO_3 and K_2CO_3, more methoxyphenols are produced [80].

In an attempt to improve the yields of phenols, rather than deoxygenating them to aromatic hydrocarbons, Nair and Vinu [55] used oxide catalysts such as TiO_2, CeO_2 and ZrO_2, and reported that, with TiO_2, the yield of guaiacols doubled as compared to non-catalytic fast pyrolysis of alkali lignin. Due to low Lewis acidity, the primary intermediates from lignin, mainly guaiacols and simple phenols, are stabilized in the pores, which effectively prevents the repolymerization reactions to form char. In non-catalytic pyrolysis, the primary phenolic intermediates are repolymerized in the melt phase, which results in high yield of char. The oxide catalysts are expected to promote the formation of hydroxyl radicals via thermal activation, which then take part in free radical reactions to form guaiacol and its derivatives.

Catalyst acidity and pore structure are the main factors that affect the catalytic efficiency of the zeolites. Zeolites with large pores are favored for higher yield of liquid products, because large pore sizes permit dimeric/trimeric macromolecules to pass through them, thus leading to effective upgradation. The acidity of the zeolite depends on both the framework of the crystalline structure as well as the overall Si/Al ratio. The Brønsted acidity aids in effective cracking, deoxygenation, coking and aromatization reactions. The existing studies using zeolites unequivocally prove that aromatic hydrocarbons including monoaromatics like BTEX (benzene, toluene, ethylbenzene and xylenes) and polymaromatics like naphthalane and methyl naphthalenes are the key products. While a number of studies have been performed using lignin and model compounds, the development of a structure-activity relationship is lacking in the literature. Relating the type of functional unit of lignin or its model compound with the activity exhibited by the catalyst can provide valuable information for the design of better catalysts with composite functions in lignin valorization. This requires synergetic combination of experimental catalysis, surface science and molecular modeling of catalysts. Scaling relationships have been developed to elucidate the reactivity trends of binding energies of different intermediates over different catalyst surfaces [81]. Such relationships are required for transformations of lignin intermediates over different surfaces.

3.2.3 *Fast Pyrolysis of Lignin Model Compounds*

3.2.3.1 Monomers

Several studies have already shown that guaiacol is one of the major products from pyrolysis of lignin. Dorrestijn and Mulder [82] investigated the radical-induced decomposition of guaiacol at different temperatures from (407 to 517) °C, and found that cumene acts as a radical scavenger. The major products were catechol, 2-hydroxybenzaldehyde, o-cresol, 2-ethylphenol, phenol, CH_4 and CO. Further, they found that increase in phenol production might be due to the conversion of 2-hydroxybenzaldehyde to phenol through hydrogen abstraction and decarbonylation. Asmadi et al. [83] studied the thermal reactions of guaiacol and syringol as lignin model aromatic nuclei. The experiments were conducted in a closed ampoule reactor in the temperature range of (400–600) °C and residence times of (40–600) s. The main products obtained from guaiacol pyrolysis were catechol, o-cresol, phenol, benzofuran, 2-ethylphenol, and 2,4-xylenol.

Shin et al. [84] investigated fast pyrolysis of vanillin using molecular beam mass spectrometry to detect gas phase products. The experiments were conducted in a spiral quartz tube inside a straight quartz tube in the temperature range of (500–800) °C, and residence time of (0.3–0.6) s. They found that the conversion of vanillin is low at 500 °C, while complete conversion occurred at 650 °C. The main products from vanillin pyrolysis were catechol, 4-hydroxybenzaldehyde, guaiacol, 3,4-dihydroxybenzaldehyde, 5-formylsalicyaldehyde at 650 °C. Further, they observed the formation of aromatic compounds such as benzene, naphthalene, and anthracene at a higher temperature of 800 °C. Liu et al. [85] studied the pyrolysis of three different guaiacyl type monomeric compounds viz., vanillin, vanillic acid and vanillyl alcohol. They evaluated the single point energies of different compounds involved in the decomposition mechanism of vanillin, vanillic acid and vanillyl alcohol using density functional theory (DFT) approach. They reported the decomposition of vanillin at different temperatures, (400–600) °C, and found that the major products were guaiacol and 5-formylsalicyaldehyde. At higher temperatures, the formation of 2-hydroxybenzaldehyde, 1,2-benzenediol, 2-methylphenol and 2-ethylphenol was observed.

3.2.3.2 Dimers

Chen et al. [86] studied the pyrolysis of β-O-4-type lignin dimer, viz., 1-(4-methoxyphenyl)-2-(2-methoxyphenoxy) ethanol via both experimental and theoretical DFT approach. The experiments were conducted in the temperature range of (300–800) °C. The pyrolysates observed at a low temperature of 300 °C were guaiacol and 4-methoxystyrene. These products were formed by the homolytic cleavage of $C_β$-O bond. At moderate temperatures, both homolytic and concerted reactions take place to form carbonyl-containing aromatics such as

2-hydroxybenzaldehyde, 4-methoxybenzaldehyde and 4-hydroxyacetophenone. At high temperatures, the primary pyrolysates undergo secondary decomposition to form phenols and mono aromatics.

Jiang et al. [87] studied the thermal reactions of β-1-type lignin dimer compound, viz., 1,2-bis-(3,5-dimethoxyphenyl) propane-1,3-diol. The experiments were conducted in an analytical Pyroprobe® pyrolyzer at 800 °C with a residence time of 20 s. The major pyrolysates were 3,5-dimethoxybenzaldehyde, 1,3-dimethoxybenzene, 3-hydroxy-5-methoxybenzaldehyde, 3,5-dimethoxybenzyl alcohol, 3-methoxybenzaldehyde, and 1,3-dimethoxy-5-vinylbenzene. 3,5-Dimethoxybenzaldehyde was the major product, accounting for more than 50% of the total products. The homolytic cleavage of C_{α}-C_{β} bond is shown to be the key reaction in pyrolysis of all β-1-type lignin dimers, which is due to its low bond dissociation energy.

3.3 Theoretical Studies of Lignin Model Compounds

The fundamental understanding of reaction chemistry and pathways is vital for better design, development and optimization of fast pyrolysis reactors for large-scale production of fine chemicals from lignocellulosic biomass and lignin. Mechanistic models have the unique ability to determine the specific species or reaction pathway responsible for the observed products. While a reasonable understanding of fast pyrolysis mechanism of cellulose and hemicellulose is achieved after the works of Broadbelt and co-workers [88–91], lignin pyrolysis chemistry is more complicated than that of the carbohydrate components of biomass. Mechanistic modeling of fast pyrolysis of lignin is challenging owing to the large number of bond types, the variety of species and their inter-conversions occurring through a number of elementary reactions. Lignin model compounds such as monomeric and dimeric phenols find value in fundamental studies that attempt to build a step-by-step and comprehensive understanding of lignin fast pyrolysis.

3.3.1 Lignin Monomers

The monomeric lignin model compounds are the primary pyrolysates from lignin. Some studies have investigated the fast pyrolysis mechanism of catechol [92, 93], guaiacol [93–95], syringol [96] and vanillin [97]. Bond dissociation energy (BDE) provides the foundation for the development of a kinetic model. BDE is a measure of the bond strength. Smaller the BDE, better is the chance for bond cleavage. BDEs (in kcal mol^{-1}) of homolytic cleavage of different bonds in selected lignin model compounds are shown in Fig. 3.9 [25, 98–100]. BDEs are usually calculated using the expression, $BDE\ (kcal\ mol^{-1}) = H_{298}(R) + H_{298}(X) - H_{298}(R - X)$, where, H ($R$), H ($X$), and H ($R - X$) refer to the enthalpies of product free radicals, R and X,

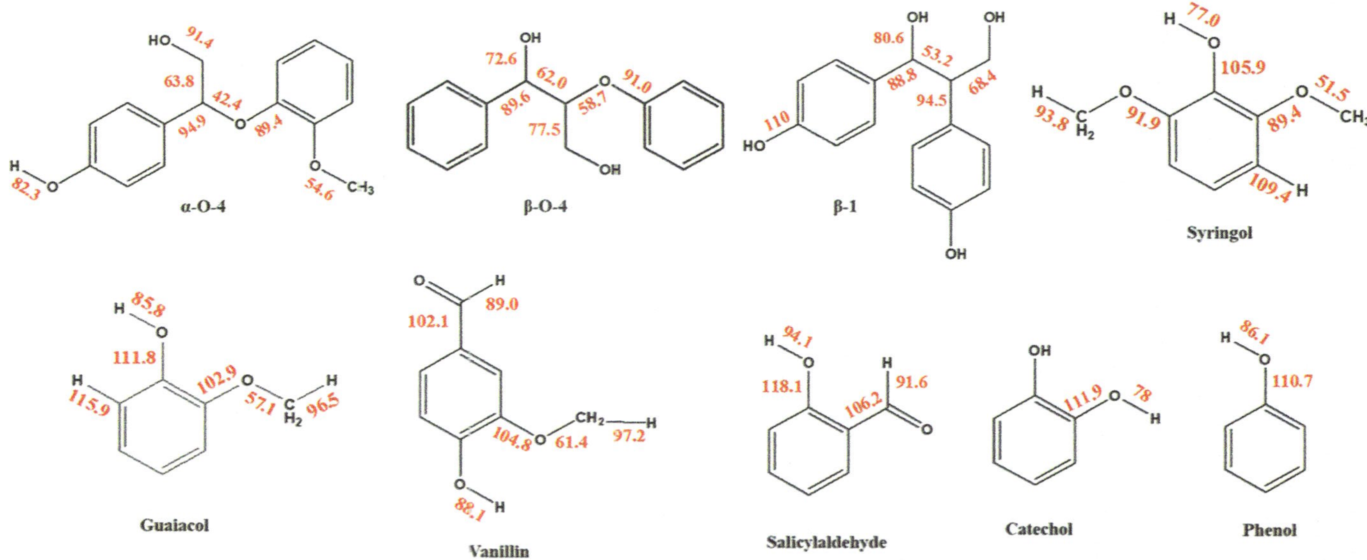

Fig. 3.9 Homolytic cleavage of different lignin model compounds and their corresponding bond dissociation energies (kcal mol^{-1}) (Redrawn with permission from [25, 98, 99, 100]. Copyright © 2014, 2015 Elsevier; Copyright © 2019 Royal Society of Chemistry). *BDEs of guaiacol, salicylaldehyde, catechol and phenol were calculated at the M06-2X/6-311+G(d,p) level, vanillin and syringol are calculated at the M06-2X/6-31++G(d,p) and B3LYP/6-31++G(d,p) level, respectively, and those of lignin dimer model compounds (β-O-4 and α-O-4) are calculated at the B3LYP/6-31+G(d), and β-1 is calculated at the M06-2X/6-31+G(d,p)

and the reactant, $R - X$, respectively, at 298.15 K and 1 atm. For bond fission or homolytic cleavage reactions, the Arrhenius activation energy (E_a) is equivalent to the magnitude of the BDE, and these are related by $E_a = \text{BDE} - \text{RT}$. For other reaction types, enthalpy, entropy and Gibbs free energy changes associated with the transition state complex are obtained from electronic structure theories, and transition state theory is used to determine the rate constant at different temperatures. The rate constants of the reverse reactions are usually calculated using the equilibrium constant, which is determined using the Gibbs free energy change associated with the reaction.

Addition of functional groups at specific positions of an aromatic ring alters the BDE of a specific bond type. For example, addition of hydroxyl group at the ortho position of phenol decreases the BDE of O–H by 8 kcal mol^{-1}. Figure 3.9 provides a comparison of BDEs of methoxy group in guaiacol and vanillin. The homolytic cleavage of O-CH_3 bond is always a facile step due to its low BDE ranging from 57.1 kcal mol^{-1} in guaiacol to 61.4 kcal mol^{-1} in vanillin. The comparison of BDEs of the hydroxyl groups in phenol, catechol, salicylaldehyde, guaiacol and vanillin is also provided in Fig. 3.9.

The BDE for O-H bond cleavage is found to be the lowest for catechol (78 kcal mol^{-1}), while the highest BDE is observed in salicylaldehyde (94.1 kcal mol^{-1}). The C_α–O bond has the lowest BDE (42.4 kcal mol^{-1}) in thermal decomposition of α-O-4 lignin dimer model compound. The BDEs of different lignin linkages follow the trend: C_α–O < O–CH_3 < C_α–C_β < O–H < O–$C_{aromatic}$ < C_β–OH < C_α–$C_{aromatic}$. The BDE of C_β–O bond, 58.7 kcal mol^{-1}, is the lowest followed by C_α–C_β bond, 62.0 kcal mol^{-1}, in thermal decomposition of β-O-4 lignin dimer model compounds. In homolytic cleavage of β-1 type lignin dimer model compounds, the lowest BDE is found for C_α-C_β bond (53.2 kcal mol^{-1}), followed by C_β–C_γ bond (68.4 kcal mol^{-1}).

Catechol is one of the secondary products from lignin pyrolysis. Altarawneh et al. [92] studied the unimolecular decomposition of catechol, and reported that, at high temperatures, both concerted and intermolecular water elimination reactions occur from the two hydroxyl groups, which result in the formation of cyclopenta-2,4-dien-1-ylidnemethanone. Cavallotti et al. [93] proposed ipso-addition reactions of catechol to form phenol and cresol. Liu et al. [94] proposed reaction pathways for guaiacol pyrolysis involving both homolytic and concerted reactions to describe the formation of o-quinonemethide intermediate, which is a precursor for the formation of polyaromatic hydrocarbons and char.

Pelucchi et al. [99] reported the major reaction classes for the decomposition of guaiacol, which include chain initiation and chain propagation (ipso-addition reactions and H-abstraction reactions) as shown in Fig. 3.10. Catechol formation can be attributed to successive H-abstraction reactions of the phenoxy-phenol radical formed in the chain initiation reaction. The products of ipso-addition reactions include cresol, anisole, 2-methyl-anisole and phenol. The frequency factor and activation energy values of the ipso-addition reactions fall in the range $(1.2–1.5) \times 10^{13}$ cm^3 mol^{-1} s^{-1} and (5–7) kcal mol^{-1}, respectively. Salicylaldehyde is obtained via decomposition of guaiacol through H-abstraction and radical rearrangement reactions. The rate

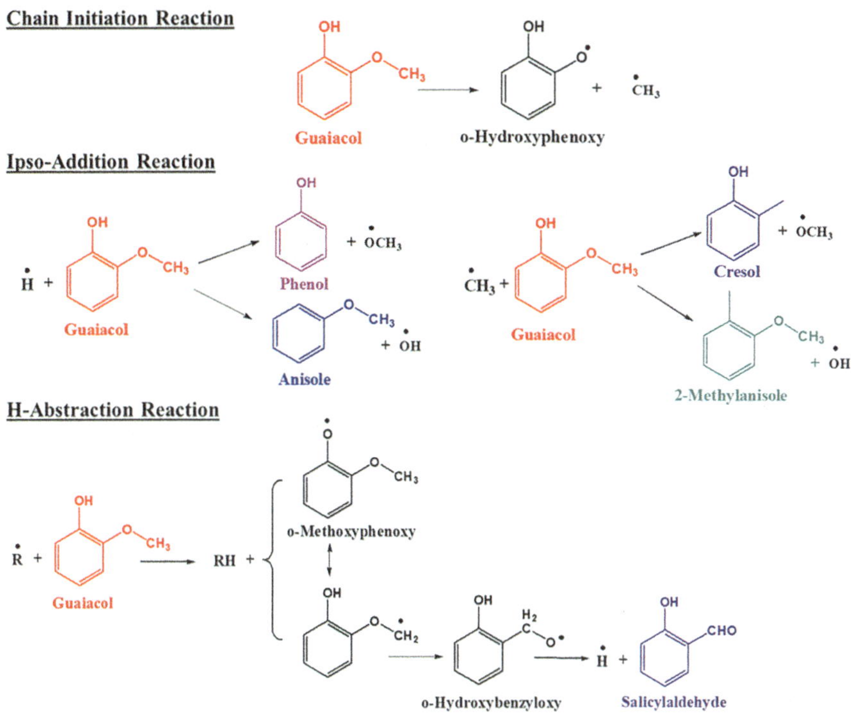

Fig. 3.10 Major reaction families in guaiacol decomposition (Redrawn with permission from [99]. Copyright © 2019 Royal Society of Chemistry)

constants of the general H-abstraction reactions usually depend on the type of hydrogen to be abstracted and the properties of the abstracting radical. The removal of H-atoms depends mainly on the strength of the corresponding O–H and C–H bonds.

Vanillin contains different functional groups such as hydroxy, formyl and methoxy, which are commonly found in the structure of lignin. Concerted reactions are dominant at high temperatures due to their high activation energy. Hydrogen abstraction reactions are shown to be the key for the formation of major products from pyrolysis of vanillin. Britt et al. [58] investigated the fast pyrolysis of vanillin, and found that guaiacol was the major product at low temperatures, (400–500) °C, while salicylaldehyde was the major product at high temperatures, (600–700) °C. This means that, at elevated temperatures, the transformation of methoxy to aldehyde functionality occurs via radical-induced rearrangement of methoxy group [58].

Liu et al. [85] studied the pyrolysis of three different monomeric compounds, viz., vanillin, vanillic acid and vanillyl alcohol, using quantum mechanics. They evaluated the single point energies of different intermediates involved in the pyrolysis of vanillin, vanillic acid and vanillyl alcohol. The major products in the temperature range of (400–600) °C were shown to be guaiacol and

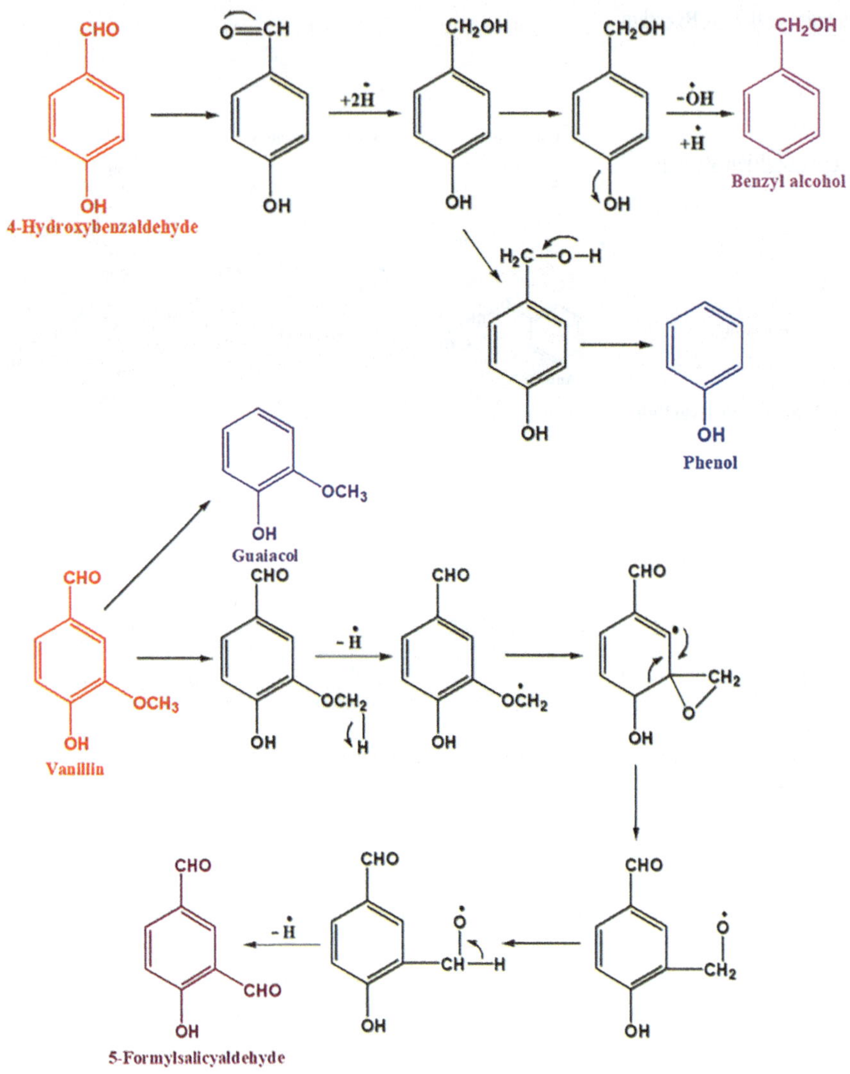

Fig. 3.11 Proposed reaction mechanism of pyrolysis of 4-hydroxybenzaldehyde and vanillin (Redrawn with permission from [99]. Copyright © 2019 Royal Society of Chemistry)

5-formylsalicyaldehyde, while at higher temperatures, 2-hydroxybenzaldehyde, 1,2-benzenediol, 2-methylphenol and 2-ethylphenol were produced. Benzaldehydes are also the major products from biomass fast pyrolysis. Akazawa et al. [101] performed pyrolysis of various benzaldehydes such as 4-hydroxybenzaldehyde, vanillin, syringaldehyde and veratrumaldehyde in Py-GC/MS. They proposed reaction mechanism for the formation of secondary products, which are shown in Fig. 3.11. The most common pyrolysis reactions include homolytic elimination of methyl group from methoxy group, radical rearrangement, addition of various radicals and recombination reactions. Huang et al. [102] reported the formation of CO_2, CO and CH_4 via decarboxylation, decarbonylation and demethylation reactions, respectively, from various lignin model compounds phenyl (p-hydroxyphenyl, guaiacyl and syringyl) formic acid, phenyl (p-hydroxyphenyl, guaiacyl and syringyl) acetaldehyde and concerted reactions of lignin model compounds.

3.3.2 Lignin Dimers

A number of studies have focused on the decomposition of lignin dimers with β-O-4 [25, 86, 103–106], α-O-4 [25, 106, 107], and β-1 [108] linkages. Huang et al. [98] proposed reaction pathways for pyrolysis of β-O-4-type lignin dimer model compound viz., 1 (1-phenyl-2-phenoxy-1,3-propanediol), which involves both homolytic and concerted reactions. They reported that concerted reactions dominate the homolytic reactions at low temperatures, whereas at high temperatures the homolytic reactions are important.

In another study, Chen et al. [86] proposed three possible reaction pathways, viz., homolysis of the C_β–O bond, homolysis of C_α–C_β bond and C_β–O concerted reaction, for pyrolysis of β-O-4 type lignin dimer model compound, as shown in Fig. 3.12. At low temperature (ca. 300 °C), the initial step involves the homolytic cleavage of C_β–O bond due to its low bond dissociation energy (53 kcal mol^{-1}). At moderate temperatures, carbonyl-containing compounds such as 2-hydroxybenzaldehyde, 4-methoxybenzaldehyde and 4-hydroxyacetophenone were formed via homolytic and concerted reactions. At high temperatures, the primary pyrolysates undergo secondary decomposition to form simple phenols and aromatic hydrocarbons.

Kim et al. [107] studied the pyrolysis mechanism of methoxy substituted α-O-4 lignin dimer model compounds. They observed that the reactivity towards ether cleavage is enhanced due to the addition of methoxy substitution in the ortho position of the aromatic ring of the α-O-4 lignin dimer. They also reported that large number of free radicals are produced from methoxy substituted α-O-4 lignin dimer. These radicals were formed not only via ether cleavage, but also by the cleavage of methoxy group.

Partharasathi et al. [109] evaluated the bond dissociation energies of a wide variety of linkages present in lignin using DFT approach. The mechanism of decomposition of phenethyl phenyl ether (PPE), a dimer with β-O-4 linkage, with various substituents, is well studied via experiments, quantum chemical computations and kinetic modeling [58, 110, 111]. Huang et al. [98] studied the pyrolysis of PPE by DFT, and reported that the major reaction channels for the formation of styrene and phenol involved concerted reactions. In another study, Elder and Beste [110] showed via quantum chemical and DFT studies that the concerted retro-ene fragmentation mechanism involves a lower reaction barrier compared to bond fission reaction. The effect of hydroxy functional groups on the decomposition of β-ether

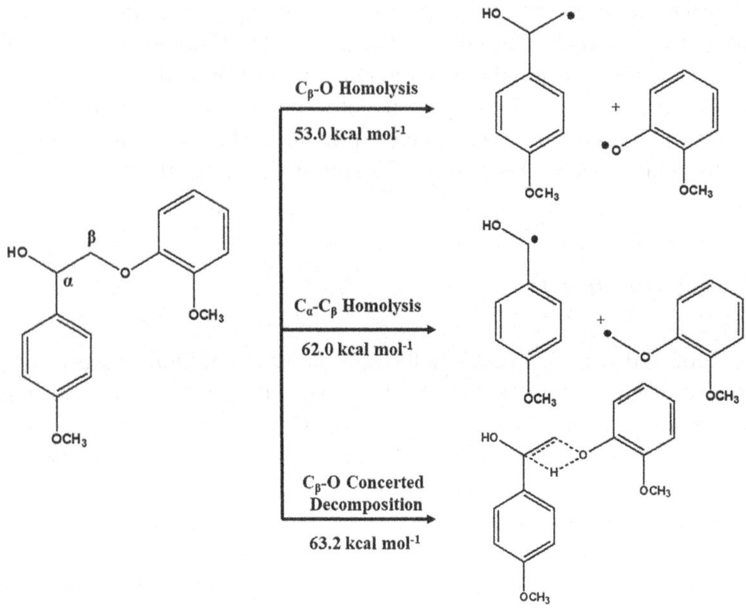

Fig. 3.12 Initial pyrolysis reactions of β-O-4-type lignin dimer model compound. The data corresponds to calculations performed using M06-2X method at 6–311++G(d,p) level of theory. (Redrawn with permission from [86]. Copyright © 2015 Elsevier)

cleavage using lignin dimer model compounds were studied by Kawamoto et al. [112]. They observed that the reactivity toward β-ether cleavage increased with the addition of hydroxyl groups in the lignin dimer model compound. In another study, Kawamoto et al. [113] studied the reactivity of major linkages present in lignin such as β-O-4, α-O-4, β-1 and biphenyl. They reported that the reactivity is higher for β-O-4 and α-O-4 linkages compared to β-1 and biphenyl linkages.

Jiang et al. [87] studied the pyrolysis of β-1-type lignin dimer compound, viz., 1,2-bis-(3,5-dimethoxyphenyl) propane-1,3-diol, by both experiments and theory. The homolytic cleavage of C_α-C_β bond is the major initiation reaction in pyrolysis of all β-1-type lignin dimers, owing to its low BDE. They proposed four reaction pathways, which include homolysis of the C_α-C_β bond, and three different concerted reaction pathways, as shown in Fig. 3.13. The major products obtained from the first two concerted pathways include 3,5-dimethoxybenzaldehyde, 3-hydroxy-5-methoxybenzaldehyde, 3-methoxybenzaldehyde, and 1,3-dimethoxy-5-vinylbenzene. The major products obtained from the homolytic reaction pathway include 1,3-dimethoxybenzene and 3,5-dimethoxybenzyl alcohol. The concerted pathway 3 hardly took place because of its high activation energy.

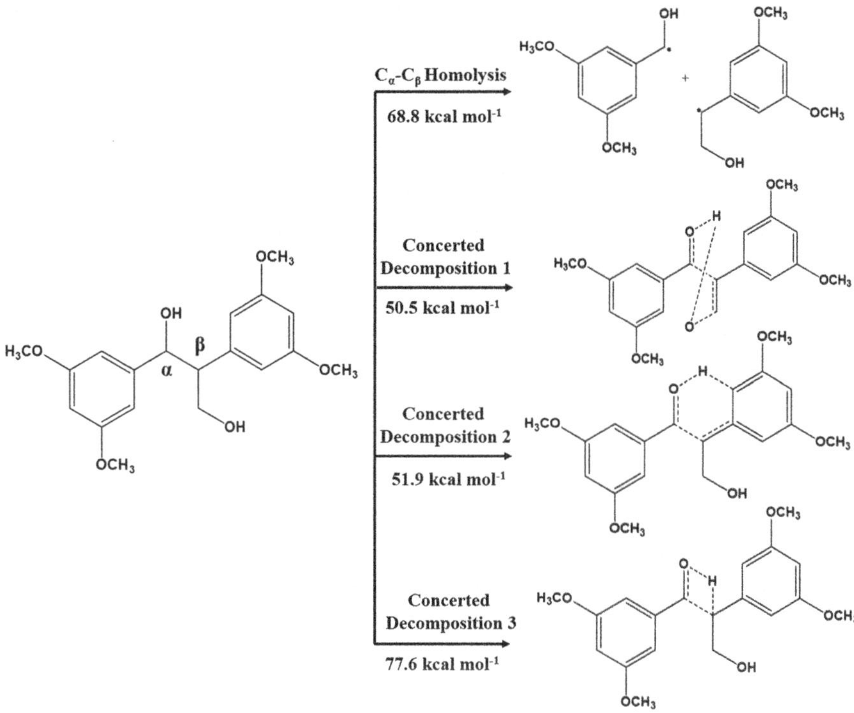

Fig. 3.13 Initial decomposition reactions involved in β-1-type lignin dimer model compound (Redrawn with permission from [87]. BioResources)

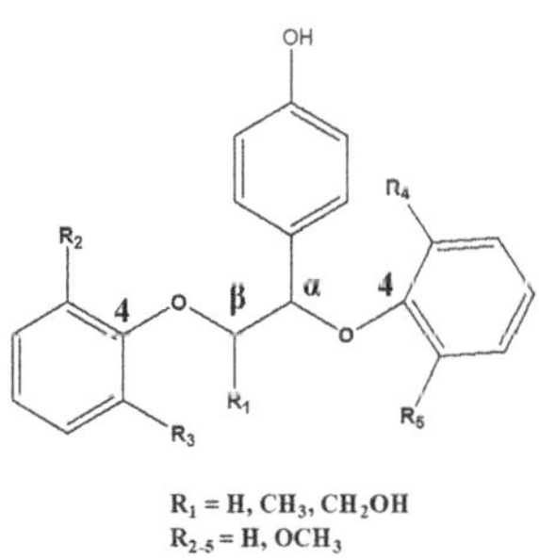

Fig. 3.14 Structure of a lignin trimer model compound containing α-O-4 and β-O-4 links

3.3.3 *Trimers and Above*

Jiang et al. [114] investigated the homolytic bond dissociation energies of various substituent groups at different positions on the C_α–O and C_β–O bonds using density functional theory methods at M06-2X level of theory with a 6-31++g (d,p) basis set. The general structure of lignin trimer model compound is shown in Fig. 3.14. It was shown that methoxy substitution at R_2 or R_3 has a minor effect on the BDE of C_β-O bond compared to the methoxy groups at R_2 and R_3. This is possibly due to the steric hindrance induced by the methoxy groups at R_2 and R_3.

Beste and Buchanan [115] investigated the effect of methoxy groups on the BDE of C_β–O bond in β-O-4 type lignin dimer model compound. It was found that the methoxy groups at R_2 and R_3 positions can reduce the BDE of C_β–O bond by 5.5 kcal mol^{-1}, and an addition of methoxy group can lead to 9 kcal mol^{-1} reduction of BDE. Similarly, single methoxy substituent at R_4 or R_5 has a minor effect on the BDE of C_α–O bond compared to the two methoxy groups at R_4 and R_5. Kim et al. [107] reported that the o-methoxy group on the aromatic ring can improve the reactivity toward the C_α–O homolysis, and lower the BDE of C_α–O bond during pyrolysis of α-O-4-type lignin dimer model. Parthasarathi et al. [109] and Huang et al. [116] reported that the BDE of C_α–O bond gradually decreases when the H atoms are replaced by methoxy groups on the phenyl ring adjacent to the oxygen of the ether bond. Interestingly, the BDE of C_α–O bond decreases regardless of the length of the oligomer (i.e., dimer or trimer), when there is an o-methoxy group on the phenyl ring adjacent to the α-O-4 ether bond. Thus, phenolic monomers, dimers, and trimers serve as valuable lignin model compounds to understand the key transformations occurring in lignin during fast pyrolysis process. A comprehensive understanding of the chemistry and reaction kinetics paves the way for selectively deriving chemicals and fuel molecules from lignin and lignin tars via catalytic approach. Furthermore, there is also an immense need to expand the knowledge by using different model compounds containing different substituents and linkages.

3.4 Deriving a Composition Model of Lignin

While the existing theoretical studies shed valuable insights on the chemistry of transformation of lignin sub-structures, extending the mechanistic understanding to the lignin macromolecule in the form of a kinetic model requires the description of the structure of lignin to a reasonable extent. As shown in Tables 3.1 and 3.2, the elemental composition and building blocks of lignin vary based on its source. This means, any description of the composition of lignin should satisfy the elemental composition, the percentage of monomeric units, i.e., the H:G:S ratio, and the way these are linked by different bond types, to a reasonable extent. Few studies have attempted to develop a composition model of lignin, which could be utilized subsequently in a kinetic model.

One of the early works in this direction is that of Faravelli et al. [117], who proposed lumping schemes to describe lignin. Based on a thorough analysis of the elemental composition of lignins, they proposed three lignin reference species containing different amounts of C, H and O. The C-rich lignin (LIG-C) was $C_{17}H_{17}O_5$, O-rich lignin (LIG-O) was $C_{16}H_{11}O_6(OCH_3)_4$, and H-rich lignin (LIG-H) was $C_{18}H_{17}O_5(OCH_3)_4$. LIG-O and LIG-H contain methoxy groups and are more representative of lignin from hardwood, whereas LIG-C represents softwood lignin, without methoxy groups. Any real lignin was modeled as a linear combination of these three reference species.

Dussan et al. [118] recently developed a composition model of lignin by designing new pseudo-components to describe the molar distribution of H, G and S units and the overall elemental composition of lignin. The four pseudocomponents include PC1 ($C_{20}H_{22}O_{10}$), PC2 ($C_{21}H_{26}O_8$), PC3 ($C_{20}H_{24}O_7$) and PC4 ($C_{16}H_{16}O_5$). Six inputs including C, H and O content from elemental analysis, and C fraction in H, G and S monomer types from NMR (Nuclear Magnetic Resonance) spectroscopy analysis were used to compute the model-based values, and further, the error between model and experiments were minimized to arrive at the lignin structure. They found that the monoaromatic fractions were reproduced with coefficient of determination $R^2 > 0.95$, while the C, H and O fractions varied with relative errors of 3–8% within expected experimental/computational uncertainty.

The Broadlbelt group from Northwestern University devised a stochastic method to generate libraries containing diverse lignin structures [119]. They demonstrated the generation of wheat straw lignin structure by considering β-O-4, β-5 and 5-5 bond types, which together account for more than 90% of the linkages. By encoding the typical bonding patterns in a decision tree, a Markov chain of lignin molecule was constructed by the use of a Metropolis Monte Carlo algorithm. Salient characteristics of lignin such as monomer and bond distribution, molecular weight distribution and branching coefficient were validated with the experimental data using statistical criteria like χ^2 goodness-of-fit and Student's *t*-test. More importantly, the composition model predicted the distribution of aromatic hydroxyl groups and the dyadic bonding distribution, which are, otherwise, difficult to obtain from experiments.

3.5 Mechanistic Model of Lignin Pyrolysis

Once lignin is described using one or more of its key attributes such as elemental composition, monomer and bond distribution, and molecular weight, it is important to subject the molecule to pyrolysis using a set of reactions to understand the conversion, overall product distribution (in terms of volatiles, gas and char yield) and the chemical composition of the volatile and gas fraction. Ranzi and co-workers developed the first semi-detailed kinetic model of lignin pyrolysis, in which they considered the reactions of their pseudocomponents, and further gas phase conversion of the pyrolysis vapors [117]. Their kinetic scheme involved 500 elementary and lumped reactions of 100 free radical and stable molecular species. The reactions were classified under the major categories of initiation, H-abstraction, β-decomposition, radical addition, condensation, char formation, CO release and termination via radical recombination. Their model described well the overall mass loss profiles of lignin from different biomass sources that were experimentally obtained using TGA. This basic model of lignin pyrolysis was further refined and incorporated in an overall biomass pyrolysis model containing the lumped decomposition reactions of cellulose, hemicellulose, and extractives [120, 121].

Dussan et al. [118] used their lignin composition model, and applied it to the kinetic model of lignin pyrolysis developed by Ranzi et al. [121]. They showed that the model-predicted yields of volatile products and char were comparable with the experimental data at high heating rates (60 and 6000) K s^{-1}, and were better than that predicted by Ranzi and co-workers. More importantly, the monoaromatics in the volatile fraction predicted by the model matched well with the experimental data. All this was attributed to better description of lignin structure.

In a latest report, Yanez et al. [122] developed a detailed kinetic model of lignin fast pyrolysis using kinetic Monte Carlo framework. The kinetic model comprised of 4313 reactions of 1615 species. The reactions were based on the elementary transformations of the model compounds reported in the literature, and the composition model of lignin was the same as that reported by the authors. The elementary reactions include ether cleavage by bond fission, demethoxylation, demethanation, aliphatic C–O cleavage at different positions, decarboxylation, deacylation, dealkylation, aliphatic C–C cleavage, methoxyl group isomerization, alcohol oxidation, aldehyde oxidation, hydrogen addition and char formation. Fingerprints of different phenylpropanoid and coumaran structures obtained from pyrolysis of lignin were included in the model. The overall yields of gases, water and char after 10 s matched well with the experimental data of fast pyrolysis of corn stover lignin. Other major products like 2-hydroxy-3-methyl-4(propenaldehyde)-phenol and p-vinylphenol were also captured by the model. Such models are valuable because, in addition to product yields and pyrolysate composition, they provide valuable information about molecular weight decrease of lignin and the time evolution of pyrolysates. Such information at short time scales of the order of seconds cannot be obtained from analytical pyrolysis experiments, which points to the fact that advancements in experimental techniques are also required to probe the structural changes in lignin, and the time evolution of vapor phase species.

3.6 Conclusions and Perspectives for Future Work

The availability and the presence of various functional groups in lignin makes it the feedstock of choice for the synthesis of chemicals such as vanillin and phenols, and polymeric materials like resins and polymers. However, this is a challenging task due to its structural complexity. This chapter has provided a flavor of the salient reactions and products from fast pyrolysis of lignin and its model compounds. In general, the yield and composition of bio-oil from fast pyrolysis of lignin depend on the type of lignin, reaction temperature, heating rate and type of catalyst. High yields of aromatic hydrocarbons and phenols in bio-oils can be obtained from lignin via thermal cleavage of the various linkages. The optimum temperature for producing aromatic hydrocarbons from lignin via catalytic pyrolysis is shown to be (550–650) °C, whereas at the same optimal temperature, non-catalytic fast pyrolysis yields phenolic compounds as the key products. The major phenolics include guaiacol, methyl guaiacol, syringol, vanillin, catechol and phenol. The design of robust catalysts with high catalytic selectivity and better reusability with low coking tendency are essential for a bio-refinery to be economically operated at scale. The selective lignin conversion to value-added fuel molecules and chemicals depends greatly on proper choice of engineered catalysts, optimal operating conditions, and more importantly, on a reasonable understanding of reaction mechanism and kinetics.

Fundamental understanding of the mechanism of pyrolysis of lignin and the development of a comprehensive kinetic model still remains a challenge. Experimental research has its own limitations in predicting the exact reaction pathways and providing time evolution of product yields under fast pyrolysis conditions. Hence, it is imperative to research the pyrolysis chemistry of lignin using various model compounds via both experimental and computational modeling techniques. While a number of studies are available on the transformation of lignin dimers and trimers, they are focused mostly on the reactions of β-O-4 linkage. Thus, there is a need to cover a wide class of inter unit bond conversions. While this is possible using DFT modeling, experimental synthesis of different dimers and trimers is not a trivial task.

Describing the composition of lignin using mathematical models has recently emerged. It is important to expand this and validate it with multiple lignin types. An equally important task is to synthesize the kinetic information of model compound conversion obtained from a molecular model, and integrate it with a mechanistic kinetic model of lignin pyrolysis. While advancements in this direction are evident from the literature, a concerted effort from multiple research groups is essential to improve the existing models and build a comprehensive one. More research in this direction will pave the way to develop a holistic picture of lignin pyrolysis pathways.

References

1. The pulp and paper industry, New Zealand Institute of Chemistry, The Forestry Industry report, IV-Forestry-C-Pulp and Paper, http://nzic.org.nz/ChemProcesses/forestry/. Accessed Oct 2019.
2. Fengel D, Wegener G. Wood: chemistry, ultrastructure, reactions. New York: de Gruyter; 1989. https://doi.org/10.1002/pol.1985.130231112.
3. Hu TQ. Chemical modification, properties and usage of lignin. New York: Kluwer Academic/Plenum Publisher; 2002.
4. Werhan H. A process for the complete valorization of lignin into aromatic chemicals based on acidic oxidation. Ph.D. Thesis, ETH Zurich; 2013. https://doi.org/10.3929/ethz-a-009790818.
5. Harkin JM. Lignin and its uses. Madison: U.S. Department of Agriculture, Forest Service Research Note FPL-0206; 1969.
6. Zakzeski J, Bruijnincx PCA, Jongerius AL, Weckhuysen BM. Catalytic volarization of lignin for the production of renewable chemicals. Chem Rev. 2010;110:3552–99. https://doi.org/10.1021/cr900354u.
7. Belgacem MN, Blayo A, Gandini A. Organosolv lignin as a filler in inks, varnishes and paints. Ind Crop Prod. 2003;18:145–53. https://doi.org/10.1016/S0926-6690(03)00042-6.

8. Lora JH, Glasser WG. Recent industrial applications of lignin: a sustainable alternative to non-renewable materials. J Polym Environ. 2002;10:39–48. https://doi.org/10.1023/A:1021070006895.
9. Torre MJ, Moral A, Hernández MD, Cabeza E, Tijero A. Organosolv lignin for biofuel. Ind Crop Prod. 2013;45:58–63. https://doi.org/10.1016/j.indcrop.2012.12.002.
10. Beis SH, Mukkamala S, Hill N, Joseph J, Baker C, Jensen B, Stemmler EA, Wheeler MC, Frederick BG, van Heiningen A, Berg AG, DeSisto WJ. Fast pyrolysis of lignins. Bioresources. 2010;5:1408–24. https://doi.org/10.15376/biores.5.3.1408-1424.
11. Nowakowski DJ, Bridgwater AV, Elliott DC, Meier D, de Wild P. Lignin fast pyrolysis: results from an international collaboration. J Anal Appl Pyrol. 2010;88:53–72. https://doi.org/10.1016/j.jaap.2010.02.009.
12. Lazaridis PA, Fotopoulos AP, Karakoulia SA, Triantafyllidis KS. Catalytic fast pyrolysis of Kraft lignin with conventional, mesoporous and nanosized ZSM-5 zeolite for the production of alkyl-phenols and aromatics. Front Chem. 2018;6:1–22. https://doi.org/10.3389/fchem.2018.00295.
13. Park S, Jae J, Farooq A, Kwon EE, Park ED, Ha JM, Jung SC, Park YK. Continuous pyrolysis of organosolv lignin and application of biochar on gasification of high-density polyethylene. Appl Energy. 2019;255:113801. https://doi.org/10.1016/j.apenergy.2019.113801.
14. Galkin MV, Dahlstrand C, Samec JS. Mild and robust redox-neutral Pd/C-catalyzed lignol β-O-4 bond cleavage through a low energy barrier pathway. ChemSusChem. 2015;8:2187–92. https://doi.org/10.1002/cssc.201500117.
15. Lan W, Amiri MT, Hunston CM, Luterbacher JS. Protection group effects during α, γ-diol lignin stabilization promote high-selectivity monomer production. Angew Chem Int Ed. 2018;57:1356–60. https://doi.org/10.1002/anie.201710838.
16. Luo M, Lin H, Li B, Dong Y, He Y, Wang L. A novel modification of lignin on corncob-based biochar to enhance removal of cadmium from water. Bioresour Technol. 2018;259:312–8. https://doi.org/10.1016/j.biortech.2018.03.075.
17. Morgan HM Jr, Bu Q, Liang J, Liu Y, Mao H, Shi A, Lei H, Ruan R. A review of catalytic microwave pyrolysis of lignocellulosic biomass for value-added fuel and chemicals. Bioresour Technol. 2017;230:112–21. https://doi.org/10.1016/j.biortech.2017.01.059.
18. Regmi YN, Mann JK, McBride JR, Tao J, Barnes CE, Labbé N, Chmely SC. Catalytic transfer hydrogenolysis of Organosolv lignin using B-containing FeNi alloyed catalysts. Catal Today. 2018;302:190–5. https://doi.org/10.1016/j.cattod.2017.05.051.
19. Xu L, Yao Q, Zhang Y, Fu Y. Integrated production of aromatic amines and n-doped carbon from lignin via ex situ catalytic fast pyrolysis in the presence of ammonia over zeolites. ACS Sustain Chem Eng. 2017;5:2960–9. https://doi.org/10.1021/acssuschemeng.6b02542.
20. Mohan D, Pittman CU, Steele PH. Pyrolysis of wood/biomass for bio-oil: a critical review. Energy Fuel. 2006;20:848–89. https://doi.org/10.1021/ef0502397.
21. Lasa HD, Salaices E, Mazumder J, Lucky R. Catalytic steam gasification of biomass: catalysts, thermodynamics and kinetics. Chem Rev. 2011;111:5404–33. https://doi.org/10.1021/cr200024w.
22. Zhou X, Broadbelt LJ, Vinu R. Mechanistic understanding of thermo chemical conversion of polymers and lignocellulosic biomass. In: Van Geem K, editor. Advances in chemical engineering: thermochemical process engineering, vol. 49. London: Elsevier; 2016. p. 95–198. https://doi.org/10.1016/bs.ache.2016.09.002.
23. Kang S, Li X, Fan J, Chang J. Hydrothermal conversion of lignin: a review. Renew Sust Energy Rev. 2013;27:546–58. https://doi.org/10.1016/j.rser.2013.07.013.
24. Amen-Chen C, Pakdel H, Roy C. Production of monomeric phenols by thermochemical conversion of biomass: a review. Bioresour Technol. 2001;79:277–99. https://doi.org/10.1016/S0960-8524(00)00180-2.
25. Huang J, He C. Pyrolysis mechanism of α-O-4 linkage lignin dimer: a theoretical study. J Anal Appl Pyrol. 2015;113:655–64. https://doi.org/10.1016/j.jaap.2015.04.012.

26. Zakzeski J, Bruijnincx PCA, Jongerius AL, Weckhuysen BM. The catalytic valorization of lignin for the production of renewable chemicals. Chem Rev. 2010;110:3552–99.. https://pubs.acs.org/doi/10.1021/cr900354u
27. Gellerstedt G, Henriksson G. Lignins: major sources, structures and properties. Amsterdam: Elsevier; 2008. https://doi.org/10.1016/B978-0-08-045316-3.00009-0.
28. Mutturi S, Palmqvist B, Lidén G. Developments in bioethanol fuel-focused biorefineries. In: Waldron K, editor. Advances in biorefineries: biomass and waste supply chain exploitation. Cambridge: Woodhead Publishing; 2014. p. 259–302. https://doi.org/10.1533/9780857097385.1.259.
29. Capanema EA, Balakshin MY, Kadla JF. Quantitative characterization of a hardwood milled wood lignin by nuclear magnetic resonance spectroscopy. J Agric Food Chem. 2005;53:9639–49. https://doi.org/10.1021/jf0515330.
30. Chakar FS, Ragauskas AJ. Review of current and future softwood Kraft lignin process chemistry. Ind Crop Prod. 2004;20:131–41. https://doi.org/10.1016/j.indcrop.2004.04.016.
31. McKendry P. Energy production from biomass (part 2): overview of biomass. Bioresour Technol. 2002;83:37–46. https://doi.org/10.1016/S0960-8524(01)00119-5.
32. Bridgwater AV, Peacocke GVC. Fast pyrolysis processes for biomass. Renew Sust Energ Rev. 2000;1:1–73. https://doi.org/10.1016/S1364-0321(99)00007-6.
33. Van Loo S, Koppejan J. The handbook of biomass combustion and co-firing. London: Earthscan; 2010.
34. Fang X, Jia L. Experimental study on ash fusion characteristics of biomass. Bioresour Technol. 2012;104:769–74. https://doi.org/10.1016/j.biortech.2011.11.055.
35. Steen EV, Claeys M. Fisher-Tropsch catalysts for the biomass-to-liquid (BTL) process. Chem Eng Technol. 2008;31:655–60. https://doi.org/10.1002/ceat.200800067.
36. McKendry P. Energy production from biomass (part 3): gasification technologies. Bioresour Technol. 2002;83:55–63. https://doi.org/10.1016/S0960-8524(01)00120-1.
37. Balat M. Sustainable transportation fuels from biomass materials. Energy Edu Sci Technol. 2006;17:83–103.
38. Yu H, Wu Z, Chen G. Catalytic gasification characteristics of cellulose, hemicellulose and lignin. Renew Energy. 2018;121:559–67. https://doi.org/10.1016/j.renene.2018.01.047.
39. Wu C, Wang Z, Huang J, Williams PT. Pyrolysis/gasification of cellulose, hemicellulose and lignin for hydrogen production in the presence of various nickel-based catalysts. Fuel. 2013;106:697–706. https://doi.org/10.1016/j.fuel.2012.10.064.
40. Azadi P, Inderwildi OR, Farnood R, King DA. Liquid fuels, hydrogen and chemicals from lignin: a critical review. Renew Sust Energy Rev. 2013;21:506–23. https://doi.org/10.1016/j.rser.2012.12.022.
41. Mermelstein J, Millan M, Brandon NP. The interaction of biomass gasification syngas components with tar in a solid oxide fuel cell and operational conditions to mitigate carbon deposition on nickel-gadolinium doped ceria anodes. J Power Sources. 2011;196:5027–34. https://doi.org/10.1016/j.jpowsour.2011.02.011.
42. Nagel FP, Ghosh S, Pitta C, Schildhauer TJ, Biollaz S. Biomass integrated gasification fuel cell systems-concept development and experimental results. Biomass Bioenergy. 2011;35:354–62. https://doi.org/10.1016/j.biombioe.2010.08.057.
43. Tekin K, Karagöz S, Bektaş S. A review of hydrothermal biomass processing. Renew Sust Energy Rev. 2014;40:673–87. https://doi.org/10.1016/j.rser.2014.07.216.
44. Peterson AA, Vogel F, Lachance RP, Froling M, Antal JM Jr, Tester JW. Thermochemical biofuel production in hydrothermal media: a review of sub and supercritical water technologies. Energy Environ Sci. 2008;1:32–65. https://doi.org/10.1039/B810100K.
45. Zhang B, Keitz M, Valentas K. Thermochemical liquefaction of high-diversity grassland perennial. J Anal Appl Pyrol. 2009;84:18–24. https://doi.org/10.1016/j.jaap.2008.09.005.
46. Kang S, Li X, Fan J, Chang J. Classified separation of lignin hydrothermal liquefied products. Ind Eng Chem Res. 2011;50:11288–96. https://doi.org/10.1021/ie2011356.

47. Wahyudiono Sasaki M, Goto M. Recovery of phenolic compounds through the decomposition of lignin in near and supercritical water. Chem Eng Process. 2008;47:1609–19. https://doi.org/10.1016/j.cep.2007.09.001.
48. Zhao C, Jiang E, Chen A. Volatile production from pyrolysis of cellulose, hemicellulose and lignin. J Energy Inst. 2017;90:902–13. https://doi.org/10.1016/j.joei.2016.08.004.
49. Balat M, Kirtay E, Balat H. Main routes for the thermo-conversion of biomass into fuels and chemicals. Part 1: pyrolysis systems. Energy Convers Manag. 2009;50:3147–57. https://doi.org/10.1016/j.enconman.2009.08.014.
50. Maschio G, Koufopanos C, Lucchesi A. Pyrolysis, a promising route for biomass utilization. Bioresour Technol. 1992;42:219–31. https://doi.org/10.1016/0960-8524(92)90025-S.
51. Demirbas A. Partly chemical analysis of liquid fraction of flash pyrolysis products from biomass in the presence of sodium carbonate. Energy Convers Manag. 2002;43:1801–9. https://doi.org/10.1016/S0196-8904(01)00137-6.
52. Collard FX, Blin J. A review on pyrolysis of biomass constituents: mechanisms and composition of the products obtained from the conversion of cellulose, hemi-celluloses and lignin. Renew Sust Energ Rev. 2014;38:594–608. https://doi.org/10.1016/j.rser.2014.06.013.
53. Carrier M, Windt M, Ziegler B, Appelt J, Saake B, Meier D, Bridgwater A. Quantitative insights into the fast pyrolysis of extracted cellulose, hemicellulose and lignin. ChemSusChem. 2017;10:3212–24. https://doi.org/10.1002/cssc.201700984.
54. Lin X, Sui S, Tan S, Pittman CU Jr, Sun J, Zhang Z. Fast pyrolysis of four lignins from different isolation processes using Py-GC/MS. Energies. 2015;8:5107–21. https://doi.org/10.3390/en8065107.
55. Nair V, Vinu R. Production of guaiacols via catalytic fast pyrolysis of alkali lignin using titania, zirconia and ceria. J Anal Appl Pyrol. 2016;119:31–9. https://doi.org/10.1016/j.jaap.2016.03.020.
56. Asmadi M, Kawamoto H, Saka S. Thermal reactivities of catechols/pyrogallols and cresols/xylenols as lignin pyrolysis intermediates. J Anal Appl Pyrol. 2011;92:76–87. https://doi.org/10.1016/j.jaap.2011.04.012.
57. Huang X, Liu C, Huang J, Li H. Theory studies on pyrolysis mechanism of phenethyl phenyl ether. Comput Theor Chem. 2011;976:51–9. https://doi.org/10.1016/j.comptc.2011.08.001.
58. Britt PF, Buchanan AC, Cooney MJ, Martineau DR. Flash vacuum pyrolysis of methoxy-substituted lignin model compounds. J Org Chem. 2000;65:1376–89. https://doi.org/10.1021/jo991479k.
59. Windt M, Meier D, Marsman JH, Heeres HJ, de Koning S. Micro-pyrolysis of technical lignins in a new modular rig and product analysis by GC-MS/FID and GC × GC-TOFMS/FID. J Anal Appl Pyrol. 2009;85:38–46. https://doi.org/10.1016/j.jaap.2008.11.011.
60. Patwardhan PR, Brown RC, Shanks BH. Understanding the fast pyrolysis of lignin. ChemSusChem. 2011;4:1629–36. https://doi.org/10.1002/cssc.201100133.
61. Mullen CA, Boateng AA. Catalytic pyrolysis-GC/MS of lignin from several sources. Fuel Process Technol. 2010;91:1446–58. https://doi.org/10.1016/j.fuproc.2010.05.022.
62. Zhang M, Moutsoglou A. Catalytic fast pyrolysis of prairie cordgrass lignin and quantification of products by pyrolysis-gas chromatography-mass spectrometry. Energy Fuel. 2014;28:1066–73. https://doi.org/10.1021/ef401795z.
63. Ma Z, Troussard E, Bokhoven JAV. Controlling the selectivity to chemicals from lignin via catalytic fast pyrolysis. Appl Catal A Gen. 2012;423-424:130–6. https://doi.org/10.1016/j.apcata.2012.02.027.
64. Branca C, Giudicianni P, Di Blasi C. GC/MS characterization of liquids generated from low-temperature pyrolysis of wood. Ind Eng Chem Res. 2003;42:3190–202. https://doi.org/10.1021/ie030066d.
65. Hosoya T, Kawamoto H, Saka S. Secondary reactions of lignin-derived primary tar components. J Anal Appl Pyrol. 2008;83:78–87. https://doi.org/10.1016/j.jaap.2008.06.003.

66. Biagini E, Barontini F, Tognotti L. Devolatilization of biomass fuels and biomass components studied by TG/FTIR technique. Ind Eng Chem Res. 2006;45:4486–93. https://doi.org/10.1021/ie0514049.
67. Liu Q, Wang S, Zheng Y, Luo Z, Cen K. Mechanism study of wood lignin pyrolysis by using TG-FTIR analysis. J Anal Appl Pyrol. 2008;82:170–7. https://doi.org/10.1016/j.jaap.2008.03.007.
68. Kawamoto H. Lignin pyrolysis reactions. J Wood Sci. 2017;63:117–32. https://doi.org/10.1007/s10086-016-1606-z.
69. Lou R, Wu S, Lyu G. Quantified monophenols in the bio-oil derived from lignin fast pyrolysis. J Anal Appl Pyrol. 2015;111:27–32. https://doi.org/10.1016/j.jaap.2014.12.022.
70. Jiang G, Nowakowski DJ, Bridgwater AV. Effect of the temperature on the composition of lignin pyrolysis products. Energy Fuel. 2010;24:4470–5. https://doi.org/10.1021/ef100363c.
71. Zhang M, Resende FLP, Moutsoglou A. Catalytic fast pyrolysis of aspen lignin via Py-GC/MS. Fuel. 2014;116:358–69. https://doi.org/10.1016/j.fuel.2013.07.128.
72. Lee HW, Kim TH, Park SH, Jeon JK, Suh DJ, Park YK. Catalytic fast pyrolysis of lignin over mesoporous Y zeolite using Py-GC/MS. J Nanosci Nanotechnol. 2013;13:2640–6. https://doi.org/10.1166/jnn.2013.7421.
73. Li X, Su L, Wang Y, Yu Y, Wang C, Li X, Wang Z. Catalytic fast pyrolysis of Kraft lignin with HZSM-5 zeolite for producing aromatic hydrocarbons. Front Environ Sci Eng. 2012;6:295–303. https://doi.org/10.1007/s11783-012-0410-2.
74. Wang K, Kim KH, Brown RC. Catalytic pyrolysis of individual components of lignocellulosic biomass. Green Chem. 2014;16:727–35. https://doi.org/10.1039/C3GC41288A.
75. Ma Z, Custodis V, Bokhoven JAV. Selective deoxygenation of lignin during catalytic fast pyrolysis. Cat Sci Technol. 2014;4:766–72. https://doi.org/10.1039/C3CY00704A.
76. Santana JA Jr, Carvalho WS, Ataíde CH. Catalytic effect of ZSM-5 zeolite and HY-340 niobic acid on the pyrolysis of industrial Kraft lignins. Ind Crop Prod. 2018;111:126–32. https://doi.org/10.1016/j.indcrop.2017.10.023.
77. Jackson MA, Compton DL, Boateng AA. Screening heterogeneous catalysts for the pyrolysis of lignin. J Anal Appl Pyrol. 2009;85:226–30. https://doi.org/10.1016/j.jaap.2008.09.016.
78. Kim JY, Lee JH, Park J, Kim JK, An D, Song IK, Choi JW. Catalytic pyrolysis of lignin over HZSM-5 catalysts: effect of various parameters on the production of aromatic hydrocarbons. J Anal Appl Pyrol. 2015;114:273–80. https://doi.org/10.1016/j.jaap.2015.06.007.
79. Atul M, Jayant E. Pyrolysis of purified Kraft lignin in the presence of $AlCl_3$ and $ZnCl_2$. J Environ Chem Eng. 2013;1:844–9. https://doi.org/10.1016/j.jece.2013.07.026.
80. Peng C, Zhang G, Yue J, Xu G. Pyrolysis of lignin for phenols with alkaline additive. Fuel Process Technol. 2014;124:212–21. https://doi.org/10.1016/j.fuproc.2014.02.025.
81. Greeley J. Theoretical heterogeneous catalysis: scaling relationships and computational catalysts design. Annu Rev Chem Biomol Eng. 2016;7:605–35. https://doi.org/10.1146/annurev-chembioeng-080615-034413.
82. Dorrestijn E, Mulder P. The radical-induced decomposition of 2-methoxyphenol. J Chem Soc Perkin Trans. 1999;2:777–80. https://doi.org/10.1039/A809619H.
83. Asmadi M, Kawamoto H, Saka S. Thermal reactions of guaiacol and syringol as lignin model aromatic nuclei. J Anal Appl Pyrol. 2011;92:88–98. https://doi.org/10.1016/j.jaap.2011.04.011.
84. Shin EJ, Nimlos MR, Evans RJ. A study of the mechanisms of vanillin pyrolysis by mass spectrometry and multivariate analysis. Fuel. 2001;80:1689–96. https://doi.org/10.1016/S0016-2361(01)00055-2.
85. Liu C, Deng Y, Wu S, Mou H, Liang J, Lie M. Study on pyrolysis mechanism of three guaiacyl-type lignin monomeric model compounds. J Anal Appl Pyrol. 2016;118:123–9. https://doi.org/10.1016/j.jaap.2016.01.007.
86. Chen L, Ye X, Luo F, Shao J, Lu Q, Fang Y, Wang X, Chen H. Pyrolysis mechanism of β-O-4 type lignin model dimer. J Anal Appl Pyrol. 2015;115:103–11. https://doi.org/10.1016/j.jaap.2015.07.009.

87. Jiang XY, Lu Q, Ye XN, Hu B, Dong CQ. Experimental and theoretical studies on the pyrolysis mechanism of β-1-type lignin dimer model compound. Bioresources. 2016;11:6232–43. https://doi.org/10.15376/biores.11.3.6232-6243.
88. Vinu R, Broadbelt LJ. A mechanistic model of fast pyrolysis of glucose-based carbohydrates to predict bio-oil composition. Energy Environ Sci. 2012;5:9808–26. https://doi.org/10.1039/C2EE22784C.
89. Zhou X, Nolte MW, Mayes HB, Shanks BH, Broadbelt LJ. Experimental and mechanistic modeling of fast pyrolysis of neat glucose-based carbohydrates: experiments and development of a detailed mechanistic model. Ind Eng Chem Res. 2014;53:13274–89. https://doi.org/10.1021/ie502259w.
90. Burnham AK, Zhou X, Broadbelt LJ. Critical review of the global chemical kinetics of cellulose thermal decomposition. Energy Fuel. 2015;29:2906–18. https://doi.org/10.1021/acs.energyfuels.5b00350.
91. Zhou X, Li W, Mabon R, Broadbelt LJ. A mechanistic model of fast pyrolysis of hemicellulose. Energy Environ Sci. 2018;11:1240–60. https://doi.org/10.1039/C7EE03208K.
92. Altarawneh M, Dlugogorski BZ, Kennedy EM, Mackie JC. Theoretical study of unimolecular decomposition of catechol. J Phys Chem A. 2009;114:1060–7. https://doi.org/10.1021/jp909025s.
93. Cavallotti C, Cuoci A, Faravelli T, Frassoldati A, Pelucchi M, Ranzi E. Detailed kinetics of pyrolysis and combustion of catechol and guaiacol, as reference components of bio-oil from biomass. Chem Eng Trans. 2018;65:79–84. https://doi.org/10.3303/CET1865014.
94. Liu C, Zhang Y, Huang X. Study of guaiacol pyrolysis mechanism based on density functional theory. Fuel Process Technol. 2014;123:159–65. https://doi.org/10.1016/j.fuproc.2014.01.002.
95. Huang J, Li X, Wu D, Tong H, Li W. Theoretical studies on pyrolysis mechanism of guaiacol as a lignin model compound. J Renew Sust Energy. 2013;5:043112. https://doi.org/10.1063/1.4816497.
96. Huang J, Liu C, Ren L, Tong H, Li WM, Wu D. Studies on pyrolysis mechanism of syringol as lignin model compound by quantum chemistry. J Fuel Chem Technol. 2013;41:657–66. https://doi.org/10.1016/S1872-5813(13)60031-6.
97. Wang M, Liu C, Xu X, Li Q. Theoretical study of the pyrolysis of vanillin as a model of secondary lignin pyrolysis. Chem Phys Lett. 2016;654:41–5. https://doi.org/10.1016/j.cplett.2016.03.058.
98. Huang J, Liu C, Wu D, Tong H, Ren L. Density functional theory studies on pyrolysis mechanism of β-O-4 type lignin dimer model compound. J Anal Appl Pyrol. 2014;109:98–108. https://doi.org/10.1016/j.jaap.2014.07.007.
99. Pelucchi M, Cavallotti C, Cuoci A, Faravelli T, Frassoldati A, Ranzi E. Detailed kinetics of substituted phenolic species in pyrolysis bio-oils. React Chem Eng. 2019;4:490–506. https://doi.org/10.1039/C8RE00198G.
100. Ince A, Carstensen HH, Sabbe M, Reyniers MF, Marin GB. Group additive modeling of substituent effects in monocyclic aromatic hydrocarbon radicals. AIChE J. 2017;63:2089–106. https://doi.org/10.1002/aic.15588.
101. Akazawa M, Kojima Y, Kato Y. Reaction mechanisms for pyrolysis of benzaldehydes. Bull Faculty Agric Niigata Univ. 2014;67:59–65.
102. Huang J, Liu C, Tong H, Li W, Wu D. A density functional theory study on formation mechanism of CO, CO_2 and CH_4 in pyrolysis of lignin. Comput Theor Chem. 2014;1045:1–9. https://doi.org/10.1016/j.comptc.2014.06.009.
103. Beste A, Buchanan AC III. Computational study of bond dissociation enthalpies for lignin model compounds. Substituent effects in phenethyl phenyl ethers. J Org Chem. 2009;74:2837–41. https://doi.org/10.1021/jo9001307.
104. Beste A, Buchanan AC III. Substituent effects on the reaction rates of hydrogen abstraction in the pyrolysis of phenethyl phenyl ethers. Energy Fuel. 2010;24:2857–67. https://doi.org/10.1021/ef1001953.

105. Beste A, Buchanan AC III, Harrison RJ. Computational prediction of α/β selectivities in the pyrolysis of oxygen-substituted phenethyl phenyl ethers. J Phys Chem A. 2008;112:4982–8. https://doi.org/10.1021/jp800767j.
106. Choi YS, Singh R, Zhang J, Balasubramanian G, Sturgeon MR, Katahira R, Chupka G, Beckham GT, Shanks BH. Pyrolysis reaction networks for lignin model compounds: unraveling thermal deconstruction of β-O-4 and α-O-4 compounds. Green Chem. 2016;18:1762–73. https://doi.org/10.1039/C5GC02268A.
107. Kim KH, Bai X, Brown RC. Pyrolysis mechanisms of methoxy substituted α-O-4 lignin dimeric model compounds and detection of free radicals using electron paramagnetic resonance analysis. J Anal Appl Pyrol. 2014;110:254–63. https://doi.org/10.1016/j.jaap.2014.09.008.
108. Huang J, He C, Liu C, Tong H, Wu L, Wu S. A computational study on thermal decomposition mechanism of β-1 linkage lignin dimer. Comput Theor Chem. 2015;1054:80–7. https://doi.org/10.1016/j.comptc.2014.12.007.
109. Parthasarathi R, Romero RA, Redondo A, Gnanakaran S. Theoretical study of the remarkably diverse linkages in lignin. J Phys Chem Lett. 2011;2:2660–6. https://doi.org/10.1021/jz201201q.
110. Elder T, Beste A. Density functional theory study of the concerted pyrolysis mechanism for lignin models. Energy Fuel. 2014;28:5229–35. https://doi.org/10.1021/ef5013648.
111. Klein MT, Virk PS. Modeling of lignin thermolysis. Energy Fuel. 2008;22:2175–82. https://doi.org/10.1021/ef800285f.
112. Kawamoto H, Horigoshi S, Saka S. Effects of side-chain hydroxyl groups on pyrolytic β-ether cleavage of phenolic lignin model dimer. J Wood Sci. 2007;53:268–71. https://doi.org/10.1007/s10086-006-0839-7.
113. Kawamoto H, Horigoshi S, Saka S. Pyrolysis reactions of various lignin model dimers. J Wood Sci. 2007;53:168–74. https://doi.org/10.1007/s10086-006-0834-z.
114. Jiang X, Lu Q, Dong X, Hu B, Dong C. Theoretical study on the effect of the substituent groups on the homolysis of the ether bond in lignin trimer model compounds. J Fuel Chem Technol. 2016;44:335–41. https://doi.org/10.1016/S1872-5813(16)30017-2.
115. Beste A, Buchanan AC. Computational study of bond dissociation enthalpies for lignin model compounds: substituent effects in phenethyl phenyl ether. J Org Chem. 2009;74:2837–41. https://doi.org/10.1021/jo9001307.
116. Huang JB, Wu SB, Cheng H, Lei M, Liang JJ, Tong H. Theoretical study of bond dissociation energies for lignin model compounds. J Fuel Chem Technol. 2015;43:429–36. https://doi.org/10.1016/S1872-5813(15)30011-6.
117. Faravelli T, Frassoldati A, Migliavacca G, Ranzi E. Detailed kinetic modelling of the thermal degradation of lignins. Biomass Bioenergy. 2010;34:290–301. https://doi.org/10.1016/j.biombioe.2009.10.018.
118. Dussan K, Dooley S, Monaghan RFD. A model of the chemical composition and pyrolysis kinetics of lignin. Proc Combust Inst. 2019;37:2697–704. https://doi.org/10.1016/j.proci.2018.05.149.
119. Yanez AJ, Li W, Mabon R, Broadbelt LJ. A stochastic method to generate libraries of structural representations of lignin. Energy Fuel. 2016;30:5835–45. https://doi.org/10.1021/acs.energyfuels.6b00966.
120. Ranzi E, Cuoci A, Faravelli T, Frassoldati A, Migliavacca G, Pierucci S, Sommariva S. Chemical kinetics of biomass pyrolysis. Energy Fuel. 2008;22:4292–300. https://doi.org/10.1021/ef800551t.
121. Ranzi E, Debiagi PEA, Frassoldati A. Mathematical modeling of fast biomass pyrolysis and bio-oil formation. Note 1: kinetic mechanism of biomass pyrolysis. ACS Sustain Chem Eng. 2017;5:2867–81. https://doi.org/10.1021/acssuschemeng.6b03096.
122. Yanez AJ, Natarajan P, Li W, Mabon R, Broadbelt LJ. Coupled structural and kinetic model of lignin fast pyrolysis. Energy Fuel. 2018;32:1822–30. https://doi.org/10.1021/acs.energyfuels.7b03311.

Chapter 4
Pyrolysis Chemistry and Mechanisms: Interactions of Primary Components

Wei Chen, Yingquan Chen, Hanping Chen, and Haiping Yang

Abstract Global warming resulting from the use of fossil fuel has led to an urgent need to develop sustainable energy resources. Biomass pyrolysis allows conversion of biomass into valuable bio-oil, gas, and char products and at the same time through optimization of reaction conditions, allows the realization of comprehensive processing of biomass. As the component structure of natural biomass is very complicated, exploring the pyrolysis behavior of individual components is a fundamental method to understand biomass pyrolysis mechanisms. In this chapter, state-of-the-art component pyrolysis behavior and pyrolysis mechanisms are discussed in detail. Individual reaction mechanisms for cellulose, hemicellulose, and lignin, and their interactions are discussed in depth, as well as the effect of inorganic species, based on the formation mechanism of bio-oil, char and gaseous products. Research status of advanced methods of component pyrolysis is given. While there are challenges, such as detailed pyrolysis mechanisms, native component pyrolysis, interaction mechanisms, and advanced characterization methods, that are still needed. Pyrolysis technology has a bright future for creating valuable products economically from biomass.

Keywords Biomass pyrolysis mechanism · Components pyrolysis · Cellulose · Hemicellulose · Lignin · Interactions

4.1 Introduction

The year 2018 is shaping up to be one of the hottest in peak climate temperatures, with new records being set in many places (48 °C in Spain, 40 °C in Japan and South Korea, 37 °C in Northeast China, and 32 °C in the Arctic). The rise in climate temperatures is attributed to global warming caused by greenhouse gas emissions

W. Chen · Y. Chen · H. Chen · H. Yang (✉)
State Key Laboratory of Coal Combustion, School of Power and Energy Engineering, Huazhong University of Science and Technology, Wuhan, China
e-mail: yhping2002@163.com

Z. Fang et al. (eds.), *Production of Biofuels and Chemicals with Pyrolysis*, Biofuels and Biorefineries 10, https://doi.org/10.1007/978-981-15-2732-6_4

mainly coming from the burning of fossil fuels [1–3]. Thus, it is urgent to develop renewable sources of carbon to replace fossil fuels. Biomass, such as lignocellulosic biomass, is a promising alternative which could be converted into fuels and chemicals [4, 5]. Biomass resources accounts for ~11% of world primary energy consumption, however, 60% of biomass is employed in traditional ways, such as cooking and space heating [6, 7]. Modern biomass use needs to account for 26–34% of primary energy by 2100 [8]. Hence, it is necessary to explore conversion methods for efficient utilization of biomass.

Biomass pyrolysis is one of the most promising thermochemical conversion methods, as it can obtain more liquid yield (~80 wt%), and retain more energy (~70%) in bio-oil, compared with combustion and gasification [9]. Biomass pyrolysis can be seen as the first step in methods for biomass thermochemical conversion [10]. Thus, biomass pyrolysis for liquid fuels has received wide attention [11]. Biomass pyrolysis has been extensively reviewed by Wang et al. [12], Anca-Couce [6], Collard and Blin [13], Gollakota et al. [14], and Shen [15] for obtaining liquid fuels. While biochar and gas products have not been reviewed in detail, these are also important pyrolysis products obtained from biomass pyrolysis.

Biomass pyrolysis allows one to obtain bio-oil, gas, and char products at the same time through optimization of reaction conditions [16–18]. Chen et al. [18] and Yang et al. [16] obtained high bio-oil yields, and gas products with ~10 MJ/m^3, as well as char products with 22 MJ/kg, at optimum pyrolysis temperatures of (550–650) °C. There are commercial biomass pyrolysis poly-generation plants running in China [19] that continuously deal with 7740 t of biomass per year, which can be converted into 1.6 million Nm^3 of bio-gas, 0.54 million kg of tar, 1.09 million kg of wood vinegar and 1.63 million kg of biochar.

A complete understanding of biomass pyrolysis mechanisms is not only essential for elucidating biomass decomposition pathways in the correlation between physicochemical structure and reactive behavior, but also, knowledge on the mechanisms is beneficial for guiding reactor design and optimizing reaction conditions [12]. It is also beneficial to realize production of valuable products, such as high-value gas products (hydrogen, methane, and light olefins), high-value liquid chemicals (levoglucosenone, phenol, furfural, ethanol, benzene, toluene and xylene (BTX)), and high-value biochar materials (biochar used for adsorption, catalysis, and energy storage) [15, 20–25].

Many reviews have been published on biomass pyrolysis mechanisms and the interested reader is directed to these compilations. For example, Sharma et al. [26] reviewed modelling and process parameters. Zhou et al. [27] focused on the evolution mechanism of gas and liquid products during pyrolysis. Kan et al. [28] focused on the influence of pyrolysis parameters on the products and their distribution. However, biomass pyrolysis is accompanied with the evolution of biochar and release of gas and bio-oil products which is highly interconnected to the entire process. Review of the evolution and relationship mechanism of gas, bio-oil and biochar allows one to reveal many aspects of the pyrolysis mechanism.

Thus, in this chapter, the pyrolysis mechanisms of biomass are comprehensively reviewed through considering the evolving properties of small-molecular gaseous

Table 4.1 Cellulose, hemicellulose, and lignin content in lignocellulosic biomass

Feedstock	Cellulose (wt%)	Hemicellulose (wt%)	Lignin (wt%)
Sapele	40.28	16.63	33.87
Badamu	29.41	24.79	27.71
Soybean stalk	37.83	18.77	19.73
Peanut shells	31.81	14.41	4.80
Corn stalk	42.73	22.09	0.23
Eucalyptus	39.07	14.29	27.81
Straw	37.12	24.41	23.44
Walnut shells	24.52	22.37	36.90
Rice husk	38.14	18.73	36.76
Cotton stalk	36.56	15.67	31.25
Elm	42.58	21.06	24.48
Bigenli	25.09	25.39	38.83
Bamboo	40.36	23.05	28.47
Poplar	43.19	18.83	12.89
Sesame stalk	40.40	18.46	1.78
Peanut stalk	31.47	14.18	23.68
Rape stalk	34.25	16.89	15.79
Willow	35.44	14.98	29.19
Tabacco stalk	24.58	11.90	29.81
Straw	39.17	23.73	21.50

Note: The analysis standard is dry basis

products, liquid oil compounds, and solid char products. The structure of natural biomass is very complicated, for example, the main components of lignocellulosic biomass are cellulose, hemicellulose, and lignin (Table 4.1), thus exploring the reactive behavior of individual components and their interactions is a fundamental method to elucidate biomass pyrolysis mechanisms. This chapter was written with the intention of helping the reader to understand key features of biomass pyrolysis mechanisms.

4.2 Cellulose Pyrolysis Mechanism

Cellulose is a kind of macromolecular polysaccharide consisting of linear long-chain glucose units linked through β-1,4-glucosidic bonds and it is the most abundant component in lignocellulosic biomass. Cellulose content in woody biomass is ~45 wt%, while that in cotton, flax and chemical pulp can reach 80–95 wt%. The chemical formula of cellulose is $(C_6H_{10}O_5)_n$, where n is the degree of polymerization (DP), and DP is at least 9000–15,000, and glucose is the repeating unit of cellulose [29]. The glycosidic bonds linking the basic units in cellulose are not as strong as

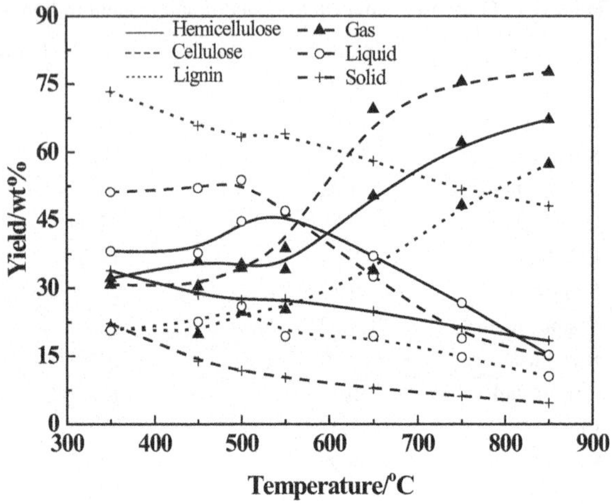

Fig. 4.1 Effect of temperature on product distribution for pyrolysis of cellulose, hemicellulose and lignin (adapted with permission from [33], Copyright © 2013 Elsevier)

other bonds, and tend to cleave under acidic conditions or high-temperatures [12]. Thus, cellulose structure degrades sharply during the pyrolysis process [12].

Yang et al. [30] found that the weight loss of cellulose pyrolysis mainly occurred in temperature range of (315–400) °C. However, the activation energy of this degradation process obtained from different studies varied significantly (166–250) kJ/mol, due to the differences in the types of cellulose, experimental conditions, and computation processes [31]. Taking into consideration of regression objective function and optimization strategy, the activation energy of slow pyrolysis of microcrystalline cellulose is about 200 kJ/mol [32].

Xin et al. [33] investigated the pyrolysis behavior of cellulose, xylan, and lignin, and found the char yield of cellulose (23–5) wt% was lower than that of xylan (25–35) wt% and lignin (50–75) wt% (Fig. 4.1) and the maximum bio-oil yield of cellulose was higher than that of xylan and lignin. Yang et al. [30] found CO_2 was the major gaseous product with secondary pyrolysis forming large amounts of CO and CH_4 at (400–600) °C, while H_2 mainly evolved at higher temperature (>600 °C).

Levoglucosan (1,6-anhydro-β-D-glucopyranose) is the major primary product from cellulose pyrolysis. Levoglucosan is generated mainly at lower temperatures during cellulose pyrolysis, and has a maximum yield at 450 °C, then it decreases and finally disappears at temperatures greater than 750 °C [33]. While levoglucosan is mainly generated through chain-end mechanism (breaking C1-O bond by transglycosylation) [34], which could further transform into levoglucosenone, 1,4:3,6-dianhydro-β-D-glucopyranose, and 1,6-anhydro-β-D-glucopyranose through dehydration and isomerization reactions, finally these anhydrosugars are further converted into furfural and hydroxymethylfurfural by dehydration reactions [35].

Xin et al. [36] investigated char evolution during cellulose pyrolysis, and found that dehydration reactions were predominant below 300 °C; decarbonylation, ring-opening and aromatization reactions became major reactions at (300–430) °C; then significant deoxygenation and condensation reactions occurred at (430–650) °C; while dehydrogenation was predominant at temperatures over 650 °C, and large aromatic systems were formed.

Yang et al. [37] explored cellulose pyrolysis mechanism with two-dimensional perturbation correlation infrared spectroscopy (2D-PCIS), which revealed that the C=O groups of cellulose changed before the aliphatic C-O-C linkages, indicating that dehydration and keto-enol tautomerism of pyran hydroxyls occurred before ring scission. Regarding solid char, the char structure was composed of low-order fused rings at temperatures around 350 °C, while it went to higher-order fused rings at temperatures over 450 °C.

The major reaction pathways of cellulose pyrolysis are shown in Fig. 4.2 [23, 34, 35, 38, 39]. In the primary step, cellulose decomposes into low-molecular compounds such as hydroxyacetaldehyde, levoglucosan, and other sugars, accompanying the release of gaseous products and the formation of char products through dehydration and decarboxylation reactions that release H_2O and CO_2.

Char contains many six-membered oxygen containing heterocycles linked by ether bonds. At increasing temperatures, levoglucosan undergoes rearrangement and hydration to form levoglucosenone, and cyclization reactions to convert into stable O-containing heterocyclic compounds such as furfural, 5-hydroxyl methyl furfural, and pyran, while some levoglucosan undergoes polymerization reactions to form char products. During this period, the char product contains large amounts of active functional groups such as -COOH, -COO-, -C=O, -OH, and C-O-C with the volatilization of char product.

For further increases in temperature, low-molecular compounds undergo decarbonylation and aromatization reactions to form large quantities of aromatics such as benzenes and naphthalenes. Further, some low-molecular compounds undergo condensation reactions to form char. The main functional groups left in char are stable -C-O-C, -OH, and -C=O through decarboxylation and decarbonylation reactions, leading to the aromatization degree of char greatly increasing. These processes would also be accompanied by the release of much CO, H_2, and C_2 compounds such as C_2H_2, C_2H_4, and C_2H_6.

4.3 Hemicellulose Pyrolysis Mechanism

Compared with cellulose, hemicellulose consists of shorter chain heteropolysaccharides, including xylose, arabinose, mannose, galactose, glucose, rhamnose, or glucuronic acid, which forms a typical amorphous branched structure with a low degree of polymerization. The content of hemicellulose in herbaceous biomass, softwood, and hardwood is (20–25) wt%, (18–23) wt%, and (10–15) wt%, respectively [12]. Obtaining native hemicellulose is extremely challenging due to its complex nature and cross-linking with cellulose and lignin, thus model compounds of hemicellulose are investigated widely, including monomeric building blocks of

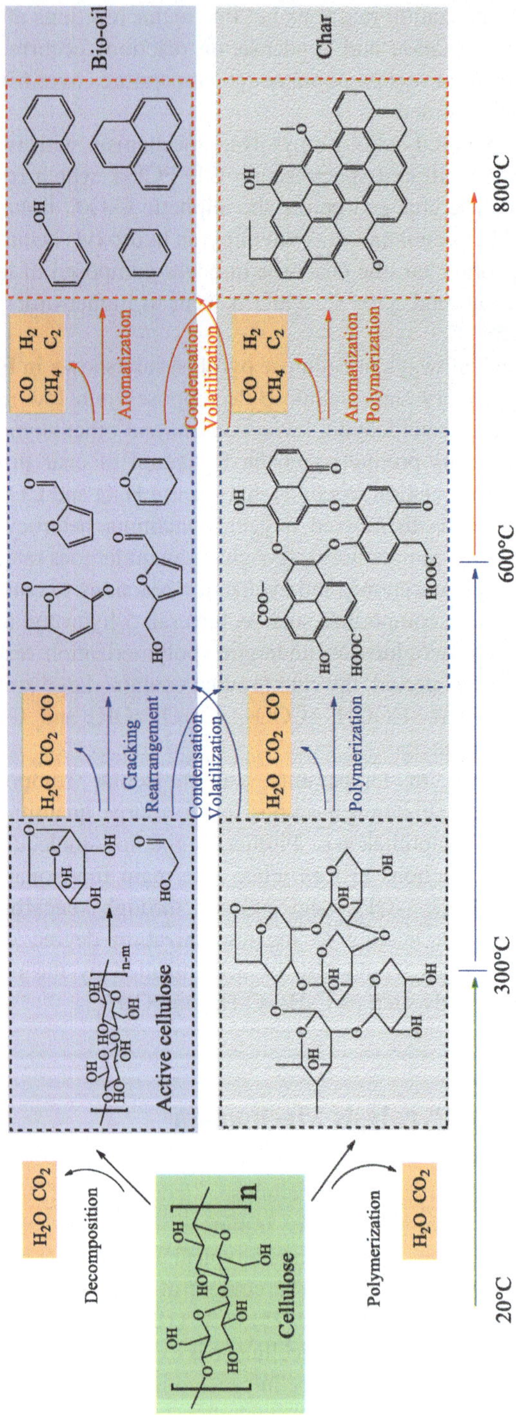

Fig. 4.2 Proposed reaction pathways of cellulose pyrolysis

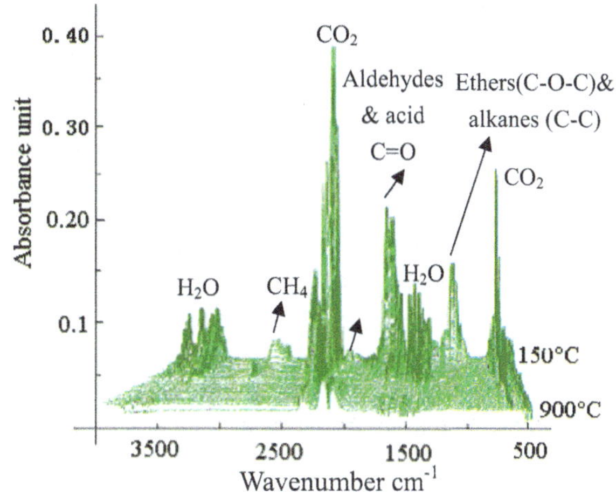

Fig. 4.3 Typical FTIR spectra of gas products from hemicellulose pyrolysis (adapted with permission from [30], Copyright © 2007 Elsevier)

hemicellulose, isolated/commercial hemicellulose polysaccharides, extracted hemicellulose, and hemicellulose as part of biomass [27, 40, 41]. Xylan is an excellent representative molecule among the possible monomeric building blocks of hemicellulose [41, 42].

Compared with cellulose, the decomposition of hemicellulose is easier, and its weight loss mainly occurs at (220–315) °C [30]. The activation energy of the main decomposition stage for hemicellulose is 95 kJ/mol, which is lower than that of cellulose (200 kJ/mol) [43].

In the pyrolysis of xylan, yields of CO and CO_2 are much higher than that of H_2 and CH_4, due to the O-acetyl groups (high content) usually at C2 position of xylan chain [44]. Yang et al. [30] found that release of gaseous products and organic compounds with C=O (1730 cm^{-1}) and C-O-C (1167 cm^{-1}) from xylan pyrolysis mainly occurred at (200–400) °C, as shown in Fig. 4.3. Carboxylic acids are the most abundant with 8.6 wt% during hemicellulose pyrolysis, followed by 3.4 wt% non-aromatic ketones, 1.4 wt% furans, 1.2 wt% sugars, and 0.87 wt% non-aromatic aldehydes [45].

However, there are differences between pyrolysis behavior of native hemicellulose and that of model compounds. Native hemicellulose pyrolysis generates more char, gaseous products, acetol and acetic acid than extracted hemicellulose pyrolysis, while giving yields less 1,2-anhydroxylopyranose, 1,2;3,4-dianhydroxylopyranose and glycolaldehyde [46]. Patwardhan et al. [42] reported that hemicellulose pyrolysis pathways mainly occurred as: depolymerization to form sugars and anhydrosugars, then dehydration to form furans and pyrans, finally ring breakages to generate O-containing small-molecular compounds [27].

Yang et al. [47] found that depolymerization and ring-opening reactions of xylan occurred at 200 °C. Furthermore, the xylan matrix structure was completely degraded at 225 °C to form numerous C=O, C=C, and C-O-C intermediates, which were further decomposed and reconstructed into small-molecule aldehydes, ketones, furans, cyclopentanes, and aromatic compounds. At lower temperatures,

(200–300) °C, the xylan char structure mainly consisted of phenyl rings and oxygen-containing cyclic compounds with aliphatic chains. Then, the molecular structure of the char rearranged to form small fused-ring compounds at (300–450) °C. The dehydrogenation and condensation reactions of xylan increased at (450–600) °C, resulting in a char structure containing a variety of fused-ring compounds, with most C=O and -CH_3 groups removed.

The major reaction pathways of hemicellulose pyrolysis are shown in Fig. 4.4 [27, 42, 45, 46, 48]. Hemicellulose pyrolysis starts with decomposition to form oligosaccharides through the breakdown of the glycosidic bonds, then further cracks into xylose with the release of H_2O and CO_2. For the char product, it contains large amounts of C=O, COOH, -OH, and C-O groups, O-containing ring structures are formed through polymerization reactions.

With the increasing of pyrolysis temperature, cracking, rearrangement, and competing reactions of xylose form many light oxygenates, such as formic acid, acetic acid, propionic acid, 1-hydroxy-2-propanone, furfural, and furan. And the char further polymerizes and forms more aromatic ring structures, with the release of much H_2O, CO_2, and CO, by the condensation of pyrolysis intermediates and volatilization of char product. At high temperatures, aromatic compounds become the main bio-oil product due to the aromatization reactions of the light oxygenates and volatilization of aromatics from char. The aromatization degree of char increases accompanying the condensation of light oxygenates, and the release of CO, H_2, and C_2.

4.4 Lignin Pyrolysis Mechanism

Lignin is currently considered as a waste and finding ways of using it as a renewable source of valuable chemicals is one of the hottest topics in research in bio-based industries [49]. In contrast to cellulose and hemicellulose, lignin is a large-molecular phenolic compound with randomly linking. Softwood lignin accounts for (25–35) wt %, while hardwood and grass lignin accounted for (20–25) wt% and (10–15) wt%, respectively [12]. Lignin has an amorphous structure with three main units (p-cournaryl, coniferyl, and sinapyl alcohol, namely p-hydroxyphenyl (H), guaiacyl (G), and syringyl (S), respectively). These aromatic monomers are linked through C-O bond such as β-O-4, α-O-4, γ-O-4, and 5-O-4 and C-C bonds such as 5-5, β-1, and β-5, with β-O-4 being the most frequent coupling linkage that occurs as (46–60) % [50–52].

Lignin is more difficult to decompose than cellulose, and its weight loss with temperature occurs over a wide range, as the activity of chemical bonds in lignin are extremely varied [30, 53]. Yeo et al. [43] reported that lignin pyrolysis, that had an activation energy of 174.4 kJ/mol, started with complex physicochemical reactions such as lignin melting with weakly bonded functional groups cracking (44–194) °C, followed by massive degradation of lignin through breaking of β-O-4 linkages (194–604) °C and cleavage of various types of bonds (604–824) °C.

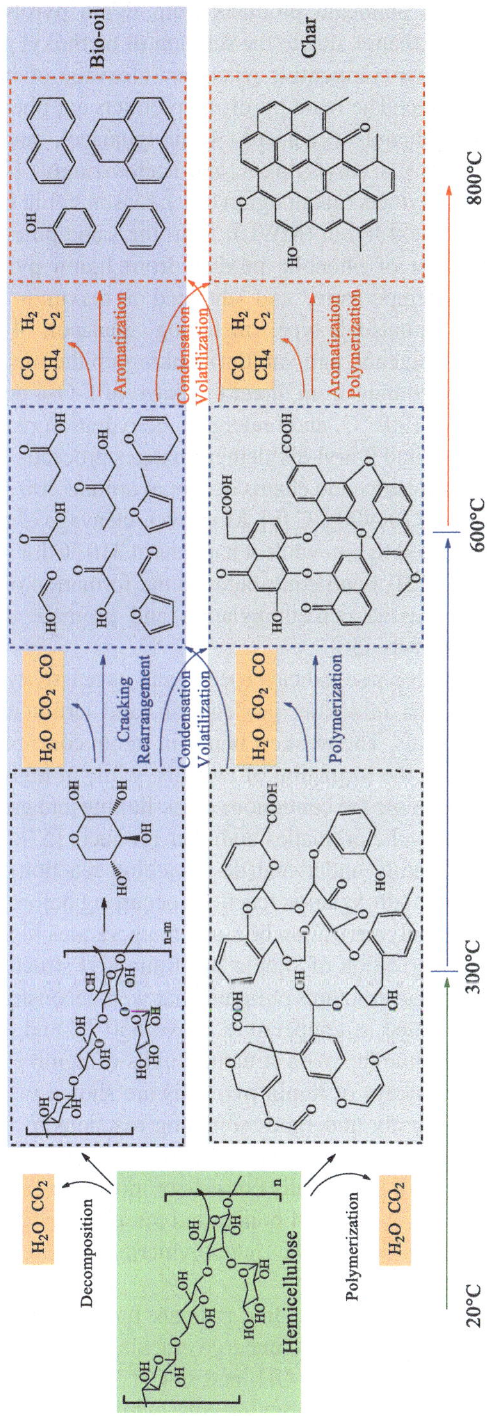

Fig. 4.4 Proposed reaction pathways for hemicellulose pyrolysis

Phenols were the most abundant products from lignin pyrolysis, which also released much CH_4 and methanol, due to the scission of methoxyl groups [54]. Lignin pyrolysis undergoes two competitive reactions: cleavage of inter-unit or alky linkages, and char formation. The main pyrolysis products are phenols (phenol and 4-vinylphenol), methoxyphenols (guaiacol, 4-vinylguaiacol, and vanillin), and dimethoxyphenols (syringol, 4-vinylsyingol, and acetosyringol) [45]. Wang et al. [55] isolated four lignin polymers (alkali lignin (AL), klason lignin (KL), organosolv lignin (OL), and milled wood lignin (MWL)) from the same pine wood pyrolysis, and found that the content of phenolic product from lignin pyrolysis increased greatly with increasing temperature, and obtained a maximum of 60–80% at 600 °C with G type (guaiacol type, including guaiacol, 4-methylguaiacol, 4-vinylguaiacol, trans-isoeugenol, and vanillin) making up the majority of the total products, due to guaiacol coming from direct breakage of β-O-4 bonds.

Lignin softens at (160–190) °C, and undergoes dehydration reactions at around 200 °C, and cracking of α- and β-aryl-alkylether linkages proceeds at (150–300) °C, after which cleavage of aliphatic side chains occurs at around 300 °C, and cleavage of C-C linkages occurs at (370–400) °C [6]. Moreover, cleavage of methoxyl groups occurs at 340 °C for softwood lignin, while it happens at 310 °C for hardwood lignin. Hemolytic cleavage of O-CH_3 bond contributes to the formation of catechol, while H-atom and CH_3-radical assist demethoxylation and promote the formation of phenol and cresol, respectively [56].

Lignin pyrolysis is a melt phase radical process, and its reactivity mainly depends on the competition between initiation, propagation, and termination reactions of different propagating radicals. The weakest bonds in lignin contribute to the formation of radical, and initiate the formation of radicals, while at higher temperatures, condensation reactions promote the continuous cross-linking and growth of aromatic structures and form poly-cyclic aromatics and char products [57]. Yang et al. [58] found that alkali lignin mainly underwent de-branching reactions at temperatures lower than 250 °C, with dehydroxylation reactions occurring before decarbonylation reactions. Alkali lignin depolymerization began at temperatures higher than 250 °C, with preferential depolymerization of simple substitute aryl structures occurring at (250–350) °C. The molecular structures of lignin char were reconstructed, with small fused-ring compounds formed as temperature above 350 °C and at (450–650) °C, high-order fused rings became the main structural units of lignin char.

The major reaction pathways of lignin pyrolysis are shown in Fig. 4.5 [13, 52, 59–65]. Lignin pyrolysis firstly undergoes softening reactions to form an intermediate liquid compound. Then the intermediate liquid compound further decomposes and generates pyrolytic lignin that usually consists of monomers, dimers, trimers, or tetramers through the cleavage of β-O-4 bonds, and these further transform into free radicals, while lignin undergoes aromatic ring polymerization reactions to form char products that release H_2O and CO_2.

As the pyrolysis reaction proceeds, free radicals further capture protons from other pyrolysis products (H donors), and undergo cracking reactions through decomposition of side chains such as C-O, -OH, and O-CH_3, to form large amounts of phenols, methoxyphenols, and dimethoxyphenols, while some free radicals form

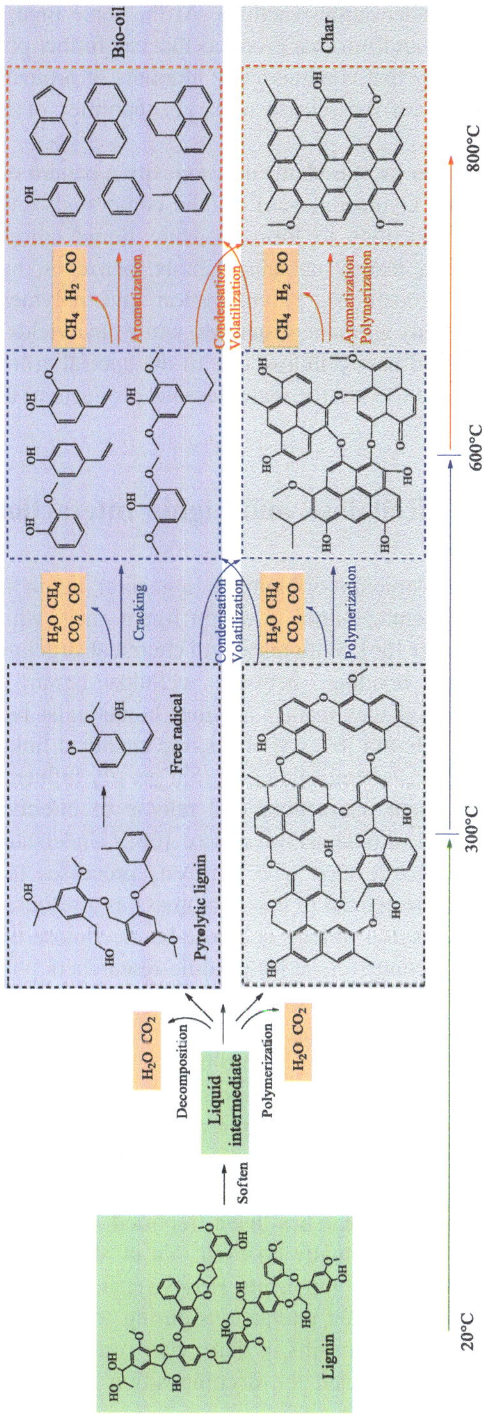

Fig. 4.5 Proposed reaction pathways of lignin pyrolysis

char products through condensation reactions. At the same time, small-molecular compounds can volatilize out from char products that can further polymerize to form more stable char structures that contains large amounts of aromatic rings, C-O-C, and -OH groups along with the release of large quantities of H_2O, CH_4, CO_2, and CO.

As temperature is further increased, the decomposition pattern of lignin is similar with that of cellulose and hemicellulose. Phenolic compounds undergo aromatization and condensation reactions to form aromatic liquid compounds and char products, and the bio-oils mainly contain phenols, benzenes, naphthalenes, and polycyclic aromatic hydrocarbons. Aromatization and polymerization of char enhance the degree of highly aromatic structures, while char yields are much higher than those of cellulose and hemicellulose due to the special aromatic structure of lignin. Then, CH_4, H_2, and CO are the main gaseous products of this step.

4.5 Cellulose, Hemicellulose, and Lignin Interactions

According to the physical structure of biomass, in general, lignin is in the outer cell wall, and cellulose and hemicellulose is within lignin shell, while hemicellulose binds among cellulose fiber [66]. With respect to chemical structure, there are large amounts of hydrogen bonding between cellulose-lignin, and cellulose-hemicellulose. Covalent linkages mainly as ether bonds exist between cellulose-lignin, and hemicellulose-lignin [67, 68]. Thus, the chemical linkage and physical arrangement among cellulose-hemicellulose-lignin greatly influences the pyrolysis products through component interactions and release of intermediates and their interactions, but not in a simple additive way [66]. Understanding interaction mechanisms would help to develop new pyrolysis processes for lignocellulosic biomass, which would be beneficial to use of biomass as a resource.

Whether there is interaction among cellulose-hemicellulose-lignin during biomass pyrolysis is always controversial [66]. Some researchers point out that interactions among three components could be negligible. Yang et al. [69] reported that pyrolysis of synthesized samples containing two or three of biomass components showed negligible interaction among cellulose-hemicellulose-lignin, and there was a linear relationship between weight loss and proportion of hemicellulose or cellulose and the residues (Fig. 4.6). Zhang et al. [66] reported that there were no significant interactions between cellulose-hemicellulose, and cellulose-lignin for both physical and native mixtures.

On the contrary, many researches have reported that interactions among cellulose-hemicellulose-lignin actually do exist. Yu et al. [70] found interactions between cellulose and other two components. Hosoya et al. [71] found that cellulose-lignin pyrolysis showed significant interactions during wood pyrolysis at 800 °C. Fan et al. [72] obtained similar results in that xylan could inhibit the cracking of cellulose, and that lignin promoted the decomposition of sugars from cellulose pyrolysis.

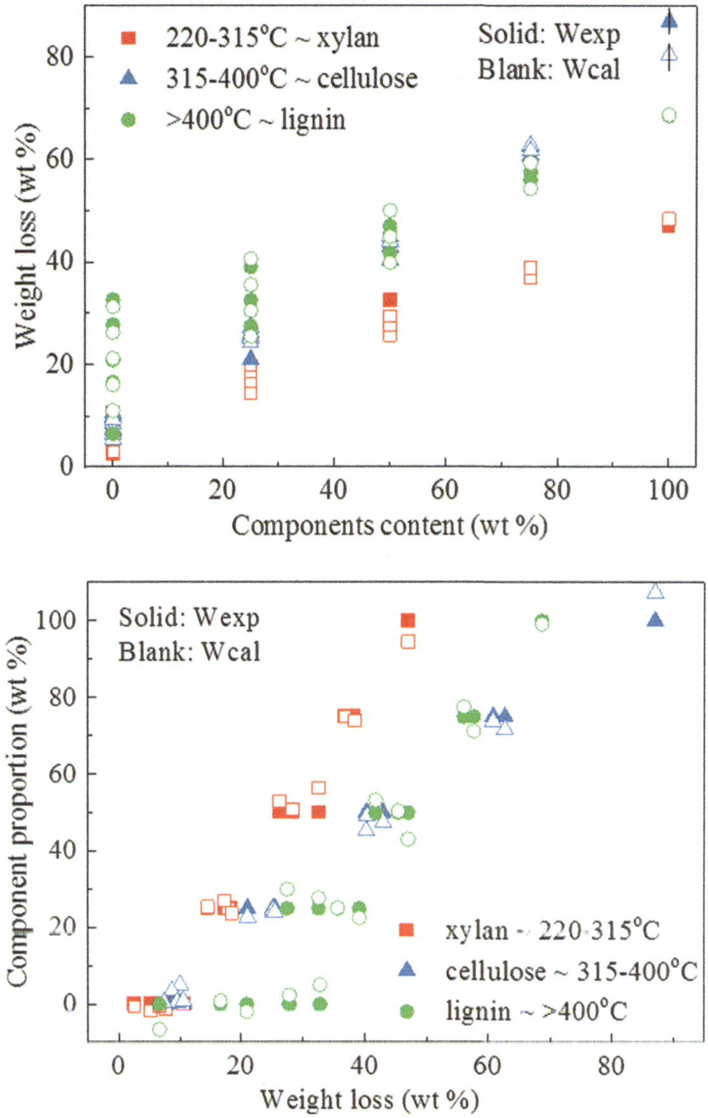

Fig. 4.6 Pyrolysis of biomass-related compounds (adapted with permission from [69], Copyright © 2006 American Chemical Society)

4.5.1 *Cellulose-Hemicellulose Interaction*

Cellulose-hemicellulose interactions increase gas yield, but decrease bio-oil yield [73] and they also greatly promote the generation of CO_2 and hemicellulose-derived products, while inhibit the formation of cellulose-derived products, especially levoglucosan [74]. Interactions of cellulose and hemicellulose also strongly promote the generation of 2,5-diethoxytetrahydrofuran and inhibit the formation of altrose, while cellulose enhances the generation of hemicellulose-derived hydroxyacetone, acetone, acetic acid, 3-buten-2-ol, 1-hydroxy-2-butanone, acetic acid and 2-furfural [74].

Kawamoto et al. [75] found that for hemicellulose, xylan promotes the formation of phenolic and non-phenolic dimers, while glucomannan inhibits the generation of non-phenolic dimers. The interactions between cellulose-xylan increase the decomposition temperature, and cellulose promotes the decomposition and aromatization of xylan [76], which is similar to results of Zhou et al. [77], who found that cellulose-xylan interaction increases the content of polycyclic aromatic hydrocarbons (PAH). Cellulose-xylan interactions greatly affect char product properties, which significantly reduces the char porosity in that the specific surface area decreased 43%, compared with calculated results of individual components by superposition [78].

4.5.2 Cellulose-Lignin Interaction

The interaction between cellulose-lignin is stronger than interactions between cellulose-hemicellulose [74]. Interactions between cellulose and lignin decrease tar yields, but increase char yields [79]. The decrease of tar may be due to cross-linking reactions between lignin-cellulose to generate H_2O and ester groups [79]. Cellulose-lignin co-pyrolysis increases the yield of CO_2 [80], which could promote the generation of cellulose-derived products (esters, aldehydes, ketones, and cyclic ketones), and lignin-derived products (phenols, guaiacols, and syringols), while they inhibit the formation of anhyrosugars, especially levoglucosan [81]. Moreover, cellulose hinders the accumulation of benzene rings generated from lignin [76], which may be due to lignin enhanced micro-explosions that decrease the residence time of cellulose derived products in liquid intermediates, which is the phase where most of dehydration reactions occur [82]. Ye et al. [83] found D-glucose interacted with homolysis radical of lignin model dimer to form a 10-numbered ring intermediate, which changed the energy barriers of pyrolytic reactions, thus promoted the formation of linear carbonyls and furans. Cellulose-xylan interactions also decrease the yields of naphthalene, acenaphthylene, and chrysene plus benzo[*a*]anthracene, while increase the yields of 1-methynaphthalene, 2-methynaphthalene, acenaphthene, fluorene, phenanthrene, anthracene, and pyrene [77].

Pyrolysis of herbaceous biomass, which is representative of native cellulose-lignin mixtures, decreases levoglucosan yield, while increases the yield of low molecular weight compounds and furans [66]. However, there were no interactions for woody biomass, perhaps due to different amounts of covalent linkages in two biomass samples (Fig. 4.7) [66]. Vapor-phase interactions between cellulose-lignin promote gas formation from cellulose-derived volatiles, while inhibit char formation from lignin-derived volatiles and greatly increase the yields of methane and

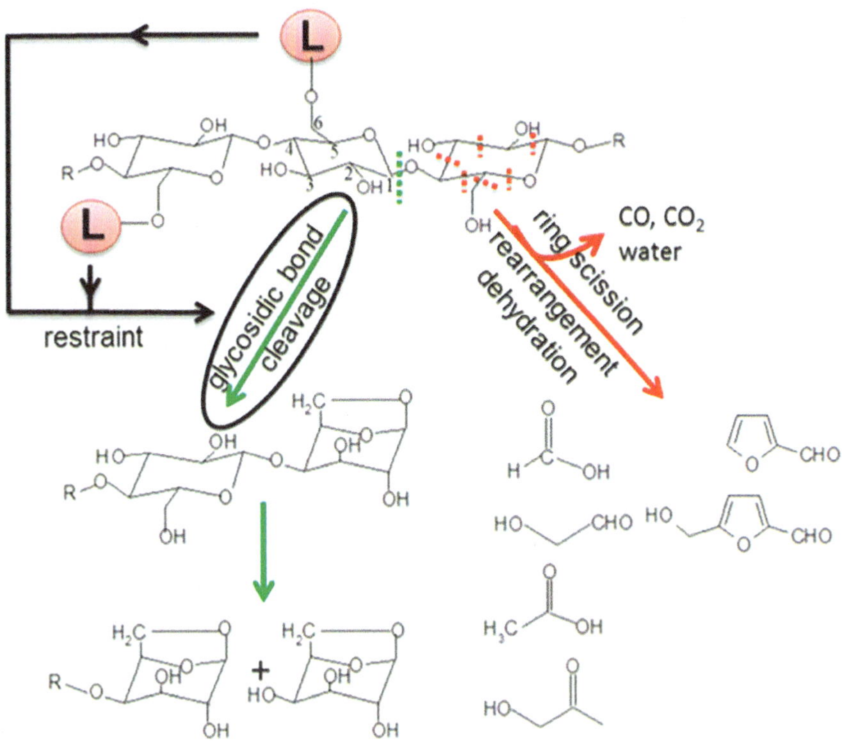

Fig. 4.7 Postulated pyrolysis mechanisms of cellulose covalently linked with lignin (L: lignin) (adapted with permission from [66], Copyright © 2015 American Chemical Society)

catechols from lignin but not o-cresols. This may be ascribed to cellulose-derived intermediates acting as H-donors, while lignin-derived intermediates act as H-acceptors [84], which is similar to conclusions of Wu et al. [81] that cellulose-lignin interaction mechanisms could mainly be H-transfer mechanisms for physical mixtures, while it is the effect of morphology and covalent bonds between them for native mixtures.

4.5.3 Hemicellulose-Lignin Interaction

Interactions between hemicellulose-lignin are much weaker than interactions of cellulose-hemicellulose and cellulose-lignin [72]. Long et al. [85] used the overlap ratio of TG curves to evaluate the interaction of biomass components quantitatively. The interactions between xylan-lignin were weak with an overlap ratio of 0.987, whereas the interactions between cellulose-lignin strongly overlap and have a ratio of 0.974, which may be related to the abundance of hydrogen bonds between hemicellulose and lignin, and strong van der Waals forces between lignin and cellulose [86].

Interactions of lignin and xylan decrease the liquid yields while increase gas yields [78] and they promote the formation of CO_2 and H_2 at temperatures of (400–700) °C and inhibit the generation of CH_4 and phenols below 500 °C and at (400–700) °C, respectively [73]. Zhao et al. [87] found that lignin largely inhibits the formation of sugars (mainly levoglucosan) from hemicellulose pyrolysis. However, hemicellulose significantly promoted lignin decomposition to produce phenolic compounds. Xylan inhibits the formation of guaiacols and syringols from lignin pyrolysis, but enhances hydroxyphenols generation [88]. Zhou et al. [77] found that the interactions between xylan and lignin decreases the content of PAHs, such as naphthalene, acenaphthylene, 1-methynaphthalene, 2-methynaphthalene, acenaphthene, and fluorene.

4.5.4 Cellulose-Hemicellulose-Lignin Interaction

In addition to interactions between two components, three component interactions (cellulose-hemicellulose-lignin) have been widely researched. Cellulose-hemicellulose-lignin interactions extend the temperature range for the formation of levoglucosan. Cellulose and lignin promote the formation of 2-furfural and acetic acid at (350–500) °C and cellulose and hemicellulose increase the content of phenol, 2,6-dimethoxy [89]. However, Liu et al. [90] showed some different results, namely, that lignin inhibits the formation of 2-furaldehyde and C=O containing compounds. Hemicellulose largely decreases levoglucosan yields, while greatly increases hydroxyacetaldehyde yields. Yu et al. [70] found that synthetic mixtures (mixing cellulose, hemicellulose, and lignin) give more tar and less char than predicted values, while actual biomass samples showed the opposite trends with less char and more tar being formed than predicted values. Furthermore, compared with synthetic mixtures, the internal porosity and channels within actual biomass, may facilitate volatile product release.

Researchers have made great efforts to explore the interactions among cellulose-hemicellulose-lignin during pyrolysis, but interaction mechanisms are still not well understood yet [12]. Moreover, the effect of other components, such pectin, which is structurally and functionally the most complex polysaccharide in plant cell walls [91], on the biomass pyrolysis mechanism still needs to be explored.

4.6 Influence of Inorganic Species on Biomass Pyrolysis

In addition to the major organic components (cellulose, hemicellulose, and lignin), biomass can contain inorganic minerals at very low concentrations and in decreasing order of abundance as Ca, K, Si, Mg, Al, S, Fe, P, Cl, Na, Mn, and Ti [92, 93]. The

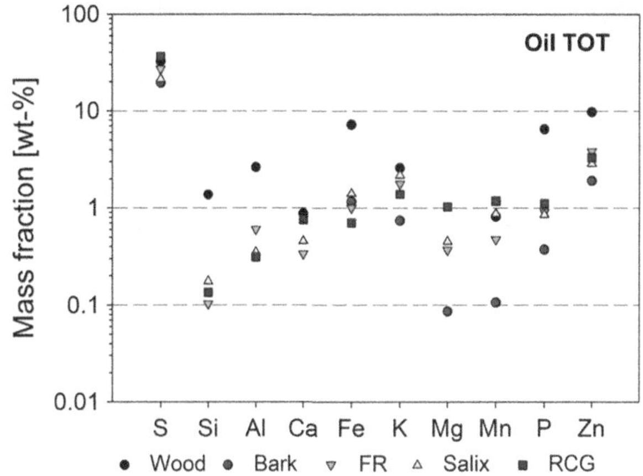

Fig. 4.8 Inorganic elements found in the bio-oil from biomass pyrolysis (adapted with permission from [96], Copyright © 2017 Elsevier)

content and distribution of these inorganic species in biomass depend on type and source. Woody biomass often contains lower inorganic content than that of agricultural residues [92]. Inorganic minerals may be present in biomass in numerous forms [94] with the simplest being free ion form that is dissolved in fluid matter within the biomass. Salts are often in structured mineral forms (e.g. NaCl, KCl). Covalent bonds between inorganic elements and organic biomass structures are seen as well [94]. Transformation mechanisms of these inorganic components is also important for understanding biomass pyrolysis mechanisms [95].

Wiinikka et al. [96] explored the fate of inorganic elements during biomass fast pyrolysis in a cyclone reactor, and found that Si, Al, Ca, Fe, K, Mg, Mn, Na, P, Ti, and Zn were mainly concentrated in char products, while much of the S is released into gaseous products during biomass pyrolysis with (1–10) wt% of Zn, K, and Fe being transformed into pyrolysis oil, while <1 wt% of other inorganic elements being in the pyrolysis oil, as shown in Fig. 4.8. Alkali earth metals (Ca, Mg), transition metals (Fe, Cu, Ni, Cd, Cr, Co, Mn, Zn), and post transition metals (Al, Pb) mostly remain in char products during biomass pyrolysis [94]. Lane et al. [97] investigated the release mechanism of inorganic elements during algae pyrolysis, and found that Cl is mainly released together with alkali metals (primarily by sublimation of NaCl) from marine algae, and in the form of HCl from freshwater algae. The S was released firstly by organic forms during biomass pyrolysis process, and then by inorganic forms during secondary oxidation and reduction of char. Furthermore, K and Na were released by volatilization of melted metal-phosphates.

Inorganic species significantly influence pyrolysis behavior and mechanisms of biomass pyrolysis. Alkali metals increase degradation rate, reduce degradation temperature, and change the reactions pathways of biomass pyrolysis [98]. Potassium impregnated in cellulose inhibits the formation of anhydrosugars and mobile protons due to its stabilization of free radicals, while hydrogen mobility and transfer are important in biomass pyrolysis mechanism, then hemolytic scissions enhancement causes formation of a more aromatic structure [99]. CH_3COOK added to cellulose and levoglucosan promotes glucosidic unit decomposition to form small-molecular compounds, such as acetic acid, hydroxyacetaldehyde, and acetol [100]. The K and Na contribute to carbonization, dehydration, decarboxylation, and demethoxylation to form modified carbonaceous structures during biomass pyrolysis [101]. Furthermore, K, Na, Ca, and Mg are able to promote competing reactions to form lower molecular weight species, and decrease levoglucosan yields [102]. It could be observed that research on the effect of nonmetals (e.g. Si, P, S, Cl) on biomass pyrolysis is rare. Thus, more work should be performed on biomass pyrolysis with additives to understand mechanism comprehensively.

4.7 Conclusions and Future Outlook

Biomass pyrolysis is a promising technology to convert biomass into valuable bio-oil, gas, and char products. By choosing appropriate reaction conditions, yields of the different products can be optimized. Pyrolysis mechanisms of individual components (cellulose, hemicellulos, lignin, and inorganic species) and interaction mechanisms of the components were reviewed in this chapter and an overview was presented on the evolution of biochar, bio-oil and gas products. Some of the challenges in developing biomass pyrolysis further can be given as follows:

Individual pyrolysis mechanisms mainly focus on cellulose and lignin, hence reaction pathways of hemicellulose, and other organic components (such as pectin [91]) should be given more attention as well as interaction mechanisms among organic and inorganic species, as well as the catalytic effect of inorganic species. It is vital to extract the native components from biomass through development of micro-depletion extraction methods that retain the original structure, special linkages, and functional groups in biomass components. Further, the combination of different size model compounds (monomer, oligomer, individual components) and biomass is very important for studying pyrolysis mechanisms.

Analysis methods are mainly thermogravimetric analysis (TGA), thermogravimetric analyzer coupled with Fourier transform infrared spectroscopy (TG-FTIR), and pyrolysis-gas chromatography/mass spectrometry-flame ionization detector (PY-GC/MS-FID). As these methods are limited in application to biomass component pyrolysis mechanisms, it is necessary to develop new in-situ analysis methods (such as pyrolysis combined with two-dimensional gas chromatography-time of flight mass spectrometer (PY-GC×GC-TOF/MS) [103], thermogravimetric analyzer combined with photoionization time-of-flight mass spectrometer (TG-PI-TOF-MS) [104], on-line single-photon ionization time-of-flight mass spectrometry (SPI-TOF-MS) [105–107], on-line near-atmospheric pressure photoionization Orbitrap mass spectrometry (nAPPI Orbitrap MS) [104], in-situ reflectance Infrared Fourier transform spectroscopy (in-situ DRIFTS) [76, 108], in-situ X-ray diffraction (in-situ XRD) [109]) for in-time and quantitative comprehensive characterization of pyrolysis gas products, bio-oil, and char, to further explore biomass pyrolysis mechanisms in depth.

To realize the biomass conversion into valuable products, catalytic pyrolysis process is necessary, and understanding catalytic pyrolysis mechanism is vital for better utilization of biomass [110]. It is also important that the integration of pyrolysis for industrial uses for bio-oil (fast pyrolysis), and the synergetic effects due to the heterogeneity of biomass or mixtures of biomass.

Acknowledgements We express great appreciation of the financial support from the National Natural Science Foundation of China (51876078, 51622604 and 51861130362), China Postdoctoral Science Foundation (2018M640696 and 2019T120664), the technical support from Analytical and Testing Center in Huazhong University of Science and Technology (http://atc.hust.edu.cn).

References

1. Grubler A, Wilson C, Bento N, Boza-Kiss B, Krey V, McCollum DL, Rao ND, Riahi K, Rogelj J, De Sterck S, Cullen J, Frank S, Fricko O, Guo F, Gidden M, Havlík P, Huppmann D, Kiesewetter G, Rafaj P, Schoepp W, Valin H. A low energy demand scenario for meeting the 1.5 °C target and sustainable development goals without negative emission technologies. Nat Energy. 2018;3:515–27. https://doi.org/10.1038/s41560-018-0172-6.
2. Rogelj J, Popp A, Calvin KV, Luderer G, Emmerling J, Gernaat D, Fujimori S, Strefler J, Hasegawa T, Marangoni G, Krey V, Kriegler E, Riahi K, Van Vuuren DP, Doelman J, Drouet L, Edmonds J, Fricko O, Harmsen M, Havlík P, Humpenöder F, Stehfest E, Tavoni M. Scenarios towards limiting global mean temperature increase below 1.5 °C. Nat Clim Change. 2018;8:325–32. https://doi.org/10.1038/s41558-018-0091-3.
3. Pachauri RK, Allen MR, Barros VR, Broome J, Cramer W, Christ R, Church JA, Clarke L, Dahe Q, Dasgupta P. Climate change 2014: synthesis report. Contribution of Working Groups I, II and III to the fifth assessment report of the Intergovernmental Panel on Climate Change, Ipcc2014.
4. Dodds DR, Gross RA. Chemicals from biomass. Science. 2007;318:1250–1. https://doi.org/10.1126/science.1146356.
5. Alonso DM, Hakim SH, Zhou S, Won W, Hosseinaei O, Tao J, Garcia-Negron V, Motagamwala AH, Mellmer MA, Huang K, Houtman CJ, Labbé N, Harper DP, Maravelias CT, Runge T, Dumesic JA. Increasing the revenue from lignocellulosic biomass: maximizing feedstock utilization. Sci Adv. 2017;3(5):e1603301. https://doi.org/10.1126/sciadv.1603301.
6. Anca-Couce A. Reaction mechanisms and multi-scale modelling of lignocellulosic biomass pyrolysis. Prog Energy Combust Sci. 2016;53:41–79. https://doi.org/10.1016/j.pecs.2015.10.002.
7. Proskurina S, Junginger M, Heinimö J, Tekinel B, Vakkilainen E. Global biomass trade for energy— Part 2: production and trade streams of wood pellets, liquid biofuels, charcoal, industrial roundwood and emerging energy biomass. Biofuels Bioprod Biorefin. 2019;13:371–87. https://doi.org/10.1002/bbb.1858.
8. Vaughan NE, Gough C, Mander S, Littleton EW, Welfle A, Gernaat DEHJ, Van Vuuren DP. Evaluating the use of biomass energy with carbon capture and storage in low emission scenarios. Environ Res Lett. 2018;13:044014. https://doi.org/10.1088/1748-9326/aaaa02.
9. Huber GW, Iborra S, Corma A. Synthesis of transportation fuels from biomass: chemistry, catalysts, and engineering. Chem Rev. 2006;106:4044–98. https://doi.org/10.1021/cr068360d.
10. Mettler MS, Vlachos DG, Dauenhauer PJ. Top ten fundamental challenges of biomass pyrolysis for biofuels. Energy Environ Sci. 2012;5:7797–809. https://doi.org/10.1039/C2EE21679E.

11. Xia Q, Chen Z, Shao Y, Gong X, Wang H, Liu X, Parker S, Han X, Yang S, Wang Y. Direct hydrodeoxygenation of raw woody biomass into liquid alkanes. Nat Commun. 2016;7:11162. https://doi.org/10.1038/ncomms11162. https://www.nature.com/articles/ncomms11162#supplementary-information
12. Wang S, Dai G, Yang H, Luo Z. Lignocellulosic biomass pyrolysis mechanism: a state-of-the-art review. Prog Energy Combust Sci. 2017;62:33–86. https://doi.org/10.1016/j.pecs.2017.05.004.
13. Collard FX, Blin J. A review on pyrolysis of biomass constituents: mechanisms and composition of the products obtained from the conversion of cellulose, hemicelluloses and lignin. Renew Sust Energy Rev. 2014;38:594–608. https://doi.org/10.1016/j.rser.2014.06.013.
14. Gollakota ARK, Reddy M, Subramanyam MD, Kishore N. A review on the upgradation techniques of pyrolysis oil. Renew Sust Energy Rev. 2016;58:1543–68. https://doi.org/10.1016/j.rser.2015.12.180.
15. Shen D, Jin W, Hu J, Xiao R, Luo K. An overview on fast pyrolysis of the main constituents in lignocellulosic biomass to valued-added chemicals: structures, pathways and interactions. Renew Sust Energy Rev. 2015;51:761–74. https://doi.org/10.1016/j.rser.2015.06.054.
16. Yang H, Liu B, Chen Y, Chen W, Yang Q, Chen H. Application of biomass pyrolytic polygeneration technology using retort reactors. Bioresour Technol. 2016;200:64–71. https://doi.org/10.1016/j.biortech.2015.09.107.
17. Chen Y, Yang H, Wang X, Chen W, Chen H. Biomass pyrolytic polygeneration system: adaptability for different feedstocks. Energ Fuel. 2016;30:414–22. https://doi.org/10.1021/acs.energyfuels.5b02332.
18. Chen Y, Yang H, Wang X, Zhang S, Chen H. Biomass-based pyrolytic polygeneration system on cotton stalk pyrolysis: influence of temperature. Bioresour Technol. 2012;107:411–8. https://doi.org/10.1016/j.biortech.2011.10.074.
19. Zhang X, Che Q, Cui X, Wei Z, Zhang X, Chen Y, Wang X, Chen H. Application of biomass pyrolytic polygeneration by a moving bed: characteristics of products and energy efficiency analysis. Bioresour Technol. 2018;254:130–8. https://doi.org/10.1016/j.biortech.2018.01.083.
20. Xu C, Arancon RAD, Labidi J, Luque R. Lignin depolymerisation strategies: towards valuable chemicals and fuels. Chem Soc Rev. 2014;43:7485–500. https://doi.org/10.1039/C4CS00235K.
21. Upton BM, Kasko AM. Strategies for the conversion of lignin to high-value polymeric materials: review and perspective. Chem Rev. 2016;116:2275–306. https://doi.org/10.1021/acs.chemrev.5b00345.
22. Liu Y, Nie Y, Lu X, Zhang X, He H, Pan F, Zhou L, Liu X, Ji X, Zhang S. Cascade utilization of lignocellulosic biomass to high-value products. Green Chem. 2019;21:3499–535. https://doi.org/10.1039/C9GC00473D.
23. Liu WJ, Jiang H, Yu HQ. Development of biochar-based functional materials: toward a sustainable platform carbon material. Chem Rev. 2015;115:12251–85. https://doi.org/10.1021/acs.chemrev.5b00195.
24. Hu W, Dang Q, Rover M, Brown RC, Wright MM. Comparative techno-economic analysis of advanced biofuels, biochemicals, and hydrocarbon chemicals via the fast pyrolysis platform. Biofuels. 2016;7:57–67. https://doi.org/10.1080/17597269.2015.1118780.
25. Chen W, Fang Y, Li K, Chen Z, Xia M, Gong M, Chen Y, Yang H, Tu X, Chen H. Bamboo wastes catalytic pyrolysis with N-doped biochar catalyst for phenols products. Appl Energy. 2020;260:114242. https://doi.org/10.1016/j.apenergy.2019.114242.
26. Sharma A, Pareek V, Zhang D. Biomass pyrolysis—A review of modelling, process parameters and catalytic studies. Renew Sust Energy Rev. 2015;50:1081–96. https://doi.org/10.1016/j.rser.2015.04.193.
27. Zhou X, Li W, Mabon R, Broadbelt LJ. A critical review on hemicellulose pyrolysis. Energ Technol. 2017;5:52–79. https://doi.org/10.1002/ente.201600327.

28. Kan T, Strezov V, Evans TJ. Lignocellulosic biomass pyrolysis: a review of product properties and effects of pyrolysis parameters. Renew Sust Energy Rev. 2016;57:1126–40. https://doi.org/10.1016/j.rser.2015.12.185.
29. French ADJC. Glucose, not cellobiose, is the repeating unit of cellulose and why that is important. 2017;24:4605–9. https://doi.org/10.1007/s10570-017-1450-3.
30. Yang H, Yan R, Chen H, Lee DH, Zheng C. Characteristics of hemicellulose, cellulose and lignin pyrolysis. Fuel. 2007;86:1781–8. https://doi.org/10.1016/j.fuel.2006.12.013.
31. Chen X, Che Q, Li S, Liu Z, Yang H, Chen Y, Wang X, Shao J, Chen H. Recent developments in lignocellulosic biomass catalytic fast pyrolysis: strategies for the optimization of bio-oil quality and yield. Fuel Process Technol. 2019;196:106180. https://doi.org/10.1016/j.fuproc.2019.106180.
32. Adenson MO, Kelley MD, Elkelany OO, Biernacki JJ, Liu YW. Kinetics of cellulose pyrolysis: ensuring optimal outcomes. Can J Chem Eng. 2018;96:926–35. https://doi.org/10.1002/cjce.23060.
33. Xin S, Yang H, Chen Y, Wang X, Chen H. Assessment of pyrolysis polygeneration of biomass based on major components: product characterization and elucidation of degradation pathways. Fuel. 2013;113:266–73. https://doi.org/10.1016/j.fuel.2013.05.061.
34. Zhang X, Yang W, Dong C. Levoglucosan formation mechanisms during cellulose pyrolysis. J Anal Appl Pyrolysis. 2013;104:19–27. https://doi.org/10.1016/j.jaap.2013.09.015.
35. Lin Y-C, Cho J, Tompsett GA, Westmoreland PR, Huber GW. Kinetics and mechanism of cellulose pyrolysis. J Phys Chem C. 2009;113:20097–107. https://doi.org/10.1021/jp906702p.
36. Xin S, Yang H, Chen Y, Yang M, Chen L, Wang X, Chen H. Chemical structure evolution of char during the pyrolysis of cellulose. J Anal Appl Pyrolysis. 2015;116:263–71. https://doi.org/10.1016/j.jaap.2015.09.002.
37. Yang H, Gong M, Hu J, Liu B, Chen Y, Xiao J, Li S, Dong Z, Chen H. Cellulose pyrolysis mechanism based on functional group evolutions by two-dimensional perturbation correlation infrared spectroscopy. Energ Fuel. 2020;34:3412–21. https://doi.org/10.1021/acs.energyfuels.0c00134.
38. Stefanidis SD, Kalogiannis KG, Iliopoulou EF, Michailof CM, Pilavachi PA, Lappas AA. A study of lignocellulosic biomass pyrolysis via the pyrolysis of cellulose, hemicellulose and lignin. J Anal Appl Pyrolysis. 2014;105:143–50. https://doi.org/10.1016/j.jaap.2013.10.013.
39. Burnham AK, Zhou X, Broadbelt LJ. Critical review of the global chemical kinetics of cellulose thermal decomposition. Energ Fuel. 2015;29:2906–18. https://doi.org/10.1021/acs.energyfuels.5b00350.
40. Zeitoun R, Pontalier PY, Marechal P, Rigal L. Twin-screw extrusion for hemicellulose recovery: influence on extract purity and purification performance. Bioresour Technol. 2010;101:9348–54. https://doi.org/10.1016/j.biortech.2010.07.022.
41. Werner K, Pommer L, Broström M. Thermal decomposition of hemicelluloses. J Anal Appl Pyrolysis. 2014;110:130–7. https://doi.org/10.1016/j.jaap.2014.08.013.
42. Patwardhan PR, Brown RC, Shanks BH. Product distribution from the fast pyrolysis of hemicellulose. ChemSusChem. 2011;4:636–43. https://doi.org/10.1002/cssc.201000425.
43. Yeo JY, Chin BLF, Tan JK, Loh YS. Comparative studies on the pyrolysis of cellulose, hemicellulose, and lignin based on combined kinetics. J Energy Inst. 2019;92:27–37. https://doi.org/10.1016/j.joei.2017.12.003.
44. Jiang G, Nowakowski DJ, Bridgwater AV. Effect of the temperature on the composition of lignin pyrolysis products. Energ Fuel. 2010;24:4470–5. https://doi.org/10.1021/ef100363c.
45. Carrier M, Windt M, Ziegler B, Appelt J, Saake B, Meier D, Bridgwater A. Quantitative insights into the fast pyrolysis of extracted cellulose, hemicelluloses, and lignin. ChemSusChem. 2017;10:3212–24. https://doi.org/10.1002/cssc.201700984.
46. Zhou X, Li W, Mabon R, Broadbelt LJ. A mechanistic model of fast pyrolysis of hemicellulose. Energy Environ Sci. 2018;11:1240–60. https://doi.org/10.1039/c7ee03208k.

47. Yang H, Li S, Liu B, Chen Y, Xiao J, Dong Z, Gong M, Chen H. Hemicellulose pyrolysis mechanism based on functional group evolutions by two-dimensional perturbation correlation infrared spectroscopy. Fuel 2020;267. https://doi.org/10.1016/j.fuel.2020.117302.
48. Shen DK, Gu S, Bridgwater AV. Study on the pyrolytic behaviour of xylan-based hemicellulose using TG-FTIR and Py-GC-FTIR. J Anal Appl Pyrolysis. 2010;87:199–206. https://doi.org/10.1016/j.jaap.2009.12.001.
49. Schutyser W, Renders T, Van den Bosch S, Koelewijn SF, Beckham GT, Sels BF. Chemicals from lignin: an interplay of lignocellulose fractionation, depolymerisation, and upgrading. Chem Soc Rev. 2018;47:852–908. https://doi.org/10.1039/C7CS00566K.
50. Zhang J, Jiang X, Ye X, Chen L, Lu Q, Wang X, Dong C. Pyrolysis mechanism of a β-O-4 type lignin dimer model compound. J Therm Anal Calorim. 2016;123:501–10. https://doi.org/10.1007/s10973-015-4944-y.
51. Jiang W, Wu S, Lucia LA, Chu J. A comparison of the pyrolysis behavior of selected β-O-4 type lignin model compounds. J Anal Appl Pyrolysis. 2017;125:185–92. https://doi.org/10.1016/j.jaap.2017.04.003.
52. Chu S, Subrahmanyam AV, Huber GW. The pyrolysis chemistry of a β-O-4 type oligomeric lignin model compound. Green Chem. 2013;15:125–36. https://doi.org/10.1039/c2gc36332a.
53. Shen D, Liu G, Zhao J, Xue J, Guan S, Xiao R. Thermo-chemical conversion of lignin to aromatic compounds: effect of lignin source and reaction temperature. J Anal Appl Pyrolysis. 2015;112:56–65. https://doi.org/10.1016/j.jaap.2015.02.022.
54. Asmadi M, Kawamoto H, Saka S. Thermal reactions of guaiacol and syringol as lignin model aromatic nuclei. J Anal Appl Pyrolysis. 2011;92:88–98. https://doi.org/10.1016/j.jaap.2011.04.011.
55. Wang S, Ru B, Lin H, Sun W, Luo Z. Pyrolysis behaviors of four lignin polymers isolated from the same pine wood. Bioresour Technol. 2015;182:120–7. https://doi.org/10.1016/j.biortech.2015.01.127.
56. Liu C, Ye L, Yuan W, Zhang Y, Zou J, Yang J, Wang Y, Qi F, Zhou Z. Investigation on pyrolysis mechanism of guaiacol as lignin model compound at atmospheric pressure. Fuel. 2018;232:632–8. https://doi.org/10.1016/j.fuel.2018.05.162.
57. Faravelli T, Frassoldati A, Migliavacca G, Ranzi E. Detailed kinetic modeling of the thermal degradation of lignins. Biomass Bioenergy. 2010;34:290–301. https://doi.org/10.1016/j.biombioe.2009.10.018.
58. Chen Y, Liu B, Yang H, Wang X, Zhang X, Chen H. Generalized two-dimensional correlation infrared spectroscopy to reveal the mechanisms of lignocellulosic biomass pyrolysis. Proc. Combust. Inst. 2019;37:3013–21. https://doi.org/10.1016/j.proci.2018.06.141.
59. Kawamoto H. Lignin pyrolysis reactions. J Wood Sci. 2017;63:117–32. https://doi.org/10.1007/s10086-016-1606-z.
60. Robichaud DJ, Nimlos MR, Ellison GB. Pyrolysis mechanisms of lignin model compounds using a heated micro-reactor. In: Schlaf M, Zhang ZC, editors. Reaction pathways and mechanisms in thermocatalytic biomass conversion II: homogeneously catalyzed transformations, acrylics from biomass, theoretical aspects, lignin valorization and pyrolysis pathways. Singapore: Springer; 2016. p. 145–71.
61. Dieguez-Alonso A, Anca-Couce A, Zobel N, Behrendt F. Understanding the primary and secondary slow pyrolysis mechanisms of holocellulose, lignin and wood with laser-induced fluorescence. Fuel. 2015;153:102–9. https://doi.org/10.1016/j.fuel.2015.02.097.
62. Zhou S, Pecha B, van Kuppevelt M, McDonald AG, Garcia-Perez M. Slow and fast pyrolysis of Douglas-fir lignin: importance of liquid-intermediate formation on the distribution of products. Biomass Bioenergy. 2014;66:398–409. https://doi.org/10.1016/j.biombioe.2014.03.064.
63. Kotake T, Kawamoto H, Saka S. Mechanisms for the formation of monomers and oligomers during the pyrolysis of a softwood lignin. J Anal Appl Pyrolysis. 2014;105:309–16. https://doi.org/10.1016/j.jaap.2013.11.018.

64. Bayerbach R, Meier D. Characterization of the water-insoluble fraction from fast pyrolysis liquids (pyrolytic lignin). Part IV: structure elucidation of oligomeric molecules. J Anal Appl Pyrolysis. 2009;85:98–107. https://doi.org/10.1016/j.jaap.2008.10.021.
65. Scholze B, Hanser C, Meier D. Characterization of the water-insoluble fraction from fast pyrolysis liquids (pyrolytic lignin): part II. GPC, carbonyl goups, and ^{13}C-NMR. J. Anal. Appl. Pyrolysis. 2001;58-59:387–400. https://doi.org/10.1016/S0165-2370(00)00173-X.
66. Zhang J, Choi YS, Yoo CG, Kim TH, Brown RC, Shanks BH. Cellulose-hemicellulose and cellulose-lignin interactions during fast pyrolysis. ACS Sustain Chem Eng. 2015;3:293–301. https://doi.org/10.1021/sc500664h.
67. Zhang N, Li S, Xiong L, Hong Y, Chen Y. Cellulose-hemicellulose interaction in wood secondary cell-wall. Model Simul Mater Sci Eng. 2015;23:085010. https://doi.org/10.1088/0965-0393/23/8/085010.
68. Terrett OM, Dupree P. Covalent interactions between lignin and hemicelluloses in plant secondary cell walls. Curr Opin Biotech. 2019;56:97–104. https://doi.org/10.1016/j.copbio.2018.10.010.
69. Yang H, Yan R, Chen H, Zheng C, Lee DH, Liang DT. In-depth investigation of biomass pyrolysis based on three major components: hemicellulose, cellulose and lignin. Energ Fuel. 2006;20:388–93. https://doi.org/10.1021/ef0580117.
70. Yu J, Paterson N, Blamey J, Millan M. Cellulose, xylan and lignin interactions during pyrolysis of lignocellulosic biomass. Fuel. 2017;191:140–9. https://doi.org/10.1016/j.fuel.2016.11.057.
71. Hosoya T, Kawamoto H, Saka S. Cellulose–hemicellulose and cellulose–lignin interactions in wood pyrolysis at gasification temperature. J Anal Appl Pyrolysis. 2007;80:118–25. https://doi.org/10.1016/j.jaap.2007.01.006.
72. Fan Y, Cai Y, Li X, Jiao L, Xia J, Deng X. Effects of the cellulose, xylan and lignin constituents on biomass pyrolysis characteristics and bio-oil composition using the Simplex Lattice Mixture Design method. Energy Convers Manag. 2017;138:106–18. https://doi.org/10.1016/j.enconman.2017.01.075.
73. Chen Y, Fang Y, Yang H, Xin S, Zhang X, Wang X, Chen H. Effect of volatiles interaction during pyrolysis of cellulose, hemicellulose, and lignin at different temperatures. Fuel. 2019;248:1–7. https://doi.org/10.1016/j.fuel.2019.03.070.
74. Wu S, Shen D, Hu J, Zhang H, Xiao R. Cellulose-hemicellulose interactions during fast pyrolysis with different temperatures and mixing methods. Biomass Bioenergy. 2016;95:55–63. https://doi.org/10.1016/j.biombioe.2016.09.015.
75. Kawamoto H, Watanabe T, Saka S. Strong interactions during lignin pyrolysis in wood – a study by in situ probing of the radical chain reactions using model dimers. J Anal Appl Pyrolysis. 2015;113:630–7. https://doi.org/10.1016/j.jaap.2015.04.009.
76. Wang J, Shen B, Kang D, Yuan P, Wu C. Investigate the interactions between biomass components during pyrolysis using in-situ DRIFTS and TGA. Chem Eng Sci. 2019;195:767–76. https://doi.org/10.1016/j.ces.2018.10.023.
77. Zhou H, Wu C, Meng A, Zhang Y, Williams PT. Effect of interactions of biomass constituents on polycyclic aromatic hydrocarbons (PAH) formation during fast pyrolysis. J Anal Appl Pyrolysis. 2014;110:264–9. https://doi.org/10.1016/j.jaap.2014.09.007.
78. Giudicianni P, Cardone G, Ragucci R. Cellulose, hemicellulose and lignin slow steam pyrolysis: thermal decomposition of biomass components mixtures. J Anal Appl Pyrolysis. 2013;100:213–22. https://doi.org/10.1016/j.jaap.2012.12.026.
79. Worasuwannarak N, Sonobe T, Tanthapanichakoon W. Pyrolysis behaviors of rice straw, rice husk, and corncob by TG-MS technique. J Anal Appl Pyrolysis. 2007;78:265–71. https://doi.org/10.1016/j.jaap.2006.08.002.
80. Volpe R, Zabaniotou AA, Skoulou V. Synergistic effects between lignin and cellulose during pyrolysis of agricultural waste. Energ Fuel. 2018;32:8420–30. https://doi.org/10.1021/acs.energyfuels.8b00767.

81. Wu S, Shen D, Hu J, Zhang H, Xiao R. Cellulose-lignin interactions during fast pyrolysis with different temperatures and mixing methods. Biomass Bioenergy. 2016;90:209–17. https://doi.org/10.1016/j.biombioe.2016.04.012.
82. Hilbers TJ, Wang Z, Pecha B, Westerhof RJM, Kersten SRA, Pelaez-Samaniego MR, Garcia-Perez M. Cellulose-Lignin interactions during slow and fast pyrolysis. J Anal Appl Pyrolysis. 2015;114:197–207. https://doi.org/10.1016/j.jaap.2015.05.020.
83. Ye X, Lu Q, Jiang X, Wang X, Hu B, Li W, Dong C. Interaction characteristics and mechanism in the fast co-pyrolysis of cellulose and lignin model compounds. J Therm Anal Calorim. 2017;130:975–84. https://doi.org/10.1007/s10973-017-6465-3.
84. Hosoya T, Kawamoto H, Saka S. Solid/liquid- and vapor-phase interactions between cellulose- and lignin-derived pyrolysis products. J Anal Appl Pyrolysis. 2009;85:237–46. https://doi.org/10.1016/j.jaap.2008.11.028.
85. Long Y, Zhou H, Meng A, Li Q, Zhang Y. Interactions among biomass components during co-pyrolysis in (macro)thermogravimetric analyzers. Korean J Chem Eng. 2016;33:2638–43. https://doi.org/10.1007/s11814-016-0102-x.
86. Youssefian S, Rahbar N. Molecular origin of strength and stiffness in bamboo fibrils. Sci Rep-UK. 2015;5:11116. https://doi.org/10.1038/srep11116.
87. Zhao S, Liu M, Zhao L, Zhu L. Influence of interactions among three Biomass components on the pyrolysis behavior. Ind Eng Chem Res. 2018;57:5241–9. https://doi.org/10.1021/acs.iecr.8b00593.
88. Liu Z, Wang L, Zhang Y, Li Y, Li Z, Cai H. Cellulose-lignin and xylan-lignin interactions on the formation of lignin-derived phenols in pyrolysis oil. Bioresources. 2017;12:4958–71. https://doi.org/10.15376/biores.12.3.4958-4971.
89. Wang S, Guo X, Wang K, Luo Z. Influence of the interaction of components on the pyrolysis behavior of biomass. J. Anal. Appl. Pyrolysis. 2011;91:183–9. https://doi.org/10.1016/j.jaap.2011.02.006.
90. Liu Q, Zhong Z, Wang S, Luo Z. Interactions of biomass components during pyrolysis: a TG-FTIR study. J Anal Appl Pyrolysis. 2011;90:213–8. https://doi.org/10.1016/j.jaap.2010.12.009.
91. Mohnen D. Pectin structure and biosynthesis. Curr Opin Plant Biol. 2008;11:266–77. https://doi.org/10.1016/j.pbi.2008.03.006.
92. Vassilev SV, Baxter D, Andersen LK, Vassileva CG. An overview of the composition and application of biomass ash. Part 1. Phase–mineral and chemical composition and classification. Fuel. 2013;105:40–76. https://doi.org/10.1016/j.fuel.2012.09.041.
93. Vassilev SV, Baxter D, Andersen LK, Vassileva CG, Morgan TJ. An overview of the organic and inorganic phase composition of biomass. Fuel. 2012;94:1–33. https://doi.org/10.1016/j.fuel.2011.09.030.
94. Leijenhorst EJ, Wolters W, Van De Beld L, Prins W. Inorganic element transfer from biomass to fast pyrolysis oil: review and experiments. Fuel Process Technol. 2016;149:96–111. https://doi.org/10.1016/j.fuproc.2016.03.026.
95. Liu WJ, Li WW, Jiang H, Yu HQ. Fates of chemical elements in biomass during its pyrolysis. Chem Rev. 2017;117:6367–98. https://doi.org/10.1021/acs.chemrev.6b00647.
96. Wiinikka H, Johansson AC, Sandstrom L, Ohrman OGW. Fate of inorganic elements during fast pyrolysis of biomass in a cyclone reactor. Fuel. 2017;203:537–47. https://doi.org/10.1016/j.fuel.2017.04.129.
97. Lane DJ, van Eyk PJ, Ashman PJ, Kwong CW, de Nys R, Roberts DA, Cole AJ, Lewis DM. Release of Cl, S, P, K, and Na during thermal conversion of algal biomass. Energ Fuel. 2015;29:2542–54. https://doi.org/10.1021/acs.energyfuels.5b00279.
98. Fahmi R, Bridgwater AV, Darvell LI, Jones JM, Yates N, Thain S, Donnison IS. The effect of alkali metals on combustion and pyrolysis of Lolium and Festuca grasses, switchgrass and willow. Fuel. 2007;86:1560–9. https://doi.org/10.1016/j.fuel.2006.11.030.
99. Yann LB, Thierry G, Sébastien L, Mohammed B, Luc D, Nicolas B, Colin S, Patrick C, Dufour A. Effect of potassium on the mechanisms of biomass pyrolysis studied using

complementary analytical techniques. ChemSusChem. 2016;9:863–72. https://doi.org/10.1002/cssc.201501560.
100. Nowakowski DJ, Jones JM. Uncatalysed and potassium-catalysed pyrolysis of the cell-wall constituents of biomass and their model compounds. J Anal Appl Pyrolysis. 2008;83:12–25. https://doi.org/10.1016/j.jaap.2008.05.007.
101. Di Blasi C, Galgano A, Branca C. Influences of the chemical state of alkaline compounds and the nature of alkali metal on wood pyrolysis. Ind Eng Chem Res. 2009;48:3359–69. https://doi.org/10.1021/ie801468y.
102. Patwardhan PR, Satrio JA, Brown RC, Shanks BH. Influence of inorganic salts on the primary pyrolysis products of cellulose. Bioresour Technol. 2010;101:4646–55. https://doi.org/10.1016/j.biortech.2010.01.112.
103. Fang Y, Li J, Chen Y, Lu Q, Yang H, Wang X, Chen H. Experiment and modeling study of glucose pyrolysis: formation of 3-hydroxy-γ-butyrolactone and 3-(2H)-furanone. Energ Fuel. 2018;32(9):9519–29. https://doi.org/10.1021/acs.energyfuels.8b01877.
104. Zhou Z, Chen X, Wang Y, Liu C, Ma H, Zhou C, Qi F, Yang J. Online photoionization mass spectrometric evaluation of catalytic co-pyrolysis of cellulose and polyethylene over HZSM-5. Bioresour Technol. 2019;275:130–7. https://doi.org/10.1016/j.biortech.2018.12.045.
105. Wang Y, Huang Q, Zhou Z, Yang J, Qi F, Pan Y. Online study on the pyrolysis of polypropylene over the HZSM-5 zeolite with photoionization time-of-flight mass spectrometry. Energ Fuel. 2015;29:1090–8. https://doi.org/10.1021/ef502529w.
106. Jia L, Le-Brech Y, Shrestha B, Frowein MB, Ehlert S, Mauviel G, Zimmermann R, Dufour A. Fast pyrolysis in a microfluidized bed reactor: effect of biomass properties and operating conditions on volatiles composition as analyzed by online single photoionization mass spectrometry. Energ Fuel. 2015;29:7364–74. https://doi.org/10.1021/acs.energyfuels.5b01803.
107. Jia LY, Raad M, Hamieh S, Toufaily J, Hamieh T, Bettahar MM, Mauviel G, Tarrighi M, Pinard L, Dufour A. Catalytic fast pyrolysis of biomass: superior selectivity of hierarchical zeolites to aromatics. Green Chem. 2017;19:5442–59. https://doi.org/10.1039/C7GC02309J.
108. Shao S, Zhang H, Xiao R, Li X, Cai Y. Evolution of coke in the catalytic conversion of biomass-derivates by combined in-situ DRIFTS and ex-situ approach: effect of functional structure. Fuel Process Technol. 2018;178:88–97. https://doi.org/10.1016/j.fuproc.2018.05.021.
109. Vazquez Thyssen V, Moreira Assaf E. Ni/CaO-SiO_2 catalysts for assessment in steam reforming reaction of acetol. Fuel. 2019;254:115592. https://doi.org/10.1016/j.fuel.2019.05.175.
110. Liu C, Wang H, Karim AM, Sun J, Wang Y. Catalytic fast pyrolysis of lignocellulosic biomass. Chem Soc Rev. 2014;43:7594–623. https://doi.org/10.1039/C3CS60414D.

[illegible]

Part II
Production of Liquid biofuels by Pyrolysis and Catalytic Pyrolysis

Chapter 5
Catalytic Upgrading of Bio-Oils into Aromatic Hydrocarbon over Highly Active Solid Catalysts

Surachai Karnjanakom, Nichaboon Chaihad, Suwadee Kongparakul, Chanatip Samart, Abuliti Abudula, and Guoqing Guan

Abstract Pyrolysis of lignocellulosic biomass is a promising method to produce bio-oils in high yields. However, the obtained bio-oils via the pyrolysis process always contain large amounts of oxygenated compounds such as phenols, ketones sugars and acids, which could make them unstable, corrosive and have low heating value. Thus, the original bio-oils need to be firstly upgraded before its application as transportation fuel. Herein, upgrading of bio-oils by using several typically highly-active solid catalysts with metal modification are introduced. Especially, the effects of pore size and acidic-basic properties of the solid catalysts on their activity, selectivity, stability and deactivation are critically reviewed. In this chapter, fundamental reaction pathways for the conversion of oxygenated compounds into aromatic hydrocarbons and coke via catalytic pyrolysis are summarized in detail along with an outline of future research needs.

Keywords Biomass · Bio-oil · Solid catalysts · Reaction pathways · Aromatic hydrocarbons

---

S. Karnjanakom
Department of Chemistry, Faculty of Science, Rangsit University, Pathumthani, Thailand
e-mail: surachai.ka@rsu.ac.th

N. Chaihad · A. Abudula
Graduate School of Science and Technology, Hirosaki University, Aomori, Japan

S. Kongparakul · C. Samart
Department of Chemistry, Faculty of Science and Technology, Thammasat University, Pathumthani, Thailand

G. Guan (✉)
Graduate School of Science and Technology, Hirosaki University, Aomori, Japan

Laboratory of Energy Conversion Engineering, Institute of Regional Innovation (IRI), Hirosaki University, Aomori, Japan
e-mail: guan@hirosaki-u.ac.jp

Z. Fang et al. (eds.), *Production of Biofuels and Chemicals with Pyrolysis*, Biofuels and Biorefineries 10, https://doi.org/10.1007/978-981-15-2732-6_5

5.1 Introduction of Biomass Pyrolysis

Nowadays, an intensive outgrowth of industrial technologies together with energy consumption is leading to the depletion of fossil resources and calamity in the environment. Investigation of renewable resources is necessary to replace traditional fossil fuels. Lignocellulosic biomass, which consists primarily of C, H and O atoms could be a future sustainable and renewable source for the production of hydrocarbon fuels and hydrogen directly and finally as a substitute for fossil resources [1–6]. Especially, lignocellulosic biomass is attractive for direct production of highly value-added chemicals. However, processes for biomass conversion into specific products are still very complex. Thermochemical conversion processes, such as pyrolysis, gasification and combustion allow a biomass conversion process to be simplified and offer effective routes for production of bio-fuels and bio-power [7]. In particular, pyrolysis, which is typically performed at temperatures of (400–600) °C under inert atmosphere, can be used to produce liquid biofuels [8, 9]. Since biomass is composed of different components such as cellulose, hemicellulose and lignin, oxygenated compounds will be generated from these components depending on both conditions and reaction mechanisms, which are also a result of the thermal processing condition and reactor design [10, 11]. Interactions of cellulose, hemicelluloses and lignin at the thermal conditions in the biomass structure during the pyrolysis are still a research topic [12]. For example, lignin-derived phenols are easily produced from the interaction between hemicellulose and lignin pyrolysis, whereas the formation of hydrocarbons from hemicellulose is inhibited simultaneously [13]. It is also reported that lignin can interact with cellulose during pyrolysis to prevent polymerization of levoglucosan derived from cellulose, leading to the reduction of char formation. Furthermore, interactions between cellulose and hemicellulose have less effects on the formation and distribution of pyrolysis products [14]. Unfortunately, direct utilization of original bio-oil for the transportation fuels is not presently possible, since it contains many oxygenated compounds such as phenols, ketones sugars and acids, which lower its heating value and make it unsuitable as a stable fuel. Therefore, additional upgrading steps are necessary to improve the quality of the bio-oil even though it will lower the energy efficiency of the entire bio-oil production process.

Reaction pathway of cellulose pyrolysis at different temperature ranges can be via dehydrogenation, depolymerization and fragmentation, which is also called as the Waterloo-mechanism [15]. That is, at a pyrolysis temperature of >250 °C with a slow heating rate, cellulose is converted into char, H_2O, CO, CO_2, etc. When the temperature is increased from (350–450) °C, anhydrosugars such as levoglucosan are released via depolymerization. Other oxygenated compounds such as ketone, aldehydes and acid also generate via fragmentation process at ~600 °C. For actual biomass, a large number of reactions occur in series such as dehydration, depolymerization, isomerization, cracking and charring during biomass pyrolysis [16–18]. The biomass pyrolysis process generally consists of three main steps: (1) initial evaporation of moisture; (2) primary decomposition; and (3) secondary decomposition and reactions via cracking and repolymerization [19].

During the primary decomposition stage at a temperature range of (200–400) °C, the decomposition of biomass results in the generation of gases such as H_2, CO, CO_2 and CH_4, bio-oils (also called tar) and char [20]. In the secondary stage, decomposition of as-generated bio-oil in the first stage and continuous decomposition of the solid matrix occur with further increasing of the reaction temperature. In general, the decomposition of hemicellulose mainly occurs at a pyrolysis temperatures range of (250–350) °C and the decomposition of cellulose takes place in the temperature range of (325–400) °C to form the main products like levoglucosan [21]. In contrast, lignin decomposition requires a higher temperature up to 900 °C [22]. Several factors such as temperature, heating rate, residence time, pressure and catalysts greatly affect the product yield and bio-oil quality [23]. For instance, a rapid heating rate favors formation of condensable liquid products. In the case of high temperature and long residence time, non-condensable gaseous products such as methane and ethane will be produced due to the occurrence of secondary decomposition. Large amounts of char can be obtained at a low temperature with a slow heating rate [24]. Hence, pyrolysis can be classified into two categories: (1) slow pyrolysis, a batch process that is carried out with a slow heating rate and a long residence time [25–27], and (2) fast pyrolysis that is carried out with a rapid heating rate which can inhibit further cracking of the pyrolysis products so that the bio-oil yield is high. Herein, the chemical kinetics and heat and mass transfer rates play crucial roles in determining the product distribution and reaction mechanisms.

The bio-oils are dark brown and free-flowing liquids and have a distinctive smoky smell [28, 29] and their physical properties are determined by the chemical composition of the biomass. Generally, bio-oils contain several hundreds of organic compounds, which can be classified into three major families: (1) small carbonyl compounds such as acetic acid, carboxylic acids, acetaldehyde, hydroxyaldehydes and hydroxyketones; (2) sugar compounds such as furfural, levoglucosan, furan and pyran compounds and (3) lignin-derived compounds such as phenols and guaiacols [30]. In addition, oligomers with molecular weights ranging from 900 to 2500 are found in bio-oil with significant amounts [31, 32]. The distribution of these compounds mostly depends on the biomass type, the pyrolysis way and whether the catalysts are present or not [33, 34]. Some properties of bio-oil, such as instability, high corrosiveness and low heating value, limit its direct use as a transportation fuel. Therefore, it is important to understand the product distribution before considering to upgrade the bio-oil. Table 5.1 compares selected properties of a bio-oil with a heavy petroleum fuel oil [35, 36]. One can see that the bio-oil has much higher oxygen and water contents than the heavy petroleum oil, which is the reason why bio-oil has a lower heating value than the petroleum oil. For example, the heating values (HHV, MJ/kg) of bio-oils derived from pyrolysis of wood are in the range of (16–19) MJ/kg, which are only about 50% of that of the petroleum oil (40 MJ/kg) even in the highest evaluation. The presence of aldehydes and phenols in the bio-oil are unstable, since they are easily transformed into macromolecules via polymerization reactions, especially under acidic conditions, resulting in an increase in bio-oil viscosity and

Table 5.1 Comparison of bio-oil properties derived from the pyrolysis of wood and heavy petroleum fuel oil

Properties	Bio-oils derived from pyrolysis of wood [35]	Heavy petroleum fuel oil [36]
Water content (wt%)	15–30	0.1
pH	2.5	–
Carbon (wt%)	54–58	85
Hydrogen (wt%)	5.7–7.0	11
Oxygen (wt%)	35–40	1.0
Nitrogen (wt%)	0–0.2	0.3
HHV (MJ/kg)	16–19	40
Viscosity (at 50 °C) (mPa s)	40–100	180
Distillation residue (wt%)	Up to 50	1

a reduction in bio-oil fluidity. In summary, bio-oils produced via pyrolysis always have high water content and contain many oxygenated compounds so that bio-oils have high acid value, high viscosity, are thermal instability, and low heating values [37]. However, although the application of bio-oils has been thus far limited by unfavorable properties, the bio-oil has promising properties such as relatively low toxicity, good lubricity and greater biodegradation than petroleum fuels. Therefore, it is important to upgrade bio-oils so that they have more advantages than fossil fuels as an energy resource.

In this chapter, recent progress in upgrading of bio-oils derived from pyrolysis of biomass using several active solid catalysts such as metal/zeolite, metal/Al_2O_3, metal/MCM-41 and metal/KIT-6 is reviewed. The effect of catalyst type, porosity, acidity-basicity and their interactions on catalytic activity, selectivity, stability and deactivation are summarized. Possible ways for increasing yields of value-added products such as benzene, toluene and xylenes (BTXs) and suppressing the formation of unwanted by-products such as polyaromatics and coke on the catalysts are discussed. Suitable spent catalyst regeneration methods are analyzed for increasing the reusability of catalysts.

5.2 Upgrading of Bio-Oil Using Solid Catalysts

Upgrading of bio-oils to aromatic hydrocarbons can be realized via thermal catalytic cracking, catalytic hydrodeoxygenation, hot vapor filtration or stabilization by adding suitable solvents. Herein, catalytic hydrodeoxygenation involves hydrotreating and catalytic vapor cracking, by which the oxygen in the bio-oil can be removed in the form of simple molecules such as H_2O, CO and CO_2 [38]. Figure 5.1 shows reactions associated with catalytic hydrodeoxygenation in bio-oil upgrading. Even though this method is one of the most promising bio-oil upgrading routes, it requires high-pressure with a large stoichiometric excess of hydrogen gas,

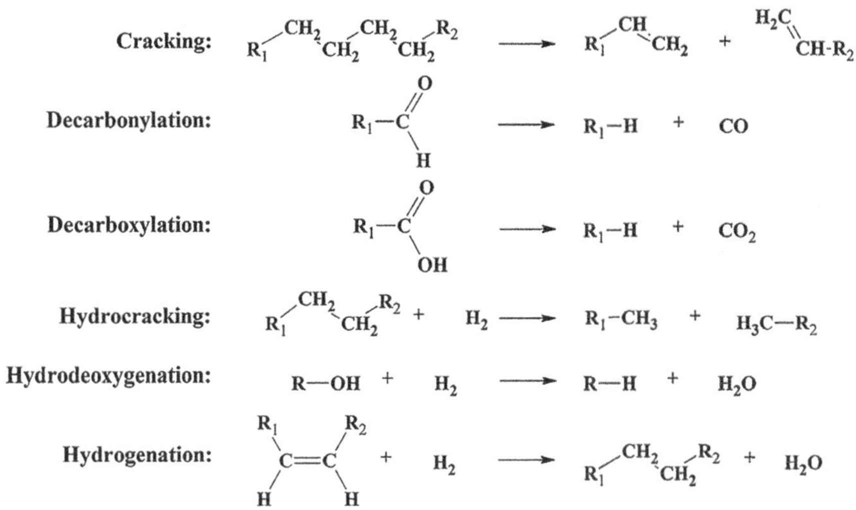

Fig. 5.1 Reactions for hydro-deoxygenation in bio-oil upgrading (adapted with permission from [38], Copyright© 2019, Elsevier)

which significantly increases the cost of upgrading [39]. Hence, it is necessary to minimize the amount of hydrogen required for deoxygenation. However, it is still a challenge to find upgrading catalysts that can work under low-pressure and at near stoichiometric H_2 conditions. Herein, fundamental understanding of the factors that favor C–O bond cleavage (and C–C bond formation) is the key to develop novel solid catalysts for bio-oil upgrading. In this section, discussion on the role of solid catalysts in deoxygenation are presented at first and then, various aspects of the developed solid acid catalysts will be summarized and discussed.

5.2.1 *Reaction Mechanisms for Bio-Oil Upgrading*

To improve the quality of bio-oils, one way is loading the solid catalysts in the downstream reactor of biomass pyrolysis unit and separately allowing the produced bio-oil to be upgraded [40]. Another way is in-situ upgrading by integrated catalytic pyrolysis, in which the catalyst is added into the pyrolysis system, so that the generated bio-oils undergo further cracking inside catalyst structure or on active sites to form monoaromatic hydrocarbons (MAHs). It should be noted that the yield of bio-oils will be reduced after the upgrading process due to co-generation of H_2O, CO and CO_2. Herein, the selectivity is also an attractive feature of catalytic pyrolysis, which is dependent on the catalyst pore size distribution and nature of the active sites [41]. Detailed reaction pathways for in-situ upgrading of bio-oils derived from the pyrolysis of biomass are shown in Fig. 5.2 [42]. One can see that the composition of bio-oils derived from the biomass pyrolysis without upgrading are very complex.

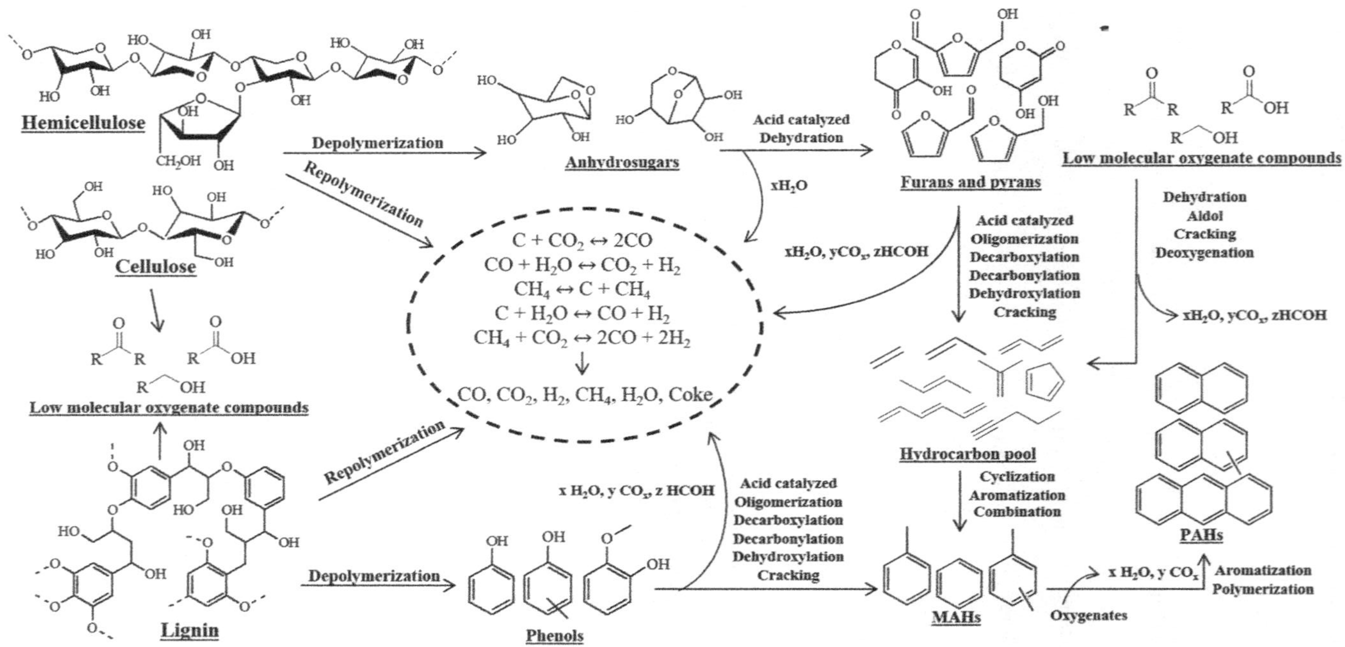

Fig. 5.2 Possible reaction mechanisms leading to formation of aromatic hydrocarbons obtained from catalytic upgrading of bio-oils derived from pyrolysis of biomass (adapted with permission from [42], Copyright© 2019 Elsevier)

Thus, selection of proper catalysts with high activity as well as selectivity is very important in the final product distribution of the bio-oil.

As shown in Fig. 5.2, for cellulose and hemicellulose, anhydrosugars are firstly formed via thermal decomposition of polymer chains that contain glycosidic bonds. During this stage, small gas-state oxygenated by-products such as CO_2, CO and H_2O are produced via cracking and depolymerization. Then, sugar compounds further undergo dehydration and re-arrangement reactions to form furans and small oxygenated compounds such as furans and pyrans [43]. Thereafter, in the presence of an acid catalyst, these intermediate oxygenated compounds are converted into aromatic hydrocarbons via aromatization. In the case of lignin, large molecules containing phenol groups are initially produced via cracking and depolymerization [19]. Subsequently, a larger amount of char is formed from lignin pyrolysis via polymerization compared with cellulose-hemicellulose pyrolysis. During catalytic upgrading, oxygenated compounds can contact active sites of the solid acid catalysts. In this case, higher acidity is needed for the cleavage of C–O and C–C bonds before the deoxygenation reactions proceed to obtain smaller hydrocarbon molecules such as BTXs. However, further aromatization and polymerization of BTXs to polycyclic aromatic hydrocarbons (PAHs) and coke can occur on catalysts with too high acidity. Therefore, it is necessary to adjust the acidity or acidic amount on the catalyst to increase the aromatic selectivity and reduce coke formation.

Ausavasukhi et al. reported a possible reaction route for catalytic deoxygenation of benzaldehyde promoted by Ga-modified ZSM-5 zeolite catalyst, in which the direct deoxygenation over Brønsted acid sites could yield benzene, but toluene could only be produced over Ga species in the presence of H_2 [44]. Experimental results also verified that no toluene was formed without the presence of hydrogen gas. Also, the ratio of benzene to toluene products was significantly increased by co-feeding water in the presence of Ga-modified HZSM-5 catalyst. The deoxygenation mechanism of acetone catalyzed by molybdenum oxide has been proposed as illustrated in Fig. 5.3 [45]. It is found that this catalyst promotes hydrogenolysis via direct scission of the C–O bond. The concentration of oxygen vacant sites is the main factor that influences catalytic performance and the strength of metal–oxygen bond within the metal oxide also affects the deoxygenation performance. Metal species with strong metal–oxygen bonds may strongly bind to oxygen on the catalyst surface, inhibiting the formation of oxygen vacant sites and promoting hydrocarbon yields. Gonzalez-Borja et al. [46] studied hydrodeoxygenation of anisole and guaiacol using monolithic Pt-Sn catalysts, and proposed transalkylation/deoxygenation pathways of anisole and guaiacol over Pt-Sn/CNF/Inconel catalysts [46]. It is found that such monolithic catalysts can provide low pressure drop even at a high reactant flow rate and as such, they are beneficial for upgrading of bio-oils at atmospheric conditions. As a result, when Pt-Sn alloy monolith catalysts are used for guaiacol and anisole deoxygenation, full deoxygenations of guaiacol and anisole to the main products of benzene, toluene and xylenes (BTXs) occur over either Pt-Sn/Inconel or Pt-Sn/CNF/Inconel catalysts. However, catalyst stability needs to be improved, since coke easily formed on the catalysts.

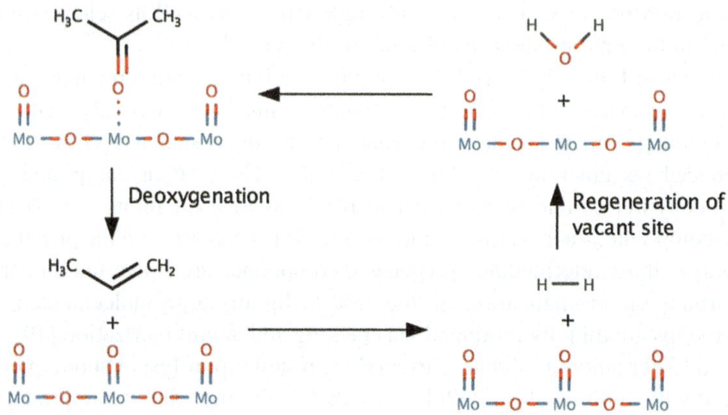

Fig. 5.3 Acetone deoxygenation over molybdenum oxide [45] (adapted with permission from [45], Copyright©2019 Royal Society of Chemistry)

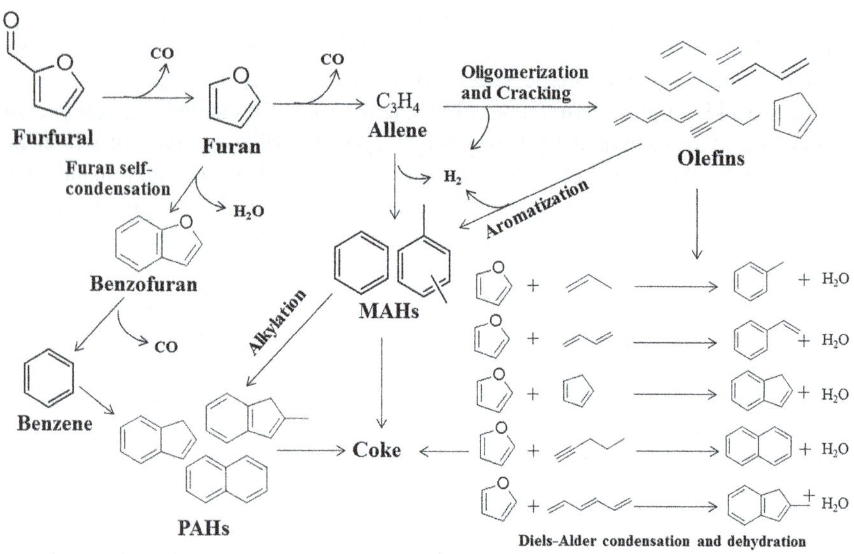

Fig. 5.4 Reaction pathways for selective conversion of furfural into aromatic hydrocarbons [47]

In our previous study [47], a possible mechanism for the catalytic conversion of model bio-oil of furfural over Mg-Cu-doped β-zeolite was proposed as shown in Fig. 5.4. Herein, the acidity of β-zeolite was tuned by various amounts of Mg and Cu species. During the reaction, furan was initially produced by decarbonylation of furfural. Since certain amounts of olefins were formed, it is considered that allene should be formed by decarbonylation of furan at first and then, the produced allene could be trapped on the acid sites of zeolite and converted into olefins. Benzofuran was also detected in the products especially at a high weight hourly space velocity (WHSV), indicating that benzofuran, which was formed by the dehydrative condensation of furan, was a possible intermediate in the formation of monocyclic aromatic hydrocarbons (MAHs). Then, these intermediates could be further transformed into aromatic, olefins and coke through decarbonylation, oligomerization, cracking, dehydrogenation, dehydration, Diels-Alder condensation, alkylation and aromatization reactions. For the Mg-Cu-doped β-zeolite catalyst, it provided Lewis acid sites or electron pair acceptors, on which hydride ions were released to promote the rate of olefin oligomerization, leading to the formation of aromatic hydrocarbons.

However, in that study [47], when the Cu loading amount was too high, higher acidity as well as bulk CuO were formed on β-zeolite, which promoted secondary reactions of aromatization and polymerization of BTXs to PAHs and coke. Therefore, the optimum acid-base properties of catalyst and interaction of metal species with zeolite support must be considered for preventing the formation of bulk copper species, which was the key to control the final aromatic product distribution. Moreover, simultaneous introduction of basic sites (MgO) on the catalyst resulting in a significant reduction of coke yield together with lower PAHs yield and higher BTXs yields. Herein, the basic MgO tuned the strong Lewis acid sites derived from extra framework Al species of β-zeolite and ionic $Cu(Cu^{2+})/CuO/Cu_2O$ composite to avoid polyaromatization and coking. Therefore, acidity of the solid catalysts for cracking, deoxygenation and aromatization should be the key to control the catalyst performance in the catalytic upgrading of real bio-oils. Figure 5.5 shows the acid properties of the parent and Cu- and Mg-loaded β-zeolite [47]. As observed in Fig. 5.5, for the parent β-zeolite, two NH_3 desorption peaks at $\leq$280 °C and $\geq$ 280 °C were attributed to weak acid sites and strong acid sites, respectively. With an increase in Cu loading, the NH_3 desorption peaks shifted to a lower temperature range, which should be the contribution of copper species, since it can be exchanged with the protons of Brønsted acid sites, resulting in an increase in Lewis acid sites. As a result, deoxygenation reactions such as dehydration, decarbonylation, decarboxylation and aromatization can be promoted, resulting in the generation of more H_2O, CO, CO_2 and aromatic hydrocarbons. The three NH_3 desorption peaks corresponding to weak, medium and strong acid sites appeared in the cases of 3 wt % and 5 wt% of Cu loading amounts. These different distributions of acid sites can be attributed to the existence of different Cu species (e.g., bulk CuO, Cu^+ and Cu^{2+}) located on the β-zeolite structure. In addition, co-loading of Mg cations with optimum loading amount at extra-framework positions was confirmed to perform as Lewis acid sites for promoting different deoxygenation reactions.

Phenolic compounds in a simulated bio-oil were unable to be converted over parent H-Beta zeolite catalyst under a hydrogen atmosphere as reported by Shafaghat et al. [43]. It is found that metal doping can effectively help to initiate conversion of bio-oil to rich hydrocarbon fuels, which could be dehydrated over the acid sites on the metal-doped zeolites to achieve oxygenated compound removal. To date, various metals have been applied to modify zeolite catalysts. Table 5.2 summarizes different metal loaded zeolite catalysts with their main reaction pathways and products. Herein, the real bio-oil can be converted into aromatics via

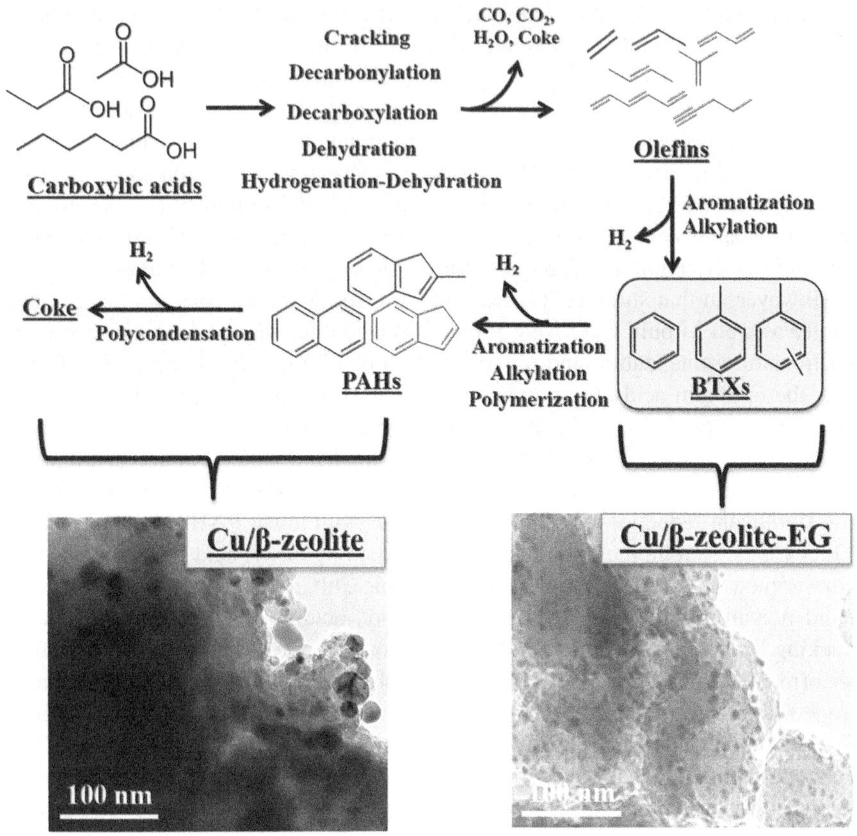

Fig. 5.5 Selective production of aromatic hydrocarbon from Cu/β-zeolite (adapted with permission from [57], Copyright© 2019 Elsevier)

hydrogenolysis using the Fe/H-Beta zeolite whereas the Ni/H-Beta zeolite catalyst shows outstanding performance for hydrogenation of benzene ring via hydrogenation-dehydration-hydrogenation to form cycloalkanes. Veses et al. [48] carefully investigated the upgrading of bio-oils over ZSM-5 doped by metals such as Mg, Ni, Cu, Ga and Sn. As expected, a significant decrease of the viscosity and oxygen content of bio-oil was found with these catalysts. They also proposed several reaction pathways for different metal modified zeolite catalysts. A large amount of phenols could be transformed into aromatic hydrocarbons via decarbonylation and oligomerization using the Ga-ZSM-5 [49] with the highest water yield obtained being with this catalyst, demonstrating high efficiency of oxygenated compound removal via hydrogenation-dehydration. For Ni-ZSM-5, its application greatly improved the aromatics formation via aromatization. The facile decarboxylation can be attributed to in-situ H_2 formation for hydrodeoxygenation reactions. For the Sn-ZSM-5, a high yield of aromatic hydrocarbons was observed together with a high water content. The decarboxylation should be the main reaction pathway since

Table 5.2 The effect of metal doping on zeolites for upgrading of bio-oil

Metal	Zeolite	Feedstock	Dominant reactions	Major products	Ref.
Fe	H-Beta	Phenolic compound	Aromatization	Aromatics	[43]
Ni			Hydrogenation	Cycloalkanes	
Ga	H-ZSM-5	Pyrolysis bio-oil	Decarbonylation, Oligomerization, Dehydration	Aromatics	[48]
Ni					
Sn					
Cu					
Mg					
Ga	H-ZSM-5	Lignocellulosic	Deoxygenation	Olefins, Aromatic	[49]
Fe	H-ZSM-5	Biomass derived bio-oil	Fast pyrolysis	Phenols	[50]
Co					
Zr				Hydrocarbons	
Zn	H-ZSM-5	Pyrolysis bio-oil; CH_4	Deoxygenation	Aromatics	[51]
Ni	H-ZSM-5	Biomass derived bio-oil	Deoxygenation	Aromatics	[52]
Sn	Hierarchical zeolite	Biomass derived bio-oil	Deoxygenation	Hydrocarbons	[53]
Cu					
Ni					
Mg					

higher CO_2 than CO yields accompanied is found [42]. In the case of Cu-ZSM-5, decarbonylation was likely promoted by copper sites rather than by acidic sites of zeolite, suggesting that most of metal cations were located on the ion exchange positions of zeolite. Mg-ZSM-5 showed remarkable difference with other metal-loaded ZSM-5 zeolites. Especially, Mg-ZSM-5 exhibited the lowest coke and polyaromatic formations among all the catalysts tested [48], suggesting the existence of a basicity property. These trends show that metal modification of a zeolite leads to different reaction routes. Li et al. [50] considered that the introduction of transition metals on the zeolite could favor the decarbonylation/decarboxylation rather than dehydration, leading to increased amounts of CO/CO_2 and less water. He et al. [51] studied in-situ catalytic upgrading of pyrolysis oil with methane over metal-loaded ZSM-5 catalyst. Compared with other transition metals (i.e., Fe, Co, Cu, Ni, Mn, Zr and Ce) loaded zeolite catalysts, the Zn/ZSM-5 gave the highest oil yield along with a high H/C atomic ratio and a low O/C atomic ratio. It is found that ZSM-5 framework mainly promoted the deoxygenation during the reaction so that the quality of bio-oil was increased. Highly dispersed Zn species mostly facilitated CH_4 activation and allowed it to be incorporated into the carbon chain of the bio-oil, which enhanced the quantity of bio-oil. As for hierarchical zeolites, the doping of metal showed great effects on the overall performance of the catalysts. Yung et al. [52] studied the upgrading of bio-oil over Ni/ZSM-5 and found that the addition of Ni on ZSM-5 greatly improved the oxygenate conversion and aromatic hydrocarbons yields. The Ni addition retained the access of bio-oil molecules to the zeolite pores and provided micropores on the external surface of ZSM-5, resulting in more specific interactions between metal centers and acidic sites and finally improved the performance in upgrading of bio-oil. Veses et al. [53] found that the impregnation of hierarchical zeolite with metals such as Sn, Cu, Ni and Mg improved the deoxygenation depth. Particularly, the use of Mg-loaded hierarchical zeolite resulted in the best fuel quality with low oxygen content and low acidity.

In one of our previous studies [54], it was found that the interaction between the metal sites and the proton sites are important when the metal-doped β-zeolite catalysts are used for bio-oil upgrading. In the case of moderate Cu loading amount, Cu species replaces some of the proton sites, leading to interactions between metal sites (isolated Cu species) so that the remaining proton sites (H^+ cation) can shift the prior route from dehydration to decarbonylation. Namely, the remaining H^+ cation promotes dehydration as well as decarboxylation of oxygenated compounds, while Cu species facilitates decarbonylation reactions [48]. Pham et al. [55] used activated carbon (AC) supported Ni_2P catalysts for atmospheric deoxygenation of waste cooking oil (WCO), and found that the proportion of oxygen atoms in the WCO sample was reduced by >85%, resulting from synergistic effects between Ni and P elements. Kaewpengkrow et al. [56] used Ce, Pd, Ru or Ni salts impregnated AC for upgrading of bio-oil derived from the pyrolysis of Jatropha waste residue. The main aromatic compounds formed were benzene, toluene, ethylbenzene and xylene (BTEX) with the highest aromatic selectivity obtained in the order of Ce/AC > Ni/AC > Pd/AC > Ru/AC. Karnjanakom et al. [57] improved the Cu dispersion on β-zeolite with ethylene glycol. It is found that smaller Cu particles without sintering can be obtained by using their method [57]. Herein, the Brønsted acid sites of the parent β-zeolite can be ion exchanged with Cu species, thus resulting in an increase of Lewis acid sites, which can promote the formation of aromatics. It should be noted that small sizes of Cu promotes BTX formation while polyaromatic hydrocarbon (PAHs) are more easily formed in the presence of larger sizes of Cu (Fig. 5.5).

As such, it can be assumed that the deoxygenation pathway can be shifted from decarboxylation (Eq. (5.1)) to decarbonylation and cracking reactions (Eqs. (5.2) and (5.3)) since a higher CO yield is observed in the case of Cu/β-zeolite-EG, which was probably results from the existence of smaller Cu particles/clusters. Thus, different Cu sites can be generated in β-zeolites after using different preparation methods for zeolite-supported Cu catalysts [57].

$$\text{Decarboxylation}: CH_3(CH_2)_4COOH \rightarrow CO_2 + C_5H_{12} \tag{5.1}$$

$$\text{Decarbonylation}: CH_3(CH_2)_4COOH \rightarrow CO + C_5H_{10} + H_2O \tag{5.2}$$

$$\text{Cracking}: CH_3(CH_2)_4COOH \rightarrow 2CO + C_4H_8 + 2H_2 \tag{5.3}$$

$$\text{Coke gasification}: C + H_2O \rightarrow CO + H_2 \tag{5.4}$$

Coke yield was clearly decreased with an increase of Cu loading in the case of β-zeolite-EG, which was opposed to the case of Cu/β-zeolite. This result is attributed to the shifting of the deoxygenation pathways, by which water was formed together with CO via decarbonylation/dehydration steps, thus promoting in-situ coke gasification, as indicated in the reaction described by Eq. (5.4). The in-situ H_2 produced enhances the hydrocracking and hydrodeoxygenation in the presence of a catalyst, and leads to an increase in alkanes, and finally to the formation of more BTX. Yung et al. [51] studied the upgrading of bio-oil over Ni/ZSM-5 and found that addition of Ni to ZSM-5 greatly improves oxygenate conversion and aromatic hydrocarbon yields. The Ni addition could retain the access of bio-oil molecules into the zeolite pores and provide micropores on the external surface of ZSM-5, resulting in more specific interactions between metal centers and acidic sites and finally improve the performance in upgrading of bio-oil.

The porous structure in solid catalysts always affects the activity and selectivity in upgrading of bio-oil. Biomass-derived feedstocks, oxygenates, and hydrocarbons with kinetic diameters, can enter into zeolite micropores, as reported in the literature [58]. Primary pyrolysis products with kinetic diameters larger than or equal to that of glucose should not diffuse into the pores of microporous materials. Thus, the pyrolysis of biomass and its primary large products/intermediates is hardly affected by the presence of microporous structures in the catalysts. Chaihad et al. [59] found that naphthalene was favored to be produced with H-ZSM-5 while more monocyclic aromatic hydrocarbons such as *p*-xylene, toluene and alkylbenzene were generated with H-beta as well as H-USY zeolites. Generally, mesoporous materials with an adjustable uniform pore size (2–15 nm) allow the interaction of large organic molecules with active sites, which could be beneficial especially for in situ catalytic upgrading of bio-oil derived from pyrolysis of biomass. Very large pores of mesoporous materials could affect the decomposition of large lignocellulosic macromolecules. For example, when mesoporous rod-like alumina was prepared by Pluronic P123 surfactant-assisted hydrothermal synthesis with different P123/Al molar ratios from 0.005 to 0.05 [60], the pore sizes of the obtained Al_2O_3 increased with an increase in P123/Al molar ratios from 0.005 to 0.05 while the surface area continuously increased until 0.01, leading to the formation of more PAHs and the reduction of MAHs to some extent. Aluminosilicate mesoporous MCM-41 are found to be good catalyst supports for catalytic cracking and deoxygenation of bio-oils [61–63]. Karnjanakom et al. [62] used Cu/MCM-41 and Cu/KIT-6 for the upgrading of bio-oil, and found that large molecules of oxygenated compounds can more easily diffuse inside the structure of KIT-6 without blockage than MCM-41, leading to the facile formation of aromatic hydrocarbons. Similar catalytic performance is observed for Cu supported KIT-6 with high loading (Fig. 5.6).

The acidity of a catalyst generated by the incorporation of metals is usually associated with the nature of the chemical linkages produced by the metal ions in the silica framework. However, larger pore sizes can lead to higher coke deposition and polyaromatics [58, 64]. For instance, Al-MCM-41 led to a higher coke deposition than ZSM-5 and USY zeolites due to its larger pore volume [65]. Different effects of pore sizes of the catalysts are observed in these experiments on the

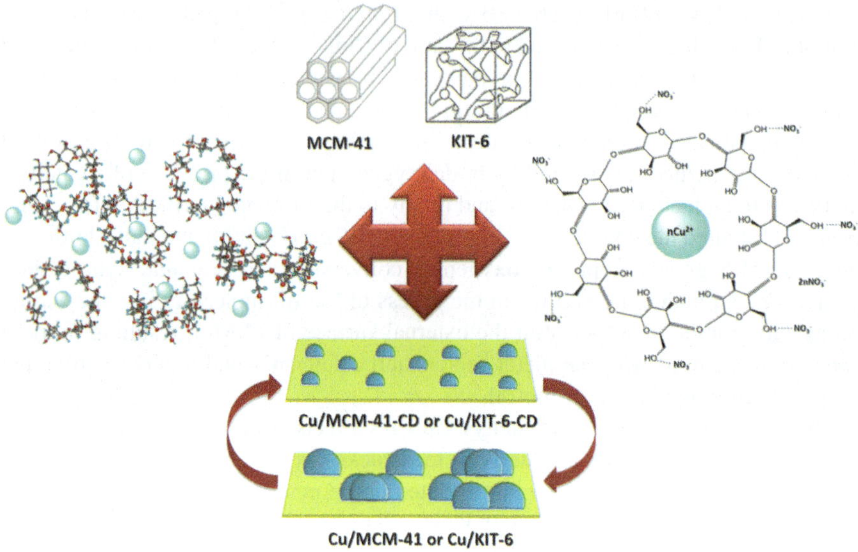

Fig. 5.6 Catalyst preparation of Cu/KIT-6 or MCM-41 by cyclodextrin (CD)-assisted impregnation method [62] (adapted with permission from [62], Copyright© 2019American Chemical Society)

decomposition of intermediate compounds formed from pyrolysis of feedstocks. It has been reported that less coke and more aromatics are obtained with ZSM-5 based catalysts since their smaller pore sizes can prevent the formation of polyaromatic compounds that can act as coke precursors [66]. Therefore, it is necessary to consider pore selectivity as well as activity during the development of solid catalysts for the upgrading of bio-oil.

5.2.2 Deactivation and Regeneration of Catalyst

There are many reasons for catalyst deactivation during bio-oil upgrading. Primarily, deactivation is associated with coke formation due to the series of reactions. Coke deposition on catalysts typically results in the catalyst deactivation and negative consequences of reducing selectivity towards the desired products. Moreover, hydrogen-rich coking species can attack the main active sites on the catalyst surface with a net consequence of catalyst deactivation. In addition, strong adsorption of phenolic compounds on the catalyst may be another reason for deactivation.

Coke deposition leads to the blockage of both active sites and micropores, which is one of the major causes of catalyst deactivation [67–69]. Carbonaceous materials or coke that deposit inside the zeolite pores hinder mass diffusion. Huang et al. [70] considered that the acid sites on solid catalysts played a significant role in the

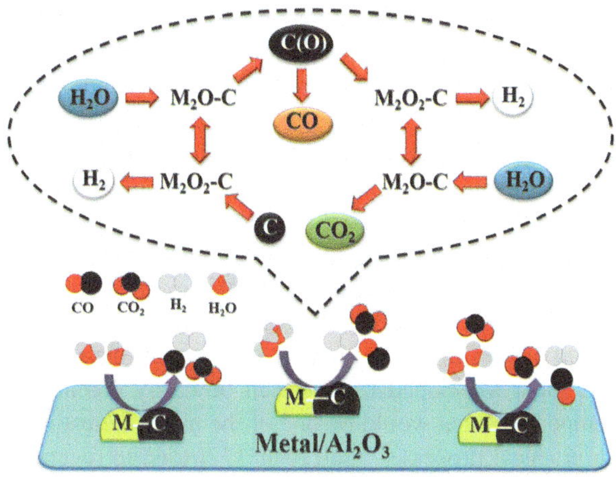

Fig. 5.7 Overall reaction for carbon elimination from metal/Al_2O_3 by AAEM-catalyzed pyrolysis of biomass [60] (adapted with permission from [60], Copyright© 2019, Royal Society of Chemistry)

formation of carbon. Proton donation from the acid sites could result in adsorption of hydrocarbon cations, and the deprotonated basic framework of the zeolite usually exhibits facilitated potential for cracking, aromatization and polymerization, leading to carbon formation. Vitolo et al. [71] regenerated spent HZSM-5 catalyst by washing it with acetone following by calcination at 500 °C for 12 h and found that the regenerated catalyst had a shorter catalyst lifetime and lower deoxygenation ability when compared with the fresh one, and this effect becomes more pronounced as a function of the regeneration cycle. Thus, persistent deactivation was evaluated as a decrease in the availability of acid sites, which decreased by 62% over 5 regeneration cycles. It can be considered that high water content in the bio-oil results in the inevitable dealumination of zeolite during cracking of bio-oil, so that the structure of zeolite itself is destroyed. Karnjanakom et al. [42] found that AAEM species in bio-oil could promote some deoxgenation reactions to remove oxygen from the bio-oil (Fig. 5.7).

Moreover, the AAEM species could assist elimination of coke on catalyst as follows [73]:

$$K_2CO_3 + 2C \rightarrow 2\,K + 3CO \tag{5.5}$$

$$K_2O - C + H_2O \rightarrow K_2O_2 - C + H_2 \tag{5.6}$$

$$\alpha K_2O_2 - C + \alpha C \rightarrow \alpha K_2O - C + \alpha CO \tag{5.7}$$

$$(1-\alpha)K_2O_2 - C + (1-\alpha)CO \rightarrow (1-\alpha)K_2O - C + (1-\alpha)CO_2 \tag{5.8}$$

$$\text{Overall reaction}: \alpha C + H_2O \rightarrow H_2 + (1-\alpha)CO_2 + (2\alpha - 1)CO \tag{5.9}$$

Here, complicated potassium intermediates such as K_2O-C and K_2O_2-C can also dominate the catalytic process. Furthermore, the two redox cycles of K_2CO_3 can promote carbon elimination, which may also be the reason why these catalysts exhibit good reusability even without regeneration. Herein, it should be noted that different metal loadings can greatly affect the amount of coke deposited on a catalyst. Moreover, it is considered that the changing of coke yield and formed coke type might be also related with the existence of metal species in the vicinity of support, which could have a synergetic effect between H^+ sites and metal/metal oxide sites in the catalyst structure. Recently, it is found that thermal coke decomposition temperature range decreased from (350–750) °C to (250–650) °C when Mg species were doped onto Al-MCM-41 for the upgrading of bio-oil, suggesting that Mg favors the delay of hard coke formation. Herein, the changing of coke species formation might be related with the existence of Mg species in the vicinity of support, which could have a synergetic effect between proton sites with metal/ metal oxide sites in the support structure for the formation of coke [58]. As a result, the spent Mg-doped catalyst can be easily regenerated by calcination at a lower temperature that would be one of the most interesting results for further application in practical processes. Guo et al. [72] tried to regenerate spent HZSM-5 at 600 °C over 12 h, but found that the regenerated catalyst resulted in an increased oxygen content in the upgraded oil as a function of the regeneration cycle when compared with the fresh one. That is, the fresh catalyst produced bio-oil with 21% oxygen, but after 5 regeneration cycles, the oxygen content increased to 30% due to a decrease in the amount of exposed active sites on the catalyst. Therefore, a suitable catalyst with high activity, selectivity and stability as well as high anti-coking ability is still needed to be developed.

5.3 Conclusions and Future Outlook

Pyrolysis of lignocellulosic biomass has potential to provide environmentally friendly bio-fuels and chemicals. However, the presence of oxygen compounds in a high amount in bio-oils is the main issue for the application as a transport fuel. In this chapter, various developed solid catalysts with relatively high activity, high selectivity and long-term stability for the conversions of the oxygenated compounds in the bio-oil to hydrocarbons were reviewed. Deoxygenation of O-containing compounds to valuable aromatic hydrocarbons has a strong dependence on the metal as well as the support properties, the preparation methods and the reaction conditions. Various metals such as Cu, Mg, Ni and Ga loaded catalysts exhibited good ability for upgrading of bio-oils. However, different metal loadings result in different catalyst deactivation and regeneration properties, and the catalyst acidity, acid site and pore size have a great effect on the formation of MAHs and PAHs.

Even though development and modification of solid catalysts for the bio-oil upgrading have made great progress, further studies are still necessary. Especially, novel catalysts with improved activity, selectivity and stability or regenerability under the upgrading conditions should be developed. Some future perspectives for catalytic upgrading of bio-oil can be summarized as follows:

- Since many kinds of complex aromatics always contain in the bio-oil and the significant deactivation of catalysts occurs due to the coke formation during the reaction, the selection of catalysts with high selectivity and long-term stability or regenerability are required for the future study. Herein, it is proposed to develop acid-base bifunctional porous catalysts and to study them more deeply. For example, using the catalyst co-loaded with transition metals and alkali metals can improve the activity and selectivity for the selective production of value-added chemicals such as BTXs together with anti-polycyclic-aromatic-hydrocarbon formation as well as anti-coking.
- Due to the highly complex nature of bio-oil usually derived from pyrolysis of real biomass, understanding the reaction pathways for conversion of each compound is highly desirable for catalyst and process screenings. Further studies on the deoxygenation mechanisms of model compounds such as acids, ketones and phenols could simplify the problem. Moreover, co-feeding of model compounds with different ratios can be considered to define the expected products and will allow one to understand the deoxygenation mechanisms in more details.
- Even though the catalytic upgrading process can be achieved via deoxygenation using solid acid catalysts, the upgraded bio-oil still contains large amounts of water. Therefore, it is suggested to solve this problem by using some new ways, for example, co-feeding of N_2 with some gases such as CH_4 to eliminate water and use the CO and H_2 via methanation reaction.
- Carbon based materials derived from char after biomass pyrolysis can be considered and applied as catalyst or catalyst support for upgrading of bio-oil. It should be noted that by using a carbon-based catalyst, catalyst deactivation might be avoided due to coke formation because of the similar properties of carbon. However, to date, the activity of carbon-based catalysts is still low when compared with commercial zeolites. Therefore, it is a challenge to improve the performance of bio-char based catalysts in the future.
- To date, effective catalysts being reported are very expensive or are too complicated to prepare that limits large-scale applications in industrial processes. Thus, abundant resources, such as calcite, dolomite, or other common materials must be further considered to modify and apply as low-cost catalysts that have sufficient performance.
- Consideration of economic, ex-situ catalytic upgrading of bio-oils using non-edible oils and waste cooking oils as feedstocks should be considered for the future.

Acknowledgments This study is supported by Hirosaki University Fund, Japan. S. Karnjanakom and N. Chaihad greatly acknowledge the Ministry of Education, Culture, Sports, Science, and Technology (MEXT) of Japan for the scholarship.

References

1. Nzihou A, Stanmore B, Lyczko N, Minh DP. The catalytic effect of inherent and adsorbed metals on the fast/flash pyrolysis of biomass: a review. Energy. 2019;170:326–37. https://doi.org/10.1016/j.energy.2018.12.174.
2. Wu KJ, Wu YL, Chen Y, Chen H, Wang JL, Yang MD. Heterogeneous catalytic conversion of biobased chemicals into liquid fuels in the aqueous phase. ChemSusChem. 2016;9:1355–85. https://doi.org/10.1002/cssc.201600013.
3. Papari S, Hawboldt K. A review on condensing system for biomass pyrolysis process. Fuel Process Technol. 2018;180:1–13. https://doi.org/10.1016/j.fuproc.2018.08.001.
4. Azadi P, Inderwildi OR, Farnood R, King DA. Liquid fuels, hydrogen and chemicals from lignin: a critical review. Renew Sust Energ Rev. 2013;21:506–23. https://doi.org/10.1016/j.rser.2012.12.022.
5. Saraeian A, Nolte MW, Shanks BH. Deoxygenation of biomass pyrolysis vapors: improving clarity on the fate of carbon. Renew Sust Energ Rev. 2019;104:262–80. https://doi.org/10.1016/j.rser.2019.01.037.
6. Yaman S. Pyrolysis of biomass to produce fuels and chemical feedstocks. Energy Convers Manag. 2004;45:651–71. https://doi.org/10.1016/S0196-8904(03)00177-8.
7. Rahman MM, Liu R, Cai J. Catalytic fast pyrolysis of biomass over zeolites for high quality bio-oil – a review. Fuel Process Technol. 2018;180:32–46. https://doi.org/10.1016/j.fuproc.2018.08.002.
8. Anis S, Zainal ZA. Tar reduction in biomass producer gas via mechanical, catalytic and thermal methods: a review. Renew Sust Energ Rev. 2011;15:2355–77. https://doi.org/10.1016/j.rser.2011.02.018.
9. Che Q, Yang M, Wang X, Chen X, Chen W, Yang Q, Yang H, Chen H. Aromatics production with metal oxides and ZSM-5 as catalysts in catalytic pyrolysis of wood sawdust. Fuel Process Technol. 2019;188:146–52. https://doi.org/10.1016/j.fuproc.2019.02.016.
10. Yang H, Yan R, Chen H, Lee DH, Zheng C. Characteristics of hemicellulose, cellulose and lignin pyrolysis. Fuel. 2007;86:1781–8. https://doi.org/10.1016/j.fuel.2006.12.013.
11. Ansari KB, Arora JS, Chew JW, Dauenhauer PJ, Mushri SH. Fast pyrolysis of cellulose, hemicellulose, and lignin: effect of operating temperature on bio-oil yield and composition and insights into the intrinsic pyrolysis chemistry. Ind Eng Chem Res. 2019;58:15838–52. https://doi.org/10.1021/acs.iecr.9b00920.
12. Caballero JA, Conesa JA, Font R, Marcilla A. Pyrolysis kinetics of almond shells and olive stones considering their organic fractions. J Anal Appl Pyrolysis. 1997;42:159–75. https://doi.org/10.1016/S0165-2370(97)00015-6.
13. Wang SR, Guo XJ, Wang KG, Luo ZY. Influence of the interaction of components on the pyrolysis behavior of biomass. J Anal Appl Pyrolysis. 2011;91:183–9. https://doi.org/10.1016/j.jaap.2011.02.006.
14. Hosoya T, Kawamoto H, Saka S. Cellulose-hemicellulose and cellulose-lignin interactions in wood pyrolysis at gasification temperature. J Anal Appl Pyrolysis. 2007;80:118–25. https://doi.org/10.1016/j.jaap.2007.01.006.
15. Boukis I. Fast pyrolysis of biomass in a circulating fluidized bed reactor. Ph.D. Thesis, University of Aston, Birmingham; 1997.
16. Vamvuka D. Bio-oil, solid and gaseous biofuels from biomass pyrolysis processes – an overview. Int J Energy Res. 2011;35:835–62. https://doi.org/10.1002/er.1804.
17. Gómez JN, Banks SW, Nowakowski DJ, Rosas JG, Cara J, Sánchez ME, Bridgwater AV. Effect of temperature on product performance of a high ash biomass during fast pyrolysis and its bio-oil storage evaluation. Fuel Process Technol. 2018;172:97–105. https://doi.org/10.1016/j.fuproc.2017.11.021.
18. Collard FX, Blin J. A review on pyrolysis of biomass constituents: mechanisms and composition of the products obtained from the conversion of cellulose, hemicelluloses and lignin. Renew Sust Energ Rev. 2014;38:594–608. https://doi.org/10.1016/j.rser.2014.06.013.

19. White JE, Catallo WJ, Legendre BL. Biomass pyrolysis kinetics: a comparative critical review with relevant agricultural residue case studies. J Anal Appl Pyrolysis. 2011;91:1–33. https://doi.org/10.1016/j.jaap.2011.01.004.
20. Fisher T, Hajaligol M, Waymack B, Kellogg D. Pyrolysis behavior and kinetics of biomass derived materials. J Anal Appl Pyrolysis. 2002;62:331–49. https://doi.org/10.1016/S0165-2370(01)00129-2.
21. Stefanidis SD, Kalogiannis KG, Iliopoulou EF, Michailof CM, Pilavachi PA, Lappas AA. A study of lignocellulosic biomass pyrolysis via the pyrolysis of cellulose, hemicellulose and lignin. J Anal Appl Pyrolysis. 2014;105:143–50. https://doi.org/10.1016/j.jaap.2013.10.013.
22. Williams PT, Besler S. The influence of temperature and heating rate on the slow pyrolysis of biomass. Renew Energy. 1996;7:233–50. https://doi.org/10.1016/0960-1481(96)00006-7.
23. Varma AK, Mondal P. Pyrolysis of sugarcane bagasse in semi batch reactor: effects of process parameters on product yields and characterization of products. Ind Crop Prod. 2017;95:704–17. https://doi.org/10.1016/j.indcrop.2016.11.039.
24. Bridgwater AV. Review of fast pyrolysis of biomass and product upgrading. Biomass Bioenergy. 2012;38:68–94. https://doi.org/10.1016/j.biombioe.2011.01.048.
25. Stamatov V, Honnery D, Soria J. Combustion properties of slow pyrolysis bio-oil produced from indigenous Australian species. Renew Energy. 2006;31:2108–21. https://doi.org/10.1016/j.renene.2005.10.004.
26. Arni S. Comparison of slow and fast pyrolysis for converting biomass into fuel. Renew Energy. 2018;124:197–201. https://doi.org/10.1016/j.renene.2017.04.060.
27. Das P, Ganesh A. Bio-oil from pyrolysis of cashew nut shellda near fuel. Biomass Bioenergy. 2003;25:113–7. https://doi.org/10.1016/S0961-9534(02)00182-4.
28. Liu RH, Shen CJ, Wu HJ, Deng CJ, Liu SY. Characterisation of bio-oil from fast pyrolysis of rice husk in fluidised bed reactor. J Energy Inst. 2011;84:73–9. https://doi.org/10.1179/014426011X12968328625397.
29. Zhang S, Zhang H, Liu X, Zhu S, Hu L, Zhang Q. Upgrading of bio-oil from catalytic pyrolysis of pretreated rice husk over Fe modified ZSM-5 zeolite catalyst. Fuel Process Technol. 2018;175:17–25. https://doi.org/10.1016/j.fuproc.2018.03.002.
30. Shuangning X, Abolghasem S. Bio-oil production and upgrading research: a review. Renew Sust Energ Rev. 2012;16:4406–14. https://doi.org/10.1016/j.rser.2012.04.028.
31. Hicks JC. Advances in C–O bond transformations in lignin-derived compounds for biofuels production. J Phys Chem Lett. 2011;2:2280–7. https://doi.org/10.1021/jz2007885.
32. Czernik S, Bridgwater AV. Overview of applications of biomass fast pyrolysis oil. Energy Fuel. 2004;18:590–8. https://doi.org/10.1021/ef034067u.
33. Cheng YT, Huber GW. Chemistry of furan conversion into aromatics and olefins over HZSM-5: a model biomass conversion reaction. ACS Catal. 2011;1:611–28. https://doi.org/10.1021/cs200103j.
34. Greenhalf CE, Nowakowski DJ, Harms AB, Titiloye JO, Bridgwater AV. A comparative study of straw, perennial grasses and hardwoods in terms of fast pyrolysis products. Fuel. 2013;108:216–30. https://doi.org/10.1016/j.fuel.2013.01.075.
35. Zhang Q, Chang J, Wang T, Xu Y. Review of biomass pyrolysis oil properties and upgrading research. Energy Convers Manag. 2007;48:87–92. https://doi.org/10.1016/j.enconman.2006.05.010.
36. Oasmaa A, Czernik S. Fuel oil quality of biomass pyrolysis oils-state of the art for the end users. Energy Fuel. 1999;13:914–21. https://doi.org/10.1021/ef980272b.
37. Lu Q, Li W, Zhu X. Overview of fuel properties of biomass fast pyrolysis oils. Energy Convers Manag. 2009;50:1376–83. https://doi.org/10.1016/j.enconman.2009.01.001.
38. Mortensen PM, Grunwaldt JD, Jensen PA, Knudsen KG, Jensen AD. A review of catalytic upgrading of bio-oil to engine fuels. Appl Catal A Gen. 2011;407:1–19. https://doi.org/10.1016/j.apcata.2011.08.046.

39. Gutierrez A, Kaila RK, Honkela ML, Slioor R, Krause AOI. Hydrodeoxygenation of guaiacol on noble metal catalysts. Catal Today. 2009;147:239–46. https://doi.org/10.1016/j.cattod.2008.10.037.
40. Bridgwater AV. Catalysis in thermal biomass conversion. Appl Catal A Gen. 1994;116:5–47. https://doi.org/10.1016/0926-860X(94)80278-5.
41. Carlson TR, Tompsett GA, Conner WC, Huber GW. Aromatic production from catalytic fast pyrolysis of biomass-derived feedstocks. Top Catal. 2009;52:241. https://doi.org/10.1007/s11244-008-9160-6.
42. Karnjanakom S, Bayu A, Hao X, Kongparakul S, Samart C, Abudula A, Guan G. Selectively catalytic upgrading of bio-oil to aromatic hydrocarbons over Zn, Ce or Ni-doped mesoporous rod-like alumina catalysts. J Mol Catal A Chem. 2016;421:235–44. https://doi.org/10.1016/j.molcata.2016.06.001.
43. Shafaghat H, Rezaei PS, Daud W. Catalytic hydrodeoxygenation of simulated phenolic bio-oil to cycloalkanes and aromatic hydrocarbons over bifunctional metal/acid catalysts of Ni/HBeta, Fe/HBeta and NiFe/HBeta. J Ind Eng Chem. 2016;35:268–76. https://doi.org/10.1016/j.jiec.2016.01.001.
44. Ausavasukhi A, Sooknoi T, Resasco DE. Catalytic deoxygenation of benzaldehyde over gallium-modified ZSM-5 zeolite. J Catal. 2009;268:68–78. https://doi.org/10.1016/j.jcat.2009.09.002.
45. Prasomsri T, Nimmanwudipong T, Román-Leshkov Y. Effective hydrodeoxygenation of biomass-derived oxygenates into unsaturated hydrocarbons by MoO_3 using low H_2 pressures. Energy Environ Sci. 2013;6:1732–8. https://doi.org/10.1039/C3EE24360E.
46. González-Borja MÁ, Resasco DE. Anisole and Guaiacol hydrodeoxygenation over monolithic Pt-Sn catalysts. Energy Fuel. 2011;25:4155–62. https://doi.org/10.1021/ef200728r.
47. Karnjanakom S, Yoshida A, Bayu A, Kurnia I, Hao X, Maneechakr P, Abudula A, Guan G. Bifunctional Mg-Cu-loaded β-zeolite: high selectivity for the conversion of furfural into monoaromatic compounds. ChemCatChem. 2018;10:3564–75. https://doi.org/10.1002/cctc.201800450.
48. Veses A, Puertolas B, Callen MS, Garcia T. Catalytic upgrading of biomass derived pyrolysis vapors over metal-loaded ZSM-5 zeolites: effect of different metal cations on the bio-oil final properties. Microporous Mesoporous Mater. 2015;209:189–96. https://doi.org/10.1016/j.micromeso.2015.01.012.
49. Cheng YT, Jae J, Shi J, Fan W, Huber GW. Production of renewable aromatic compounds by catalytic fast pyrolysis of lignocellulosic biomass with bifunctional Ga/ZSM-5 catalysts. Angew Chem Int Ed. 2012;51:1387–90. https://doi.org/10.1002/ange.201107390.
50. Li P, Li D, Yang HP, Wang XH, Chen HP. Effects of Fe-, Zr-, and Co-modified zeolites and pretreatments on catalytic upgrading of biomass fast pyrolysis vapors. Energy Fuel. 2016;30:3004–13. https://doi.org/10.1021/acs.energyfuels.5b02894.
51. He P, Shan WP, Xiao Y, Song H. Performance of Zn/ZSM-5 for in situ catalytic ppgrading of pyrolysis bio-oil by methane. Top Catal. 2016;59:86–93. https://doi.org/10.1007/s11244-015-0508-4.
52. Yung MM, Starace AK, Mukarakate C, Crow AM, Leshnov MA, Magrini KA. Biomass catalytic pyrolysis on Ni/ZSM-5: effects of nickel pretreatment and loading. Energy Fuel. 2016;30:5259–68. https://doi.org/10.1021/acs.energyfuels.6b00239.
53. Veses A, Puertolas B, Lopez JM, Callen MS, Solsona B, Garcia T. Promoting deoxygenation of bio-oil by metal-loaded hierarchical ZSM-5 zeolites. ACS Sustain Chem Eng. 2016;4:1653–60. https://doi.org/10.1021/acssuschemeng.5b01606.
54. Widayatno WB, Guan G, Rizkiana J, Yang J, Hao X, Tsutsumi A, Abudula A. Upgrading of bio-oil from biomass pyrolysis over Cu-modified β-zeolite catalyst with high selectivity and stability. Appl Catal B Environ. 2016;186:166–72. https://doi.org/10.1016/j.apcatb.2016.01.006.
55. Pham LKH, Tran TTV, Kongparakul S, Reubroycharoen P, Karnjanakom S, Guan G, Samart C. Formation and activity of activated carbon supported Ni_2P catalysts for atmospheric

deoxygenation of waste cooking oil. Fuel Process Technol. 2019;185:117–25. https://doi.org/10.1016/j.fuproc.2018.12.009.
56. Kaewpengkrow P, Atong D, Sricharoenchaikul V. Selective catalytic fast pyrolysis of Jatropha curcas residue with metal oxide impregnated activated carbon for upgrading bio-oil. Int J Hydrog Energy. 2017;42:18397–409. https://doi.org/10.1016/j.ijhydene.2017.04.167.
57. Karnjanakom S, Yoshida A, Widayatno WB, Bayu A, Kurnia I, Hao X, Maneechakr P, Abudula A, Guan G. Selective deoxygenation of carboxylic acids to BTXs over Cu/β-zeolite prepared by ethylene glycol-assisted impregnation. Catal Commun. 2018;110:33–7. https://doi.org/10.1016/j.catcom.2018.03.002.
58. Jae J, Tompsett GA, Foster AJ, Hammond KD, Auerbach SM, Lobo RF, Huber GW. Investigation into the shape selectivity of zeolite catalysts for biomass conversion. J Catal. 2011;279:257–68. https://doi.org/10.1016/j.jcat.2011.01.019.
59. Chaihad N, Karnjanakom S, Kurnia I, Yoshida A, Abudula A, Reubroycharoen P, Guan G. Catalytic upgrading of bio-oils over high alumina zeolites. Renew Energy. 2019;136:304–1310. https://doi.org/10.1016/j.renene.2018.09.102.
60. Karnjanakom S, Bayu A, Xiaoketi P, Hao X, Kongparakul S, Samart C, Abudula A, Guan G. Selective production of aromatic hydrocarbons from catalytic pyrolysis of biomass over Cu or Fe loaded mesoporous rod-like alumina. RSC Adv. 2016;6:50618–29. https://doi.org/10.1039/C6RA09431G.
61. Silva FCM, Lima MS, Neto COC, Sá JLS, Souza LD, Caldeira VPS, Santos AGD, Luz GE Jr. Catalytic deoxygenation of C18 fatty acids over HAlMCM-41 molecular sieve. Biomass Convers Biorefin. 2018;8:159–67. https://doi.org/10.1007/s13399-017-0263-9.
62. Karnjanakom S, Guan G, Asep B, Hao X, Kongparakul S, Samart C, Abudula A. Catalytic upgrading of bio-oil over Cu/MCM-41 and Cu/KIT-6 prepared by β-cyclodextrin-assisted co-impregnation method. J Phys Chem C. 2016;120:3396–407. https://doi.org/10.1021/acs.jpcc.5b11840.
63. Tran NTT, Uemura Y, Chowdhury S, Ramli A. Vapor-phase hydrodeoxygenation of guaiacol on Al-MCM-41supported Ni and Co catalysts. Appl Catal A Gen. 2016;512:93–100. https://doi.org/10.1016/j.apcata.2015.12.021.
64. Rezaei PS, Shafaghat H, Daud WMAW. Production of green aromatics and olefins by catalytic cracking of oxygenate compounds derived from biomass pyrolysis: a review. Appl Catal A Gen. 2014;469:490–511. https://doi.org/10.1016/j.apcata.2013.09.036.
65. Twaiq FA, Mohamed AR, Bhatia S. Liquid hydrocarbon fuels from palm oil by catalytic cracking over aluminosilicate mesoporous catalysts with various Si/Al ratios. Microporous Mesoporous Mater. 2003;64:95–107. https://doi.org/10.1016/j.micromeso.2003.06.001.
66. Williams PT, Horne PA. The influence of catalyst type on the composition of upgraded biomass pyrolysis oils. J Anal Appl Pyrolysis. 1995;31:39–61. https://doi.org/10.1016/0165-2370(94)00847-T.
67. Prasomsri T, To AT, Crossley S, Alvarez WE, Resasco DE. Catalytic conversion of anisole over HY and HZSM-5 zeolites in the presence of different hydrocarbon mixtures. Appl Catal B Environ. 2011;106:204–11. https://doi.org/10.1016/j.apcatb.2011.05.026.
68. Graça IS, Comparot JD, Laforge SB, Magnoux P, Lopes JM, Ribeiro MF, Ribeiro FR. Influence of phenol addition on the H-ZSM-5 zeolite catalytic properties during methylcyclohexane transformation. Energy Fuel. 2009;23:4224–30. https://doi.org/10.1021/ef9003472.
69. Shao S, Zhang H, Xiao R, Li X, Cai Y. Evolution of coke in the catalytic conversion of biomass-derivates by combined in-situ DRIFTS and ex-situ approach: effect of functional structure. Fuel Process Technol. 2018;178:88–97. https://doi.org/10.1016/j.fuproc.2018.05.021.
70. Huang J, Long W, Agrawal PK, Jones CW. Effects of acidity on the conversion of the model bio-oil ketone cyclopentanone on H-Y zeolites. J Phys Chem C. 2009;113:16702–10. https://doi.org/10.1021/jp905661w.
71. Vitolo S, Bresci B, Seggiani M, Gallo MG. Catalytic upgrading of pyrolytic oils over HZSM-5 zeolite: behaviour of the catalyst when used in repeated upgrading–regenerating cycles. Fuel. 2001;80:17–26. https://doi.org/10.1016/S0016-2361(00)00063-6.

72. Guo X, Zheng Y, Zhang B, Chen J. Analysis of coke precursor on catalyst and study on regeneration of catalyst in upgrading of bio-oil. Biomass Bioenergy. 2009;33:1469–73. https://doi.org/10.1016/j.biombioe.2009.07.002.
73. Karnjanakom S, Guan G, Asep B, Du X, Hao X, Samart C, Abudula A. Catalytic steam reforming of tar derived from steam gasification of sunflower stalk over ethylene glycol assisting prepared Ni/MCM-4. Energy Convers Manag. 2015;98:359–68. https://doi.org/10.1016/j.enconman.2015.04.007.

Chapter 6
Catalytic Pyrolysis of Lignocellulosic Biomass for Production of Liquid Biofuels

Bo Zhang, Kai Wu, Jing Zhang, Siying Zhong, and Huiyan Zhang

Abstract Catalytic fast pyrolysis (CFP) processes for conversion of lignocellulose biomass into liquid biofuels (bio-oil) have been extensively studying to meet increasing fuel demands and to address environmental issues. Direct use of crude bio-oil, however, is usually restricted due to its high content of oxygen. Therefore, reaction principles and mechanisms of the biomass catalytic pyrolysis need further reveal to develop improved bio-oil. This chapter presents a comprehensive review of the development of biomass CFP and bio-oil improving routes, including its catalysts, feedstocks, reaction methods and reactors.

Keywords Lignocellulose biomass · Catalytic pyrolysis · Bio-oil · ZSM-5 catalyst

6.1 Introduction

Lignocellulose biomass has attracted great attention from many researchers due to the low carbon release during its utilization. Fast pyrolysis, a thermochemical technique, has been developed for several decades to convert the biomass into bio-oil [1].

Fast pyrolysis, as a dominant thermochemical utilizing method, requires a moderate reaction temperature of (450–650) °C, a rapid heating rate of (10^3–10^4) K/s, and short residence time of <2 s [2–5]. The generated raw bio-oil, however, shows low quality, such as high oxygen content, decreased heating value, poor thermal stability and low corrosion resistance [6, 7], and thus needs further improving.

Catalytic fast pyrolysis (CFP) is generally regarded as a complement to reduce the oxygen content and promote the bio-oil quality [8–10], which can be implemented in-situ or ex-situ. In the in-situ CFP, biomass particles are entirely blended with the catalyst powder, and then fed into one reactor for catalytic pyrolysis

B. Zhang · K. Wu · J. Zhang · S. Zhong · H. Zhang (✉)
Key Laboratory of Energy Thermal Conversion and Control of Ministry of Education, School of Energy and Environment, Southeast University, Nanjing, China
e-mail: hyzhang@seu.edu.cn

Z. Fang et al. (eds.), *Production of Biofuels and Chemicals with Pyrolysis*, Biofuels and Biorefineries 10, https://doi.org/10.1007/978-981-15-2732-6_6

[11, 12]. Consequently, an abundant amount of catalysts are needed in the in-situ CFP to provide an adequate contact between primary pyrolytic vapor and the catalysts. The ex-situ CFP mode, on the other hand, produces better bio-oil with less catalyst needed [13, 14], in which the catalyst powder is fixed to form a catalyst fixed bed. During the ex-situ CFP, all the initial biomass pyrolytic vapor will pass through the catalyst fixed bed to ensure the fully contact. Additionally, an independent temperature control system is applicable in the catalyst fixed bed, which increases its flexibility and efficiency [15]. What is more, catalyst inactivation caused by residual char can be reduced in the ex-situ CFP system [16].

This chapter presents a comprehensive review of the progress and development of the biomass CFP. Its principles and mechanisms are illuminated in Sect. 6.2. The CFP related catalyst, feedstocks, operation routines and reactors are discussed in Sects. 6.3–6.6, respectively. Section 6.7 provides an overview of the biomass CFP.

6.2 Principle and Mechanism of Biomass Fast Pyrolysis

Biomass pyrolysis, a typical thermochemical treatment, is a thermal degradation process under anoxic or anaerobic condition, which mainly yields non-condensable light gases, liquid (bio-oil) and solid residual char (bio-char). Among various pyrolytic reactions, the highest yield rate of bio-oil (over 50 wt%) is generated by fast pyrolysis in which medium-high temperatures, a rapid heating rate, extremely short residence time, and timely condensation of pyrolytic vapor are needed. The yield rate (ω) of bio-oil can be calculated by the following Eq. (6.1):

$$\omega = \frac{m}{M} \times 100\% \tag{6.1}$$

In this equation, m means the quality of bio-oil while M means the quality of biomass feedstocks. During the fast pyrolysis, the biomass with low energy density can be transformed into the bio-oil with high energy density under continuous operations and with low cost. Furthermore, the energy-enriched bio-oil is environmental friendly and can be easily refined for further utilization in existing infrastructures. High value-added chemicals can also be extracted from the bio-oil later.

Biomass is generally composed of cellulose, hemicellulose and lignin. During the fast pyrolysis, hemicellulose and lignin will be initially depolymerized to moderate molecules, and then a portion of the molecules will be further decomposed into new compounds with low molecular weight. The pyrolysis mechanisms and transformation paths of cellulose, hemicellulose (xylan as a model compound), and lignin are summarized in Figs. 6.1 and 6.2a–c.

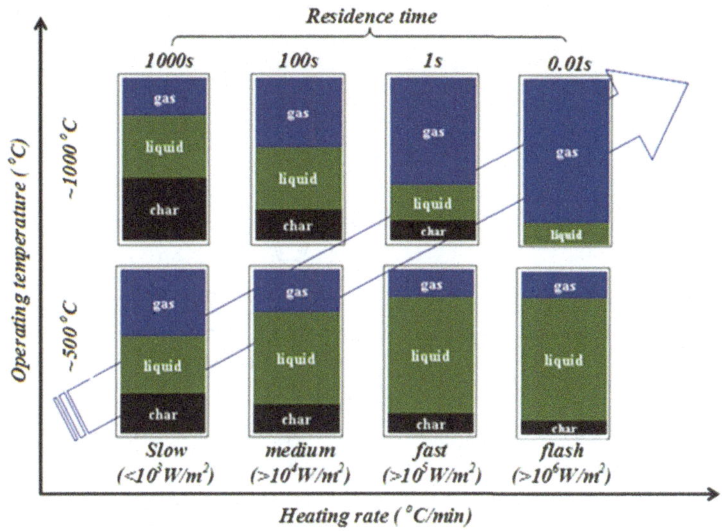

Fig. 6.1 Schematic overview of the staged decomposition concept within the thermochemical biorefinery (reprinted with permission from [17], Copyright © 2011 Elsevier)

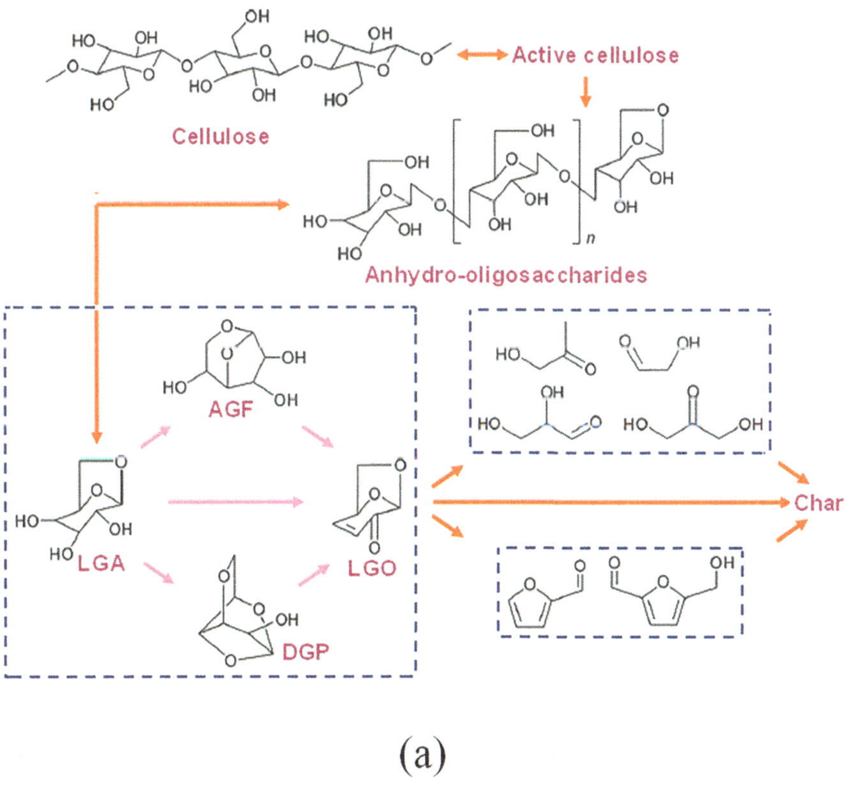

Fig. 6.2 The pyrolysis mechanisms and transformation paths of cellulose, hemicellulose, and lignin: (**a**) cellulose; (**b**) hemicellulose (xylan as a model compound); (**c**) lignin (**a** is adapted with permission from [18], Copyright © 1991 Elsevier. (**b**) is adapted with permission from [19], Copyright © 2009 Elsevier. (**c**) is adapted with permission from [20], Copyright © 2008 Elsevier)

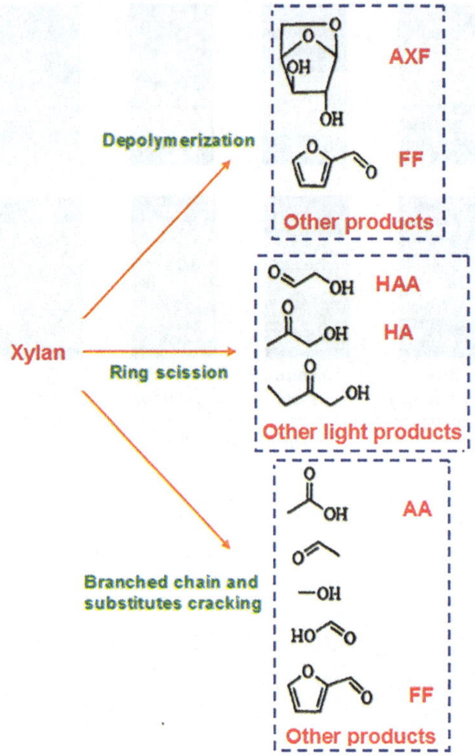

(b)

Fig. 6.2 (continued)

6.3 Catalyst

During the fast pyrolysis, many catalysts are used to reduce undesirable compounds and to obtain user-friendly bio-oil. These catalysts can be divided into three main types: inorganic minerals, metal oxides and zeolites. Their experimental conditions and results are shown in Table 6.1.

6.3.1 ZSM-5 Zeolite Catalyst

In the past decades, numerous microporous, mesoporous and macroporous catalysts (e.g. ZSM-5, ZSM-12, MCM-41, SBA-15, FSM-16, MSU, CNT and their modified derivatives) have been evaluated for biomass CFP conversion [10, 11, 28]. ZSM-5

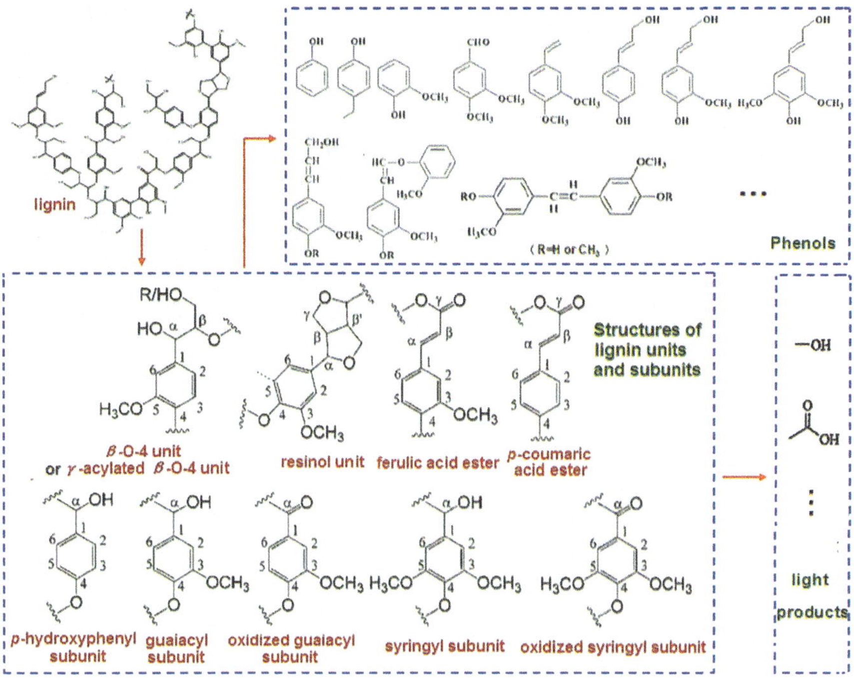

(c)

Fig. 6.2 (continued)

Table 6.1 Summary of catalysts used in CFP (*I* increased or improved, *D* decreased)

Catalyst type	Catalyst	Temperature (°C)	Bio-oil yield rates	Gas yield rates	Char yield rates	Ref
Inorganic minerals	Alkali/alkaline earth metals	480–520	I	I	I	[21]
	Transition metals	380	D	I	I	[22]
Metal oxides	Al_2O_3	500	D	I	I	[23]
	MgO	500	D	I	I	[23]
	CaO	520	I	I	I	[24]
	CaO·MgO	450	D	I	I	[25]
	ZnO	500	–	D	I	[26]
	Fe_2O_3	500	I	I	I	[26]
Zeolites	HZSM-5	400–550	D	I	I	[27]
	H-Y	400–550	D	I	I	[27]
	Na-Y	400–550	D	I	I	[27]
	Mg-Y	400–550	D	I	I	[27]
	Ca-Y	400–550	D	I	I	[27]

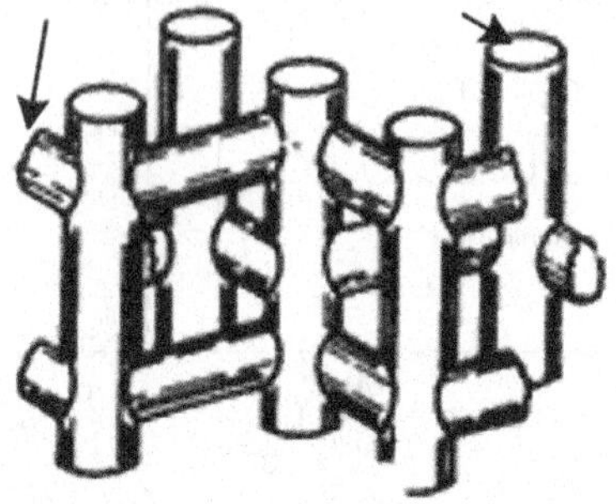

Fig. 6.3 Pore structure of ZSM-5

catalyst (zeolite) has a unique three-dimensional pore system with 10-member rings, and has internal straight channels of 0.53 nm × 0.56 nm linked to its internal zigzag channels of 0.51 nm × 0.55 nm (Fig. 6.3). Owing to the special structure, zeolite is rated the best one to favor hydrocarbon production and shows compelling deoxygenating capacity [29–32]. During the biomass CFP, oxygenated chemicals coming from biomass primary pyrolytic vapor are continually adsorbed by ZSM-5 catalysts, and then be stored within its confined pore channels and intersection cavities. What is more, the catalyst also exhibits assorted isomerization and deoxygenation responses (e.g. dehydration, dehydroxylation, decarbonylation, and decarboxylation reactions). For instance, oxygen atoms stored within zeolite are often released in the form of H_2O, CO_2, and CO [33].

Furthermore, the pore diameter of ZSM-5 catalyst is similar to dynamic diameter of benzene, toluene and xylene. Thus, ZSM-5 catalyst shows a prominent shape-selective catalytic effect on the formation of aromatic hydrocarbons during the biomass CFP.

The particular catalytic mechanism of ZSM-5 is mainly due to "hydrocarbon pool" formed within the catalyst. Oxygenated chemicals coming from biomass primary pyrolytic vapor, as mentioned above, are trapped into the inner pores of ZSM-5 after contact with the catalyst, and then be transformed into alkanes, olefins, and aromatic hydrocarbons by reacting with the "hydrocarbon pool" (Fig. 6.4) [35]. Specifically, a duplex aromatic- and olefin-based carbon pool cycle takes charge of hydrocarbon formation within ZSM-5 during biomass CFP [36]: C2 olefin (ethylene) and aromatic hydrocarbons are produced from the aromatic-based methylation catalytic cycle (aromatic carbon pool), while >C2 olefins are generated from another olefin-based one (olefin carbon pool) (Fig. 6.5). Interestingly, the formation pathway of ethylene is different from that of other olefins. Ethylene is involved in aromatic carbon pool where aromatic hydrocarbons play active catalyst roles.

During the biomass CFP, adding ZSM-5, on the other hand, has negative effects on bio-oil yield rate and promotes non-condensable light gases release because a lot of condensable pyrolytic vapor is break into non-condensable light gases.

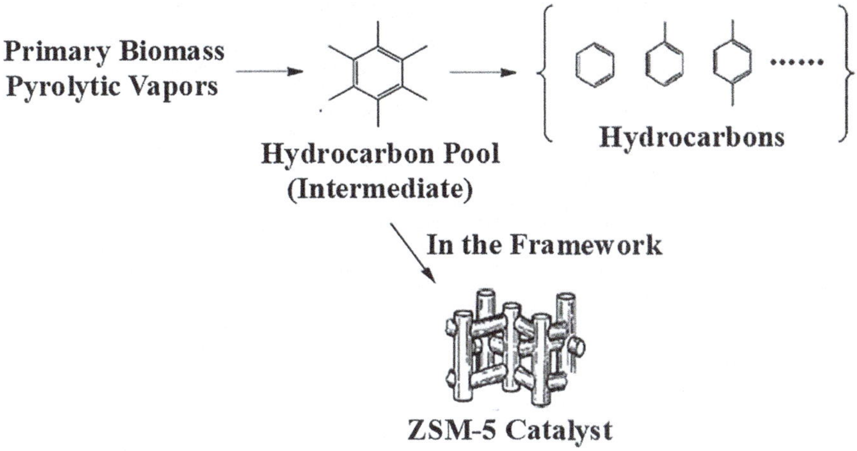

Fig. 6.4 "Hydrocarbon pool" catalytic mechanism (reprinted with permission from [34], Copyright © 2014 Elsevier)

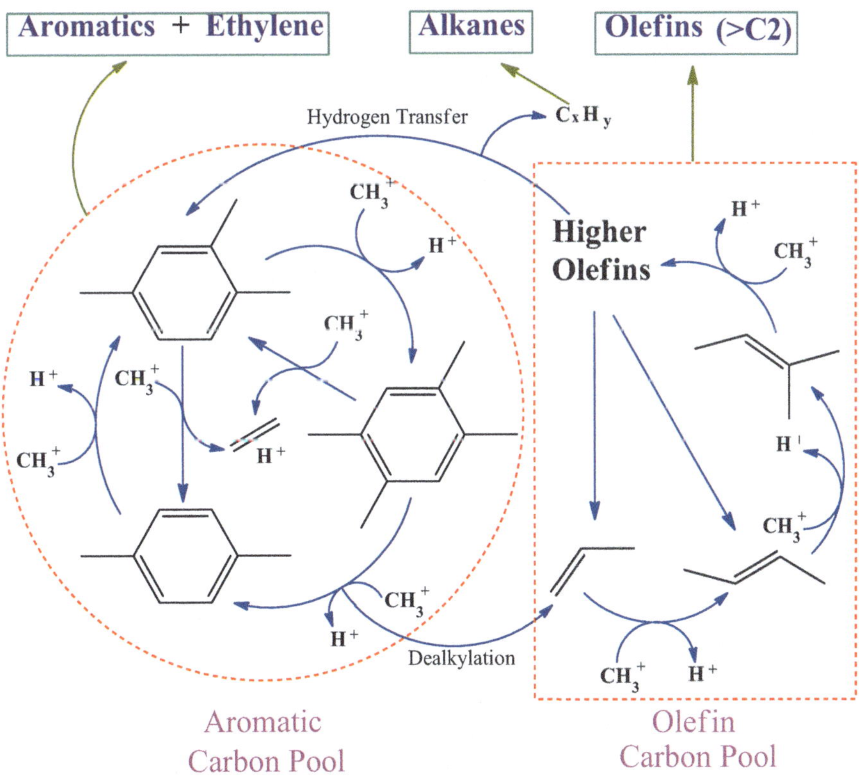

Fig. 6.5 Duplex aromatic- and olefin-based carbon pool cycles for hydrocarbon formation within ZSM-5 catalyst (reprinted with permission from [37], Copyright © 2017 Elsevier)

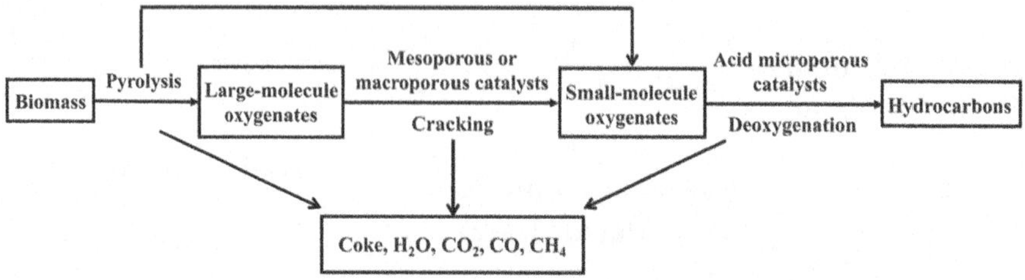

Fig. 6.6 Catalytic fast pyrolysis (CFP) of biomass to hydrocarbons over dual catalysts

6.3.2 *Dual Catalysts*

During biomass CFP, some bulky molecules, derived from acid-catalyzed polymerization, often deposit on ZSM-5 external surface to form coke [38]. The coke yield is often less than 5.0 wt% [16], but the attached coke will result in pore blockage and rapid inactivation of the catalysts. The shape-selective catalytic reactions, ultimately, will be blocked.

According to previous studies [39], large molecules can be break into smaller ones by some mesoporous and macroporous materials with excellent cracking capacity. For example, the amount of compounds with low molecular weight are remarkably increased after adding CaO as its catalyst [40]. The combination of ZSM-5 and these mesoporous/macroporous cracking materials might promote hydrocarbon production while mitigate coke deposition. Therefore, mesoporous/macroporous cracking catalysts are adopted to crack heavy components coming from biomass pyrolytic vapor into smaller oxygenated ones, and then these smaller ones are transformed into hydrocarbons via passing through a ZSM-5 fixed bed (Fig. 6.6). When the mass ratio of CaO to ZSM-5 is 1:3, high yield of bio-oil can be obtained [40]. MgO, CeO_2, ZrO_2, and MCM-41 also show the cracking ability. When they are used in combination with ZSM-5, the dual catalysts can effectively reduce coke yield rate, promote oxygen removal and facilitate hydrocarbon production during biomass CFP [41].

6.3.3 *Hierarchical Catalyst*

The mesoporous zeolite named MCM-41 shows cracking capability to break large molecules into relatively light compounds to reduce coke generation, but its deoxygenation ability is limited [42]. Therefore, MCM-41 plus ZSM-5 might be a possible solution to inhibit coke accumulation and to promote hydrocarbon formation. Zhang et al. [43] synthesized an innovative hierarchical ZSM-5/MCM-41 composite zeolite catalyst by growing a layer of MCM-41 structure onto the outer surface of ZSM-5 zeolite seeds. When primary biomass pyrolytic vapor passes through the hierarchical ZSM-5/MCM-41 composite catalyst, the majority of heavy oxygenated compounds in the vapor are cracked into smaller ones on the MCM-41 surface. These smaller compounds are, subsequently, converted into various hydrocarbons in the ZSM-5 core. There are growing attempts on the synthesis of zeolitic micro-mesoporous composite catalyst to integrate the excellent deoxygenation ability of ZSM-5 and the breaking capability of MCM-4.

For the synthesis of hierarchical ZSM-5/MCM-41 composite zeolite, hydrothermal crystallization technique is mostly used with ZSM-5 powder as original seeds in lab- and industrial-scale operation. First, parent ZSM-5 powder is alkali-leached by NaOH solution (2.0 mol/L) with mechanical stirring to form desirable aluminosilicate fragment in the solution. Second, CTAB solution (hexadecyl trimethyl ammonium bromide, 10 wt%) is added into the zeolite solution, in which the CTAB acts as MCM-41 precursor. Third, an autoclave with THE lining is used for crystallization of the mixture (110 °C, 24 h). Fourth, the heated solution is cooled to room temperature and then is adjusted to pH 8.5. Finally, the second crystallization of the solution (pH 8.5) is implemented in the autoclave again (110 °C, 24 h). Sixthly, filtration, washing, drying and calcination steps are subsequently conducted to obtain Na-ZSM-5/MCM-41 composite zeolite. Finally, Na-ZSM-5/MCM-41 is transformed into ZSM-5/MCM-41 composite catalyst through leaching, filtration, washing, drying and calcination treatment of NH_4Cl solution.

6.3.4 Modification of ZSM-5 Catalyst

Heavy compounds coming from primary biomass pyrolytic vapor will precipitate on the surface of acidic catalysts to form detrimental coke that is mainly composed of aromatic compounds with large molecular weight. For ZSM-5 catalyst, the coke is usually gathered on its external surface rather than on its internal surface due to the large size of the coke [44]. The favorable deoxygenation ability of ZSM-5 is mainly credited to its internal acid sites, whereas its external acid sites promote coke accumulation. Thus, the external acid sites of ZSM-5 should be inactivated to hinder coke gathering, while its internal acid sites are ought to be preserved to facilitate hydrocarbon formation.

Researchers have inactivated the outer acid sites of ZSM-5 catalyst via pre-coked handling, chemical vapor deposition of inert silica (SiO_2-CVD), and EDTA chemical modification. Accordingly, the modified ZSM-5 catalysts are referred to as PC-ZSM-5, SiO_2-ZSM-5 and EDTA-ZSM-5, respectively [45]. For PC-ZSM-5 catalyst, pre-arranged coke is deposited on its outer surface to partly passivate its external acid sites and to adjust its external acidity and competence. For SiO_2-ZSM-5 catalyst, an ingenious method is conducted to modify ZSM-5 by using tetra-ethyl-orthosilicate (TEOS) as the modifying precursor. When ZSM-5 catalyst is impregnated into the TEOS solution, TEOS with large molecule size cannot enter into the narrow intracrystalline pores of ZSM-5. During calcination of the TEOS-impregnated ZSM-5, SiO_2 will gradually accumulate on the external acid sites of the ZSM-5. Meanwhile, the internal acid sites of the ZSM-5 are unaffected. Therefore, the silanization of ZSM-5 external surface, evidently, retards the coke formation while remains its catalytic capacity.

Furthermore, there is a close relationship between the acid sites of ZSM-5 and its framework aluminums. The external acid sites, correspondingly, can be adjusted via outer framework aluminums. Ethylene diamine tetraacetie acid (EDTA), a chelating complex, has been proposed to remove framework aluminums in ZSM-5 to improve its anti-coking ability. Owing to its large size, EDTA cannot enter into the inner channels of ZSM-5 catalyst, and thus the external acid sites of ZSM-5 can be selectively inactivated via EDTA chemical modification, while its internal acid sites are not altered.

6.4 Catalytic Fast Co-Pyrolysis (Co-CFP)

Another critical issue for biomass CFP process is the relatively low hydrocarbon yield rate in bio-oil due to the limited hydrogen available in traditional biomass feedstocks. To determine the relative hydrogen content in various biomass feedstocks, a new parameter named effective hydrogen index (EHI) has been proposed [46], which is defined in the followed equation:

$$\mathrm{EHI} = \frac{\mathrm{H} - 2\mathrm{O} - 3\mathrm{N} - 2\mathrm{S}}{\mathrm{C}} \tag{6.2}$$

This definition assumes that O, N and S atoms are completely transformed into hydrides, and then be used for hydrocarbon production. The EHI value of conventional lignocellulosic biomass, unfortunately, is often around 0–0.3, which means biomass lacks hydrogen in general [47]. Thus, ZSM-5 zeolite is needed during the pyrolysis of biomass with EHI below 1.0 [46].

The hydrogen-deficient problem can be solved via co-feeding the biomass with hydrogen-enriched materials to rise the total EHI value of the feedstocks, which is called catalytic fast co-pyrolysis (co-CFP) technique [48]. As reported, synergy between the defective biomass and hydrogen-enriched materials is observed and verified during the co-CFP process. Experimental results showed that higher hydrocarbon yield rate can be received from the co-CFP than its theoretical value (synergistic effect 1). The definition of theoretical hydrocarbon yield in co-CFP is:

$$Y_{\text{theoretical}} = \omega_B Y_B + \omega_{\text{HRM}} Y_{\text{HRM}} \;\; (\omega\mathrm{B} + \omega\mathrm{HRM} = 1) \tag{6.3}$$

In this equation, ω_B is the mass fraction of biomass in total mixture feedstocks; ω_{HRM} is the mass fraction of hydrogen-enriched materials in total mixture feedstocks; Y_B is the hydrocarbon yield in bio-oil from the CFP of biomass alone; Y_{HRM} is the hydrocarbon yield in bio-oil from the CFP of hydrogen-enriched materials alone. What is more, co-CFP of biomass and hydrogen-enriched materials provides

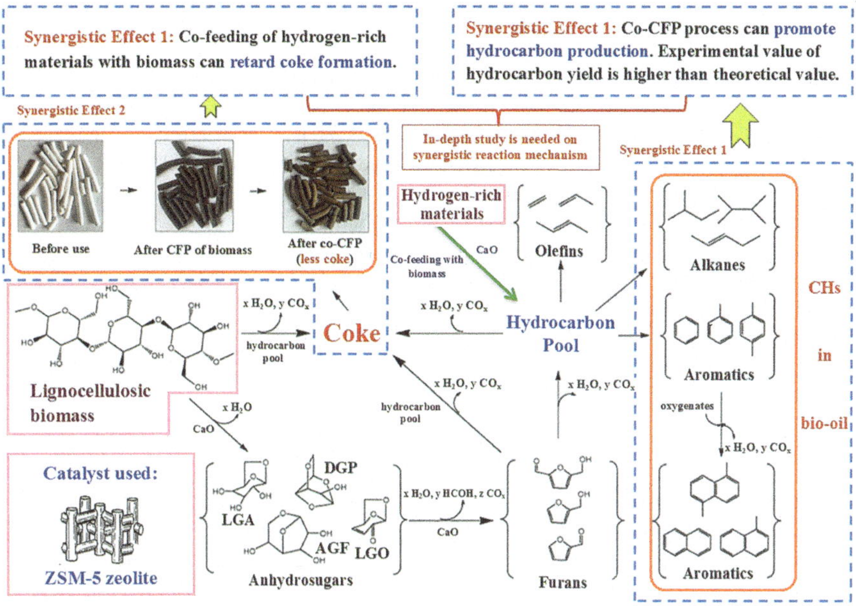

Fig. 6.7 Synergy between biomass feedstocks and hydrogen-enriched materials during co-CFP (reprinted with permission from [37], Copyright © 2017 Elsevier)

the ability to retard the coke generation when comparing with biomass used alone. Therefore, catalyst inactivation can be delayed and the catalytic shape-selective effect can be promoted (synergistic effect 2). These results are in good agreement with those of previous studies [49] and these synergistic effects are summarized in Fig. 6.7.

Co-CFP of biomass and high-density polyethylene (HDPE) is shown in Table 6.2. The total peak area and the relative content of hydrocarbons of III are larger than these of IV, which reveals a great synergistic effect between biomass and HDPE. HDPE can provide hydrogen atoms during the process and increase the yield rate of hydrocarbon production. Oxygenated compounds in the biomass fast pyrolysis vapor, on the other hand, can promote the chain scission and cracking of HDPE. From Table 6.3, the presence of hydrogen-enriched materials can reduce the formation of coke.

6.5 Ex-Situ CFP Process

CFP configurations can be divided into in-situ and ex-situ according to the pyrolytic vapor and catalyst contact methods [50]. Figure 6.8 shows the process of the two different configurations.

Table 6.2 Comparison of CFP of biomass, HDPE, and their mixtures (Py-GC/MS experiments)

	I	II	III	IV
Experiments	CFP of biomass	CFP of HDPE	CFP of their mixtures	
Feedstocks	0.50 mg powdered corn stalk samples	0.50 mg powdered HDPE samples	0.25 mg powdered corn stalk samples + 0.25 mg powdered HDPE samples, fully mixed	$\frac{I+II}{2}$
Total peak area ($\times 10^9$)	2.88	12.14	9.55	7.51
Relative content of hydrocarbons (%)	59.1	100.0	84.9	79.5

CFP temperature: 600 °C

Table 6.3 Coke formation on ZSM-5 from co-CFP in Py-GC/MS

Feedstocks	Corn stalk (mg)	1.00	1.00	1.00	1.00	1.00
	HDPE (mg)	0	0.10	0.20	0.30	0.40
Coke amount on ZSM-5 (mg)		0.046	0.042	0.039	0.036	0.032

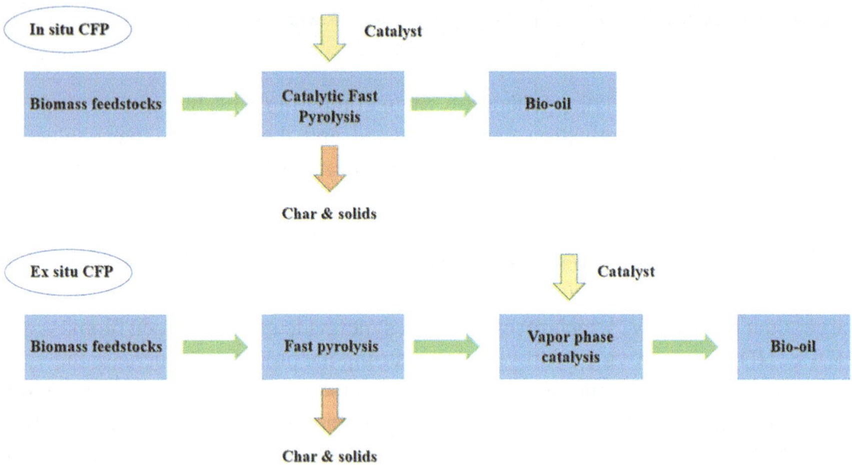

Fig. 6.8 In-situ and ex-situ catalytic configurations

In the in-situ CFP, biomass feedstocks and catalyst powder are fully blended in a CFP reaction chamber at the beginning, and then large pyrolysis fragments are immediately decomposed to reduce repolymerization and secondary char formation [51]. A large quantity of catalyst is needed in this manner due to the method and the short reaction time; in-situ CFP needs a high ratio of catalyst to biomass. Moreover, the process of biomass pyrolysis and catalytic conversion can only be conducted

Table 6.4 Comparison of in-situ and ex-situ configurations [52–55]

Configuration	Bio-oil quality	Reaction degree	Temperature	Bio-oil yield
In-situ CFP	Low	Inadequate	Fixed	High
Ex-situ CFP	High	Adequate	Flexible	Low

under the same temperature because the reaction occurs in one step and within one reactor.

For the ex-situ CFP, biomass feedstocks and catalysts are set in two separate reactors, and thus the biomass pyrolysis and the catalytic conversion can be separately regulated. Pre-treatment on the pyrolysis vapor can be conducted to improve bio-oil quality and to realize catalyst-saving in the ex-situ CFP manner [16]. With the decrease of residence time on the catalyst bed, light oxygenates can easily enter into the pore of the catalyst, and then be transformed into hydrocarbons. What is more, char generated during the ex-situ CFP can also be easily handled as a promising by-product.

Both of in-situ and ex-situ CFP can improve bio-oil quality [52], but they have many differences. Güngör et al. [13] found that bio-char, during in-situ CFP, is adsorbed on the surface of the catalysts due to the fully mixture of biomass and catalysts, which causes pore blockage and inactivation of the catalysts. In ex-situ CFP, its bio-char shows limited negative impacts on its catalysts. Furthermore, larger amount of aromatics are generated during the in-situ catalytic pyrolysis than the ex-situ reaction [53]. Ex-situ configuration removes more oxygen compounds and shows better deoxygenating capacity than in-situ one [54], which attributes to an independent temperature control system in the ex-situ CFP. The catalysts of the ex-situ CFP can be pre-heated before contact with its vapor. Huang et al. noted that the ex-situ shows a higher techno-economic uncertainty than the in-situ due to the increased cost [54]. The main differences between in-situ and ex-situ configurations are listed in the following table (Table 6.4) [52–55].

6.6 CFP Reactors

A typical biomass CFP system consists of feed pretreatment unit, biomass feeding apparatus, CFP reactor, gas-solid separation device, liquid condensing unit, bio-oil collector and reservoir vessel. To realize high bio-oil yield rate, CFP reactor ought to fulfill the following requirements. First, sufficient and consecutive heat input with great fuel efficiency is needed to maintain the endothermic CFP reactions and to match its rapid heat exchange rate. Second, the primary pyrolytic vapor shows short residence time within the reactor to inhibit secondary cracking reactions. Third, limited non-condensable gas with small molecules is produced to maximize the bio-oil yield rate. Various CFP reactors are being developed during the past decades and are generally divided into two types, fluidized bed reactors and non-fluidized bed reactors, which depends on whether its carrier gas is introduced into the reactor or not. Fluidized bed reactors mainly include bubbling fluidized bed, spouted fluidized bed, circulating fluidized bed and entrained-flow bed; non-fluidized bed reactors comprise fixed bed, ablative vortex flow reactor, rotating cone reactor, screw reactor and vacuum moving bed reactor.

Some innovative CFP reactor designs are gradually emerging in these few years. Zhang et al. [56] drafted a kind of reactor termed internally interconnected fluidized bed (IIFB) that possesses cost-saving and convenient attributes. Within its reaction chamber, its char and coke are burnt to support its biomass CFP running and to realize catalysts regeneration. The key characteristic of the IIFB system is a direct connection of a biomass CFP area and an annular combustion zone. A cross-section outline of the system is depicted in Fig. 6.9. IIFB provides a comprehensive solution to the high energy needed and the catalyst inactivation during the continuous biomass conversion. A comparison of major CFP reactors is summarized in Table 6.5.

A unique and advanced microwave-assisted CFP (MACFP) technique has aroused interest due to its flexibility, manageability, effectiveness, uniformity, and time-saving features [67]. Compared with traditional electrical heating ways,

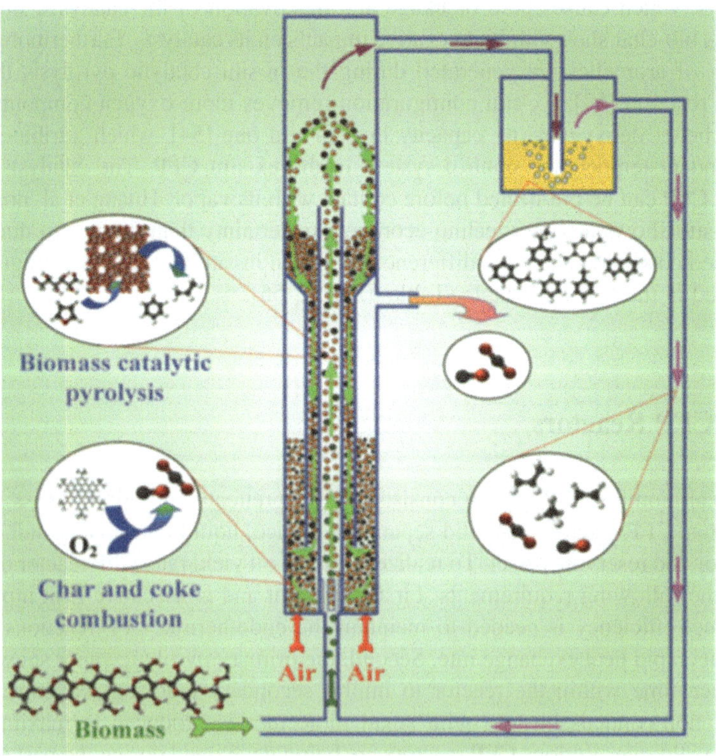

Fig. 6.9 A cross-section outline of IIFB system (adapted with permission from [57], Copyright © 2011 Elsevier)

Table 6.5 A comparison of major reactors

Type	Reactor model	Advantages	Disadvantages	Ref
Fluidized bed	Bubbling fluidized bed	Simple structure, easy to operate, long service life, temperature can be controlled	Require small particles, high operating energy consumption, hard to generate vapor	[58, 59]
	Spouted fluidized bed	High heat transfer efficiency, simple structure, low pressure decrease, short residence time	Limits for particle size, high requirement for bed, hard to generate vapor	[60]
	Circulating fluidized-bed	Processes large biomass feedstock, can conduct large-scale experiment	Hard to control and operate, high comsumption, complex separation and condensation equipment	[61]
	Entrained flow-model	No solid heat carrier, simple structure	Large air flow, large equipment, poses char handling challenges, low bio-oil yields	[62]
Non-fluidized bed	Ablative reactor configuration	Processes large biomass feedstock, no gas requirement, compact device structure, less costly	High consumption, limits to wall heat transfer and reaction, unsuitable to large-scale experiment, complex in nature	[63]
	Rotating cone configuration	No gas requirement, compact device structure, low operation temperature	Complex system, high requirement for materials	[64]
	Helical configuration	Good adaptability of materials, no gas requirement, compact device structure, low operation temperature	Complex system, low efficiency of heat, low product yield, easy to be blocked	[65]
	Vacuum pyrolysis reactor	Good adaptability of materials, no gas requirement, short residence time	High consumption, high cost, low product yield and quality	[66]

microwave-induced heating method produces uniform internal heating, thus, various biomass particle sizes and non-fluidized bed reactors are applicable to the MACFP system [68]. Biomass feedstocks, on the other hand, are difficult to be rapidly heated by microwave due to the low dielectric loss factor of lignocellulose. However, carbonaceous materials with high radiation absorption capacity, namely "microwave absorbents", can be added into the feedstocks to promote microwave energy absorption, with such absorbents being materials such as activated carbon, graphite, bio-char and silicon carbide (SiC) [16]. Therefore, the assembled biomass feedstocks, during the MACFP process, are heated to desirable high temperature via the indirect and direct microwave heating methods simultaneously. Figure 6.10 shows an outline of a typical biomass MACFP reactor, in which SiC is used as the microwave absorbent.

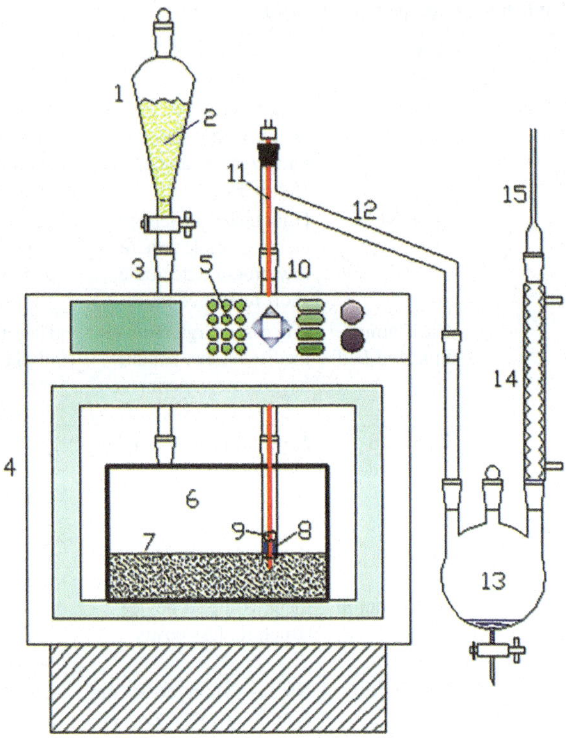

Fig. 6.10 Schematic of a typical MACFP reactor. (1) Semi-continuous biomass feeder; (2) biomass particles; (3) inlet connector; (4) microwave oven; (5) microwave control panel; (6) reactor with two necks; (7) microwave absorbent (SiC) bed; (8) catalyst fixed bed; (9) quartz wool; (10) outlet connector; (11) thermocouple; (12) connector; (13) bio-oil collector; (14) condenser; (15) connection to a vacuum pump to draw the pyrolysis vapors out of quartz reactor (reprinted with permission from [69], Copyright © 2016 Elsevier)

6.7 Conclusions and Future Work

Biomass CFP is an ideal technology to produce bio-oil. The performance and inactivation of its catalyst have aroused great interests but are still being technological bottlenecks. Its catalytic mechanisms also require in-depth reveal. The synergistic mechanisms of co-CFP process between biomass and hydrogen-enriched materials need further work. Additionally, development of highly advanced CFP reactor system should be taken into consideration.

References

1. Li R, Zhong ZP, Jin BS, Zheng AJ. Selection of temperature for bio-oil production from pyrolysis of algae from lake blooms. Energy Fuel. 2012;26:2996–3002. https://doi.org/10.1021/ef300180r.
2. Yanik J, Kornmayer C, Saglam M, Yüksel M. Fast pyrolysis of agricultural wastes: characterization of pyrolysis products. Fuel Process Technol. 2007;88:942–7. https://doi.org/10.1016/j.fuproc.2007.05.002.

3. Czernik S, Bridgwater AV. Overview of applications of biomass fast pyrolysis oil. Energy Fuel. 2004;18:590–8. https://doi.org/10.1021/ef034067u.
4. French R, Czernik S. Catalytic pyrolysis of biomass for biofuels production. Fuel Process Technol. 2010;91:25–32. https://doi.org/10.1016/j.fuproc.2009.08.011.
5. Encinar JM, González JF, González J. Fixed-bed pyrolysis of *Cynara cardunculus* L. product yields and compositions. Fuel Process Technol. 2000;68(3):209–22. https://doi.org/10.1016/S0378-3820(00)00125-9.
6. Jiang XX, Zhong ZP, Ellis N, Wang Q. Aging and thermal stability of the mixed product of the ether-soluble fraction of bio-oil and bio-diesel. Chem Eng Technol. 2011;34(5):727–36. https://doi.org/10.1002/ceat.201000441.
7. Zhang B, Zhong ZP, Ding K, Cao YY, Liu ZC. Catalytic upgrading of corn stalk fast pyrolysis vapors with fresh and hydrothermally treated HZSM-5 catalysts using Py-GC/MS. Ind Eng Chem Res. 2014;53(24):9979–84. https://doi.org/10.1021/ie404426x.
8. Shao SS, Zhang HY, Xiao R, Shen DK, Zheng J. Comparison of catalytic characteristics of biomass derivates with different structures over ZSM-5. Bioenerg Res. 2013;6:1173–82. https://doi.org/10.1007/s12155-013-9303-x.
9. Liu HL, Ma XQ, Li LJ, Hu ZF, Guo PS, Jiang YH. The catalytic pyrolysis of food waste by microwave heating. Bioresour Technol. 2014;106:45–50. https://doi.org/10.1016/j.biortech.2014.05.020.
10. Dickerson T, Soria J. Catalytic fast pyrolysis: a review. Energies. 2013;6:514–38. https://doi.org/10.3390/en6010514.
11. Adam J, Antonakou E, Lappas A, Stocker M, Nilsen MH, Bouzga A, Hustad JE, Oye G. *In situ* catalytic upgrading of biomass derived fast pyrolysis vapours in a fixed bed reactor using mesoporous materials. Microporous Mesoporous Mater. 2006;96:93–101. https://doi.org/10.1016/j.micromeso.2006.06.021.
12. Imran A, Bramer EA, Seshan K, Brem G. High quality bio-oil from catalytic flash pyrolysis of lignocellulosic biomass over alumina-supported sodium carbonate. Fuel Process Technol. 2014;127:72–9. https://doi.org/10.1016/j.fuproc.2014.06.011.
13. Güngör A, Önenc S, Ucar S, Yanika J. Comparison between the "one-step" and "two-step" catalytic pyrolysis of pine bark. J Anal Appl Pyrolysis. 2012;97:39–48. https://doi.org/10.1016/j.jaap.2012.06.011.
14. Williams PT, Nugranad N. Comparison of products from the pyrolysis and catalytic pyrolysis of rice husks. Energy. 2000;25:493–513. https://doi.org/10.1016/S0140-6701(02)85315-2.
15. Fan LL, Chen P, Zhou N, Liu SY, Zhang YN, Liu YH, Wang YP, Omar MM, Peng P, Addy M, Cheng YL, Ruan R. *In-situ* and *ex-situ* catalytic upgrading of vapors from microwave-assisted pyrolysis of lignin. Bioresour Technol. 2018;247:851–8. https://doi.org/10.1016/s0960-8524(99)00085-1.
16. Zhang B, Zhong ZP, Xie QL, Liu SY, Ruan R. Two-step fast microwave-assisted pyrolysis of biomass for bio-oil production using microwave absorbent and HZSM-5 catalyst. J Environ Sci. 2016;45:240–7. https://doi.org/10.1016/j.jes.2015.12.019.
17. De Wild PJ, Den Uil H, Reith JH, Kiel JHA, Heeres HJ. Biomass valorisation by staged degasification A new pyrolysis-based thermochemical conversionoption to produce value-added chemicals from lignocellulosic biomass. J Anal Appl Pyrolysis. 2009;85:124–33. https://doi.org/10.1016/j.jaap.2008.08.008.
18. Radlein D, Piskorz J, Scott DS. Fast pyrolysis of natural polysaccharides as a potential industrial process. J Anal Appl Pyrolysis. 1991;19:41–63. https://doi.org/10.1016/0165-2370(91)80034-6.
19. Shen DK, Gu S, Bridgwater AV. Study on the pyrolytic behaviour of xylan-based hemicellulose using TG–FTIR and Py–GC–FTIR. J Anal Appl Pyrolysis. 2010;87:199–206. https://doi.org/10.1016/j.jaap.2009.12.001.
20. Bayerbach R, Meier D. Characterization of the water-insoluble fraction from fast pyrolysis liquids (pyrolytic lignin). Part IV: structure elucidation of oligomeric molecules. J Anal Appl Pyrolysis. 2009;85:98–107. https://doi.org/10.1016/j.jaap.2008.10.021.

21. Trendewicz A, Evans R, Dutta A. Evaluating the effect of potassium on cellulose pyrolysis reaction kinetics. Biomass Bioenergy. 2015;74:15–25. https://doi.org/10.1016/j.biombioe.2015.01.001.
22. Di Blasi C, Branca C, Galgano A. Modifications in the thermicity of the pyrolysis reactions of ZnCl2-loaded wood. Ind Eng Chem Res. 2015;54(51):12741–9. https://doi.org/10.1021/acs.iecr.5b03694.
23. Stefanidis SD, Kalogiannis KG, Iliopoulou EF. In-situ upgrading of biomass pyrolysis vapors: catalyst screening on a fixed bed reactor. Bioresour Technol. 2011;102(17):8261–7. https://doi.org/10.1016/j.biortech.2011.06.032.
24. Lin Y, Zhang C, Zhang M. Deoxygenation of bio-oil during pyrolysis of biomass in the presence of CaO in a fluidized-bed reactor. Energ Fuels. 2011;24(10):5686–95. https://doi.org/10.1021/ef1009605.
25. Veses A, Aznar M, Martínez I. Catalytic pyrolysis of wood biomass in an auger reactor using calcium-based catalysts. Bioresour Technol. 2011;162:250–8. https://doi.org/10.1016/j.biortech.2014.03.146.
26. Torri C, Reinikainen M, Lindfors C. Investigation on catalytic pyrolysis of pine sawdust: catalyst screening by Py-GC-MIP-AED. J Anal Appl Pyrolysis. 2010;88(1):7–13. https://doi.org/10.1016/j.jaap.2010.02.005.
27. Imran A, Bramer E, Seshan K, Brem G. Catalytic flash pyrolysis of biomass using different types of zeolite and online vapor fractionation. Energies. 2016;9(3):187. https://doi.org/10.3390/en9030187.
28. Wang SR, Dai GX, Yang HP, Luo ZY. Lignocellulosic biomass pyrolysis mechanism: a state-of-the-art review. Prog Energy Combust Sci. 2017;62:33–86. https://doi.org/10.1016/j.pecs.2017.05.004.
29. Foster AJ, Jae J, Cheng YT, Huber GW, Lobo RF. Optimizing the aromatic yield and distribution from catalytic fast pyrolysis of biomass over ZSM-5. Appl Catal A Gen. 2012;423:154–61. https://doi.org/10.1016/j.apcata.2012.02.030.
30. Yang HP, Coolman R, Karanjkar P, Wang HY, Dornath P, Chen HP, Fan W, Conner WC, Mountziaris TJ, Huber G. The effects of contact time and coking on the catalytic fast pyrolysis of cellulose. Green Chem. 2017;19:286–97. https://doi.org/10.1039/c6gc02239a.
31. Maneffa M, Priecel P, Lopez-Sanchez JA. Biomass-derived renewable aromatics: selective routes and outlook for p-xylene commercialization. ChemSusChem. 2016;9:2736–48. https://doi.org/10.1002/cssc.201600605.
32. Hassan E, Elsayed I, Eseyin A. Production high yields of aromatic hydrocarbons through catalytic fast pyrolysis of torrefied wood and polystyrene. Fuel. 2016;174:317–24. https://doi.org/10.1016/j.fuel.2016.02.031.
33. Zhang HY, Xiao R, Huang H, Xiao G. Comparison of non-catalytic and catalytic fast pyrolysis of corncob in a fluidized bed reactor. Bioresour Technol. 2009;100(3):1428–34. https://doi.org/10.1016/j.biortech.2008.08.031.
34. Zhang B, Zhong ZP, Ding K, Song ZW. Production of aromatic hydrocarbons from catalytic co-pyrolysis of biomass and high density polyethylene: analytical Py-GC/MS study. Fuel. 2015;139:622–8. https://doi.org/10.1016/j.fuel.2014.09.052.
35. Dahl IM, Kolboe SJ. On the reaction mechanism for hydrocarbon formation from methanol over SAPO-34: 2. Isotopic labeling studies of the co-reaction of propene and methanol. J Catal. 1996;161:304–9. https://doi.org/10.1006/jcat.1996.0188.
36. Tau LM, Fort AW, Bao SQ, Davis BH. Methanol to gasoline: 14C tracer studies of the conversion of methanol/higher alcohol mixtures over ZSM-5. Fuel Process Technol. 1990;26:209–19. https://doi.org/10.1016/0378-3820(90)90006-e.
37. Zhang B, Zhong ZP, Chen P, Ruan R. Microwave-assisted catalytic fast co-pyrolysis of *Ageratina adenophora* and kerogen with CaO and ZSM-5. J Anal Appl Pyrolysis. 2017;127:246–57. https://doi.org/10.1016/j.jaap.2017.07.027.

38. Carlson TR, Jae J, Lin YC, Tompsett GA, Huber GW. Catalytic fast pyrolysis of glucose with HZSM-5: the combined homogeneous and heterogeneous reactions. J Catal. 2010;270:110–24. https://doi.org/10.1016/j.jcat.2009.12.013.
39. Zhang HY, Xiao R, Jin BS, Xiao GM, Chen R. Biomass catalytic pyrolysis to produce olefins and aromatics with a physically mixed catalyst. Bioresour Technol. 2013;140:256–62. https://doi.org/10.1016/j.biortech.2013.04.094.
40. Lu Q, Zhang ZF, Dong CQ, Zhu XF. Catalytic upgrading of biomass fast pyrolysis vapors with nano metal oxides: an analytical Py-GC/MS study. Energies. 2010;3:1805–20. https://doi.org/10.3390/en3111805.
41. Ding K, He AX, Zhong DX, Fan LL, Liu SY, Wang YP, Liu YH, Chen P, Lei HW, Ruan R. Improving hydrocarbon yield *via* catalytic fast co-pyrolysis of biomass and plastic over ceria and HZSM-5: an analytical pyrolyzer analysis. Bioresour Technol. 2018;268:1–8. https://doi.org/10.1016/j.biortech.2018.07.108.
42. Casoni AI, Nievas ML, Moyano EL, Alvarez M, Diez A, Dennehy M, Volpe MA. Catalytic pyrolysis of cellulose using MCM-41 type catalysts. Appl Catal A Gen. 2016;514:235–40. https://doi.org/10.1016/j.apcata.2016.01.017.
43. Zhang B, Zhong ZP, Li T, Xue ZY, Wang XJ, Ruan R. Biofuel production from distillers dried grains with solubles (DDGS) co-fed with waste agricultural plastic mulching films via microwave-assisted catalytic fast pyrolysis using microwave absorbent and hierarchical ZSM-5/MCM-41 catalyst. J Anal Appl Pyrolysis. 2018;130:1–7. https://doi.org/10.1016/j.jaap.2018.02.007.
44. Carlson TR, Cheng YT, Jae J, Huber GW. Production of green aromatics and olefins by catalytic fast pyrolysis of wood sawdust. Energy Environ Sci. 2011;4:145–61. https://doi.org/10.1039/C0EE00341G1.
45. Zhang B, Zhong ZP, Song ZW, Ding K, Chen P, Ruan R. Optimizing anti-coking abilities of zeolites by ethylene diamine tetraacetie acid modification on catalytic fast pyrolysis of corn stalk. J Power Sources. 2015;300:87–94. https://doi.org/10.1016/j.jpowsour.2015.09.075.
46. Chen NY, Degnan TF, Koenig LR. Liquid fuel from carbohydrates. Chem Tech. 1986;16(8):506–11.
47. Valle B, Aramburu B, Santiviago C, Bilbao J, Gayubo AG. Upgrading of bio-oil in a continuous process with dolomite catalyst. Energy Fuel. 2014;28:6419–28. https://doi.org/10.1021/ef501600f.
48. Dorado C, Mullen CA, Boateng AA. H-ZSM5 catalyzed co-pyrolysis of biomass and plastics. ACS Sustain Chem Eng. 2014;2(2):301–11. https://doi.org/10.1021/sc400354g.
49. Liu SY, Xie QL, Zhang B, Cheng YL, Liu YH, Chen P, Ruan R. Fast microwave-assisted catalytic co-pyrolysis of corn stover and scum for bio-oil production with CaO and HZSM-5 as the catalyst. Bioresour Technol. 2016;204:164–70. https://doi.org/10.1016/j.biortech.2015.12.085.
50. Wan S, Wang Y. A review on ex situ catalytic fast pyrolysis of biomass. Front Chem Sci Eng. 2014;8(3):280–94. https://doi.org/10.1007/s11705-014-1436-8.
51. Galadima A, Muraza O. In situ fast pyrolysis of biomass with zeolite catalysts for bioaromatics/gasoline production: a review. Energ Convers Manag. 2015;105:338–54. https://doi.org/10.1016/j.enconman.2015.07.078.
52. Wang S, Ru B, Lin H. Pyrolysis mechanism of hemicellulose monosaccharides in different catalytic processes. Chem Res Chin Univ. 2014;30(5):848–54. https://doi.org/10.1007/s40242-014-4019-9.
53. Park YK, Jung JS, Jae J, Hong SB, Watanabe A, Kim Y-M. Catalytic fast pyrolysis of wood plastic composite over microporous zeolites. Chem Eng J. 2019;377:119742. https://doi.org/10.1016/j.cej.2018.08.128.
54. Nguyen TS, Zabeti M, Lefferts L, Brem G, Seshan K. Catalytic upgrading of biomass pyrolysis vapours using faujasite zeolite catalysts. Biomass Bioenergy. 2012;48:100–10. https://doi.org/10.1016/j.biombioe.2012.10.024.

55. Duan D, Wang Y, Dai L, Ruan R, Zhao Y, Fan L, Tayier M, Liu Y. Ex-situ catalytic co-pyrolysis of lignin and polypropylene to upgrade bio-oil quality by microwave heating. Bioresour Technol. 2017;241:207–13. https://doi.org/10.1016/j.biortech.2017.04.104.
56. Zhang HY, Zheng J, Xiao R, Shen DK, Jin BS, Xiao GM, Chen R. Co-catalytic pyrolysis of biomass and waste triglyceride seed oil in a novel fluidized bed reactor to produce olefins and aromatics integrated with self-heating and catalyst regeneration processes. RSC Adv. 2013;3:5769–74. https://doi.org/10.1039/c3ra40694f.
57. Zhang HY, Xiao R, Wang DH, Cho J, He GY, Shao SS, Zhang JB. Hydrodynamics of a novel biomass autothermal fast pyrolysis reactor: solid circulation rate and gas bypassing. Chem Eng J. 2012;181-182:685–93. https://doi.org/10.1016/j.cej.2011.12.057.
58. Bridgwater AV. Fast pyrolysis reactors worldwide. PyNe. 2010;27(1):8–20.
59. Wu CZ, Yin XL, Yuan ZH. The development of bioenergy technology in China. Energy. 2010;35(11):4445–50. https://doi.org/10.1016/j.energy.2009.04.006.
60. Epsterin N, Grace JR. Spouted and spout-flmd beds: fundamentals and applications. Cambridge: Cambridge University Press; 2011. https://doi.org/10.1017/CBO9780511777936.011.
61. Bridgwater AV. Biomass for energy. J Sci Food Agric. 2006;86(12):1755–68. https://doi.org/10.1002/jsfa.2605.
62. Hayes RD. Biomass pyrolysis technology and products: a Canadian viewpoint. In: Pyrolysis oils from biomass. Washington: American Chemical Society; 1988. https://doi.org/10.1021/bk-1988-0376.ch002.
63. Galadima A, Muraza O. In situ fast pyrolysis of biomass with zeolite catalysts for bioaromatics/gasoline production: a review. Energy Convers Manag. 2015;105:338–54. https://doi.org/10.1016/j.enconman.2015.07.078.
64. Li F, Choudhari MM, Paredes P. GÃķrtler instability and its control via surface suction over an axisymmetric cone fluid dynamics conference 3069; 2018.
65. Diebold JP, Czernik S. Additives to lower and stabilize the viscosity of pyrolysis oils during storage. Energy Fuel. 1997;11(5):1081–91. https://doi.org/10.1021/ef9700339.
66. Zhang W, Yuan C, Xu J. Beneficial synergetic effect on gas production during co-pyrolysis of sewage sludge and biomass in a vacuum reactor. Bioresour Technol. 2015;183:255–8. https://doi.org/10.1016/j.biortech.2015.01.113.
67. Yu F, Deng SB, Chen P, Liu YH, Wan YQ, Olson A, Kittelson D, Ruan R. Physical and chemical properties of bio-oils from microwave pyrolysis of corn stover. Appl Biochem Biotechnol. 2007;136-140:957–70. https://doi.org/10.1007/s12010-007-9111-x.
68. Lam SS, Liew RK, Cheng CK, Chase HA. Catalytic microwave pyrolysis of waste engine oil using metallic pyrolysis char. Appl Catal B Environ. 2015;176-177:601–17. https://doi.org/10.1016/j.apcatb.2015.04.014.
69. Zhang B, Tan GW, Zhong ZP, Ruan R. Microwave-assisted catalytic fast pyrolysis of spent edible mushroom substrate for bio-oil production using surface modified zeolite catalyst. J Anal Appl Pyrolysis. 2017;123:92–8. https://doi.org/10.1016/j.jaap.2016.12.022.

Part III
Production of Liquid Biofuels with Microwave Pyrolysis

Chapter 7
Microwave-Assisted Pyrolysis of Biomass: An Overview

Jiby Kurian and G. S. Vijaya Raghavan

Abstract Biomass is considered as an important resource for the production of biofuels and bioproducts through conventional and advanced thermochemical and biochemical technologies. Among the thermochemical technologies, pyrolysis is a quick and efficient method to produce solid, liquid, and gaseous fuels from biomass. Microwave-assisted pyrolysis has many advantages over conventional pyrolysis methods. Volumetric heating obtained though the supply of microwaves leads to the rapid initiation of biomass pyrolysis. Microwave-assisted pyrolysis has high energy efficiency and feedstock with high moisture content can be directly pyrolyzed with this technique. Computer-aided modeling and simulation tools have been applied in the development of equipment and process for the microwave-assisted pyrolysis of biomass. The modeling of process has helped to elucidate the heat and mass transfer mechanism during the microwave-assisted pyrolysis, and the optimised models have been validated using laboratory-scale experiments. Microwave-assisted pyrolysis has been investigated in the production of fuels and biproducts from various agricultural and forest residues. The studies have shown that the composition of feedstock as well as the process conditions have significant effects on the yield and composition of pyrolysis products. Additionally, microwave-absorbers have been investigated to increase the efficiency of the microwave-assisted pyrolysis. There are many challenges to be addressed in the further development of microwave-assisted pyrolysis. The high capital cost and heterogeneity in the heating of complex materials like biomass are to be resolved. These challenges should be addressed to make this technique scalable and suitable for remote and under-developed locations.

Keywords Microwaves-assisted pyrolysis · Liquid biofuels · Biooil · Modelling and simulation · Microwave-absorbers · Reaction kinetics · Scale-up of technology

---

J. Kurian · G. S. V. Raghavan (✉)
Department of Bioresource Engineering, McGill University, St. Anne-de-Bellevue, QC, Canada
e-mail: jiby.kurian@mail.mcgill.ca; vijaya.raghavan@mcgill.ca

Z. Fang et al. (eds.), *Production of Biofuels and Chemicals with Pyrolysis*, Biofuels and Biorefineries 10, https://doi.org/10.1007/978-981-15-2732-6_7

7.1 Introduction

Converting biomass to fuels and materials is actively pursued as a method of carbon management and addressing the escalating demand for energy and materials. Conversion of biomass into fuels can be achieved through numerous combinations of resources, conversion processes, and end products [1]. The feedstocks for biofuel production include agricultural residues, bioenergy crops, forest residues, industry wastes, and municipal wastes [2]. These biomass resources can be converted into fuels through thermochemical and biochemical routes (e.g., fermentation). The thermochemical routes have high efficiency for the production of energy from biomass [3]. Thermochemical routes of biofuel production include combustion, pyrolysis, and gasification [1].

Among the methods to convert biomass into fuels and materials, pyrolysis is relatively quick and produces a wide range of products. Solid, liquid, and gaseous fuels can be produced through the pyrolysis of biomass. In general, pyrolysis is defined as the thermal breakdown of complex compounds into simpler molecules at elevated temperatures in an inert atmosphere [4]. The activation energy required for elevating the temperature of feedstock from room temperature to the reaction temperature needs to be supplied [5]. Different technologies are available for the supply of activation energy for the pyrolysis of biomass into fuels and materials. Conventional pyrolysis process provides this energy through conduction, convection and radiation mechanisms. Conventional pyrolysis process requires long duration for elevating the temperature of biomass from room temperature to the peak temperature of pyrolysis. Conventional methods of supplying thermal energy into biomass result in poor heating characteristics in the core of the biomass, whereas in advanced energy transfer method using microwaves, the biomass is heated volumetrically, and heat generates inside the biomass [1] and provides targeted heating so that the quality of fuel products can be improved with the use of effective catalysts [4]. Microwave-assisted pyrolysis has been applied on biomass materials such as bamboo sawdust [6], Douglas fir sawdust pellet [7], larch wood block [8], corn stover [9], rice straw [10], rice husk [11], sugarcane bagasse [12], waste paper [13], prairie cordgrass [14], wheat straw [15], and pine wood sawdust [16, 17] to name a few for the production of biofuels. The following sections briefly discuss the mechanism of microwave-assisted pyrolysis of lignocellulosic biomass, developments in this subject, use of microwave absorbers in the pyrolysis, modeling and simulation of microwave-assisted pyrolysis of biomass, reaction kinetics and mass transfer studies, and the challenges in the scaling up of microwave-assisted pyrolysis from laboratory scale to industry scale.

7.2 Mechanism of Microwave-Assisted Pyrolysis of Biomass

Microwaves are electromagnetic waves with wavelengths between 1 mm and 1 m and frequency of 300 MHz to 300 GHz. The electromagnetic frequencies that are used for telecommunication systems such as cellular phones, radar, and satellite television communications are also present within these frequency range. The 915 MHz, 2.45 GHz, 5.8 GHz, 22.0 GHz, 24.12 GHz, 28 GHz, and 60 GHz frequencies of microwaves can be used for the processing of dielectric materials. However, among the microwave frequencies, 915 MHz and 2.45 GHz are widely applied for scientific, medical and industrial uses. These frequencies provide uniform electric field distribution in the microwave cavity and higher power dissipation in the dielectric materials [18].

The energy carried by a specific frequency of electromagnetic wave can be estimated using the Planck's equation:

$$E = h\nu \tag{7.1}$$

where, E is the energy in joule, h is the Planck's constant (6.626×10^{-34} J s), and ν is the frequency (s^{-1} or hertz) of the electromagnetic wave [19].

The important parameters that influence the effectiveness of microwave-assisted pyrolysis of biomass include the microwave power absorbed (P) and the depth (D) of microwave penetrated into the biomass. These parameters are highly dependent on the dielectric properties of the biomass material. The microwave power (P) absorbed per unit volume of biomass can be estimated using the following equation:

$$P = \sigma|E|^2 = 2\pi\nu\varepsilon_0\, \varepsilon''|E|^2 = 2\pi\nu\varepsilon_0\varepsilon' tan\delta|E|^2 \tag{7.2}$$

where, E is the magnitude of the internal electric field, ε'' is the dielectric loss factor, ε_0 is the permittivity of free space, ν is the microwave frequency, σ is the total effective conductivity, ε' is the dielectric constant, and $tan\delta$ is the energy dissipation factor (loss tangent) [20].

The dielectric properties (ε'', ε', and $tan\delta$) of the material significantly influence the level of microwave power absorbed by the material. The dielectric constant (ε') is the polarization ability of a material under the electromagnetic field. The dielectric loss factor (ε'') is the ability of the material to dissipate the absorbed energy as heat. The structural and compositional properties of the material influence its dielectric properties [19].

The relation between the dielectric properties of materials at specific temperatures and under specific frequencies can be represented as follows:

$$tan\delta = \frac{\varepsilon''}{\varepsilon'} \tag{7.3}$$

Materials with higher *tanδ* values absorb significantly more electromagnetic energy than the materials with lower *tanδ* values [19]. The absorbed microwave power is converted into heat within the material and the extent of this heat production can be estimated by using the following equation:

$$\frac{\Delta T}{\Delta t} = \frac{2\pi \nu \varepsilon_0 \, \varepsilon'' |E|^2}{\rho C_p} \tag{7.4}$$

where, T is the temperature, t is the process time, ρ is the density, and C_p is the heat capacity of the material [20].

The dielectric properties of the material influence the depth of microwave penetration into the material. The materials with higher ε_r' and *tanδ* values allow less penetration of microwaves of specific wavelengths into them. The depth of penetration of microwaves into the materials can be estimated using the following equation:

$$D = \frac{3\lambda_0}{8.686\pi \, tan\delta \left(\frac{\varepsilon'}{\varepsilon_0}\right)^{\frac{1}{2}}} \tag{7.5}$$

where, D is the depth of penetration of microwaves at which the incident power is reduced by one half and λ_0 is the incident wavelength [20]. The microwaves undergo reflection, refraction, interference, diffraction, and polarization in the biomass [21].

The interaction of materials with the electric field of electromagnetic waves is measured in terms of the dielectric response of the material. It is important to know the dielectric properties of materials for the effective use of microwave-assisted heating technique [22]. Depending on type of material and its temperature, each material has different capacities to interact with and absorb microwaves. Depending on the depth of penetration of microwaves and interaction with the electromagnetic fields of microwaves, materials can be classified as microwave opaque, microwave transparent, and microwave-absorbers. The microwave opaque materials reflect the microwaves without allowing any penetration of microwaves. Examples are metal conductors that largely reflect microwaves and are not effectively heated by microwaves. The microwave transparent materials are also called as low dielectric loss materials and they allow the microwaves to pass through them without any interaction between the material and microwaves. Examples of microwave transparent materials include glass, paper, and most of the plastic materials. Microwave-absorbers are materials with high dielectric loss properties, and they absorb microwave energy and convert the electromagnetic energy into thermal energy to a certain degree depending on the dielectric loss factor of the material. Examples of microwave-absorbers are water, nitric acid, and the biological materials containing polar compounds [19]. Multiphase materials like biomass are mixed absorbers, where some components are high dielectric loss materials while some are with low dielectric loss factors. This property of composite materials can be utilised in the

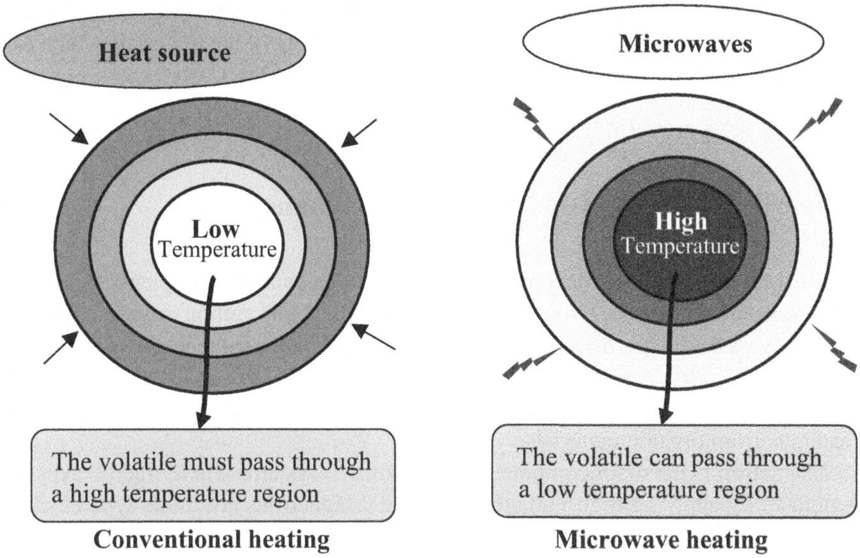

Fig. 7.1 Schematic diagram of temperature distribution and heat and mass transfer mechanisms in the conventional and microwave heating of biomass. Adapted with permission from [8], Copyright © 2004, Elsevier

selective heating process, which is not achievable with conventional heating methods [20].

In microwave-assisted heating, heat is generated inside the biomass due to the molecular friction due to the ionic migration and rotation of dipolar molecules caused by microwaves. Molecular rotation occurs in materials containing polar molecules with electrical dipole moment. The alignment of these dipole molecules will fluctuate in an electromagnetic field along with the oscillation of the fields of electromagnetic waves. Since the electric field of 2.45 GHz microwaves alternate 2.45×10^9 times per second, the dipole molecules align and realign 4.9×10^9 times per second causing continuous friction between the molecules [19]. These rotations cause friction between the molecules and heat energy will be generated within the material. The heat energy will be transferred between the atoms and molecules and also it will be transferred from the inside to the outside of the material leading to volumetric, fast and uniform heating of the material [23, 24]. A general schematic of the difference in temperature distribution, heat transfer, and mass transfer in conventional and microwave-assisted heating methods is shown in Fig. 7.1.

Different components of biomaterials absorb microwaves at different levels and causes selective heating with hotspots in heterogeneous materials. Studies have shown that the highest temperature and heating rate of microwave-assisted pyrolysis of biomass increased with the increase in polysaccharide content of biomass [1]. Polysaccharide contents are decomposed to volatile compounds while lignin molecules decomposed to char. Also, the dielectric properties of biomass changes with the increase in temperature and different components of biomass decompose to char at different rates. Char heats up faster than biomass components under microwave pyrolysis [25]. This can lead to nonuniform heating of the materials which can be addressed with the addition of microwave-absorbers [25].

7.3 Microwave-Assisted Pyrolysis of Biomass

In 1968, Fu et al. reported the production of hydrogen cyanide, acetylene, cyanogen, low molecular weight hydrocarbons, hydrogen, and carbon oxides through microwave-assisted pyrolysis of high-volatile bituminous coal in the presence of nitrogen [27, 26]. As shown in Fig. 7.2 [28], microwave-assisted pyrolysis is actively investigated in the recent decades as a process to produce energy and materials from organic materials.

Magnetron is the main component of microwave systems where microwaves are generated by applying high voltage electricity. Magnetrons are diode-type electron tubes containing the anode, the cathode, the antenna, and high-power permanent magnets. Microwaves are passed from the microwave generator to the application chamber through wave guides. The conversion efficiency of electrical energy to 915 MHz microwaves is about 85% and to 2.45 GHz is about 50% [16]. Compared to 2.45 GHz, 915 MHz microwaves penetrate deeply into the materials. The microwave application chamber confines the microwaves in the system and the rotation of biomass increases the absorption of microwaves. The alternating electric and magnetic fields of microwaves is used for generating heat inside biomass [30]. A schematic of the system for microwave-assisted pyrolysis of biomass is shown in Fig. 7.3.

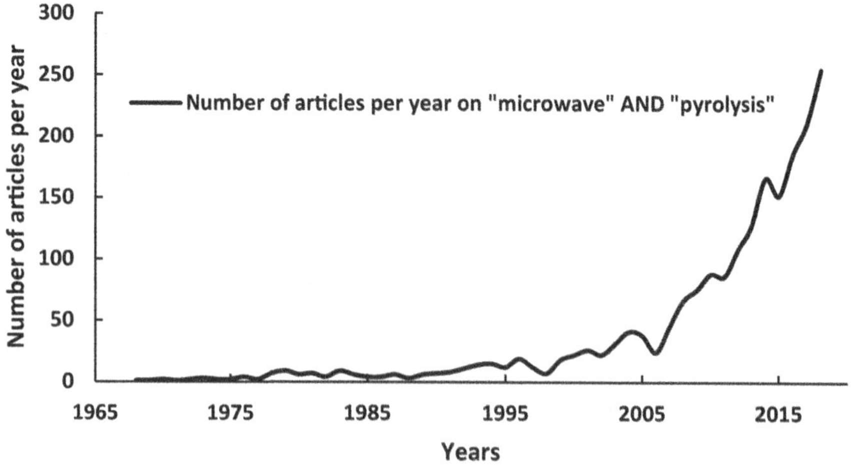

Fig. 7.2 Number of articles over the years on microwave-assisted pyrolysis of organic materials [28]

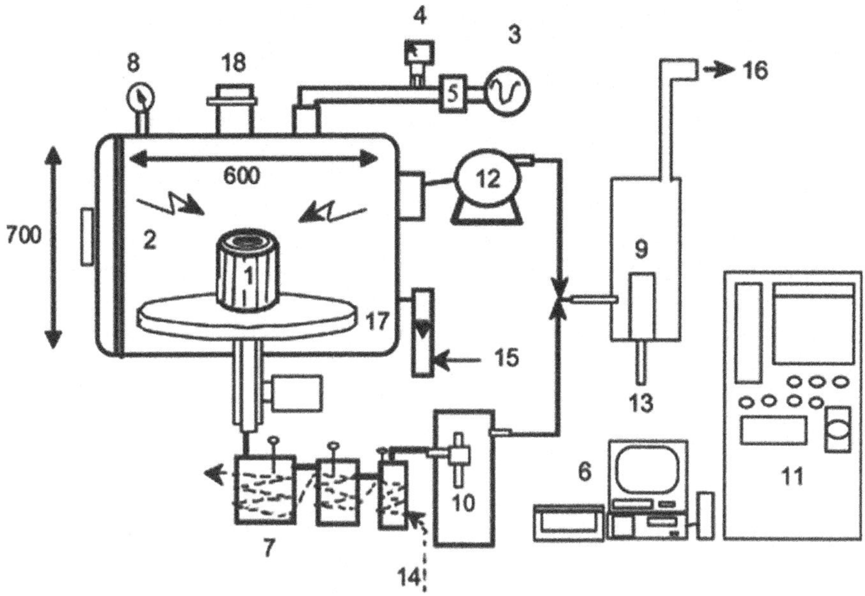

Fig. 7.3 Microwave-assisted pyrolysis reactor. 1: Sample; 2: Reactor (width: 600 mm; depth: 600 mm; height: 700 mm); 3: Microwave power generator; 4: Power monitor; 5: Isolator; 6: Computer; 7: Collection tanks; 8: Pressure gauge; 9: Afterburner; 10: Aspirator pump; 11: Control panel; 12: Blower; 13: Kerosene; 14: Condenser; 15: Nitrogen gas; 16: Exhaust gas; 17: Rotating table; and 18: Safety shutoff. Reprinted with permission from [8], Copyright © 2004, Elsevier

Heating of biomass under microwave-assisted method is faster than in conventional pyrolysis method. Studies have shown that the temperature of lignocellulosic agricultural residues has reached 320–530 °C within 10 min when the microwave power levels were 300–500 W. Increase in the microwave power level used increases the gas production and decreases the production of solid and liquid fuels from biomass [30]. Also, the activation energy for microwave pyrolysis is lower than that of conventional pyrolysis, and therefore the reaction kinetics of microwave pyrolysis is different from that of conventional pyrolysis [16]. Some of the examples of microwave-assisted pyrolysis of biomass resources to produce biofuels are given hereafter.

Miura et al. [8] have investigated the microwave-assisted pyrolysis of wood and concluded that the temperature distribution and heat and mass transfer mechanisms are different from that observed in conventional pyrolysis method. The microwave-assisted pyrolysis is controllable to produce the desired product and to prevent the further pyrolysis of the products formed. The biochar obtained through microwave-assisted pyrolysis had specific surface area of about 450 m^2/g, which is higher than that of biochar produced through conventional methods. Also, the biochar produced in the microwave-assisted pyrolysis contained less amount of tar inside its pores. The researchers have reported that electric power consumption is inversely proportional to the diameter of the wood block used, which shows that a larger wood block requires less electrical power per unit weight than by a wood block with smaller diameter [8].

Microwave-assisted pyrolysis has been investigated for the conversion of distillers dried grain with solubles (DDGS) to bio-oil, syngas, and biochar. The effects of factors such as reaction temperature, reaction time, and microwave power input on the yield of biofuels were analyzed. About 26–50% of the feedstock was converted into biofuels, and the biooils obtained had an energy content of about 28 MJ/kg which is about 66% of the heating value of gasoline. The biooils from DDGS contained many important aliphatic and aromatic hydrocarbons and many of them are also found in regular unleaded gasoline [31]. Similarly, studies have shown that the biooil produced through the microwave-assisted pyrolysis of wet sewage sludge contained over 70 different industrially valuable chemicals such as benzene, pyridine, decane, limonene, toluene, ethylbenzene, styrene, palmitic acid, hexadecanamide, and cholestane [32]. The calorific value of biooil produced from wet sewage sludge was about 36 MJ/kg. The high calorific value of this biooil makes it useful as a fuel [33].

The product distribution from the microwave-assisted pyrolysis of biomass materials is influenced by type of biomass, feedstock weight, particle size, microwave power level, reaction temperature, reaction time, product vapor residence time, reactor design, and the heating rate [16]. Microwave-assisted pyrolysis of soapnut seed powder with particle size of about $\leq$2 mm at 7 W/g microwave power for 20 min under nitrogen atmosphere produced 52% yield of biooil samples with about 22 MJ/kg calorific value and a moisture content of 10–15%. The peak temperature of pyrolysis was 528 °C. The high moisture content shows that the biooil needs to be upgraded to be used as fuel in internal combustion engines [34].

Microwave-assisted pyrolysis of corn stover with different particle sizes (0.5–4 mm) shows that, thermochemical reactions occur rapidly in larger particles than in smaller particles [9]. Milled office paper (printed and photocopied) wastes were converted into biooil through low temperature (<200 °C) microwave-assisted pyrolysis. About 42% yield of biooil was obtained, and it contained 19% as organic phase compounds and 23% as aqueous phase compounds. The organic phase biooil mainly contains aromatic compounds and aqueous phase biooil mainly contains carbohydrate sugars [13].

Microwave-assisted pyrolysis of soybean straw under CO_2 atmosphere was studied to produce H_2 rich syngas. The yield of syngas was about 55% which was higher than that obtained from the pyrolysis under N_2 atmosphere (45% yield) at 1000 W microwave power. The concentration of H_2 plus CO in the syngas was about 84% with relatively higher concentration of CO than H_2. This study has shown that increase in microwave power up to 800 W increases the production of biooil with faster attainment of the highest temperature for pyrolysis. Further increase in microwave power decreases the yields of biooil and biochar and increases the production of syngas [35].

Microwave-assisted pyrolysis of dried seaweed *Laminaria digitata* and its extract was investigated to produce biooil without adding any microwave-absorbers. The input energy requirement was in the range of 1.84–2.83 kJ/g and it pyrolyzed

55–70% of the feedstock in 200 s. The biooil yield was about 15% from the seaweed and 5% from its extract. The biooil from seaweed and its extract contained mainly nitrogenous and carbohydrate compounds such as L-Proline, 1-methyl-5-oxo, methylester and lacked phenolic compounds. The low yield and the absence of phenolic compounds in biooil make this feedstock not a favorable source of bioenergy [36].

7.4 Microwave-Assisted Pyrolysis of Biomass with Microwave-Absorbers

Dry biomass is a poor absorber of microwave, and therefore microwave-absorbers are often added to increase the microwave absorption rate and thereby to reduce the microwave power requirement [4]. Some microwave-absorbers (e. g., K_3PO_4, clinoptilolite, and bentonite) also act as catalyst for enhancing the efficiency of pyrolysis. These catalysts improve the quality of biooil by reducing viscosity and acidity and also increase the concentration of hydrocarbon products [4]. A few of the studies on the use of microwave-absorbers for the enhanced microwave-assisted pyrolysis are discussed in this section.

The doping of biomass with char formed through pyrolysis can enhance the heat generation and distribution in the microwave-assisted pyrolysis. Zhao et al. have investigated the effects of silicon carbide (SiC) on the microwave-assisted pyrolysis of wheat straw. The silicon carbide was used to heat the microwave reactor and the pipeline parts for preventing the condensation of volatile compounds as tars on these parts and thus to extend the lifetime of the system. Also, the collection of volatile compounds as biofuels enhanced the overall efficiency of biomass pyrolysis [37].

Catalytic microwave-assisted pyrolysis of biomass for the production of biofuels and phenols has been investigated by using Douglas fir as the biomass and activated carbon as the catalyst. The biooils produced contained high concentrations of phenols (38%) and phenolic compounds (66%). High concentration of long chain fatty acid esters was obtained in the presence of zinc powder as catalyst. The high concentration of phenols and fatty acid esters make the biooil suitable for the partial substitution of gasoline fuels and as feedstock for the synthesis of industrial materials. However, the oxygenates in biooil needs to be separated and the product should be purified and upgraded before the industrial uses [38].

Microwave-assisted pyrolysis of sugarcane bagasse in the presence of metal oxides (NiO, CuO, CaO, and MgO) was studied by Kuan et al. and the results have shown that the use of different metal oxides leads to different product distribution pattern. The addition of metal oxides increased the reaction rate and influenced the kinetic parameters [12]. Similarly, the use of iron (Fe) and cobalt (Co) particles as microwave-absorbers have increased the yield of H_2 and CH_4 gases in the microwave-assisted pyrolysis of sugarcane bagasse. It also shown to produce biochar nanoparticles and carbon nanotubes, with average diameter in the range of (20–120) nm, during the pyrolysis of biomass [39].

Microwave-assisted pyrolysis has been applied to produce fuels from sewage sludge of waste water treatment plants. Graphite and sewage sludge char were used as microwave-absorbers in the microwave-assisted pyrolysis of wet sewage sludge. Use of microwave-absorbers helped in achieving about 1040 °C as the peak temperature of pyrolysis within 6 min of the process. The heating rate obtained in microwave-assisted pyrolysis was about 200 °C/min which is significantly higher than the highest heating rate of 74.3 °C/min obtained in an electric furnace pyrolysis system. Also, studies have shown that microwave-assisted pyrolysis produced higher quantity of syngas from wet sewage sludge than by the conventional pyrolysis method [33]. The quality of biooil produced from sewage sludge was not affected by the type of microwave-absorber used. Importantly, microwave-assisted pyrolysis produced less amount of hazardous polycyclic aromatic hydrocarbons (PAH) than by the conventional pyrolysis method [33].

Rice husk char and rice husk char with the metallic (Ni, Fe, and Cu) catalysts were investigated for improving the product quality of the microwave-assisted pyrolysis of rice husk. The rice husk char with metallic catalysts improved the microwave absorption capability, heating rate, and the highest temperature reached of pyrolysis. These catalysts convert tar and heavy organic compounds into simpler organic compounds [11].

Oil palm shells and fibers were converted into biooil, biochar, and syngas through the microwave-assisted pyrolysis with biochar as microwave-absorber. Presence of microwave-absorbers can reduce the consumption of energy, time, and resources for microwave-assisted pyrolysis of biomass. Additionally, direct pyrolysis of large particles of biomass is possible in microwave-assisted pyrolysis method which can reduce the energy required for grinding and drying of biomass [40]. Coconut activated carbon has also been used as a microwave-absorber during the pyrolysis of oil palm shell for increasing the pyrolysis temperature, and the biooil produced was rich in high energy components such as phenol and 1,1-dimethyl hydrazine. Uniform distribution of microwave-absorbers in the biomass leads to uniform heating of the feedstock and reduced the hotspot formation. It can improve the biomass conversion efficiency and biooil composition profile [41].

Microwave-assisted pyrolysis of switchgrass in the presence of a mixture of catalysts was investigated to understand the effect of catalyst mixtures on the rates of microwave heating and biomass decomposition. The formation of coke on the surface of catalysts was investigated to determine the role of biomass decomposition product on further enhancement of microwave absorption and pyrolysis. A mixture of bentonite ($Al_2O_{34}SiO_2H_2O$), potassium phosphate tribasic (K_3PO_4), and clinoptilolite ($(K, Ca, Na)_2O\text{-}Al_2O_3\text{-}10SiO_2\text{-}6H_2O$) at different ratio was mixed with switchgrass for microwave-assisted pyrolysis. K_3PO_4 and clinoptilolite are ionic compounds and have good ability for absorbing microwaves. Bentonite is a natural aluminum phyllosilicate which has low capacity for microwave absorption but with high thermal conductivity (1.15 W/m K). It was shown that pure switchgrass powder had low capacity for the absorption of microwaves, and the highest temperature reached was 160 °C after 30 min of microwave treatment. Microwaves will be absorbed by the catalysts with high microwave absorption capacity and heated faster than biomass and the heat will be transferred to biomass through conduction. The vapours formed from the pyrolysis of biomass will be cracked by the catalysts to different products. The coke deposited on the surface of catalysts have significant effect on the microwave heating rate and heating behaviour of biomass. Oxygenated coke decreases the microwave absorption, and graphitic carbon has high ability for the absorption of microwaves and it further improves the efficiency of microwave-assisted pyrolysis of biomass [4]. In a similar study, the use of biochar with iron (Fe) as a catalyst and microwave-absorber in the microwave-assisted pyrolysis of torrefied corn cob has shown the production of biooil rich in phenolic compounds [42]. These results are useful in the development of systems for the microwave-assisted pyrolysis of biomass for biofuel production [4].

7.5 Modeling and Simulation of Microwave-Assisted Pyrolysis of Biomass

Mathematical modelling and simulation of microwave-assisted pyrolysis of biomass has been reported by many researchers [29, 44, 25]. Recent developments in computer technology help in modelling microwave-assisted pyrolysis of biomass through different methods including the finite-difference time-domain (FDTD) and the finite element method (FEM) [25]. Finite element analysis with Maxwell's equations has been applied to simulate the energy distribution inside a microwave oven and also in a complex material like biomass. The Maxwell's equations can be written as follows:

$$\nabla \times \overline{H} = \frac{\partial \overline{D}}{\partial t} + \overline{I} \tag{7.6}$$

where, $\overline{H}$ represents the magnetic field vector, $\overline{D}$ represents the electric flux density vector, and $\overline{I}$ represents the current density vector. Eq. (7.6) represents Ampere's law which states that the change in magnetic field around a closed circuit is equal to the net electric current in the circuit. The relationship between the electric field and magnetic field in a closed circuit is given by Faraday's law as shown below:

$$\nabla \times \overline{E} = -\frac{\partial \overline{B}}{\partial t} \tag{7.7}$$

where, $\overline{E}$ represents the electric field vector, and $\overline{B}$ represents the magnetic flux density vector. Gauss's laws state the boundary conditions for the electromagnetic fields in a closed circuit.

$$\nabla \cdot \overline{D} = \rho_v \tag{7.8}$$

$$\nabla \cdot \overline{B} = 0 \tag{7.9}$$

where, ρ_v represents the electric volume charge density. These equations show that the net magnetic flux in a region of the circuit is zero and the net electric flux is dependent on the charge in that region. An in-depth understanding of electromagnetic theory is required for the design of effective microwave systems for the processing of materials. Finite element analysis can be used to couple electromagnetic heating, combustion, and heat and mass transfer in microwave-assisted pyrolysis of biomass [23]. The detailed partial differential equations of the mathematical model for the microwave-assisted pyrolysis of biomass can be found in articles by Santaniello et al. [43], Dutta et al. [29], and Halim and Swithenbank [25].

The differences in heating dynamics of microwave-assisted pyrolysis and conventional pyrolysis of biomass were mathematically simulated by Santaniello et al. and good agreement between the predicted and actual data on the mass loss during pyrolysis was obtained. It was shown that low microwave power and thicker sample materials lead to faster pyrolysis of biomass and that microwave heating is more thermally efficient than conventional heating at similar input power setting [43]. In a similar study, the finite element model of microwave-assisted pyrolysis of biomass added with char and graphite as microwave-absorbers was developed by Dutta et al. and the equations were used for simulation of the process using COMSOL Multiphysics® software (COMSOL Inc., USA). The simulation results have shown that the presence of microwave-absorbers enhance the heat transfer within the biomass and char particles were better than graphite in enhancing the microwave heating of biomass [29].

The effects of different heating methods for the biomass pyrolysis and the production of biofuels through microwave-assisted pyrolysis were simulated using software packages such as Aspen Plus™ and Aspen HYSYS® (Aspen Tech Inc., USA). Production of methanol from sugarcane bagasse using microwave-assisted pyrolysis was modelled and simulated using Aspen HYSYS®. The pyrolysis reactor was designed and the effects of process parameters such as temperature (0–700) °C, nitrogen gas flow rate (0–600) cm^3/min), heating rate (0–35) K/min, and biomass moisture content (0–9) % on the chemical reactions involved in the process were considered. The energy balance of the process was analysed. The simulation studies have shown that the pyrolysis temperature, nitrogen gas flow rate, and heating rate have significant effects on the pyrolysis process and methanol yield. The moisture content of biomass has no significant impact on the microwave-assisted pyrolysis process. The simulation results were verified with the data from laboratory-scale experiments and it was shown that about 6.42% yield of methanol was obtained from sugarcane bagasse [44]. However, this study has not considered the effect of microwave power and pyrolysis time on the yield of biofuel.

The modelling and simulation of microwave-assisted pyrolysis of *Acacia Nilotica*, *Calophyllum inophyllum* seed, rice husk, and Bael shell for the production of biooil was conducted using Aspen Plus™ software package. This software package allows the user to develop and test a chain of chemical processes having different physical, chemical, and biological parameters. The proximate and ultimate composition of biomass was used as input data for the estimation of biooil yield. Among the feedstocks tested, the seeds of *Calophyllum inophyllum* produce the highest (48%) yield of biooil when pyrolyzed at 500 °C with microwave energy. The simulation has shown that the yield of biooil decreased with the increase in pyrolysis temperature in the range of (500–700) °C [45].

Pyrolysis of gumwood with conventional electrical heating and microwave-assisted heating for the production of H_2 was modelled and simulated using Aspen Plus™ software package. The results were verified with the data from laboratory-scale experiments. The results have shown that the computer aided modelling and simulation tools can help in detailed evaluation of the process and determining the energy efficiency and mass flow at different levels of operation [46].

Halim and Swithenbank [25] have used COMSOL Multiphysics® software to study the microwave-assisted heating of biomass and the electromagnetic field distribution inside the reactor. They investigated the effects of mode stirrer, location of biomass inside the microwave cavity, waveguide position, dielectric properties of biomass, and the presence of microwave-absorbers. The presence of cold and hot spots during the heating of biomass leads to heterogeneity in the temperature distribution. Placing waveguide at the bottom of microwave cavity results in poor distribution of electric field. Side-fed microwave energy and mode stirrer improve the distribution of electric field inside the microwave cavity. The location of biomass inside the microwave cavity influence the electric field distribution and heating of biomass. The optimum bed size of biomass was found at 50 mm height for the highest absorption of microwaves. These modelling and simulation efforts are useful in further developing the computational tools incorporating complex mass transfer mechanisms and chemical reactions [25].

7.6 Reaction Kinetics and Mass-Energy Balance of Microwave-Assisted Pyrolysis of Biomass

Studies on the pyrolysis reaction kinetics would help in better understanding of the reaction pathways and predict the behaviour of biomass materials at different operating conditions. It will further help in designing more efficient pyrolysis reactors and improve the product quality [4]. Biomass pyrolysis is a very complex process and can not be described by single step reaction model. The components of biomass decompose at different rate and temperatures. The mass loss rate curves show overlapping peaks and the kinetics of decomposition of individual component is challenging [2]. Generally, the pyrolysis of biomass can be represented as:

$$\text{Biomass} \rightarrow \text{volatile matter} + \text{solid residue}$$

The rate of decomposition of biomass during pyrolysis is given as:

$$\frac{d\propto}{dt} = k(T)f(\propto) \quad (7.10)$$

where, T is the absolute temperature, $k(T)$ is the rate constant dependent on temperature, $\propto$ is the fractional conversion of biomass at time t, $f(\propto)$ is a function dependent on the reaction mechanism. The fractional conversion of biomass is defined as:

$$\propto = \frac{(w_0 - w)}{(w_0 - w_\infty)} \quad (7.11)$$

where, w_0, w, and w_∞ are the initial weight, weight at time t, and final weight of biomass sample [6].

The rate constant $k(T)$ can be calculated according to the Arrhenius equation:

$$k(T) = A\exp\left(-\frac{E}{RT}\right) \quad (7.12)$$

where, A is the pre – exponential or the frequency factor (s – 1), E is the activation energy (J mol^{-1}), and R is the universal gas constant (8.314 J mol^{-1} K^{-1}).

The reaction kinetics of microwave-assisted pyrolysis of biomass have been investigated to estimate the activation energy required and the preexponential factor of the pyrolysis. The thermal decomposition of biomass at a constant heating rate can be expressed by using Arrhenius equation:

$$\frac{d\propto}{dT} = \frac{A}{\beta}\exp\left(-\frac{E}{RT}\right)(1-\propto)^n \quad (7.13)$$

where, β is the heating rate (Ks^{-1}). The equation can be integrated as:

$$\ln\left[\frac{-\ln(1-\propto)}{T^2}\right] = \ln\left[\frac{AR}{\beta E}\left(1 - \frac{2RT}{E}\right)\right] - \frac{E}{RT} \text{ (for n = 1)} \quad (7.14)$$

$$\ln\left[\frac{1-(1-\propto)^{1-n}}{T^2(1-n)}\right] = \ln\left[\frac{AR}{\beta E}\left(1 - \frac{2RT}{E}\right)\right] - \frac{E}{RT} \text{ (for n} \neq 1) \quad (7.15)$$

If $\frac{2RT}{E} \ll 1$, Eqs. (7.13) and (7.14) can be simplified as:

Table 7.1 Kinetic parameters observed for microwave-assisted pyrolysis of biomass

Biomass	*E* (kJ mol^{-1})	*A* (1/s)	Ref.
Prairie cordgrass	3.1	2.26	[14]
Sugarcane bagasse	18.9	0.18	[12]
Sugarcane bagasse with NiO (10% wt)	20.2	0.25	
Sugarcane bagasse with CuO (10% wt)	20.9	0.29	
Sugarcane bagasse with CaO (10% wt)	21.4	0.34	
Sugarcane bagasse with MgO (10% wt)	20.0	0.23	
Douglas fir sawdust pellet	16.5	15.3	[7]
Rice straw	27.1	1.06	[10]
Rice straw (50–250) W microwave	6.82	4.98×10^{-3}	[1]
Rice straw (250–500) W microwave	21.37	2.47×10^{-1}	
Moso bamboo sawdust	24.5	5.27	[6]

$$\ln\left[\frac{-\ln(1-\propto)}{T^2}\right] = \ln\left[\frac{AR}{\beta E}\right] - \frac{E}{RT}\ (\text{for } n = 1) \tag{7.16}$$

$$\ln\left[\frac{1-(1-\propto)^{1-n}}{T^2(1-n)}\right] = \ln\left[\frac{AR}{\beta E}\right] - \frac{E}{RT}\ (\text{for } n \neq 1) \tag{7.17}$$

The *E* and *A* can be determined from the plot of $\ln\left[\frac{-\ln(1-\propto)}{T^2}\right]$ versus $\frac{1}{T}$ when n = 1, or $\ln\left[\frac{1-(1-\propto)^{1-n}}{T^2(1-n)}\right]$ versus $\frac{1}{T}$ when n ≠ 1, that result in a straight line with $-\frac{E}{R}$ as the slope and $\ln\left[\frac{AR}{\beta E}\right]$ as the intercept [16].

Studies have shown that microwave-assisted pyrolysis has lower activation energy and pre-exponential factor than that of conventional pyrolysis processes. Some examples of the kinetic parameter values of microwave-assisted pyrolysis of lignocellulosic biomass is listed in Table 7.1.

Investigations on the microwave-assisted pyrolysis of biomass such as corn stover, rice straw, rice husk, sugarcane bagasse, sugarcane peel, coffee grounds, and bamboo leaves have shown that the highest temperature of pyrolysis is highly correlated with the combustible content of the feedstocks. When the combustible content of feedstocks increased from 88 wt% to 96 wt%, the heating rate increased from 108 °C/min to 135 °C/min and the highest temperature of pyrolysis increased from 495 °C to 539 °C. Microwave power level applied (300, 400, 500) W influence the heating rate (51–140) °C and the highest temperature of heating (346–551) °C achieved. The heating rate and the highest temperature of pyrolysis of bamboo leaves was the lowest among the feedstocks tested. Use of biomass such as corn stover, sugarcane peel and rice straw with higher concentration of combustible contents, decreased the input energy for the pyrolysis. Microwave-assisted pyrolysis results in higher mass conversion to fuels than by the conventional pyrolysis methods. At 500 W, the average yields of solid, liquid, and gaseous fuels from these feedstocks was (20 ± 2) wt%, (44 ± 2) wt%, and (36 ± 4) wt%, respectively.

The biooil was primarily composed of aliphatic, phenolic, and polycyclic aromatic compounds. The kinetic parameters of microwave-assisted pyrolysis were found to be different at lower and higher pyrolysis temperatures. Reaction rates were higher by one order of magnitude at higher temperature of pyrolysis than at lower temperature of pyrolysis. The rate constant for microwave-assisted pyrolysis was much higher with lower activation energy and pre-exponential factor than that of conventional electrical heating pyrolysis method. These studies have shown that microwave-assisted pyrolysis would be industrially more viable with the use of biomass feedstocks containing higher concentration of easily combustible contents such as cellulose and hemicellulose [1].

Studies have analysed the input energy required for the microwave-assisted pyrolysis of biomass for the production of biofuels. The electricity consumption under experimental conditions for the microwave-assisted pyrolysis of wheat straw and corn stover bales was estimated at (0.58–0.65) kWh per kg of biomass. The minimum power of microwaves required for the microwave-assisted pyrolysis was 0.37 kW per kg of biomass. The energy consumption increases with the increase in microwave power and processing time applied. This energy consumption decreased with the increase in mass of biomass used. About 42% of the total energy required is lost as heat and in the conversion of electricity to microwaves. Use of syngas to provide the activation energy required for microwave-assisted biomass pyrolysis can reduce the electricity consumption by 55% [47].

The energy requirement for the microwave-assisted pyrolysis of wheat straw includes for drying of the biomass and activation of materials for the reactions. The amount of this energy can be estimated as the sum of energy for the evaporation of moisture during the drying stage and the evaporation of organic compounds during the pyrolysis stage. A unit operating at 6000 kg wheat straw per hour needs 474 kJ energy for the drying of biomass and 1302 kJ energy for the microwave-assisted pyrolysis. The total energy input for the biooil production is estimated at 1.78 MJ and the energy output value of biooil is 8.98 MJ, based on 20.2% yield and a low energy value of 16,000 kJ per kg biooil. This shows about five times return on energy input in the form of biooil as the product of pyrolysis. Additional energy return in the form of biochar and syngas confirms the favourable energy balance of the microwave-assisted pyrolysis process [15].

The energy content of syngas produced through the microwave-assisted pyrolysis of gumwood at (600–800) °C was found to be 23% higher than that of the syngas produced through conventional pyrolysis at similar temperatures. About 15% higher concentration of H_2 (120 g H_2/kg gumwood) was obtained in microwave-assisted pyrolysis process. The yield of biochar was higher in the microwave-assisted pyrolysis process. However, the yield of biooil was found to be lower in the microwave-assisted pyrolysis than in the conventional pyrolysis process. The overall efficiency of microwave-assisted pyrolysis process was 13.5% higher than that of the conventional pyrolysis process for gumwood [46].

7.7 Challenges in the Scaling Up of Microwave-Assisted Pyrolysis of Biomass

Though the microwave-assisted pyrolysis technology has advantages over the conventional techniques, there are many challenges need to be addressed for the commercial success of this technology. Large-scale systems for the microwave-assisted pyrolysis of biomass are relatively expensive to build and operate [48]. Most of the studies on microwave-assisted pyrolysis of biomass were on batch operation mode and many of these investigations were using microwave-absorbers for preventing hotspot formation and increasing the efficiency of the process. The volume of biomass processed in these investigations was very low and the large-scale operations increased the heterogeneity of the heating. The penetration depth of microwaves into large volume of biomass was very low. These problems lead to the lack of control on the biomass pyrolysis as well as on the quality of biofuels produced. The dielectric properties of the biomass feedstock vary with the change in temperature and it influences the microwave-assisted pyrolysis of biomass. Evaporation of moisture from the biomass feedstock decreases the dielectric properties of biomass and reduce the heating rate beyond 100 °C. The dielectric properties of biomass feedstock are not changed significantly in the temperature range of (100–500) °C [49]. Formation of char particles at (500–600) °C significantly increases the dielectric properties of biomass and leads to temperature runaway in the process. This effect is difficult to control and leads to secondary pyrolysis reactions such as gasification of char and reformation of valuable biooil components. Use of microwave-absorbers reduce the temperature threshold for the temperature runaway effect, and therefore the use of microwave-absorbers may not be effective at industrial scale for biofuel production. Also, the use of microwave-absorbers reduces the penetration depth of microwaves into biomass and leads to heating heterogeneity in the process [50].

Continuous mode operation of microwave-assisted pyrolysis requires the design and integration of specific systems for the addition of biomass and separation of biofuels produced. The systems should prevent the loss of microwaves and volatile components into the surrounding environment. The materials used for the parts that come in contact with microwaves should be transparent to microwaves, mechanically robust, and thermally stable at high temperatures of biomass pyrolysis. Basalt fibers, ceramic, and non-coated glass fibers are reported as suitable to make parts for the handling of biomass in microwave-assisted pyrolysis systems. However, the design of the systems needs to consider the change in structure, particle size, and composition of biomass feedstock [50, 48].

The environmental impacts and sustainability of microwave-assisted pyrolysis of biomass need to be fully assessed. Life cycle assessment (LCA) is a useful tool to analyse the environmental impacts of the microwave-assisted pyrolysis of biomass and the products formed. LCA considers the different phases in the microwave-assisted pyrolysis of biomass including biomass planting, harvesting, transporting, building and operating the pyrolysis plant, product processing, product use, as well as the demolition and recycling of pyrolysis plant [52]. It gives information on the potential impacts such as soil erosion, biodiversity change, global warming, acidification, and eutrophication. The values are estimated in terms of CO_2, SO_2, and PO_2 equivalents. Studies have shown that the pretreatment such as size reduction and drying of biomass for pyrolysis has the highest environmental impacts potential. Use of exhaust gases for the drying of biomass would reduce the greenhouse gas emission of the process. Also, the ability of microwave-assisted process to process wet biomass can reduce the global warming potential of the pyrolysis process. Use of fossil fuel derived electricity in the pyrolysis process leads to the highest effects on greenhouse gas emissions and global warming. Similarly, the use of nitrogen gas has significant effects on the greenhouse gas emissions [51].

Economic viability of microwave-assisted pyrolysis of biomass is yet to be analysed in commercial scale [48, 52]. Economic analyses based on data from laboratory-scale microwave-assisted pyrolysis of Douglas fir pellets have shown positive economic outcomes when using portable small-scale systems for the production of biooil rich in aromatic hydrocarbons. Capital cost required, yield of biooil, and the selling price of biooil are the important variables that significantly impact the economic viability of microwave-assisted pyrolysis process. The cost of microwave systems is much higher than the conventional heating systems. However, there would be smaller number of unit operations with low operating cost in the microwave-assisted pyrolysis process than in the conventional pyrolysis process [51].

Distributed systems for the microwave-assisted pyrolysis has been proposed for the in-situ processing of biomass [53, 54]. The syngas produced can be used for the heating of biomass, water, and working space, and it can also be used for cooking food. Use of syngas in the pyrolysis process can reduce the overall consumption of electrical energy for the microwave-assisted pyrolysis of biomass. Studies have shown high energy efficiency of microwave-assisted pyrolysis in comparison with conventional pyrolysis process. The energy recovery efficiency is calculated as the ratio between energy content of biofuel and the sum of energy required for pyrolysis plus the energy contained in the biomass. Microwave-assisted pyrolysis of wheat straw has resulted in an energy recovery efficiency of about 79.8% whereas the energy recovery was 99% in the case of microwave-assisted pyrolysis of coffee hulls [48]. Nevertheless, these values are obtained based on laboratory-scale studies and the actual energy efficiency of industry-scale microwave-assisted pyrolysis of biomass is yet to be analysed. These studies need to be conducted with different feedstock types, system designs, processing capacities, product options, and operational techniques [51, 53].

Microwave-assisted pyrolysis has many unique advantages that can be explored for future applications [55]. Despite the challenges in scale-up of the technology, microwave-assisted pyrolysis has been proposed to be implemented on average-sized farms for the production of biofuels through the pyrolyze crop residues. The capital cost of microwave-assisted pyrolysis system is significantly lower than that for cellulosic ethanol plants and are portable to be transported between the farms. Overcoming the technical and environmental challenges for the implementation of farm-level microwave-assisted pyrolysis systems can benefit rural communities for additional employment and income generation [51, 53].

7.8 Conclusions

Biomass holds great potential for meeting the increasing demand for energy and materials. The microwave-assisted pyrolysis of biomass is a promising technology for the production of biofuels and bioproducts. Use of catalytic microwave-absorbers greatly enhance the efficiency of microwave-assisted pyrolysis of biomass. Quality of biooil produced through microwave-assisted pyrolysis has been improved with the use of catalysts. Compared to conventional pyrolysis methods, microwave-assisted pyrolysis method has high energy efficiency and reduced reaction time. However, there is a large gap between the technical advantages provided by the microwave-assisted pyrolysis of biomass and its commercial implementation. Modeling and simulation tools with capabilities to predict the complex mass and energy transfer and chemical reactions can be helpful in determining the performance of this technology at different levels of operation. Multidisciplinary studies are essential for the successful integration of the various processes involved in the microwave-assisted pyrolysis of biomass, and to evaluate the technical, economical, and environmental aspects of this technology. Further research is required for the successful commercial production of biofuels through microwave-assisted pyrolysis technology.

Acknowledgement The authors have acknowledged the support from The *Natural Sciences and Engineering Research Council* of Canada (*NSERC*) for the research on microwave-assisted pyrolysis of biomass.

References

1. Huang YF, Chiueh PT, Kuan WH, Lo SL. Microwave pyrolysis of lignocellulosic biomass: heating performance and reaction kinetics. Energy. 2016;100:137–44. https://doi.org/10.1016/j.energy.2016.01.088.
2. Kurian JK, Nair GR, Hussain A, Raghavan GSV. Feedstocks, logistics and pre-treatment processes for sustainable lignocellulosic biorefineries: a comprehensive review. Renew Sust Energ Rev. 2013;25:205–19. https://doi.org/10.1016/j.rser.2013.04.019.
3. McKendry P. Energy production from biomass (part 2): conversion technologies. Bioresour Technol. 2002;83(1):47–54. https://doi.org/10.1016/S0960-8524(01)00119-5.
4. Mohamed BA, Ellis N, Kim CS, Bi X. Microwave-assisted catalytic biomass pyrolysis: effects of catalyst mixtures. Appl Catal B Environ. 2019;253:226–34. https://doi.org/10.1016/j.apcatb.2019.04.058.
5. White JE, Catallo WJ, Legendre BL. Biomass pyrolysis kinetics: a comparative critical review with relevant agricultural residue case studies. J Anal Appl Pyrolysis. 2011;91(1):1–33. https://doi.org/10.1016/j.jaap.2011.01.004.

6. Dong Q, Xiong Y. Kinetics study on conventional microwave pyrolysis of moso bamboo. Bioresour Technol. 2014;171:127–31. https://doi.org/10.1016/j.biortech.2014.08.063.
7. Ren S, Lei H, Wang L, Bu Q, Chen S, Wu J, Julson J, Ruan R. Biofuel production and kinetics analysis for microwave pyrolysis of Douglas fir sawdust pellet. J Anal Appl Pyrolysis. 2012;94:163–9. https://doi.org/10.1016/j.jaap.2011.12.004.
8. Miura M, Kaga H, Sakurai A, Kakuchi T, Takahashi K. Rapid pyrolysis of wood block by microwave heating. J Anal Appl Pyrolysis. 2004;71(1):187–99. https://doi.org/10.1016/S0165-2370(03)00087-1.
9. Lei H, Ren S, Julson J. The effects of reaction temperature and time and particle size of corn stover on microwave pyrolysis. Energy Fuels. 2009;23(6):3254–61. https://doi.org/10.1021/ef9000264.
10. Huang YF, Chiueh PT, Kuan WH, Lo SL. Microwave pyrolysis of rice straw: products, mechanism, and kinetics. Bioresour Technol. 2013;142:620–4. https://doi.org/10.1016/j.biortech.2013.05.093.
11. Zhang S, Dong Q, Zhang L, Xiong Y. High quality syngas production from microwave pyrolysis of rice husk with char-supported metallic catalysts. Bioresour Technol. 2015;191:17–23. https://doi.org/10.1016/j.biortech.2015.04.114.
12. Kuan WH, Huang YF, Chang CC, Lo SL. Catalytic pyrolysis of sugarcane bagasse by using microwave heating. Bioresour Technol. 2013;146:324–9. https://doi.org/10.1016/j.biortech.2013.07.079.
13. Zhang Z, Macquarrie DJ, De Bruyn M, Budarin VL, Hunt AJ, Gronnow MJ, Fan J, Shuttleworth PS, Clarka JH, Matharu AS. Low-temperature microwave-assisted pyrolysis of waste office paper and the application of bio-oil as an Al adhesive. Green Chem. 2015;17(1):260–70. https://doi.org/10.1039/C4GC00768A.
14. Zhou R, Lei H, Julson J. The effects of pyrolytic conditions on microwave pyrolysis of prairie cordgrass and kinetics. J Anal Appl Pyrolysis. 2013;101:172–6. https://doi.org/10.1016/j.jaap.2013.01.013.
15. Budarin VL, Clark JH, Lanigan BA, Shuttleworth P, Breeden SW, Wilson AJ, Macquarrie DJ, Milkowski K, Jones J, Bridgeman T, Ross A. The preparation of high-grade bio-oils through the controlled, low temperature microwave activation of wheat straw. Bioresour Technol. 2009;100(23):6064–8. https://doi.org/10.1016/j.biortech.2009.06.068.
16. Huang YF, Chiueh PT, Lo SL. A review on microwave pyrolysis of lignocellulosic biomass. Sustain Environ Res. 2016;26:103–9. https://doi.org/10.1016/j.serj.2016.04.012.
17. Chen MQ, Wang J, Zhang MX, Chen MG, Zhu XF, Min FF, Tan ZC. Catalytic effects of eight inorganic additives on pyrolysis of pine wood sawdust by microwave heating. J Anal Appl Pyrolysis. 2008;82(1):145–50. https://doi.org/10.1016/j.jaap.2008.03.001.
18. Menezes RR, Souto PM, Kiminami RHGA. Microwave fast sintering of ceramic materials. In: Lakshmanan A, editor. Sintering of ceramics: new emerging techniques. Rijeka: InTech; 2012. p. 1–26. ISBN: 978-953-51-0017-1.
19. Mello PA, Barin JS, Guarnieri RA. Microwave heating. In: Flores ÉMDM, editor. Microwave-assisted sample preparation for trace element analysis. Waltham: Elsevier; 2014. p. 59–75. ISBN: 978-0-444-59420-4.
20. Clark DE, Folz DC, West JK. Processing materials with microwave energy. Mater Sci Eng A. 2000;287(2):153–8. https://doi.org/10.1016/S0921-5093(00)00768-1.
21. Ziherl S, Bajc J, Čepič M. Refraction and absorption of microwaves in wood. Eur J Phys. 2013;34:449. https://doi.org/10.1088/0143-0807/34/2/449.
22. Leonelli C, Veronesi P. In: Fang Z, Smith Jr RL, Qi X, editors. Microwave reactors for chemical synthesis and biofuels preparation. Production of biofuels and chemicals with microwave, biofuels and biorefineries 3. Dordrecht: Springer; 2015. p. 17–40. https://doi.org/10.1007/978-94-017-9612-5_2.
23. Dutta B. Development and optimization of pyrolysis biochar production systems towards advanced carbon management. PhD Dissertation, McGill University, Montreal; 2014. https://escholarship.mcgill.ca/concern/theses/9z903318r.

24. Clark DE, Sutton WH. Microwave processing of materials. Annu Rev Mater Sci. 1996;26:299–331. https://doi.org/10.1146/annurev.ms.26.080196.001503.
25. Halim SA, Swithenbank J. Simulation study of parameters influencing microwave heating of biomass. J Energy Inst. 2019;92(4):1191–212. https://doi.org/10.1016/j.joei.2018.05.010.
26. American Chemical Society. Discharge produces hydrocarbons from coal. Chem Eng News. 1968;46(4):34–5. https://doi.org/10.1021/cen-v046n004.p034.
27. Fu YC, Blaustein BD, Sharkey AG. Reaction of coal with nitrogen in a microwave discharge. Fuel. 1972;51(4):308–11. https://doi.org/10.1016/0016-2361(72)90009-9.
28. Elsevier BV. Scopus; 2019. https://www.scopus.com. Accessed 16 Aug 2019.
29. Dutta B, Dev SRS, Raghavan VGS. Finite element modeling of selective heating in microwave pyrolysis of lignocellulosic biomass. Prog Electromagnetics Res B. 2013;56:1–24. https://doi.org/10.2528/PIERB13082502.
30. Huang YF, Chiueh PT, Kuan WH, Lo SL. Effects of lignocellulosic composition and microwave power level on the gaseous product of microwave pyrolysis. Energy. 2015;89:974–81. https://doi.org/10.1016/j.energy.2015.06.035.
31. Lei H, Ren S, Wang L, Bu Q, Julson J, Holladay J, Ruan R. Microwave pyrolysis of distillers dried grain with solubles (DDGS) for biofuel production. Bioresour Technol. 2011;102 (10):6208–13. https://doi.org/10.1016/j.biortech.2011.02.050.
32. Domínguez A, Menéndez JA, Inguanzo M, Pís JJ. Investigations into the characteristics of oils produced from microwave pyrolysis of sewage sludge. Fuel Process Technol. 2005;86 (9):1007–20. https://doi.org/10.1016/j.fuproc.2004.11.009.
33. Domínguez A, Menéndez JA, Inguanzo M, Pís JJ. Production of bio-fuels by high temperature pyrolysis of sewage sludge using conventional and microwave heating. Bioresour Technol. 2006;97(10):1185–93. https://doi.org/10.1016/j.biortech.2005.05.011.
34. Mathiarasu A, Pugazhvadivu M. Production of bio-oil from soapnut seed by microwave pyrolysis. National Conference on Recent Advances in Fuel Cells and Solar Energy, Karaikal, U.T of Puducherry, India. IOP Conference Series: Earth Environmental Science; 2019. https://doi.org/10.1088/1755-1315/312/1/012022.
35. Xin S, Guo L, Lifang L, Wei T, Xu Q. Hydrogen-rich gas production from soybean straw via microwave pyrolysis under CO2 atmosphere. Energy Sources, Part A. 2019. https://doi.org/10.1080/15567036.2019.1676330.
36. Kostas ET, Williams OSA, Duran-Jimenez G, Tapper AJ, Cooper M, Meehan R, Robinson JP. Microwave pyrolysis of Laminaria digitata to produce unique seaweed-derived bio-oils. Biomass Bioenergy. 2019;125:41–9. https://doi.org/10.1016/j.biombioe.2019.04.006.
37. Zhao X, Wang M, Liu H, Li L, Ma C, Song Z. A microwave reactor for characterization of pyrolyzed biomass. Bioresour Technol. 2012;104:673–8. https://doi.org/10.1016/j.biortech.2011.09.137.
38. Bu Q, Lei H, Ren S, Wang L, Zhang Q, Tang J, Ruan R. Production of phenols and biofuels by catalytic microwave pyrolysis of lignocellulosic biomass. Bioresour Technol. 2012;108:274–9. https://doi.org/10.1016/j.biortech.2011.12.125.
39. Debalina B, Reddy RB, Vinu R. Production of carbon nanostructures in biochar, bio-oil and gases from bagasse via microwave assisted pyrolysis using Fe and Co as susceptors. J Anal Appl Pyrolysis. 2017;124:310–8. https://doi.org/10.1016/j.jaap.2017.01.018.
40. Salema AA, Ani FN. Microwave induced pyrolysis of oil palm biomass. Bioresour Technol. 2011;102(3):3388–95. https://doi.org/10.1016/j.biortech.2010.09.115.
41. Mushtaq F, Abdullah TAT, Mat R, Ani FN. Optimization and characterization of bio-oil produced by microwave assisted pyrolysis of oil palm shell waste biomass with microwave absorber. Bioresour Technol. 2015;190:442–50. https://doi.org/10.1016/j.biortech.2015.02.055.
42. Dai L, Zeng Z, Tian X, Jiang L, Yu Z, Wu Q, Wang Y, Liu Y, Ruan R. Microwave-assisted catalytic pyrolysis of torrefied corn cob for phenol-rich bio-oil production over Fe modified bio-char catalyst. J Anal Appl Pyrolysis. 2019;143:104691. https://doi.org/10.1016/j.jaap.2019.104691.

43. Santaniello R, Galgano A, Blasi CD. Coupling transport phenomena and tar cracking in the modeling of microwave-induced pyrolysis of wood. Fuel. 2012;96:355–73. https://doi.org/10.1016/j.fuel.2012.01.040.
44. Arul M, Dineshkumar M, Ramanathan A. Aspen HYSYS simulation of biomass pyrolysis for the production. National Conference on Recent Advances in Fuel Cells and Solar Energy, Karaikal, U.T of Puducherry, India. IOP Conference Series: Earth and Environmental Science; 2019. https://doi.org/10.1088/1755-1315/312/1/012015.
45. Dineshkumar M, Shrikar B, Ramanathan A. Development of computer aided modelling and optimization of microwave pyrolysis of biomass by using ASPEN plus. National Conference on Recent Advances in Fuel Cells and Solar Energy, Karaikal, U.T of Puducherry, India. IOP Conference Series: Earth and Environmental Science; 2019. https://doi.org/10.1088/1755-1315/312/1/012006.
46. Parvez AM, Wu T, Afzal MT, Mareta S, He T, Zhai M. Conventional and microwave-assisted pyrolysis of gumwood: a comparison study using thermodynamic evaluation and hydrogen production. Fuel Process Technol. 2019;184:1–11. https://doi.org/10.1016/j.fuproc.2018.11.007.
47. Zhao X, Zhang J, Song Z, Liu H, Li L, Ma C. Microwave pyrolysis of straw bale and energy balance analysis. J Anal Appl Pyrolysis. 2011;92:43–9. https://doi.org/10.1016/j.jaap.2011.04.004.
48. Asomaning J, Haupt S, Chae M, Bressler DC. Recent developments in microwave-assisted thermal conversion of biomass for fuels and chemicals. Renew Sust Energ Rev. 2018;92:642–57. https://doi.org/10.1016/j.rser.2018.04.084.
49. Beneroso D, Albero-Ortiz A, Monzó-Cabrera J, Díaz-Morcillo A, Arenillas A, Menéndez JA. Dielectric characterization of biodegradable wastes during pyrolysis. Fuel. 2016;172:146–52. https://doi.org/10.1016/j.fuel.2016.01.016.
50. Beneroso D, Monti T, Kostas ET, Robinson J. Microwave pyrolysis of biomass for bio-oil production: scalable processing concepts. Chem Eng J. 2017;316:481–98. https://doi.org/10.1016/j.cej.2017.01.130.
51. Iribarren D, Peters JF, Dufour J. Life cycle assessment of transportation fuels from biomass pyrolysis. Fuel. 2012;97:812–21. https://doi.org/10.1016/j.fuel.2012.02.053.
52. Wang L, Lei H, Ruan R. Techno-economic analysis of microwave-assisted pyrolysis for production of biofuels. In: Fang Z, Smith Jr RL, Qi X, editors. Production of biofuels and chemicals with microwave. Dordrecht: Springer; 2015. p. 251–63. https://doi.org/10.1007/978-94-017-9612-5_12.
53. Doucet J, Laviolette JP, Farag S, Chaouki J. Distributed microwave pyrolysis of domestic waste. Waste Biomass Valoriz. 2014;5:1–10. https://doi.org/10.1007/s12649-013-9216-0.
54. Sun J, Wang W, Yue Q. Review on microwave-matter interaction fundamentals and efficient microwave-associated heating strategies. Materials. 2016;9(4):231. https://doi.org/10.3390/ma9040231.

Chapter 8
From Waste to Chemicals: Bio-Oils Production Through Microwave-Assisted Pyrolysis

Mattia Bartoli, Luca Rosi, and Marco Frediani

Abstract In the last decades, sustainable chemical productions have gained remarkable attention due to the increased accountability for environmentally issues leading to an intensification on the use of biomasses based platforms. Biorefinery has been prooved as a sound approach for the conversion of biomasses to raw chemicals integrating several processes ranging from fermentation to thermochemical treatments. Among them, pyrolysis represents a valuable tool for biomass conversion: as a matter of fact, high-quality oil with proprieties like a fuel can be recovered by fast processes run on lignocellulosic biomasses using different reactors designs and process parameters. In the same field, an innovative approach is represented by the use of microwaves as a heating source for pyrolytic conversion. Microwave (MW) induces very fast and volumetric heating but usually require a susceptor able to adsorb MW and dissipate heat after the interaction with the electromagnetic field. Microwave-assisted pyrolysis has been largely used for biomass conversion with a particular emphasis on the production of liquid fractions also known as bio-oils. It finds also a lot of applications as a source of chemicals such as pesticides, mixtures for chemical treatments or employed after purification procedures (i.e. extraction or a fractionating process to isolate the more interesting compounds present, such as acetic acid, levoglucosane, aromatics, and furans.

Keywords Microwave pyrolysis · Biomass conversion · Waste plastic · Waste management · Biofuels · Bio-oils

M. Bartoli
Department of Applied Science and Technology (DISAT), Politecnico di Torino, Turin, Italy

L. Rosi · M. Frediani (✉)
Department of Chemistry "Ugo Schiff", University of Florence, Sesto Fiorentino (Florence), Italy
e-mail: marco.frediani@unifi.it

Z. Fang et al. (eds.), *Production of Biofuels and Chemicals with Pyrolysis*, Biofuels and Biorefineries 10, https://doi.org/10.1007/978-981-15-2732-6_8

8.1 Introduction

During the last century, anthropogenic activities have deeply affected the worldwide environment [1]. Rise of mankind industrialization led to the massive consumption of non-renewable resources such as fossil fuels together with the increment of pollution [2]. Waste streams generated by human conglomerates represent both an unneglectable environmental issue and a loss of resources. In 2017, up to $4 * 10^8$ ton/y of plastics were produced but only 18 wt% was recycled [3]. Nowadays, the unrecycled plastic materials are disposed mainly through incineration and landfilling increasing the environmentally risks [4]. Besides, oil supplies cannot sustain worldwide demand endlessly even considering the not conventional oil sources [5]. Biomass may represent an interesting alternative to oil-based platforms because they are largely available and yearly renewable [6]. Nonetheless, all the current production technologies are based on oil-derived feedstocks and to shift to bio-based one is a hard task. The use of biomass-derived feedstock involves two main issues that are the low production yields and high oxygen content [6]. An interesting solution to recycle oil derived materials and convert biomasses is represented by pyrolysis. It is a thermochemical treatment at high temperature in an oxygen-free atmosphere [7]. Pyrolytic treatments induce the cracking of polymeric materials and produce three fractions: a solid, a liquid and a gas [8]. Pyrolytic treatments may be employed to convert several plastic waste streams into raw materials [9], solid and liquid fuels [10]. Furthermore, biomass has been used for the conversion into useful chemicals [11] and high-quality carbon [12]. Despite this, both pyrolytic treatments of plastic and biomasses present unneglectable issues. The main drawback of plastic pyrolysis is represented by the complexity of the reactors [13, 14] while the pyrolysis of biomasses produces a complex mixture of products hard to purify and dry [15, 16]. Nonetheless, the use of waste stream derived biomasses (i.e. agricultural and forestry residues, rotten foods, cooked oils) increase the sustainability of the overall process. In the last 20 years, an interesting alternative approach based on microwave has gained a lot of interest [17]. Microwave-assisted pyrolysis (MAP) is a very sound and simple way to process a great variety of materials under several different conditions [18]. In the next section are reported the main results achieved in the MAP of waste biomass together with an exhaustive description of the interactions of microwave (MW) with the material showing the main mechanisms involved.

8.2 Interaction Between MW and Materials

MW is an alternating electromagnetic radiation with frequencies from 300 MHz to 300 GHz, which means typical energies from 1.2 μeV to 1.2 meV [19]. The radiation can interact with materials mainly in three different ways, depending on their peculiarities: (1) through dipole reorientation or (2) Maxwell-Wagner polarization and through (3) simple electric conduction.

Table 8.1 Brief comparison between thermal pyrolysis and MAP with highlight of process differences

Thermal pyrolysis	MAP
Unselective heating and "wall effect"	Fast and selective heating
High energy demand	Reduction of energy consumption
Expensive units	Cheap units
Simply scalability	Complex scalability
No additives are required	MW adsorbent are generally required

According to their electric properties, materials could be classified as dielectric (i.e. water [20], silica [21], alumina [22]) or conductive (i.e. salts [23], metals [24] and metal oxides [25]) that display different heating mechanisms under MWs irradiation. In a perfect dielectric material, heating is caused by the rapid dipole reorientation of the permanent or induced "r" dipoles, randomly distributed within the material and due to the magnetic field induced by the radiation that changes at the frequency of the radiation itself. On the contrary, conductive materials are heated under MW irradiation through a different mechanism, called conduction loss, caused by the resistance of the material itself to the flux of electricity generated by the movement of *free* electrons and induced by the oscillating electric field.

Motasemi et al. [26] studied the MW dielectric properties of hay during MAP at two frequencies (915 and 2450) MHz from room temperature up to 975 K in an inert environment. This finding has confirmed the three stages in the modification of the materials. The first one was the drying happening in the temperature range from room temperature to 475 K, the second was the proper MAP from 475 K to 525 K and the last was the biochar formation up to 975 K. The dielectric properties were found to decrease during drying and pyrolysis stages, while increased significantly during biochar formation as shown by MWs absorption capability of the hay, it improved significantly after pyrolysis process while the MWs penetration decreased. The general advantages of MAP upon thermal pyrolysis are summarized in Table 8.1 and mainly consisted in the reduced processing time, energy consumption and to the possibility to used as MW adsorber a catalysts to improve the quality of bio-oils.

8.3 Microwave-Assisted Pyrolysis (MAP) of the Biomass

8.3.1 A General Introduction About Lignocellulosic Biomasses

Lignocellulosic biomasses are commonly identified as a great number of materials from different sources. Among them, woody ones are the most available renewable and natural resources [27]. The global amount of lignocellulosic biomass is presently estimated to be 1.24×10^{15} kg, among which 80% is attributed to woody ones. Wood is exploited as raw material for structural timber [28], sawn wood [29], furniture [30] and pulp [31] but its use for energy production in industrialized countries is presently limited to pellets [32].

Wood is generally defined as the inner tissue of stems, branches, and roots of perennial plants and it is classified like hardwood and softwood Hardwood means wood from dicot angiosperm trees while softwood means from gymnosperm trees. The main components of wood are cellulose, hemicellulose, and lignin but it contains many other organic and inorganic compounds.

The amount of wood components is affected by several parameters such as the tree specie [33] and among each species by the composition of the soil [34], the climate [35], and the harvest period [36].

MAP of biomasses has been studied by several authors to obtain high-quality bio-oils thanks to fast heating at relatively low temperatures (500–700 K) followed by rapid quenching of the volatile products formed [37–39]. Furthermore, MAP can be performed without finely milling of the feedstock because MW allows the volumetric heating of the sample [40], furthermore the presence of water inside the material may enhance the heating rate and mitigate the temperature reached during pyrolysis avoiding advanced cracking degradation. Even if biomasses can absorb MW, the addition of MW absorbers significantly affects the rate of the process and the quality of the products formed [41].

MAP of biomasses was deeply studied and the main mechanisms can be rationalized in a few different steps. The first is the release of moisture from the feedstock, increasing the surface area and improving the pore structure, which favors a quick release of volatiles and minimized char-catalyzed secondary cracking. The three components were pyrolyzed at different reaction temperatures connected with their thermal stability and the presence of ash in the sample. Pyrolysis of cellulose takes place at (430–470) K, hemicellulose at (470–600) K while lignin at (600–1000) K. During this process other reactions such as dehydration of the sample, pyrolysis of the volatiles present, formation of levoglucosan from cellulose [42] or formation of substituted aromatic rings from lignin [43] take place. The last step is the occurrence of secondary reactions among the intermediates present with the formation of furans, small organic molecules (*i.e* acetic acid, hydroxypropanone), and char [44].

8.3.2 MAP of Cellulose

Cellulose is a fibrous water-insoluble polymer, which is found in the protective cell walls of trees, particularly in stalks, stems, trunks and all woody portions of plant tissues [45]. Cellulose is the most abundant wood component, up to 40% of the total weight in a common wood having linear long chains of *D*-glucose units linked by β-(1,4) glycosidic bonds as shown in Fig. 8.1.

The average molecular weight of native cellulose is ranging from 1.800 kDa to 4000 kDa [46]. In the industrial pulping process, cellulose is degraded reaching an average molecular weight between 180 kDa and 1000 kDa [47].

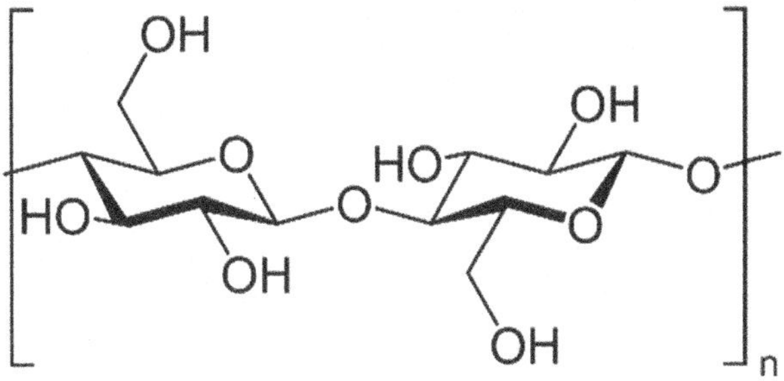

Fig. 8.1 Simplified linear cellulose structure

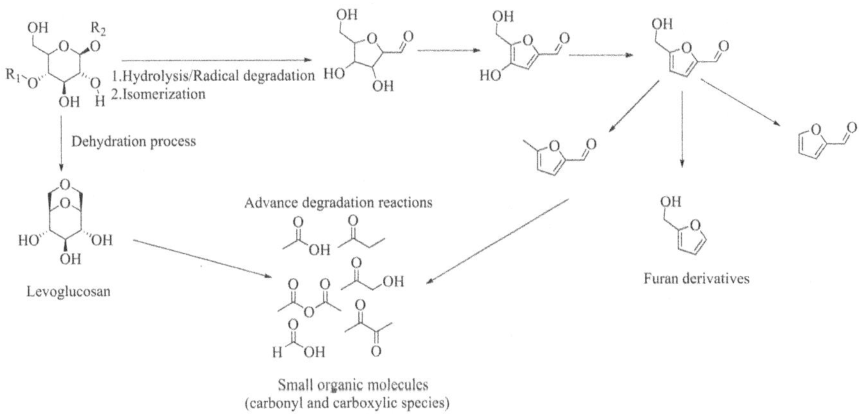

Fig. 8.2 Main reaction pathways of cellulose pyrolysis [42, 60]

Cellulose is one of the most useful raw materials for the textile industry as cotton [48]. It may also be modified to produce viscose and rayon [49], biofuels [50] and paper [51]. Furthermore, cellulose is currently used for a lot of fine applications like filler for polymers [52, 53] drugs [54], stationary chiral phase for both chromatography with liquid and gas [55], environmental materials for building insulation [56], fire retardants [57] or smokeless gunpowder [58].

Cellulose MAP shows the typical reaction pathway of thermochemical cracking processes as summarized in Fig. 8.2 [59].

Anhydrosugars are the most interesting chemicals produced during cellulose MAP. Miura et al. [61] reported the MW conversion of cellulose in a large-scale MW reactor with the large production of levoglucosan up to 2.6 wt% based on dry wood weight. Furthermore, compounds such as levoglucosenone, mannosan, galactosan, and xylosan, were detected in significant amounts.

Contrary to traditional thermal cracking, MAP can operate at significantly reduced temperatures. Al Shra'ah et al. [62] performed a MAP of amorphous cellulose using a lab-scale microwave synthesis system at low temperatures (up to 455 K). They reached a production of bio-oil up to 47 wt% showing significantly higher yields of levoglucosan if compared with conventional pyrolysis run at 673 K.

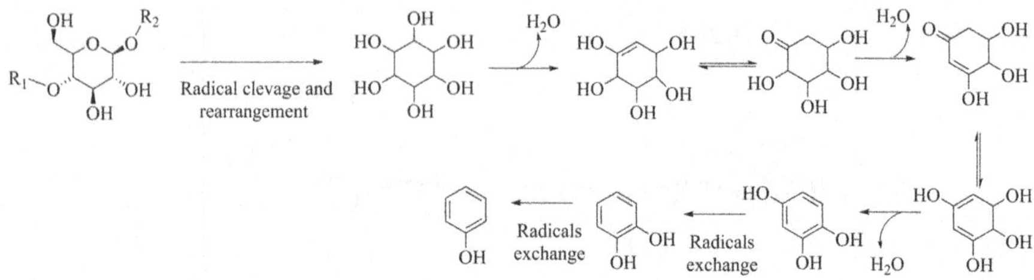

Fig. 8.3 Reaction pathway for aromatics production from cellulose

Similarly, Bartoli et al. [63] pyrolyzed α-cellulose in a multimode batch reactor using several MW absorbers. Authors showed the strong influence of MW susceptor moving from large gasification using carbon (gas yield up to 53.8 wt%) to a high yield of bio-char using Al_2O_3 (up to 64.1 wt%). Using Fe as MW absorber, the bio-oil yield reached up to 37.6 wt%. Furthermore, a high concentration of levoglucosan (up to 133.9 mg/mL) together with acetic acid, acetic anhydride, 1-hydroxy-2-propanone, formic acid and furfural were obtained using graphite as MW absorber. Interesting the authors detected some aromatics formed according to the mechanism reported in Fig. 8.3.

Wang et al. [64] magnified the production of aromatics from cellulose using a catalytic system based on modified HZSM-5. The authors reported the use of metal-doped zeolites (Fe/HZSM-5, Ni/HZSM-5, and Fe–Ni/HZSM-5) to promote the conversion of sugars and anhydrosugars into phenols and low molecular compounds.

Waste streams of cellulose were used as feedstock for MAP. Zhang et al. [65] used paper de-inking residue for MAP at relatively low temperatures (less than 475 K) to achieve interesting results. Furthermore, Undri et al. [66] used the cellulose rich multilayered packaging for producing two phases bio-oils with a products analogue product distribution of those derived from direct pyrolysis of cellulose.

Bio-oils are not the only products achievable through cellulose MAP. Solid residue known as biochar is a very attractive material [67]. Cellulose was used by Zhang et al. [68] as a starting material for the production of hydrochar through the MAP approach. Hydrochar produced showed chemical properties comparable with those produced using hydrothermal carbonization. Furthermore, Omoriyekomwan et al. [69] use cellulose as a direct precursor for the synthesis of carbon nanotubes at low temperatures (up to 875 K).

8.3.3 *MAP of Hemicellulose*

Cellulose is not the only one polysaccharide derived from wood and used for MAP conversions. Hemicellulose decomposition showed unneglectable different products distribution compared to cellulose in thermal pyrolysis. Patwardhan et al. [70] reported a hemicellulose degradative mechanism that highlighted the relevance of xylan degradation with the formation of xylose upon laevoglucose in the early stage of degradation. Considering MAP processes, Li et al. [71] used a pressurized batch MW reactor for the pyrolytic conversion of hemicellulose in the absence of any external microwave absorber. These authors achieved the production of bio-oils up to 21 wt% reach in aldehydes, alkenes, phenols, polyaromatic hydrocarbons, cyclic ketones, and furans.

Generally, MAP of neat hemicellulose is not well described such as cellulose due the poor economical feasibility of its isolation and conversion. Anyhow, it has been studied through temperature modelling [72] in order to evaluate it effect on production of gaseous fraction upon the others [73]. Classical thermal investigations showed hemicellulose decomposition in the temperature range from 625 K to 675 K [74].

8.3.4 MAP of Lignin

Lignin is a cross-linked biopolymer containing several aromatic units, which are linked together by at least ten different C–C and C–O bonds [75] and it may represent up to 30 wt% of the wood weight. Lignin structure derives in part from monomers and oligomers of the three primary monolignols known as *p*-coumaryl, coniferyl, and sinapyl alcohols (Fig. 8.4).

The great number of different monomer combinations induces a great variability into the lignin structure. Moreover, the final irregular structure of lignin also arises from the last step of its biosynthesis that is a random recombination of phenoxy radicals as reported by Felby et al. [76]. Commonly, lignin can be isolated as pure, sulphonate or Kraft derivatives of biomass feedstock, mainly as a residue [77] of pulp and mills industry or from bioethanol production [78]. Kraft lignin gained an addressable global market [79] thanks to its many applications. Indeed it is used as low energy solid fuels [80] currently employed as filler or dispersant in

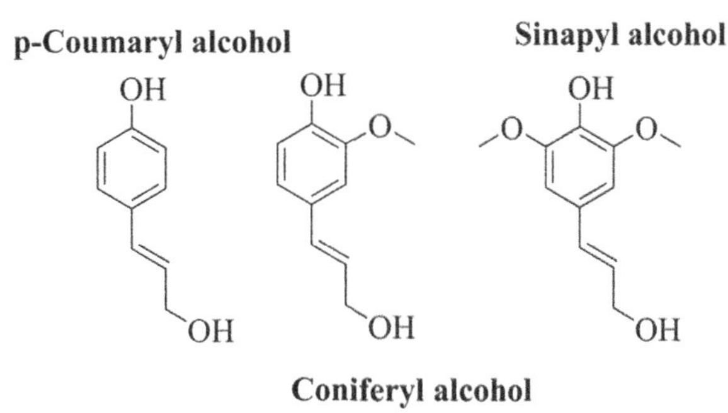

Fig. 8.4 The three aromatic sub-units of lignin

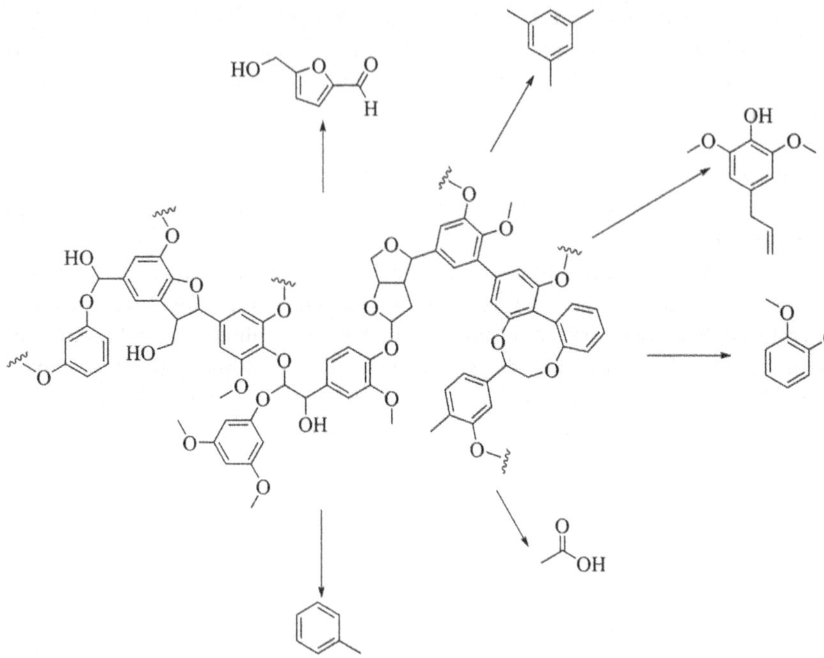

Fig. 8.5 Schematic representation of lignin degradation during MAP with the main products formed

high-performance conglomerate [81], for water treatment [82] and chemicals productions [83].

During pyrolysis, lignin can undergo various degradation pathways as reported by several authors [84–86] and summarized in Fig. 8.5.

Fan et al. [87] studied MAP of lignin using several MW absorbers such as silicon carbide, activated carbon and biochar. The use of silicon carbide promoted three distinct heating stages corresponding to the three main stages of lignin decomposition. The first step is represented by the cleavage of the functional groups leading to low molecular weight products. Afterwards, aryl moieties could undergo to radical rearrangement and degradation. The last step is the more complex and it is characterized by the self condensation of products formed in the previous stage. This mechanism is highly sensible to a lot of factors such as phenol concentration and inorganic traces [88]. Furthermore, a high loading silicon carbide promoted the formation of alkyl phenols and syngas. Both biochar and activated carbon used as MW adsorber promoting a high formation of syngas (up to 70 wt%) if compared with silicon carbide.

Carbonaceous MW absorbers (activated carbons, charcoal and graphite) were used by Yerrayya et al. [89] to improve the yield and selectivity of phenols in the bio-oil using a batch reactor. They proved that the increment of MW absorbers enhanced the bio-oil up to 66 wt% with selectivity to phenols up to 90%.

Also, the MAP of Kraft lignin was deeply investigated. Farag et al. [90] and the yield and composition of bio-oils were compared with those of conventional pyrolysis of Kraft lignin. They proved that MAP enhanced the selectivity towards chemicals in the bio-oils formed.

Furthermore, several upgraded MAP processes have been studied in order to obtain better quality lignin-derived bio-oils. Duan et al. [91] studied the MW co-pyrolytic conversion of a mixture of lignin and polypropylene improving the quality of bio-oils. Similar results were achieved by Fan et al. [92] using catalytic co-pyrolysis of lignin and low-density polyethylene with HZSM-5 and MgO. Curiously, the total amount of aromatics was increased with increasing of polyethylene content. Additionally, using a lignin/polyethylene ratio of 2-methoxy functionalities were completely removed from phenols. Furthermore, aromatics increased and alkylated phenols decreased with increasing of HZSM-5/MgO ratio.

The same authors reported a further improvement of this process using an ex-situ catalytic upgrading of lignin-derived bio-oils over HZSM-5 [93]. They reached the highest selectivity of alkyl phenols using a 0.2 catalyst/lignin ratio.

8.3.5 *MAP of Woody Biomass*

MAP of woody biomasses leads to the production of hundreds of compounds classified in several families [94]. An exhaustive bio-oils analysis is quite challenging and several authors merely run a qualitative investigation [95–97]. The complexity of the problem discourages the simple realization of multiples calibration curves both for the great number of the compounds present and their difficult univocal identification. Recently, few authors proposed methodologies based on a theoretical calculation of response factors of identified compounds leading to a simultaneous qualitative and quantitative analysis [98–100].

Despite these issues, the study on fractions from MAP of woody biomasses has harvested great attention. Huang et al. [101] studied the effect of both maximum temperature and heating rate on MAP outputs using several biomasses (i.e. rice straw, rice husk, corn stover, sugarcane bagasse, sugarcane peel, waste coffee grounds, bamboo leaves) using the results of the chromatographic area for a quantification. A more detailed study was reported by Gao et al. [102] on the distribution of polychlorinated polycyclic aromatic hydrocarbons (PPAHs). PPAHs are formed through chlorination of aromatic moieties in high temperature low oxygen atmosphere in the presence of inorganic or organic chlorine sources [103, 104]. All of this conditions are satisfied during pyrolysis of biomasses and identification and quantification of PPAHs is a relevant task due their toxicity [105]. This study successfully identified polychlorinated dibenzo-p-dioxins, dibenzofurans, and naphthalenes in all the fractions produced. The estimated concentrations ranged in the nanomolar scale. This was an example of the possibility to focus on the classical characterization of a small set of compounds. In a close study, Ravikumar et al. [106] described both the composition and the high heating values of the liquid fraction from the MAP of several biomasses.

A more comprehensive study about all the properties of bio-oils from biomass MAP was proposed by Martín et al. [107]. In this research work, several Mediterranean wasted biomasses underwent MAP conversion where both liquid and gas fractions have been characterized by gas chromatography/mass spectrometry (GC/MS) and through the collection of their rheological properties.

Similarly, Rajasekhar and Vinu [108] were able to produce a good qualitative estimation of the bio-oils compositions produced from MW induced co-pyrolysis of high ash Indian coal and rice husk finding the evidence of cross-reactions.

Anyway, the bio-oils composition is not the only challenge related to the MAP of biomasses. Reactors realization, the great variability of the feedstock, the geographical factors, they are all key points for all the process.

Scale-up and realization of large MW reactors are then quite challenging. Mutsengerere et al. [18] reported a summa of all factors that affect the bio-oils production, identifying pyrolysis temperature, MW power, feedstock particle size, type of MW adsorber as equally important for the quality and the yields of bio-oils. Other parameters such as the pyrolysis retention time, the type of purge gas and the flow rate have less influence on the bio-oil yield. Liu et al. [109] compared the efforts of MW and classical heating on biomass pyrolysis in a traditional fluidized bed auger reactor showing lower carbon dioxide production during MAP.

Salema et al. [110] used a reactor equipped with an overhead stirrer for MAP of oil palm shell to produce high-phenol containing bio-oil and they compared these results with a stationary batch oven. They attribute the better results obtained to a higher interaction with the MWs of the suspended particles rather than bulky ones.

Wang and co-workers [111] realized and innovative continuous reactor for fast biomass MAP and claimed a bio-oils yield around 30 wt% lower than the traditional fast pyrolysis units.

Another class of reactor is based on the interaction between reactor metallic walls with feedstock during MAP. A clear example was described by Hussain et al. [112] using an aluminum coil reactor or in the work of Bashir et al. [113] where an up-graded bio-oil was produced.

Beneroso et al. [114] provide an overview of the scalable processing concepts for building MW reactors on the industrial scale. Many points on that summary represent the core of the work of Salema et al. [115]. They developed a batch reactor operating at 2.45 GHz able to process a large quantity of wood briquette. Experimental outputs showed the good bio-oils yields, up to 40 wt% even with a poor heating value (around 3 MJ/kg).

All the scale-up processes are possible thanks to a method optimization of all MAP parameters. The enlightening work using the response surface methodology was a very strong approach for both simply biomass [116] or mixed feedstocks [117] for MAP modeling.

About the influence of the feedstock, a general investigation on standardized pellets was reported by Undri et al. [118]. In this study, the authors processed pine wood using different MW susceptors producing a two-phase bio-oil. Moreover,

Halim et al. [119] described similar MAP experiments using a Malaysian wood pellets using two fixed temperatures, (775 and 1075) K. They demonstrated the possibility to orientate the MAP from the production of bio-oil and a slightly porous biochar to syngas production together with a very porous biochar moving from lower to higher temperature. A scale-up of the process, using pellets and a bench-scale fixed bed MW reactor, was described by Nhuchhen et al. [120]. MAP outputs showed a complex interaction between all the process factors (i.e. feedstock loading, MW power).

Pristine poplar wood was used by Bartoli et al. [38] to study the behavior of both stump-roots and leaves from short rotation coppice of poplar using a multimode batch MW reactor, investigating several parameters. The particle size, the effect of carbonaceous and metallic MW susceptor were analyzed and experimental outputs showed different effects. The use of iron and wood chips promoted the formation of gas and biochar while the use of poplar powder and carbonaceous MW susceptor induce a larger bio-oil formation. These authors showed the consistency of these results testing several Mediterranean lignocellulosic biomasses such as Vitis vinifera [121] and olive cuts [122]. In all cases, it was proved that a reach radical environment provided by MW susceptors such as iron or sodium hydroxide depleted the formation of bio-oils that were reached in water and advance degradation compounds such as acetic acid and furans.

Moving away from the Mediterranean area, several crops residue can be easily used as feedstock MAP. Oil palm fibers represent a largely available feedstock that is generally disposed of through incineration. Abas and co-workers [123] optimized the MAP process based on the conversion of oil palm fiber for the production of liquid containing an abundance of phenolic compounds. They claimed a bio-oil yield up to 41 wt% composed of 73% of aromatic with gallic acid concentration up to 2.6 wt%. Mushtaq et al. [124] used solid oil palm shells mixed with MW absorber based on coconut activated carbon for the massive production of bio-oil. They achieved a bio-oils with phenols content ranging from 35% to 54% of the total chromatographic area. The bio-oils also contained 1,1-dimethylhydrazine in a very high amount (up to 21%). The presence of this compound was attributed to the synergistic effects of optimized MAP parameters.

Residues from fast-growing crops are a good source for MAP. Bartoli et al. [125] used residues from Arundo donax cultivation for MW pyrolytic conversion achieving a large biochar production (up to 63 wt%) using rhyzomes while using leaves bio-oils yield reached up 41 wt%. In these experiments, bio-oils were collected as one-phase dark brown liquids containing a huge amount of aromatics, acetic acid, furanes, and levoglucosan up to a concentration of 47.6 g/L.

Another cane type plant largely used for MAP is bamboo. Dong et al. [126] studied MAP of Moso bamboo mixed with bamboo biochar as an MW absorber. Authors claimed the production of bio-oils reach in acetic acid and phenol with the total contents ranging from 73% to 83%. Furthermore, they produce a gas fraction comparable with syngas suggesting the positive effect of biochar as MW susceptor in MAP processes. Also, the effect of iron as a high oxidational state cationic specie improves the outputs of bamboo MAP [127] with a significant shift from the bio-oil to syngas production with traces of methane formation. This effect was magnified by the simultaneous presence of iron and activated carbon that induce a syngas production of up to 81% [128].

Alternatively, to traditional catalytic upgrading, several authors explored the treatment of feedstock before the MAP. Tarves et al. [129] use a simple hot water extractive methodology to pretreat of shrub willow, showing that using the same temperature during MAP conditions the yield of bio-oil, bio-char and gases were similar between pretreated and untreated willow. Nonetheless, the bio-oil from treated willow contained a higher concentration of aromatic hydrocarbons and lesser acetic acid and levoglucosan amounts. Another pretreatment approach is based on the impregnation of biomasses with inorganic compounds highly interacting with MW. An acid pretreatment was described by Feng et al. [130] using formic acid claiming an increment of bio-oil yields together with a significant decrement of biochar formation after MAP.

An approach based on the use of potassium hydroxide was described by Grycova et al. [131] claimed an advance bio-oil cracking together with an in situ activation of biochar. Authors claimed an appreciable improvement of surface area reaching a value of up to 530 m^2/g. and a pore volume of 333 mm^3/g. These morphological properties match with the requirements of high performances adsorbed for environmental remediation [132, 133].

Specific salts like copper or zinc sulfate could promote the formation of levoglucosenon during the MAP of ash juniper waste [134].

Wan et al. [135] study the addition of several metal oxides, salts, and solid acids to corn stover and aspen wood during their MAP conversions, showing that this approach could improve bio-oil yields and particular simplify the composition in the presence of chloride salts.

Recently, Bartoli et al. [122] reported interesting observation about the effect of NaOH as MW adsorber during the MAP of olive pruning. Authors clearly described how NaOH promote a very efficient degradation of the biomass with a solid residue of around 2.1 wt% and at the same time a large acetic acid production. This evidence suggested that the formation of advance cracking production took place in gas phase instead of in the reactor bulk.

A mixture of salts with a high content of phosphates has been proved as a very effective methodology for the magnification of bio-oils [136–138]. This effect seems to be ascribed to the cracking ability of phosphates groups as claimed by many researchers [139–141].

Another immense field of bio-oils upgrading is represented by the use of zeolites [142]. The most used.

Used is ZSM-5 [92, 143–152]. The effect of this catalyst can be exploited by the addition of basic oxides, such as calcium oxide, leading to a reduce biochar formation. Liquid fractions from Map could be upgraded using HZSM-5 with enrichment in aromatic together a depletion of oxygen content. Bio-oils obtained from this kind of MAP processes are closer to fuel marketplace standards than others.

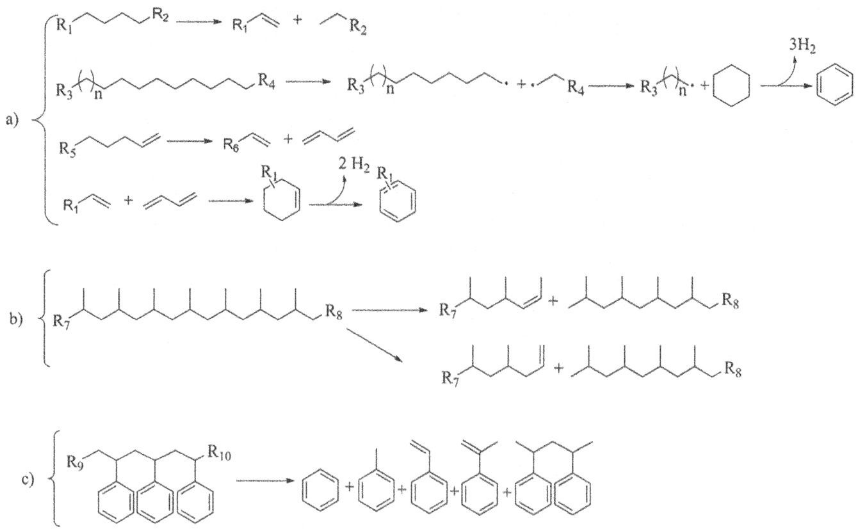

Fig. 8.6 Pyrolytic degradation pathways of (**a**) poly(ethylene), (**b**) poly(propylene) and (**c**) poly (styrene)

Furthermore, complex inorganic species cold efficiently used as reported by Zhang et al. [153]. In this research, the authors used a two-step fast MAP in which Ce-doped γ-Al_2O_3/ZrO_2 was used as an ex-situ reactor catalyst. Findings demonstrated an improvement of hydrocarbon fraction in the bio-oils recovered.

Another efficient process to obtain upgraded bio-oils is represented by the co-pyrolytic approach. Considering the huge amount of highly oxygenated compounds, the simultaneous cracking of biomasses in the presence of a polyolefin lead to the formation of hydrocarbon-rich bio-oils as the consequence of reaction pathways showed in Fig. 8.6 for the pyrolytic degradation polyolefins.

MAP of polyolefins is promising approach for the production of hydrocarbon rich fuel-like fractions as enlighten by several research studies [154–159]. The co-pyrolytic approach lead to a general bio-oil improvement enriching it with high valuable hydrocarbons. This kind of process could represent an efficient way to convert bio-oil into a fuel-like mixtures.

Chen et al. [160] reported the MW induced co-pyrolysis of high-density poly (ethylene) and waste newspaper. This research showed the synergistic effect of this process with an increment of bio-oil up to 32 wt% together with a decrement in viscosity and total acid number.

Zhao et al. [161] described a fast MW co-pyrolysis of bamboo and poly(propylene) achieving a bio-oil yield up to 62 wt%. with an appreciable improvement of aromatics and naphthenic amount.

Also, poly(styrene) or poly(propylene) mixed with biomasses was used as described by Suriapparo et al. [162] for the production of high-quality bio-oil having a heating value up to 42 MJ/kg and an aromatic concentration up to 54 wt%.

8.3.6 *MAP of Non-Lignocellullosic Biomasses*

MAP of lignocellulosic biomasses is definitively one of the most studied biomass conversion. Despite this, bio-oils are characterized by an oxygen content that largely exceeds the requirements for fuels and its reduction is a hard task partially fulfill through catalytic upgrading [163] or fractional condensation [164, 165].

Consequently, alternative biomass feedstocks were studied trying to produce a high-quality bio-oil [166] useful as a fuel, however fatty acids pyrolysis represent the hottest solution due to the composition of the bio-oil formed. A natural fatty acid is composed of a carboxylic function bonded to a long saturated or unsaturated alkyl chain and sometimes to a polyalcohol that undergoes a pyrolytic degradation accordingly to the scheme reported in Fig. 8.7.

The radical cleavage of the carboxylic group during the pyrolysis of fats represents a very important step that leads to the formation of linear, cyclic and aromatics hydrocarbons. Bio-oils produced from these feedstocks are very close to the properties required by the fuel market.

Even if the main body of fatty acids is well defined the alkyl chains could have different degrees of unsaturation that affect the distribution of the products. Wang

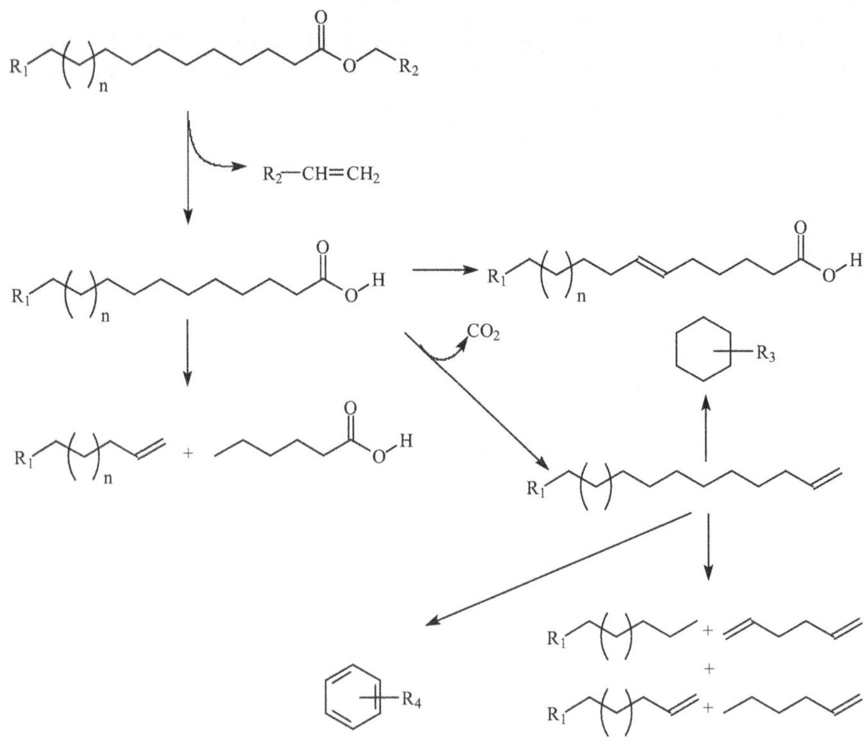

Fig. 8.7 Main pyrolytic degradation pathway of fatty acids esters and main products formed

et al. [167] studied this behavior during MAP of fatty acids salts (sodium stearate, sodium oleate, and sodium linoleate). Findings proved the production alkadienes, cycloalkanes, cycloalkenes, alkynes and aromatics from unsaturated fatty acids while they are contained in a very low amount in bio-oils produced from MAP of saturated fatty.

The other task related to MAP of fatty acids is represented by the presence of glycerol derived from triglycerides. Despite the high oxygen content, this molecule could be effectively converted into fuel components using MAP as shown by Ng et al. [168].

Rapeseed oil represents another feedstock for sustainable MAP conversions as proved by Omar et al. [169]. They compared the efforts of MAP with the outputs of traditional pyrolysis in a temperature range from 775 K to 875 K producing an interesting amount of diesel-like liquid fraction composed by aromatics, long-chain alkanes, cyclo-alkanes, alkenes and cyclo-alkenes; acids and other heavy oxygenated products. Compared with traditional pyrolysis, MAP improved the formation of aromatics in the liquid fraction.

In the last years, algae have harvested a huge interest due to their high content of lipids [170]. A lot of studies reported the MAP conversion of microalgae to hydrocarbon mixture [171–173] and their use in MW induced co-pyrolytic process from the production of hydrocarbon reach bio-oils [174–176]. Also in the MAP of algae, catalytic upgrading was performed mainly using zeolite materials for the improvement of aromatic fractions [177–179].

8.4 Conclusions and Future Outlook

The data reported in this overview on the MAP of biomass showed its feasibility for the production of bio-oils. The great variability of feedstocks represents a very large platform for the orientation of the MAP to the production of different kinds of bio-oils ranging from anhydrosugars to aromatic mixtures. The innovative studies on upgrading open an endless way to the development of a target-oriented MW pyrolytic process while the consolidation of co-pyrolytic approaches lead the way of a large plethora of waste management.

Furthermore, the MAP of fats represents an unneglectable source of renewable hydrocarbons usefully as fuel like mixtures.

Nonetheless, the efficient use of MAP is far to come. The main undressed issues are related to the addition of a MW absorbers and to the product selectivity. Many studies investigated the possible solutions but a final and comprehensive answer has not yet proposed.

Anyhow, MAP still remains a solid platforms to produce chemical several chemical stream that do not require further purifications (i.e. furans, aromatics).

Considering all the aspects, MAP represents a ground-breaking technology that could lead to biomass conversion development in the near future.

References

1. Karl TR, Trenberth KE. Modern global climate change. Science. 2003;302(5651):1719–23.
2. Hill MK. Understanding environmental pollution. Cambridge: Cambridge University Press; 2010.
3. Qualman D. Global plastics production, 1917 to 2050. https://www.darrinqualman.com/global-plastics-production/. Accessed 7 Jan 2020.
4. Geyer R, Jambeck JR, Law KL. Production, use, and fate of all plastics ever made. Sci Adv. 2017;3(7):e1700782. https://doi.org/10.1126/sciadv.1700782.
5. De Castro C, Miguel LJ, Mediavilla M. The role of non conventional oil in the attenuation of peak oil. Energy Policy. 2009;37(5):1825–33. https://doi.org/10.1016/j.enpol.2009.01.022.
6. Artz J, Palkovits R. Cellulose-based platform chemical: the path to application. Curr Opin Green Sustainable Chem. 2018;14:14–8. https://doi.org/10.1016/j.cogsc.2018.05.005.
7. Scheirs J, Kaminsky W. Feedstock recycling and pyrolysis of waste plastics. Chichester: Wiley; 2006.
8. Ferrero G, Maniatis K, Buekens A, Bridgwater A. Pyrolysis and gasification. London: Elsevier; 1989.
9. Faravelli T, Pinciroli M, Pisano F, Bozzano G, Dente M, Ranzi E. Thermal degradation of polystyrene. J Anal Appl Pyrolysis. 2001;60(1):103–21. https://doi.org/10.1016/S0165-2370(00)00159-5.
10. Kunwar B, Cheng H, Chandrashekaran SR, Sharma BK. Plastics to fuel: a review. Renew Sust Energ Rev. 2016;54:421–8. https://doi.org/10.1016/j.rser.2015.10.015.
11. Bridgwater AV, Meier D, Radlein D. An overview of fast pyrolysis of biomass. Org Geochem. 1999;30(12):1479–93. https://doi.org/10.1016/S0146-6380(99)00120-5.
12. Weber K, Quicker P. Properties of biochar. Fuel. 2018;217:240–61. https://doi.org/10.1016/j.fuel.2017.12.054.
13. Kaminsky W. Thermal recycling of polymers. J Anal Appl Pyrolysis. 1985;8:439–48. https://doi.org/10.1016/0165-2370(85)80042-5.
14. Aguado R, Olazar M, Gaisán B, Prieto R, Bilbao J. Kinetic study of polyolefin pyrolysis in a conical spouted bed reactor. Ind Eng Chem Res. 2002;41(18):4559–66. https://doi.org/10.1021/ie0201260.
15. Laurent E, Pierret C, Keymeulen O, Delmon B. Hydrodeoxygenation of oxygenated model compounds: simulation of the hydro-purification of bio-oils. In: Bridgewater AV, editor. Advances in thermochemical biomass conversion. Dordrecht: Springer; 1993. p. 1403–14, https://link.springer.com/content/pdf/10.1007%2F978-94-011-1336-6_110.pdf.
16. Wang S, Gu Y, Liu Q, Yao Y, Guo Z, Luo Z, Cen K. Separation of bio-oil by molecular distillation. Fuel Process Technol. 2009;90(5):738–45. https://doi.org/10.1016/j.fuproc.2009.02.005.
17. Undri A, Rosi L, Frediani M, Frediani P. Microwave heating. In: Microwave pyrolysis of polymeric materials. London: IntechOpen; 2011. https://doi.org/10.5772/24008.
18. Mutsengerere S, Chihobo C, Musademba D, Nhapi I. A review of operating parameters affecting bio-oil yield in microwave pyrolysis of lignocellulosic biomass. Renew Sust Energ Rev. 2019;104:328–36. https://doi.org/10.1016/j.rser.2019.01.030.
19. Pozar DM. Microwave engineering. Hamilton: Wiley; 2009.
20. Rosenkranz PW. Water vapor microwave continuum absorption: a comparison of measurements and models. Radio Sci. 1998;33(4):919–28. https://doi.org/10.1029/98RS01182.
21. Guo X, Deng Y, Gu D, Che R, Zhao D. Synthesis and microwave absorption of uniform hematite nanoparticles and their core-shell mesoporous silica nanocomposites. J Mater Chem. 2009;19(37):6706–12. https://doi.org/10.1039/B910606E.
22. Clark DE, Folz DC, West JK. Processing materials with microwave energy. Mater Sci Eng A. 2000;287(2):153–8. https://doi.org/10.1016/S0921-5093(00)00768-1.

23. Wu R, Zhou K, Yang Z, Qian X, Wei J, Liu L, Huang Y, Kong L, Wang L. Molten-salt-mediated synthesis of SiC nanowires for microwave absorption applications. CrystEngComm. 2013;15(3):570–6. https://doi.org/10.1039/C2CE26510A.
24. Ni Z, Masel RI. Rapid production of metal–organic frameworks via microwave-assisted solvothermal synthesis. J Am Chem Soc. 2006;128(38):12394–5. https://doi.org/10.1021/ja0635231.
25. Kingman S, Rowson N. Microwave treatment of minerals—a review. Miner Eng. 1998;11(11):1081–7. https://doi.org/10.1016/S0892-6875(98)00094-6.
26. Motasemi F, Afzal MT, Salema AA. Microwave dielectric characterization of hay during pyrolysis. Ind Crop Prod. 2014;61:492–8. https://doi.org/10.1016/j.indcrop.2014.07.046.
27. Pepke E. Global, wood markets: consumption, production and trade. Preseted at International Forestry and Global Issues, Nancy, FR, 18 May 2010.
28. Thelandersson S, Larsen HJ. Timber engineering. Hamilton: Wiley; 2003.
29. Parikka M. Global biomass fuel resources. Biomass Bioenergy. 2004;27(6):613–20. https://doi.org/10.1016/j.biombioe.2003.07.005.
30. Kaplinsky R, Memedovic O, Morris M, Readman J. The global wood furniture value chain: what prospects for upgrading by developing countries. UNIDO Sectoral Studies Series Working Paper; 2003.
31. Smook GA. Handbook for pulp & paper technologists. Montreal: Canadian Pulp and Paper Association; 1992.
32. Peksa-Blanchard M, Dolzan P, Grassi A, Heinimö J, Junginger M, Ranta T, Walter A. Global wood pellets markets and industry: policy drivers, market status and raw material potential. IEA Bioenergy Task 40; 2007.
33. Lamlom S, Savidge R. A reassessment of carbon content in wood: variation within and between 41 North American species. Biomass Bioenergy. 2003;25(4):381–8. https://doi.org/10.1016/S0961-9534(03)00033-3.
34. Demeyer A, Nkana JV, Verloo M. Characteristics of wood ash and influence on soil properties and nutrient uptake: an overview. Bioresour Technol. 2001;77(3):287–95. https://doi.org/10.1016/S0960-8524(00)00043-2.
35. Chave J, Muller-Landau HC, Baker TR, Easdale TA, Steege H, Webb CO. Regional and phylogenetic variation of wood density across 2456 neotropical tree species. Ecol Appl. 2006;16(6):2356–67. https://doi.org/10.1890/1051-0761(2006)016[2356:RAPVOW]2.0.CO;2.
36. Pereira H. Variability in the chemical composition of plantation eucalypts (Eucalyptus globulus Labill.). Wood Fiber Sci. 2007;20(1):82–90.
37. Undri A, Zaid M, Briens C, Berruti F, Rosi L, Bartoli M, Frediani M, Frediani P. Bio-oil from pyrolysis of wood pellets using a microwave multimode oven and different microwave absorbers. Fuel. 2015;153:464–82. https://doi.org/10.1016/j.fuel.2015.02.081.
38. Bartoli M, Rosi L, Giovannelli A, Frediani P, Frediani M. Bio-oil from residues of short rotation coppice of poplar using a microwave assisted pyrolysis. J Anal Appl Pyrolysis. 2016;119:224–32. https://doi.org/10.1016/j.jaap.2016.03.001.
39. Bartoli M, Rosi L, Giovannelli A, Frediani P, Frediani M. Pyrolysis of a-cellulose in a microwave multimode batch reactor. J Anal Appl Pyrolysis. 2016;120:284–96. https://doi.org/10.1016/j.jaap.2016.05.016.
40. Metaxas AC, Meredith RJ. Industrial microwave heating, vol. 4. London: Peter Peregrinus; 1983.
41. Macquarrie DJ, Clark JH, Fitzpatrick E. The microwave pyrolysis of biomass. Biofuels Bioprod Biorefin. 2012;6(5):549–60. https://doi.org/10.1002/bbb.1344.
42. Zhang X, Yang W, Dong C. Levoglucosan formation mechanisms during cellulose pyrolysis. J Anal Appl Pyrolysis. 2013;104:19–27. https://doi.org/10.1016/j.jaap.2013.09.015.
43. Kotake T, Kawamoto H, Saka S. Mechanisms for the formation of monomers and oligomers during the pyrolysis of a softwood lignin. J Anal Appl Pyrolysis. 2014;105:309–16. https://doi.org/10.1016/j.jaap.2013.11.018.

44. Asomaning J, Haupt S, Chae M, Bressler DC. Recent developments in microwave-assisted thermal conversion of biomass for fuels and chemicals. Renew Sust Energ Rev. 2018;92:642–57. https://doi.org/10.1016/j.rser.2018.04.084.
45. O'sullivan AC. Cellulose: the structure slowly unravels. Cellulose. 1997;4(3):173–207. https://doi.org/10.1023/a:1018431705579.
46. Sjostrom E. Wood chemistry: fundamentals and applications. London: Elsevier; 2013.
47. Kraemer EO. Molecular weights of celluloses and cellulose derivates. Ind Eng Chem. 1938;30 (10):1200–3. https://doi.org/10.1021/ie50346a023.
48. Myers D, Stolton S. Organic cotton: from field to final product. London: Intermediate Technology; 1999.
49. Goetze K. Viscose rayon production (Russian translation). Khimiya, Moscow; 1972, p. 234.
50. Carroll A, Somerville C. Cellulosic biofuels. Annu Rev Plant Biol. 2009;60:165–82. https://doi.org/10.1146/annurev.arplant.043008.092125.
51. Kline JE. Paper and paperboard: manufacturing and converting fundamentals, A pulp & paper book. San Francisco: Miller Freeman Publications; 1982.
52. Kunioka M, Inuzuka Y, Ninomiya F, Funabashi M. Biobased contents of biodegradable poly (ε-caprolactone) composites polymerized and directly molded using aluminium triflate from caprolactone with cellulose and inorganic filler. Macromol Biosci. 2006;6(7):517–23. https://doi.org/10.1002/mabi.200600037.
53. Dufresne A, Vignon MR. Improvement of starch film performances using cellulose microfibrils. Macromolecules. 1998;31(8):2693–6. https://doi.org/10.1021/ma971532b.
54. Zhang W, Zhang X, Liang M, Lu C. Mechanochemical preparation of surface-acetylated cellulose powder to enhance mechanical properties of cellulose-filler-reinforced NR vulcanizates. Compos Sci Technol. 2008;68(12):2479–84. https://doi.org/10.1016/j.compscitech.2008.05.005.
55. Ishikawa A, Shffiata T. Cellulosic chiral stationary phase under reversed-phase condition. J Liq Chromatogr Relat Technol. 1993;16(4):859–78. https://doi.org/10.1080/10826079308020939.
56. Breum N, Schneider T, Jørgensen O, Rasmussen TV, Eriksen SS. Cellulosic building insulation versus mineral wool, fiberglass or perlite: installer's exposure by inhalation of fibers, dust, endotoxin and fire-retardant additives. Ann Occup Hyg. 2003;47(8):653–69. https://doi.org/10.1093/annhyg/meg090.
57. Camino G, Costa L, Martinasso G. Intumescent fire-retardant systems. Polym Degrad Stab. 1989;23(4):359–76. https://doi.org/10.1016/0141-3910(89)90058-X.
58. Kurtz EF. Gunpowder substituted composition and method. US patent 4,497,676; 1985.
59. Budarin VL, Clark JH, Lanigan BA, Shuttleworth P, Macquarrie DJ. Microwave assisted decomposition of cellulose: a new thermochemical route for biomass exploitation. Bioresour Technol. 2010;101(10):3776–9. https://doi.org/10.1016/j.biortech.2009.12.110.
60. Zhang X, Li J, Yang W, Blasiak W. Formation mechanism of levoglucosan and formaldehyde during cellulose pyrolysis. Energy Fuel. 2011;25(8):3739–46. https://doi.org/10.1021/ef2005139.
61. Miura M, Kaga H, Yoshida T, Ando K. Microwave pyrolysis of cellulosic materials for the production of anhydrosugars. J Wood Sci. 2001;47(6):502–6. https://doi.org/10.1007/BF00767906.
62. Al Shra'ah A, Helleur R. Microwave pyrolysis of cellulose at low temperature. J Anal Appl Pyrolysis. 2014;105:91–9. https://doi.org/10.1016/j.jaap.2013.10.007.
63. Grath MC, Thomas E, Chan W, Hajaligol G, Mohammad R. Low temperature mechanism for the formation of polycyclic aromatic hydrocarbons from the pyrolysis of cellulose. J Anal Appl Pyrolysis. 2003;66(1-2):51–70. https://doi.org/10.1016/S0165-2370(02)00105-5.
64. Wang W, Wang M, Huang J, Tang N, Dang Z, Shi Y, Zhaohe M. Microwave-assisted catalytic pyrolysis of cellulose for phenol-rich bio-oil production. J Energy Inst. 2018;92:1997. https://doi.org/10.1016/j.joei.2018.10.012.

65. Zhang Z, Matharu AS. Chapter 23 - Thermochemical valorization of paper deinking residue through microwave-assisted pyrolysis. In: Bhaskar T, Pandey A, Mohan SV, Lee D-J, Khanal SK, editors. Waste biorefinery. Amsterdam: Elsevier; 2018. p. 671–92. https://doi.org/10.1016/B978-0-444-63992-9.00023-9.
66. Undri A, Rosi L, Frediani M, Frediani P. Fuel from microwave assisted pyrolysis of waste multilayer packaging beverage. Fuel. 2014;133:7–16. https://doi.org/10.1016/j.fuel.2014.04.092.
67. Nanda S, Dalai AK, Berruti F, Kozinski JA. Biochar as an exceptional bioresource for energy, agronomy, carbon sequestration, activated carbon and specialty materials. Waste Biomass Valoriz. 2016;7(2):201–35. https://doi.org/10.1007/s12649-015-9459-z.
68. Zhang J, An Y, Borrion A, He W, Wang N, Chen Y, Li G. Process characteristics for microwave assisted hydrothermal carbonization of cellulose. Bioresour Technol. 2018;259:91–8. https://doi.org/10.1016/j.biortech.2018.03.010.
69. Omoriyekomwan JE, Tahmasebi A, Zhang J, Yu J. Mechanistic study on direct synthesis of carbon nanotubes from cellulose by means of microwave pyrolysis. Energy Convers Manag. 2019;192:88–99. https://doi.org/10.1016/j.enconman.2019.04.042.
70. Patwardhan PR, Brown RC, Shanks BH. Product distribution from the fast pyrolysis of hemicellulose. ChemSusChem. 2011;4(5):636–43. https://doi.org/10.1002/cssc.201000425.
71. Li T, Remón J, Shuttleworth PS, Jiang Z, Fan J, Clark JH, Budarin VL. Controllable production of liquid and solid biofuels by doping-free, microwave-assisted, pressurised pyrolysis of hemicellulose. Energy Convers Manag. 2017;144:104–13. https://doi.org/10.1016/j.enconman.2017.04.055.
72. Namazi AB, Allen DG, Jia CQ. Probing microwave heating of lignocellulosic biomasses. J Anal Appl Pyrolysis. 2015;112:121–8. https://doi.org/10.1016/j.jaap.2015.02.009.
73. Huang Y-F, Chiueh P-T, Kuan W-H, Lo S-L. Effects of lignocellulosic composition and microwave power level on the gaseous product of microwave pyrolysis. Energy. 2015;89:974–81. https://doi.org/10.1016/j.energy.2015.06.035.
74. Yang H, Yan R, Chen H, Zheng C, Lee DH, Liang DT. In-depth investigation of biomass pyrolysis based on three major components: hemicellulose, cellulose and lignin. Energy Fuel. 2006;20(1):388–93. https://doi.org/10.1021/ef0580117.
75. Hatakeyama H, Hatakeyama T. Lignin structure, properties, and applications. In: Abe A, Dusek K, Kobayashi S, editors. Biopolymers. Berlin: Springer; 2009. p. 1–63.
76. Felby C, Nielsen B, Olesen P, Skibsted L. Identification and quantification of radical reaction intermediates by electron spin resonance spectrometry of laccase-catalyzed oxidation of wood fibers from beech (Fagus sylvatica). Appl Microbiol Biotechnol. 1997;48(4):459–64. https://doi.org/10.1007/s002530051080.
77. Chakar FS, Ragauskas AJ. Review of current and future softwood kraft lignin process chemistry. Ind Crop Prod. 2004;20(2):131–41. https://doi.org/10.1016/j.indcrop.2004.04.016.
78. Pye EK. Industrial lignin production and applications. In: Biorefineries-industrial processes and products: status quo and future directions. Weinheim: Wiley; 2008. p. 165–200. https://doi.org/10.1002/9783527619849.ch22.
79. Lebo SE, Gargulak JD, McNally TJ. Lignin. In: Kirk-Othmer encyclopedia of chemical technology. New York: Wiley; 2000. https://doi.org/10.1002/0471238961.12090714120914.a01.pub2.
80. Berlin A, Balakshin M. Chapter 18 - Industrial lignins: analysis, properties, and applications. In: Gupta VK, Kubicek MGTP, Xu JS, editors. Bioenergy research: advances and applications. Amsterdam: Elsevier; 2014. p. 315–36. https://doi.org/10.1016/B978-0-444-59561-4.00018-8.
81. Li Y, Zhu H, Yang C, Zhang Y, Xu J, Lu M. Synthesis and super retarding performance in cement production of diethanolamine modified lignin surfactant. Constr Build Mater. 2014;52:116–21. https://doi.org/10.1016/j.conbuildmat.2013.09.024.
82. Li Z, Kong Y, Ge Y. Synthesis of porous lignin xanthate resin for Pb2+ removal from aqueous solution. Chem Eng J. 2015;270:229–34. https://doi.org/10.1016/j.cej.2015.01.123.

83. Bu Q, Lei H, Zacher AH, Wang L, Ren S, Liang J, Wei Y, Liu Y, Tang J, Zhang Q, Ruan R. A review of catalytic hydrodeoxygenation of lignin-derived phenols from biomass pyrolysis. Bioresour Technol. 2012;124:470–7. https://doi.org/10.1016/j.biortech.2012.08.089.
84. Kawamoto H. Lignin pyrolysis reactions. J Wood Sci. 2017;63(2):117. https://doi.org/10.1007/s10086-016-1606-z.
85. Farag S, Fu D, Jessop PG, Chaouki J. Detailed compositional analysis and structural investigation of a bio-oil from microwave pyrolysis of kraft lignin. J Anal Appl Pyrolysis. 2014;109:249–57. https://doi.org/10.1016/j.jaap.2014.06.005.
86. Farag S, Kouisni L, Chaouki J. Lumped approach in kinetic modeling of microwave pyrolysis of kraft lignin. Energy Fuel. 2014;28(2):1406–17. https://doi.org/10.1021/ef4023493.
87. Fan L, Song H, Lu Q, Leng L, Li K, Liu Y, Wang Y, Chen P, Ruan R, Zhou W. Screening microwave susceptors for microwave-assisted pyrolysis of lignin: comparison of product yield and chemical profile. J Anal Appl Pyrolysis. 2019;142:104623. https://doi.org/10.1016/j.jaap.2019.05.012.
88. Brebu M, Vasile C. Thermal degradation of lignin—a review. Cellul Chem Technol. 2010;44(9):353.
89. Yerrayya A, Suriapparao DV, Natarajan U, Vinu R. Selective production of phenols from lignin via microwave pyrolysis using different carbonaceous susceptors. Bioresour Technol. 2018;270:519–28. https://doi.org/10.1016/j.biortech.2018.09.051.
90. Farag S, Mudraboyina BP, Jessop PG, Chaouki J. Impact of the heating mechanism on the yield and composition of bio-oil from pyrolysis of kraft lignin. Biomass Bioenergy. 2016;95:344–53. https://doi.org/10.1016/j.biombioe.2016.07.005.
91. Duan D, Wang Y, Dai L, Ruan R, Zhao Y, Fan L, Tayier M, Liu Y. Ex-situ catalytic co-pyrolysis of lignin and polypropylene to upgrade bio-oil quality by microwave heating. Bioresour Technol. 2017;241:207–13. https://doi.org/10.1016/j.biortech.2017.04.104.
92. Fan L, Chen P, Zhang Y, Liu S, Liu Y, Wang Y, Dai L, Ruan R. Fast microwave-assisted catalytic co-pyrolysis of lignin and low-density polyethylene with HZSM-5 and MgO for improved bio-oil yield and quality. Bioresour Technol. 2017;225:199–205. https://doi.org/10.1016/j.biortech.2016.11.072.
93. Fan L, Chen P, Zhou N, Liu S, Zhang Y, Liu Y, Wang Y, Omar MM, Peng P, Addy M, Cheng Y, Ruan R. In-situ and ex-situ catalytic upgrading of vapors from microwave-assisted pyrolysis of lignin. Bioresour Technol. 2018;247:851–8. https://doi.org/10.1016/j.biortech.2017.09.200.
94. Garcia-Perez M, Chaala A, Pakdel H, Kretschmer D, Roy C. Characterization of bio-oils in chemical families. Biomass Bioenergy. 2007;31(4):222–42. https://doi.org/10.1016/j.biombioe.2006.02.006.
95. Ingram L, Mohan D, Bricka M, Steele P, Strobel D, Crocker D, Mitchell B, Mohammad J, Cantrell K, Pittman CU Jr. Pyrolysis of wood and bark in an auger reactor: physical properties and chemical analysis of the produced bio-oils. Energy Fuel. 2007;22(1):614–25. https://doi.org/10.1021/ef700335k.
96. Lu Q, Yang X-l, Zhu X-f. Analysis on chemical and physical properties of bio-oil pyrolyzed from rice husk. J Anal Appl Pyrolysis. 2008;82(2):191–8. https://doi.org/10.1016/j.jaap.2008.03.003.
97. Özbay N, Pütün AE, Pütün E. Structural analysis of bio-oils from pyrolysis and steam pyrolysis of cottonseed cake. J Anal Appl Pyrolysis. 2001;60(1):89–101. https://doi.org/10.1016/S0165-2370(00)00161-3.
98. Djokic MR, Dijkmans T, Yildiz G, Prins W, Van Geem KM. Quantitative analysis of crude and stabilized bio-oils by comprehensive two-dimensional gas-chromatography. J Chromatogr A. 2012;1257:131–40. https://doi.org/10.1016/j.chroma.2012.07.035.
99. Bartoli M, Rosi L, Frediani M, Frediani P. A simple protocol for quantitative analysis of bio-oils through gas-chromatography/mass spectrometry. Eur J Mass Spectrom. 2016;22(4):199–212. https://doi.org/10.1255/ejms.1432.

100. Undri A, Abou-Zahid M, Briens C, Berruti F, Rosi L, Bartoli M, Frediani M, Frediani P. A simple procedure for chromatographic analysis of pyrolysis bio-oils. J Anal Appl Pyrolysis. 2015;114:208–21. https://doi.org/10.1016/j.jaap.2015.05.019.
101. Huang Y-F, Chiueh P-T, Kuan W-H, Lo S-L. Product distribution and heating performance of lignocellulosic biomass pyrolysis using microwave heating. Energy Procedia. 2018;152:910–5. https://doi.org/10.1016/j.egypro.2018.09.092.
102. Gao Q, Budarin VL, Cieplik M, Gronnow M, Jansson S. PCDDs, PCDFs and PCNs in products of microwave-assisted pyrolysis of woody biomass – distribution among solid, liquid and gaseous phases and effects of material composition. Chemosphere. 2016;145:193–9. https://doi.org/10.1016/j.chemosphere.2015.11.110.
103. Stieglitz L. Selected topics on the de novo synthesis of PCDD/PCDF on fly ash. Environ Eng Sci. 1998;15(1):5–18. https://doi.org/10.1089/ees.1998.15.5.
104. Born JG, Mulder P, Louw R. Fly ash mediated reactions of phenol and monochlorophenols: oxychlorination, deep oxidation, and condensation. Environ Sci Technol. 1993;27 (9):1849–63. https://doi.org/10.1021/es00046a013.
105. Harvey RG. Polycyclic aromatic hydrocarbons: chemistry and carcinogenicity. New York: CUP Archive; 1991.
106. Ravikumar C, Senthil Kumar P, Subhashni SK, Tejaswini PV, Varshini V. Microwave assisted fast pyrolysis of corn cob, corn stover, saw dust and rice straw: experimental investigation on bio-oil yield and high heating values. Sustain Mater Technol. 2017;11:19–27. https://doi.org/10.1016/j.susmat.2016.12.003.
107. Martín MT, Sanz AB, Nozal L, Castro F, Alonso R, Aguirre JL, González SD, Matía MP, Novella JL, Peinado M, Vaquero JJ. Microwave-assisted pyrolysis of Mediterranean forest biomass waste: bioproduct characterization. J Anal Appl Pyrolysis. 2017;127:278–85. https://doi.org/10.1016/j.jaap.2017.07.024.
108. Rajasekhar Reddy B, Vinu R. Microwave-assisted co-pyrolysis of high ash Indian coal and rice husk: product characterization and evidence of interactions. Fuel Process Technol. 2018;178:41–52. https://doi.org/10.1016/j.fuproc.2018.04.018.
109. Liu Y, Ran C, Siddiqui AR, Mao X, Kang Q, Fu J, Deng Z, Song Y, Jiang Z, Zhang T, Dai J. Pyrolysis of textile dyeing sludge in fluidized bed and microwave-assisted auger reactor: comparison and characterization of pyrolysis products. J Hazard Mater. 2018;359:454–64. https://doi.org/10.1016/j.jhazmat.2018.07.055.
110. Salema AA, Ani FN. Microwave-assisted pyrolysis of oil palm shell biomass using an overhead stirrer. J Anal Appl Pyrolysis. 2012;96:162–72. https://doi.org/10.1016/j.jaap.2012.03.018.
111. Wang Y, Zeng Z, Tian X, Dai L, Jiang L, Zhang S, Wu Q, Wen P, Fu G, Liu Y, Ruan R. Production of bio-oil from agricultural waste by using a continuous fast microwave pyrolysis system. Bioresour Technol. 2018;269:162–8. https://doi.org/10.1016/j.biortech.2018.08.067.
112. Hussain K, Bashir N, Hussain Z, Sulaiman SA. Cement catalyzed conversion of biomass into upgraded bio-oil through microwave metal interaction pyrolysis in aluminum coil reactor. J Anal Appl Pyrolysis. 2018;129:37–42. https://doi.org/10.1016/j.jaap.2017.12.006.
113. Bashir N, Hussain K, Hussain Z, Naz MY, Ibrahim KA, Abdel-Salam NM. Effect of metal coil on product distribution of highly upgraded bio-oil produced by microwave-metal interaction pyrolysis of biomass. Chem Eng Process Process Intensif. 2018;130:140–7. https://doi.org/10.1016/j.cep.2018.05.024.
114. Beneroso D, Monti T, Kostas ET, Robinson J. Microwave pyrolysis of biomass for bio-oil production: scalable processing concepts. Chem Eng J. 2017;316(Supplement C):481–98. https://doi.org/10.1016/j.cej.2017.01.130.
115. Salema AA, Afzal MT, Bennamoun L. Pyrolysis of corn stalk biomass briquettes in a scaled-up microwave technology. Bioresour Technol. 2017;233:353–62. https://doi.org/10.1016/j.biortech.2017.02.113.

116. Jamaluddin MA, Ismail K, Mohd Ishak MA, Ab Ghani Z, Abdullah MF, Safian MTU, Idris SS, Tahiruddin S, Mohammed Yunus MF, Mohd Hakimi NIN. Microwave-assisted pyrolysis of palm kernel shell: optimization using response surface methodology (RSM). Renew Energy. 2013;55:357–65. https://doi.org/10.1016/j.renene.2012.12.042.
117. Kadlimatti HM, Raj Mohan B, Saidutta MB. Bio-oil from microwave assisted pyrolysis of food waste-optimization using response surface methodology. Biomass Bioenergy. 2019;123:25–33. https://doi.org/10.1016/j.biombioe.2019.01.014.
118. Undri A, Abou-Zaid M, Briens C, Berruti F, Rosi L, Bartoli M, Frediani M, Frediani P. Bio-oil from pyrolysis of wood pellets using a microwave multimode oven and different microwave absorbers. Fuel. 2015;153:464–82. https://doi.org/10.1016/j.fuel.2015.02.081.
119. Halim SA, Swithenbank J. Characterisation of Malaysian wood pellets and rubberwood using slow pyrolysis and microwave technology. J Anal Appl Pyrolysis. 2016;122:64–75. https://doi.org/10.1016/j.jaap.2016.10.021.
120. Nhuchhen DR, Afzal MT, Dreise T, Salema AA. Characteristics of biochar and bio-oil produced from wood pellets pyrolysis using a bench scale fixed bed, microwave reactor. Biomass Bioenergy. 2018;119:293–303. https://doi.org/10.1016/j.biombioe.2018.09.035.
121. Bartoli M, Rosi L, Giovannelli A, Frediani P, Passaponti M, Frediani M. Microwave assisted pyrolysis of crop residues from Vitis vinifera. J Anal Appl Pyrolysis. 2018;130:305–13. https://doi.org/10.1016/j.jaap.2017.12.018.
122. Bartoli M, Rosi L, Giovannelli A, Frediani P, Frediani M. Characterization of bio-oil and bio-char produced by low-temperature microwave-assisted pyrolysis of olive pruning residue using various absorbers. Waste Manag Res. 2019;38:213. https://doi.org/10.1177/0734242X19865342.
123. Abas FZ, Ani FN, Zakaria ZA. Microwave-assisted production of optimized pyrolysis liquid oil from oil palm fiber. J Clean Prod. 2018;182:404–13. https://doi.org/10.1016/j.jclepro.2018.02.052.
124. Mushtaq F, Abdullah TAT, Mat R, Ani FN. Optimization and characterization of bio-oil produced by microwave assisted pyrolysis of oil palm shell waste biomass with microwave absorber. Bioresour Technol. 2015;190:442–50. https://doi.org/10.1016/j.biortech.2015.02.055.
125. Bartoli M, Rosi L, Giovannelli A, Frediani P, Frediani M. Production of bio-oils and bio-char from Arundo donax through microwave assisted pyrolysis in a multimode batch reactor. J Anal Appl Pyrolysis. 2016;122:479–89. https://doi.org/10.1016/j.jaap.2016.10.016.
126. Dong Q, Li H, Niu M, Luo C, Zhang J, Qi B, Li X, Zhong W. Microwave pyrolysis of moso bamboo for syngas production and bio-oil upgrading over bamboo-based biochar catalyst. Bioresour Technol. 2018;266:284–90. https://doi.org/10.1016/j.biortech.2018.06.104.
127. Dong Q, Li X, Wang Z, Bi Y, Yang R, Zhang J, Luo H, Niu M, Qi B, Lu C. Effect of iron(III) ion on moso bamboo pyrolysis under microwave irradiation. Bioresour Technol. 2017;243 (Supplement C):755–9. https://doi.org/10.1016/j.biortech.2017.07.009.
128. Dong Q, Niu M, Bi D, Liu W, Gu X, Lu C. Microwave-assisted catalytic pyrolysis of moso bamboo for high syngas production. Bioresour Technol. 2018;256:145–51. https://doi.org/10.1016/j.biortech.2018.02.018.
129. Tarves PC, Serapiglia MJ, Mullen CA, Boateng AA, Volk TA. Effects of hot water extraction pretreatment on pyrolysis of shrub willow. Biomass Bioenergy. 2017;107:299–304. https://doi.org/10.1016/j.biombioe.2017.10.024.
130. Feng Y, Li G, Li X, Zhu N, Xiao B, Li J, Wang Y. Enhancement of biomass conversion in catalytic fast pyrolysis by microwave-assisted formic acid pretreatment. Bioresour Technol. 2016;214:520–7. https://doi.org/10.1016/j.biortech.2016.04.137.
131. Grycova B, Pryszcz A, Lestinsky P, Chamradova K. Influence of potassium hydroxide and method of carbonization treatment in garden and corn waste microwave pyrolysis. Biomass Bioenergy. 2018;118:40–5. https://doi.org/10.1016/j.biombioe.2018.07.022.

132. Inyang MI, Gao B, Yao Y, Xue Y, Zimmerman A, Mosa A, Pullammanappallil P, Ok YS, Cao X. A review of biochar as a low-cost adsorbent for aqueous heavy metal removal. Crit Rev Environ Sci Technol. 2016;46(4):406–33. https://doi.org/10.1080/10643389.2015.1096880.
133. Cha JS, Park SH, Jung S-C, Ryu C, Jeon J-K, Shin M-C, Park Y-K. Production and utilization of biochar: a review. J Ind Eng Chem. 2016;40:1–15. https://doi.org/10.1016/j.jiec.2016.06.002.
134. Choi J, Nam H, Capareda SC. Effect of metal salts impregnation and microwave-assisted solvent pretreatment on selectivity of levoglucosenone and levoglucosan from vacuum pyrolysis of ashe juniper waste. J Environ Chem Eng. 2019;7(1):102796. https://doi.org/10.1016/j.jece.2018.11.041.
135. Wan Y, Chen P, Zhang B, Yang C, Liu Y, Lin X, Ruan R. Microwave-assisted pyrolysis of biomass: catalysts to improve product selectivity. J Anal Appl Pyrolysis. 2009;86(1):161–7. https://doi.org/10.1016/j.jaap.2009.05.006.
136. Mohamed BA, Ellis N, Kim CS, Bi X. Microwave-assisted catalytic biomass pyrolysis: effects of catalyst mixtures. Appl Catal B Environ. 2019;253:226–34. https://doi.org/10.1016/j.apcatb.2019.04.058.
137. Mohamed BA, Kim CS, Ellis N, Bi X. Microwave-assisted catalytic pyrolysis of switchgrass for improving bio-oil and biochar properties. Bioresour Technol. 2016;201:121–32. https://doi.org/10.1016/j.biortech.2015.10.096.
138. Shang H, Lu R-R, Shang L, Zhang W-H. Effect of additives on the microwave-assisted pyrolysis of sawdust. Fuel Process Technol. 2015;131:167–74. https://doi.org/10.1016/j.fuproc.2014.11.025.
139. Chitnis GK, Herbst JA; ExxonMobil Oil Corp. Cracking catalysts containing phosphate treated zeolites, and method of preparing the same. US patent 5,110,776; 1992.
140. Emig G, Hofmann H. Action of zirconium phosphate as a catalyst for the oxydehydrogenation of ethylbenzene to styrene. J Catal. 1983;84(1):15–26. https://doi.org/10.1016/0021-9517(83)90081-7.
141. Swift HE, Stanulonis JJ, Reynolds EH; Gulf Research, Development Co. Alumina-aluminum phosphate-silica-zeolite catalyst. US patent 4,228,036; 1980.
142. Vitolo S, Seggiani M, Frediani P, Ambrosini G, Politi L. Catalytic upgrading of pyrolytic oils to fuel over different zeolites. Fuel. 1999;78(10):1147–59. https://doi.org/10.1016/S0016-2361(99)00045-9.
143. Zhang B, Zhong Z, Xie Q, Liu S, Ruan R. Two-step fast microwave-assisted pyrolysis of biomass for bio-oil production using microwave absorbent and HZSM-5 catalyst. J Environ Sci. 2016;45:240–7. https://doi.org/10.1016/j.jes.2015.12.019.
144. Zhang B, Zhong Z, Li T, Xue Z, Wang X, Ruan R. Biofuel production from distillers dried grains with solubles (DDGS) co-fed with waste agricultural plastic mulching films via microwave-assisted catalytic fast pyrolysis using microwave absorbent and hierarchical ZSM-5/MCM-41 catalyst. J Anal Appl Pyrolysis. 2018;130:1–7. https://doi.org/10.1016/j.jaap.2018.02.007.
145. Zhang B, Zhong Z, Chen P, Ruan R. Microwave-assisted catalytic fast co-pyrolysis of Ageratina adenophora and kerogen with CaO and ZSM-5. J Anal Appl Pyrolysis. 2017;127:246–57. https://doi.org/10.1016/j.jaap.2017.07.027.
146. Zhang B, Zhong Z, Chen P, Ruan R. Microwave-assisted catalytic fast pyrolysis of biomass for bio-oil production using chemical vapor deposition modified HZSM-5 catalyst. Bioresour Technol. 2015;197:79–84. https://doi.org/10.1016/j.biortech.2015.08.063.
147. Zhang B, Zhang J, Zhong Z, Zhang Y, Song M, Wang X, Ding K, Ruan R. Conversion of poultry litter into bio-oil by microwave-assisted catalytic fast pyrolysis using microwave absorbent and hierarchical ZSM-5/MCM-41 catalyst. J Anal Appl Pyrolysis. 2018;130:233–40. https://doi.org/10.1016/j.jaap.2018.01.002.
148. Zhang B, Tan G, Zhong Z, Ruan R. Microwave-assisted catalytic fast pyrolysis of spent edible mushroom substrate for bio-oil production using surface modified zeolite catalyst. J Anal Appl Pyrolysis. 2017;123:92–8. https://doi.org/10.1016/j.jaap.2016.12.022.

149. Wang J, Zhong Z, Song Z, Ding K, Deng A. Modification and regeneration of HZSM-5 catalyst in microwave assisted catalytic fast pyrolysis of mushroom waste. Energy Convers Manag. 2016;123:29–34. https://doi.org/10.1016/j.enconman.2016.06.024.
150. Sun J, Wang K, Song Z, Lv Y, Chen S. Enhancement of bio-oil quality: metal-induced microwave-assisted pyrolysis coupled with ex-situ catalytic upgrading over HZSM-5. J Anal Appl Pyrolysis. 2019;137:276–84. https://doi.org/10.1016/j.jaap.2018.12.006.
151. Liu S, Xie Q, Zhang B, Cheng Y, Liu Y, Chen P, Ruan R. Fast microwave-assisted catalytic co-pyrolysis of corn stover and scum for bio-oil production with CaO and HZSM-5 as the catalyst. Bioresour Technol. 2016;204:164–70. https://doi.org/10.1016/j.biortech.2015.12.085.
152. Dai L, Fan L, Duan D, Ruan R, Wang Y, Liu Y, Zhou Y, Yu Z, Liu Y, Jiang L. Production of hydrocarbon-rich bio-oil from soapstock via fast microwave-assisted catalytic pyrolysis. J Anal Appl Pyrolysis. 2017;125:356–62. https://doi.org/10.1016/j.jaap.2017.03.003.
153. Zhang B, Zhong Z, Li T, Xue Z, Ruan R. Bio-oil production from sequential two-step microwave-assisted catalytic fast pyrolysis of water hyacinth using Ce-doped γ-Al2O3/ZrO2 composite mesoporous catalyst. J Anal Appl Pyrolysis. 2018;132:143–50. https://doi.org/10.1016/j.jaap.2018.03.006.
154. Undri A, Rosi L, Frediani M, Frediani P. Upgraded fuel from microwave assisted pyrolysis of waste tire. Fuel. 2014;115:600–8. https://doi.org/10.1016/j.fuel.2013.07.058.
155. Undri A, Rosi L, Frediani M, Frediani P. Efficient disposal of waste polyolefins through microwave assisted pyrolysis. Fuel. 2014;116:662–71. https://doi.org/10.1016/j.fuel.2013.08.037.
156. Undri A, Frediani M, Rosi L, Frediani P. Reverse polymerization of waste polystyrene through microwave assisted pyrolysis. J Anal Appl Pyrolysis. 2014;105:35–42. https://doi.org/10.1016/j.jaap.2013.10.001.
157. Frediani P, Rosi L, Frediani M, Undri A, Occhialini S, Meini S. Production of hydrocarbons from pyrolysis of tyres Search Results Web results European patent PCT/IB2012/050748; 2012.
158. Frediani P, Rosi L, Frediani M, Undri A, Occhialini S; Cooperativa autotrasportatori fiorentini CAF-società cooperativa arl. Production of hydrocarbons from copyrolysis of plastic and tyre material with microwave heating. US patent application 14/007,236; 2014.
159. Bartoli M, Rosi L, Frediani M, Undri A, Frediani P. Depolymerization of polystyrene at reduced pressure through a microwave assisted pyrolysis. J Anal Appl Pyrolysis. 2015;113:281–7. https://doi.org/10.1016/j.jaap.2015.01.026.
160. Chen W, Shi S, Zhang J, Chen M, Zhou X. Co-pyrolysis of waste newspaper with high-density polyethylene: synergistic effect and oil characterization. Energy Convers Manag. 2016;112:41–8. https://doi.org/10.1016/j.enconman.2016.01.005.
161. Zhao Y, Wang Y, Duan D, Ruan R, Fan L, Zhou Y, Dai L, Lv J, Liu Y. Fast microwave-assisted ex-catalytic co-pyrolysis of bamboo and polypropylene for bio-oil production. Bioresour Technol. 2018;249:69–75. https://doi.org/10.1016/j.biortech.2017.09.184.
162. Suriapparao DV, Boruah B, Raja D, Vinu R. Microwave assisted co-pyrolysis of biomasses with polypropylene and polystyrene for high quality bio-oil production. Fuel Process Technol. 2018;175:64–75. https://doi.org/10.1016/j.fuproc.2018.02.019.
163. Mortensen PM, Grunwaldt J-D, Jensen PA, Knudsen K, Jensen AD. A review of catalytic upgrading of bio-oil to engine fuels. Appl Catal A Gen. 2011;407(1–2):1–19. https://doi.org/10.1016/j.apcata.2011.08.046.
164. Gooty AT, Li D, Briens C, Berruti F. Fractional condensation of bio-oil vapors produced from birch bark pyrolysis. Sep Purif Technol. 2014;124:81–8. https://doi.org/10.1016/j.seppur.2014.01.003.
165. Westerhof RJ, Brilman DWF, Garcia-Perez M, Wang Z, Oudenhoven SR, van Swaaij WP, Kersten SR. Fractional condensation of biomass pyrolysis vapors. Energy Fuel. 2011;25(4):1817–29. https://doi.org/10.1021/ef2000322.

166. Dai L, Wang Y, Liu Y, Ruan R, Yu Z, Jiang L. Comparative study on characteristics of the bio-oil from microwave-assisted pyrolysis of lignocellulose and triacylglycerol. Sci Total Environ. 2019;659:95–100. https://doi.org/10.1016/j.scitotenv.2018.12.241.
167. Wang Y, Dai L, Shan S, Zeng Q, Fan L, Liu Y, Ruan R, Zhao Y, Zhou Y. Effect of unsaturation degree on microwave-assisted pyrolysis of fatty acid salts. J Anal Appl Pyrolysis. 2016;120:247–51. https://doi.org/10.1016/j.jaap.2016.05.012.
168. Ng J-H, Leong SK, Lam SS, Ani FN, Chong CT. Microwave-assisted and carbonaceous catalytic pyrolysis of crude glycerol from biodiesel waste for energy production. Energy Convers Manag. 2017;143:399–409. https://doi.org/10.1016/j.enconman.2017.04.024.
169. Omar R, Robinson JP. Conventional and microwave-assisted pyrolysis of rapeseed oil for bio-fuel production. J Anal Appl Pyrolysis. 2014;105:131–42. https://doi.org/10.1016/j.jaap.2013.10.012.
170. Du Z, Li Y, Wang X, Wan Y, Chen Q, Wang C, Lin X, Liu Y, Chen P, Ruan R. Microwave-assisted pyrolysis of microalgae for biofuel production. Bioresour Technol. 2011;102(7):4890–6. https://doi.org/10.1016/j.biortech.2011.01.055.
171. Zhang R, Li L, Tong D, Hu C. Microwave-enhanced pyrolysis of natural algae from water blooms. Bioresour Technol. 2016;212:311–7. https://doi.org/10.1016/j.biortech.2016.04.053.
172. Hu Z, Ma X, Chen C. A study on experimental characteristic of microwave-assisted pyrolysis of microalgae. Bioresour Technol. 2012;107:487–93. https://doi.org/10.1016/j.biortech.2011.12.095.
173. Budarin VL, Zhao Y, Gronnow MJ, Shuttleworth PS, Breeden SW, Macquarrie DJ, Clark JH. Microwave-mediated pyrolysis of macro-algae. Green Chem. 2011;13(9):2330–3. https://doi.org/10.1039/C1GC15560A.
174. Xie Q, Addy M, Liu S, Zhang B, Cheng Y, Wan Y, Li Y, Liu Y, Lin X, Chen P, Ruan R. Fast microwave-assisted catalytic co-pyrolysis of microalgae and scum for bio-oil production. Fuel. 2015;160:577–82. https://doi.org/10.1016/j.fuel.2015.08.020.
175. Dai M, Xu H, Yu Z, Fang S, Chen L, Gu W, Ma X. Microwave-assisted fast co-pyrolysis behaviors and products between microalgae and polyvinyl chloride. Appl Therm Eng. 2018;136:9–15. https://doi.org/10.1016/j.applthermaleng.2018.02.102.
176. Duan D, Ruan R, Lei H, Liu Y, Wang Y, Zhang Y, Zhao Y, Dai L, Wu Q, Zhang S. Microwave-assisted co-pyrolysis of pretreated lignin and soapstock for upgrading liquid oil: effect of pretreatment parameters on pyrolysis behavior. Bioresour Technol. 2018;258:98–104. https://doi.org/10.1016/j.biortech.2018.02.119.
177. Wang Y, Wu Q, Duan D, Ruan R, Liu Y, Dai L, Zhou Y, Zhao Y, Zhang S, Zeng Z. Ex-situ catalytic upgrading of vapors from fast microwave-assisted co-pyrolysis of Chromolaena odorata and soybean soapstock. Bioresour Technol. 2018;261:306–12. https://doi.org/10.1016/j.biortech.2018.04.042.
178. Wang Y, Tian X, Zeng Z, Dai L, Zhang S, Jiang L, Wu Q, Yang X, Liu Y, Zhang B, Yu Z, Wen P, Fu G, Ruan R. Catalytic co-pyrolysis of Alternanthera philoxeroides and peanut soapstock via a new continuous fast microwave pyrolysis system. Waste Manag. 2019;88:102–9. https://doi.org/10.1016/j.wasman.2019.03.029.
179. Borges FC, Xie Q, Min M, Muniz LAR, Farenzena M, Trierweiler JO, Chen P, Ruan R. Fast microwave-assisted pyrolysis of microalgae using microwave absorbent and HZSM-5 catalyst. Bioresour Technol. 2014;166:518–26. https://doi.org/10.1016/j.biortech.2014.05.100.

Part IV
Production of Bio-Chemicals by Pyrolysis

Chapter 9
Integrating Biomass Pyrolysis with Microbial Conversion Processes to Produce Biofuels and Biochemicals

Tharaka Rama Krishna C. Doddapaneni and Timo Kikas

Abstract Biomass is considered to be one of the most promising sources of renewable resources when it comes to producing biofuels and biochemicals. Several technologies have been developed to assist in transforming biomass into useful products, pyrolysis being one of them. However, pyrolysis products have not yet become competitive alternatives to fossil fuel resources. They still require further processing to improve their fuel characteristics and to produce chemicals. On the other hand, due to advancements in industrial biotechnology, the microbial conversion of biomass has rapidly been evolving in recent times. Thanks to this, integrating pyrolysis into microbial conversion processes is provoking a great deal of interest. This would allow the further transformation of pyrolysis products into high value biofuels and biochemicals. In this chapter, an overview of possible approaches to integrate pyrolysis and microbial processes is presented. Furthermore, the opportunities and challenges involving microbial conversion of pyrolysis products are summarised. In particular, the inhibitory effects of pyrolysis oil and possible detoxification methods are discussed. Finally, the future direction of research is highlighted.

Keywords Pyrolysis · Pyrolysis oil · Microbial conversion · Pyrolysis oil fermentation · Anaerobic digestion · Detoxification · Microbial fuel cell · Algae

9.1 Introduction

To mitigate the consequences of global warming, EU is aiming to reduce greenhouse gas emissions by 40% to the 1990 level by 2030 under the stipulations of the EU 2030 climate and energy framework [1]. The well-established fact is that fossil fuels are mainly responsible for release of greenhouse gas emissions, such as CO_2,

T. R. K. C. Doddapaneni (✉) · T. Kikas
Chair of Biosystems Engineering, Institute of Technology, Estonian University of Life Sciences, Tartu, Estonia
e-mail: tharaka.doddapaneni@emu.ee

Z. Fang et al. (eds.), *Production of Biofuels and Chemicals with Pyrolysis*, Biofuels and Biorefineries 10, https://doi.org/10.1007/978-981-15-2732-6_9

methane, and nitrous oxide [2]. Therefore, to be able to meet these environmental targets, fossil fuels need to be replaced with renewable energy sources [1]. Today, the energy mix has become much more diversified thanks to the rapid uptake and inclusion of solar and wind energy. However, chemicals and transportation fuels are still being produced in large quantities from fossil fuels [3]. Biomass has been considered as a feasible alternative to fossil fuel resources when it comes to producing fuels and chemicals [4].

Biomass needs to be processed through many approaches to convert it into useful products. In general, these processes can be grouped into thermochemical and biochemical processes. Thermochemical processes include combustion, pyrolysis, gasification, torrefaction, and hydrothermal liquefaction. The biochemical processes include fermentation, anaerobic digestion, and microbial electrolysis. Among the other available thermochemical conversion processes, pyrolysis is interesting because it is capable of producing multiple products, such as solid char, and others in gaseous and liquid forms. Most interesting is the liquid product that is commonly known as pyrolysis oil, which can be further converted into liquid biofuels, chemicals, and other products, such as plastics [5].

Pyrolysis is the thermochemical degradation of organic matter in an inert environment. In general, pyrolysis is carried out within a temperature range of (450–700) °C. Depending on the heating rate, pyrolysis is divided into either slow pyrolysis or fast pyrolysis. The heating rate for slow pyrolysis is within the range of (1–50) °C/min and for fast pyrolysis within the range of (1–100) °C/s. Slow pyrolysis favours the biochar production and fast pyrolysis favours the pyrolysis oil production [5].

In the early days, interest in pyrolysis began with the production of biochar through the means of slow pyrolysis. However, today's processing has seen a shift towards fast pyrolysis thanks to the considerable range of applications that are available for pyrolysis oil. Although pyrolysis technology is well-established, creating a fuel source for the transportation industry by using pyrolysis oil is something that has not yet been competitively realised. Pyrolysis oil contains several different components, such as water, sugars, acids, phenolics, furans, aldehydes, and ketones [5, 6]. All of these make pyrolysis oil corrosive and oxidative, with a low heating value, and unstable [5]. Therefore, pyrolysis oil needs to be further processed to improve its fuel characteristics, for example the hydrogenation of pyrolysis oil over catalysts to hydrocarbons [6]. These treatments serve to increase production costs and ultimately reduce the cost competitiveness of pyrolysis oil when compared to fossil-based fuels. Due to its complexity, separating organic compounds from pyrolysis oil is also somewhat challenging. At the same time, biochar, which is produced by means of slow pyrolysis, is also not yet cost-competitive against coal [7].

On the other hand, the applications of biotechnology to the conversion of biomass into valuable bio-products is rapidly developing. Some of the processes i.e. ethanol production through fermentation and biogas production through anaerobic digestion have already been demonstrated at industrial scale [8]. The biochemical conversion processes require lower operating temperatures compared to thermochemical conversion processes, and also permit a greater degree of product flexibility [9] At the same time the composition of biomass hydrolysates and pyrolysis oil are comparable. Therefore, integrating pyrolysis with microbial conversion processes is a highly interesting field, and one that could create new process configurations to produce multiple products.

In the available literature, various approaches have been studied in terms of integrating pyrolysis with microbial processes. The primary focus has been given to the microbial conversion of pyrolysis oil. However, some studies also focus on the application of biochar in microbial processes. In this chapter, an overview on the application of pyrolysis products, such as pyrolysis oil and biochar in microbial processes is presented. A brief overview of microbial conversion of biomass is presented. Furthermore, a summary of the opportunities and challenges that are faced when it comes to the microbial conversion of pyrolysis products are discussed. Particularly noteworthy in the discussion are the inhibitory effects of the pyrolysis oil and possible detoxification methods that can be applied to it. Finally, future opportunities are also highlighted.

9.2 An Overview on Microbial Conversion of Biomass into Biofuels and Biochemicals

The microbial conversion of biomass into biofuels and chemicals usually refers to the biochemical conversion processes. This biochemical processing of biomass involves several process steps, such as biomass pre-processing, pre-treatment, hydrolysis, fermentation, and product separation.

Biomass initially needs to be pre-processed and the steps involved in this process include debarking, chipping, and size reducing (which involves milling and grinding). The aim of this stage is the reduction of the biomass size from log to a particle size of (100–200) μm to increase the specific surface area for further processing steps [10]. Because of its reluctant nature, biomass needs to be pre-treated to convert its complex carbohydrates into simple sugars for their better utilisation [11]. Later, the pre-treated biomass is hydrolysed by adding either enzymes or chemicals. These pre-treatment methods can be classified as mechanical, chemical, or biological. Due to the complex structure involved, the lignocellulosic biomass requires extensive pre-treatment such as, explosive and chemical treatment. The end product of the pre-treatment stage could be a mixture of solid and liquid products. The liquid product is mainly a mixture of hemicellulose and lignin-derived products. The solid product could consist of cellulose alone or cellulose and lignin [12]. Later, the pre-treated biomass is hydrolysed to convert complex biomass carbohydrates into fermentable simple sugars through the process of enzymatic hydrolysis [13].

These pre-treatment processes destroy the biomass structure and increase cellulose accessibility. However, these processes also produce compounds, such as furfural, 5-hydroxymethylfurfural (5-HMF), organic acids, and phenolics that are toxic to enzyme and micro-organism activity during the subsequent operations

[14]. However, these inhibitory compounds can be removed or reduced by means of detoxification methods [14]. The detoxified biomass hydrolysates are further processed through microbial processes, such as fermentation and anaerobic digestion to produce biofuels and biochemicals using micro-organisms [9].

9.3 Integrating Biomass Pyrolysis with Microbial Processes

Biomass pyrolysis can be integrated with microbial processes in different approaches as presented in Fig. 9.1. The pyrolysis oil can be fermented to produce ethanol, lipids, chemicals (succinic acid and citric acid). Pyrolysis oil can also be used in anaerobic digestion to produce methane and in microbial electrolysis to produce hydrogen. Furthermore, the pyrolysis oil can be used for algae cultivation to produce lipids. On the other hand, biochar can be used in microbial fuel cells as electrode material and in anaerobic digestion as a stabilizing agent.

9.3.1 Pyrolysis Oil as a Feedstock for Biochemical Processes

9.3.1.1 The Properties and Composition of Pyrolysis Oil

The properties and composition of pyrolysis oil varies significantly depending on feedstock properties and operating conditions, such as temperature, heating rate, pressure, reactor type, and condensing unit [5]. The yield of pyrolysis oil varies between 30–70 wt% depending on the operating conditions. Pyrolysis oil contains hundreds of different components (over 300 of them). These compounds can generally be grouped into acids, alcohols, furans, phenols, sugars, esters, aromatics,

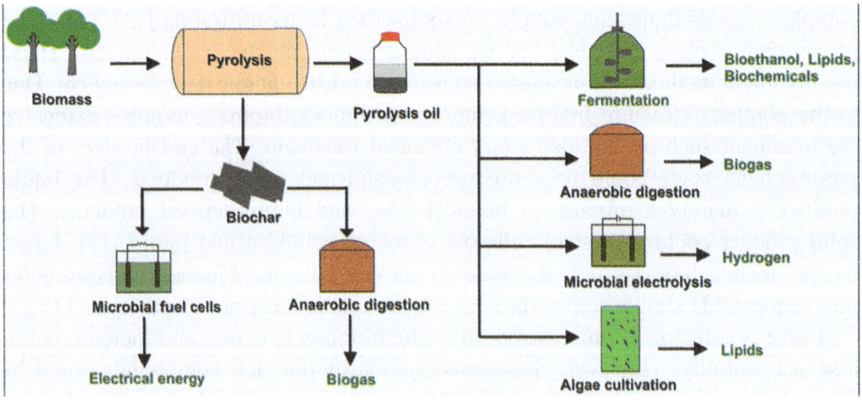

Fig. 9.1 Different approaches to integrate pyrolysis with microbial processes

aldehydes, and ketones [15]. The water content in pyrolysis oil varies between 15 and 30 wt%. The water in fresh pyrolysis oil comes from feedstock moisture and through dehydration reactions during pyrolysis [16]. The reaction water in the pyrolysis oil accounts for around 12 wt% [5]. If the water content is higher than 30 wt%, a phase change occurs in pyrolysis oil [17]. During the phase change, pyrolysis oil is separated into water soluble organics (aqueous liquid) and water insoluble heavy compounds. The water soluble fractions are derived mainly from biomass carbohydrates. The water insoluble fractions are derived from lignin, which are commonly known as pyrolytic lignin [15]. The concentration of water soluble and insoluble compounds varies depending on the cellulose, hemicellulose and lignin content in the biomass. For example, the water insoluble content of wood pyrolysis oil varies between 32 and 35 wt% (dry basis) and the same for straw pyrolysis oil varies between 17 and 28 wt% (dry basis) [17]. The reason could be the reduced lignin content in the straw. Levoglucosan, one of the major products in the pyrolysis oil, is formed through the depolymerisation of cellulose and thereby, having a yield that depends on the cellulose content in the biomass, and also on the experimental conditions [18]. According to [15], levoglucosan could also act as an intermediate product in the formation of hydroxyacetone, 5-HMF, and hydroxyacetaldehyde. Hydroxyacetaldehyde is another one of the major compounds that are present in biomass pyrolysis oils. Pyrolysis oil also contains, (5–12) wt% of organic acids [15].

Oxygen content of the fresh pyrolysis oil is in the range of (30–40) wt%, which is higher than that of conventional fossil fuels. The high oxygen content is mainly due to the high water content and the high concentration of oxygenated compounds [5]. This is also the reason for the low heating value of the pyrolysis oil and the immiscibility with petroleum derived fuels [5]. In general, the heating value of pyrolysis oil is in the range of (15–20) MJ/kg [5]. The viscosity of pyrolysis oil varies between (25–100) m^2/s at 40 °C [5].

One of the challenges faced in the use of pyrolysis oil relates to ageing reactions. During storage at room temperature, pyrolysis oil undergoes several reactions and forming high molecular weight compounds and water. An example of this is esterification between hydroxyl and carbonyl groups. These ageing reactions increase the viscosity and molecular weight of pyrolysis oil. The char particles that are present in pyrolysis oil accelerate the ageing reactions. However, these reactions can be controlled by mixing alcohols, such as ethanol and methanol into pyrolysis oil [17].

9.3.1.2 A Comparative Analysis Between Biomass Hydrolysates and Pyrolysis Oil

In the conventional biochemical conversion processes, the biomass hydrolysates are generally used for microbial conversion. Therefore, a comparative analysis between biomass hydrolysates and pyrolysis oil could aid in understanding the feasibility of the microbial conversion of pyrolysis oils. Limited studies are available on this subject, [4] previously presented a comparative analysis between the enzymatic hydrolysis and the fast pyrolysis of cellulose to produce sugars. Comparatively, the biomass pyrolysis oil also contains the same group of compounds with varying concentrations.

As sugars are the main carbon source in the most of the microbial processes, understanding the total sugar concentrations in biomass hydrolysate and pyrolysis oil is important. The sugar fraction in biomass hydrolysates contains mainly glucose, xylose, arabinose, galactose, and mannose. The total amount of sugars in the biomass hydrolysates varies between (10–60) g/L and the concentration of these compounds depends upon the type of biomass [19]. For example, acid hydrolysis of aspen biomass resulted in a glucose yield of 45 g/L while under the same operating conditions the glucose yield for birch hydrolysis was 18.7 g/L [19]. Another study [20] reported that xylose represented 75% of total sugars in the compositions of wheat straw acid hydrolysates composition. The reason for this could be higher hemicellulose content and low lignin content in wheat straw.

In contrast to biomass hydrolysates, anhydrosugars, such as levoglucosan (1,6-anhydro-β-D-glucopyranose) and cellobiosan are the major sugar compounds in the pyrolysis oil [18]. However, these compounds can further be converted into glucose by means of dilute acid hydrolysis in a way similar to enzymatic hydrolysis in pre-treated biomass hydrolysates. The levoglucosan concentration in pyrolysis oil varies between (5–80) percent carbon (C%) respective to initial cellulose content and depending on the operating conditions [21, 22]. Previously, [4] reported levoglucosan yield of 61.5 wt% during the fast pyrolysis of cellulose at 500 °C. Cellulose crystallinity is also an influencing parameter on levoglucosan yield. For example, [23] reported a levoglucosan yield of 70 wt% of the original cellulose for the microcrystalline cellulose pyrolysis, which is much higher than the results from similar studies. Levoglucosan yield also reduces with increasing temperature, for example its yield reduced from 64.3 wt% at 400 °C to 48.2 wt% at 600 °C during cellulose pyrolysis [4]. The sugar fraction in the pyrolysis oil is mostly water soluble (>96%) [24]. By adding the water, pyrolysis oil can be separated into aqueous and oil rich fractions. Total sugar levels present in the aqueous fraction of pine wood pyrolysis oil is about 13 wt%. In that, levoglucosan accounts for 3.4 wt%. Under the same operating conditions, the total concentration of sugars in the aqueous fraction increased to 21.8 wt% for acid leached pine wood while levoglucosan accounted for 10% [25]. The glucose content in pyrolysis oil varied between (55–69) mg/g of pyrolysis oil [24]. The same source, [24] also reported a glucose content in eucalyptus pyrolysis oil that increased from (0.5 to 43.1) mg/g of pyrolysis oil with the acid hydrolysis treatment of the aqueous fraction of pyrolysis oil. In the total glucose that was produced, levoglucosan and cellobiosan contributed between 45–62% and 16–27%, respectively [24].

The organic acids that are commonly present in biomass hydrolysates are acetic acid, formic acid, levulinic acid, and uronic acid. Both, acetic and uronic acids are formed from hemicellulose [26], while formic acid and levulinic acid are formed from the sugars of both, cellulose and hemicellulose [14, 26]. The acetic acid is formed through the hydrolysis of acetyl groups that are present in hemicellulose. The formic acid and levulinic acid are formed from the furan aldehydes through intermediate reactions [26]. When compared to other acids, acetic acid concentration is much higher in biomass hydrolysates. The acetic acid concentration varied with the pre-treatment and biomass type. For example, the acetic acid concentration in the hydrolysates of sugar cane bagasse, which were treated with liquid hot water and sulphuric acid was at 3.4 g/L and 2.48 g/l, respectively [27]. In case of pyrolysis oil, the acetic acid, formic acid, propionic acid, and butyric acid are the major acids present in the pyrolysis oil. Similar to the biomass hydrolysates, acetic acid concentration (5 wt% for pine biomass) is higher than other acids in the pyrolysis oil.

The other major compounds commonly present both, in biomass hydrolysates and pyrolysis oil are furfural, 5-HMF, guaiacol and coniferyl aldehyde.

9.3.2 Fermentation of Pyrolysis Oil Fractions

As discussed previously (Sect. 9.3.1.1), pyrolysis oil is complex in nature. Among the several compounds that are present in pyrolysis oil, anhydrosugars (levoglucosan and cellobiosan) are primarily interesting in the perspective of the biological conversion of pyrolysis oil [18]. The other compounds, such as oranic acids , are also interesting to produce chemicals and fuels through microbial conversion [28, 29]. Pyrolysis oil can be used in microbial processes in different approaches as presented in Fig. 9.2. These include (1) direct fermentation of crude/original pyrolysis oil (2) fermentation of separated pyrolytic sugars, such as levoglucosan and (3) fermentation of acid hydrolysed pyrolytic sugars, such as glucose. The direct fermentation of pyrolysis oil eliminates the pre-processing stage; however, pyrolysis oil is toxic to microbes at higher concentrations. Although most micro-organisms are not capable of utilizing the levoglucosan directly, there are some micro-organisms that have been identified, which are capable of metabolizing it. For example, fungal

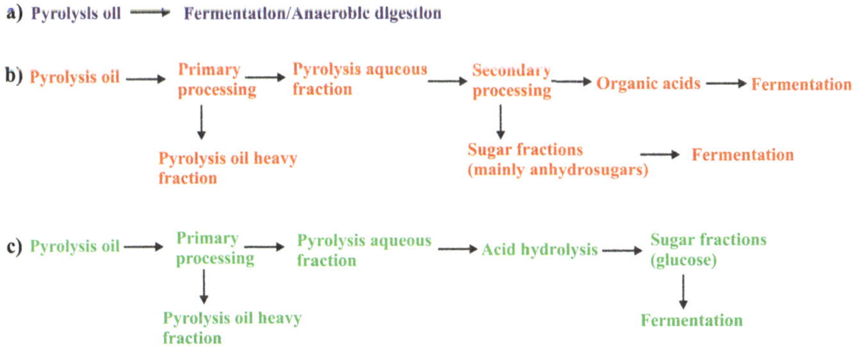

Fig. 9.2 Different approaches for the fermentation of pyrolysis oil fractions (**a**) direct utilization (**b**) fermentation of separated organic acids and sugar fractions (**c**) fermentation of acid hydrolysed sugar fractions

strains—*Aspergillus niger*, *Phanerochaete chrysosporium*, *Penicillium chrysogenum*, and yeasts—*Candida utilis*, *Cryptococcus laurentii*, *Cryptococcus flavescens* [18]. A list of such organisms can be found at [18]. The prokaryotic microorganisms can metabolize the levoglucosan through a process of three steps, such as dehydrogenation intramolecular hydrolysis and NADH- dependent reduction using NAD+ as a cofactor [18, 28]. Whereas eukaryotic microorganisms convert levoglucosan in a different biochemical pathway. Initially, the levoglucosan is converted to glucose-6-phosphate and ADP in the presence of ATP + Mg^{2+} by levoglucosan kinase (LGK). Then, glucose-6-phosphate is further converted to pyruvate following the glycolytic pathway. Later, pyruvate is decarboxylased into acetaldehyde by means of pyruvate decarboxylase. Finally, acetaldehyde is further converted into the desired products, such as ethanol by alcohol dehydrogenase or lactate through lactate dehydrogenase [18, 30]. On the other hand, there are several microbial strains that can utilise glucose as a carbon source, which can go on to produce different chemicals and fuel products. This means that pyrolytic sugars can be further converted into glucose through the process of acid hydrolysis and can be further utilised in microbial processes. Technically, different products can be produced through the microbial conversion of pyrolysis oil. These include, ethanol, citric acid, lipids, PHA, succinic acid, 1,2-propanediol and styrene, hydrogen, and methane.

Prosen et al. [31] studied the ability of microbial strains (eight yeast strains and four fungal strains) to utilise the pyrolysis oil as a carbon and energy source. Three different pyrolysis oil fractions—the aqueous fraction separated after mixing the pyrolysis oil and water at 1:2 ratio, the activated carbon treated aqueous fraction, and the fraction that was hydrolysed with sulphuric acid (2% v/v) were all used in those experiments. The fungal strains were not able to grow on the original aqueous fraction. In contrast, five yeast strains were able to grow on all three pyrolytic fractions. At the same time, no growth of strains *C. albidus* and *S. salmonicolor* was noted on any pyrolysate samples. This study shows that fungal and yeast strains can metabolise the pyrolysis oil fractions. However, aqueous pyrolysis oil is more toxic (due to lignin derived water soluble aromatics) to fungal strains than to yeast strains. Interestingly, some of the yeast strains were not able to metabolize even the detoxified and hydrolysed pyrolysate. Therefore, the microbial strain selection and/or preadaptation is another crucial aspect in the microbial conversion of pyrolysis oils. As an example, Table 9.1 provides a list of different microbial products produced from pyrolysis oil fractions.

9.3.2.1 Bioethanol Production

Previously several researchers have successfully demonstrated that ethanol could be produced from pyrolysis oil through laboratory-scale experiments. As an example, previously reported ethanol yields for pyrolytic sugars were at 70 mg/L (pine biomass; aqueous fraction; 0.1 wt% sugars), 4 mg/L (poplar pine biomass; aqueous fraction; 0.1 wt% sugars), 14.2 g/L (waste cotton; aqueous fraction hydrolysate). In

Table 9.1 Fermentation products that have been produced from biomass derived pyrolysis oil fractions

Biomass-pyrolysis oil fraction	Strain	Product	Productivity/yield	Reference
Loblolly pine-hydrolysed pyrolysis oil aqueous phase	*S. pastorianus ATCC 234*	Ethanol	0.4 g ethanol/g glucose (80% of theoretical yield)	[32]
Pine-pyrolytic sugars	*E. coli KO11 + lgk*	Ethanol	70 mg/L	[33]
Waste cotton-acid-hydrolyzed pyrolysate	*S. cerevisiae 2.399*	Ethanol	0.43 g ethanol/g glucose	[34]
Chicken beds-pyrolysis oil		PHA	0.098 g of PHA/ g of cell dry weight	[35]
Red oak- pyrolytic sugars	*E. coli LJS1*	Styrene	37 mg/L	[36]
Douglas fir- pyrolytic levoglucosan aqueous phase	*R. glutinis*	Lipids (mainly palmitic and oleic acids)	0.78 g/L	[37]
Demineralized pinewood-pyrolytic sugars	*R. diobovatum*	Lipids	26.9 mg/L-h	[38]
Rice husk-aqueous fraction of pyrolysis oil	*E. coli MG 165*	succinic acid	2.42 g/L	[39]
Cotton cellulose-pyrolytic sugars	*A. niger CBX-209*	citric acid	1.9 g/L	[40]

addition, [41] demonstrated the ethanol production from pyrolytic sugars with 7 L batch fermentation with a 3 L working volume. These studies showed that the productivity of ethanol production was increased from 0.92 g/L-h for shake flasks to 1.32 g/L-h for fermenter.

Biomass type also influences the ethanol yield from pyrolysis oil. For example, under the same operating conditions, the ethanol yields for pine and poplar biomass derived pyrolysis sugar solutions were 70 mg/L and 4 mg/L, respectively [33]. The main reason could be the increased concentration of aromatic hydrocarbons, such as anisoles in poplar derived pyrolytic sugar solutions. In another study, [34] observed that *S. cerevisiae* adapted to pyrolysis sugar hydrolysates for 12 times, produced an ethanol yield that was 47% higher when compared to the results from initial experiments i.e. without pre-adaptation [34]. Identifying optimum concentrations of pyrolysis oil or pyrolysis sugar hydrolysates in the fermentation medium is important in terms of process feasibility. Previously, [42] observed that pyrolytic sugar concentration higher than 0.25 wt% showed a severe inhibition to ethanol production. Similarly, [33] also observed the same results, no ethanol production being observed when pyrolytic sugar concentration was higher than 0.25 wt%. In both the studies, the strain *E.coli* ko11+lgk was used. In another study [32], pyrolysis sugar hydrolysates in concentration higher than 10% (v/v) in YPD medium completely inhibited the yeast growth. A concentration higher than 5% (v/v) also showed significant inhibition. The concentration of yeast cells in the fermentation medium also revealed significant effect on pyrolysis sugars fermentation. For example, [41] presented final ethanol concentration for 1% inoculum and 7% inoculum that were within the same range i.e. 7.47 and 7.75 g/L, respectively. However, the fermentation time was reduced from twenty-one hours for 1% inoculum to twelve hours for 7% inoculum.

Based on previous studies, it can be concluded that pyrolysis oil is highly inhibitory to ethanol producing microorganisms. Thus, pre-processing of pyrolysis oil is crucial for achieving ethanol yields that are comparable with conventional processes. Pre-processing steps for pyrolysis oil include its separation into aqueous and organic fractions, separation and/or concentration of pyrolytic sugars, and detoxification of pyrolysis oil fractions.

9.3.2.2 Polyhydroxyalkanoates (PHAs) Production

PHA (polyhydroxyalkanoates), a group of polyesters that are used to synthesise bioplastics is an interesting compound in a view of environment and market demand. PHA can be produced from pyrolysis oil through different approaches as presented in Fig. 9.3. Fidalgo et al. [35] studied the microbial PHA production from pyrolytic sugars that were extracted from chicken bed derived pyrolysis oil. Maximum PHA content of 9.8% (0.098 g of PHA/ g of cell dry weight) with a monomer composition of 75% hydroxybutyrate (HB) and 25% hydroxyvalerate (HV) was gained when

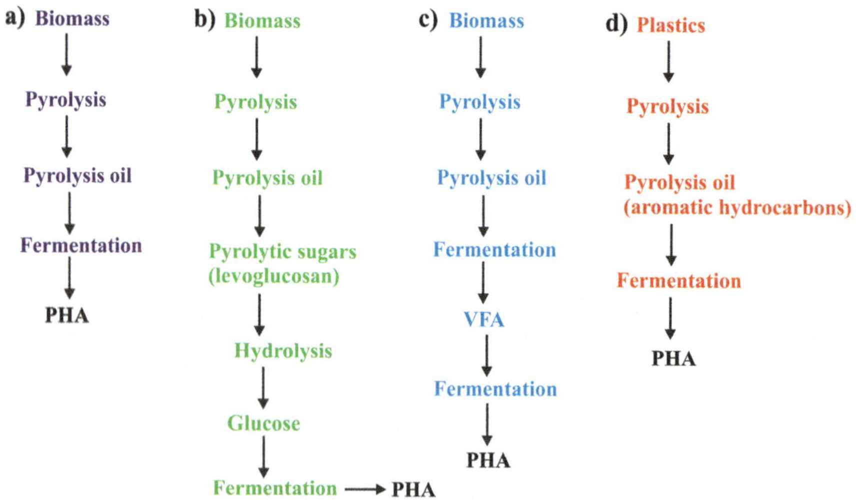

Fig. 9.3 Different approaches of polyhydroxyalkanoates (PHA) production from pyrolysis oil fractions (**a**) direct fermentation (**b**) fermentation of hydrolyzed sugars (**c**) fermentation of pyrolysis oil to volatile fatty acids (VFA) and later fermentation of VFA to PHA (two-step process) (**d**) fermentation of plastics derived pyrolysis oils

untreated pyrolysis oil was used. PHA can also be produced from pyrolysis oil in a two-step process. Initially, pyrolysis oil can be used as a substrate to produce volatile fatty acids (VFA) using mixed microbial cultures and later, these VFA can be fermented to produce PHA (refer Fig. 9.3 for details). Anaerobic fermentation of the pyrolysis oil using mixed cultures resulted in the accumulation of VFA, such as acetic acid (62.84 Cmmol/L), propionic acid (32.15 Cmmol/L) and butyric acid (73.50 Cmmol/L). The concentration of these compounds is several times higher than for the original pyrolysis oil: acetic acid (19.89 Cmmol/L), propionic acid (5.83 Cmmol/L) and butyric acid (8.12 Cmmol/L). Later, the fermented pyrolysis oil (accumulated VFA) was used as a substrate to produce PHA. The maximum PHA content of 16.83% (0.1683 g of PHA/g of cell dry weight) was observed for the two-step process. The increased PHA yield, from 9.8% for a single step process to 16.83% for the two-step process, shows that two-step process is beneficial when it comes to producing PHA from pyrolysis oil.

In another study, [36] examined the production of styrene from pure levoglucosan and pyrolytic sugars (4 g/L of levoglucosan) produced from red oak derived pyrolysis oil. Styrene production from pyrolytic sugar, glucose, and levoglucosan yielded similar results i.e. 38, 42, and 37 mg/L at 5 g/L sugars concentration in the fermentation medium respectively. As the styrene producing strain *Escherichia coli NST74* is not capable of metabolizing levoglucosan due to the lack of LGK, an engineered strain *Escherichia coli LJS1* was used in the same study.

Other than biomass, pyrolysis oil that was produced from waste plastics pyrolysis was also used as a substrate for microbial PHA production. Guzik et al. [43] analysed the ability of PHA production from polyethylene-(PE)-derived pyrolysis oil for 23 different strains and observed maximum PHA accumulation with *Pseudomonas aeruginosaPAO-1* i.e. 25 wt% cell dry weight. In another study [44], polystyrene was converted to PHA by using *Pseudomonas putida CA-3*. The yield was 62.5 mg of PHA per 1 gm of polystyrene derived pyrolysis oil (82.8 wt% of styrene). When the process was scaled up to a capacity of 7.5 L, the PHA yield increased by 1.6 times. PHA can be produced from pyrolysis oil without prior detoxification. However, for higher yield levels a two-step process could be beneficial.

9.3.2.3 Lipids Production

Traditionally, biodiesel is produced from the vegetable oils, through transesterification process. In another approach, lipids that have properties equal to vegetable oils can be produced through microbial fermentation. Later, these lipids can be further converted into biodiesel. Previously, sugar-based feedstocks have been studied for lipid production using oleaginous yeasts and bacterial strains [45]. Similarly, pyrolytic sugars could also be a feasible feedstock for lipids production. For example, [37] reported production of lipids that were equivalent to vegetable oils (39 wt% of saturated fatty acids and 60 wt% of unsaturated fatty acids) using the strain *Rhodotorula glutinis* and pure levoglucosan as a carbon source. In another study, [38] reported lipid production rate in the range of (7.3–50.4) mg/L-h for *Rhodosporidium diobovatum* from pine derived pyrolytic sugars. The maximum lipid yields of 0.078 g/L and 0.061 g/L were reported for sugars from switch grass pyrolysis oil for *Rhodococcus opacus PD630* and *Rhodococcus opacus DSM 1069*, respectively [46].

Although lipid accumulation with pure levoglucosan is comparable to that of glucose, the cell growth and lipid accumulation is significantly inhibited with original pyrolytic sugars. Previously, [37] studied the feasibility of producing lipids from Douglas fir derived pyrolytic sugars without pre-hydrolysis stage. The results showed that even after the activated carbon treatment (detoxification), the cell growth was just 0.74 g/L while, at the same operating conditions, the cell growth with pure levoglucosan was 6.8 g/L. The authors did observe significant improvement in cell growth with sequential detoxification methods (ethyl acetate extraction-evaporation-activated carbon treatment). In another study, [46] reported maximum cell growth of (0.74 and 1.45) g/L for the pyrolysis aqueous fraction and glucose, respectively.

Although the cell growth is restricted with pyrolysis sugars, the lipid accumulation for glucose and pyrolysis aqueous fraction follows the same pattern. Wei et al. [46] observed maximum lipid content of 26.9 wt% and 25.8 wt% with glucose and pyrolysis aqueous fraction, respectively for the *Rhodococcus opacus PD630* strain. The same observed for *Rhodococcus opacus DSM 1069* strain which produced lipid content of 21.6 wt% for glucose and 22 wt% for pyrolysis aqueous fraction. Interestingly, the palmitic (C16:00) and stearic acid (C18:00) accumulation increased in the case of the pyrolysis aqueous fraction compared with glucose as a carbon source. These compounds are considered as fatty acids better suited for biodiesel applications [46].

The carbon to nitrogen (C/N) ratio in the fermentation medium plays a vital role in the accumulation of lipids in oleaginous yeasts and filamentous fungi. In a reduced or low nitrogen environment, any excess carbon flow results in the accumulation of intra-cellular citric acid. Later, citric acid is converted into acetyl-coA by ATP-citrate lyase (citric acid clenage). Then, acetyl—CoA is further transformed into fatty acids [30, 47]. When considering this outcome, the low nitrogen content of the pyrolysis oil could make it a feasible form of feedstock for microbial lipid production. Wood derived pyrolysis oil contains carbon content of 56% and nitrogen content of 0.1% [45].

In addition to pyrolytic sugars, the carboxylic acids (acetic acid, formic acid, and propionic acid) that are present in pyrolysis oils have also been considered as an ingredient in the production of lipids [45]. The carboxylic acids that were present in the waste wood derived pyrolysis oil after treatment and detoxification was used for lipids production with *Cryptococcus curvatus*, *Rhodotorula glutinis*, and *Lipomyces starkeyi* [45]. The authors reported 6.9 g/L of cell growth and 2.2 g/L of lipid accumulation for pyrolytic aqueous phase (acetate concentration of 20 g/L). These values are closely following the yield for model acetate fermentation i.e. 8 g/L cell growth, 3.3 g/L lipid, 41.3 wt% lipid accumulation. This study [45], showed that carboxylic acids present in the pyrolysis oil could also be a considerable source of feedstock for microbial processes.

9.3.2.4 Other Microbial Products Produced from Pyrolysis Oil

Attempts have also been made to produce succinic acid, malic acid, fumaric acid, citric acid and 1, 2-propanediol from pyrolysis oil in laboratory scale studies. Wang et al. [39] studied the production of succinic acid production using the aqueous phase of rice husk derived pyrolysis oil (separated by adding water at a ratio 1:20) using the strain *E. coli MG 165*. The authors observed no succinic acid production when only pyrolysis oil was used as a substrate. However, succinic acid yield of 2.42 g/L was observed when salts and glucose were added (M9 mineral salt + glucose + 20% pyrolysis aqueous phase). The succinic acid production in the absence of glucose was 0.38 g/L. In another study, [48] observed succinate yield of 0.25 g/g at 2% (w/v) concentration of levoglucosan in the fermentation medium for the engineered strain *Corynebacterium glutamicum*.

The production of malic acid with *A. oryzae DSM1863* and fumaric acid production with *R. delemar DSM 095* was also studied using wheat straw derived pyrolysis oil [49]. The authors observed significant effect of pyrolysis oil concentration in fermentation medium. For example, when pyrolysis oil concentration increased from 0.5% to 2%, the fumaric acid yield reduced from 26 g/L to 0 g/L. The citric acid production of 1.9 and 15.4 g/L were observed for cotton cellulose derived pyrolysis oil using *A. niger CBX-209* for original pyrolysis oil and activated carbon treated pyrolysis oil, respectively [40]. The 1,2-propanediol yield of 18.3 mM was reported when using engineered *Corynebacterium glutamicum* for wheat straw pyrolysis water [50].

9.3.3 Microbial Electrolysis of Pyrolysis Oil

Microbial electrolysis is another approach to produce high value energy carrier, such as hydrogen from pyrolysis oil. The aqueous fraction of the pyrolysis oil can be used as a substrate in the microbial electrolysis to produce hydrogen and the produced hydrogen can be further used for the hydrogenation of the organic rich fraction to produce fuel quality bio-oil [51]. Previously, Lewis et al. [51] demonstrated the hydrogen production through microbial electrolysis using the aqueous fraction of pyrolysis oil derived from switchgrass. The authors observed a maximum hydrogen production rate of 4.3 L H_2/L anode volume-day at a loading of 10 g COD/L-anode volume-day. Current densities varied in the range of (1.2–4.5) A/m^2. In another study, Park et al. [52] used neutralized and phase separated organic fraction of the switch grass derived pyrolysis oil in microbial electrolysis and reported the peak hydrogen production rate and current densities as 4.3 L H_2/L anode volume-day and 5.3 A/m^2, respectively. In case of pyrolysis oil derived from catalytic pyrolysis of pine sawdust, a maximum hydrogen production of 5.8 L H_2/L anode volume-day at an organic loading rate of 50 g/L-anode volume-day was observed [53]. In the same study [53], interestingly, the authors observed the COD removal rate of 60% at all the organic loading rates. At the same time, the columbic efficiency decreased with increased loading rate. The reason for this phenomena could be that the conversion of pyrolysis oil aqueous fraction in microbial electrolysis through exoelectrogenesis is limited and conversion continues through fermentation and other pathways [53].

The hydrogen productivity varied depending on the biomass used for pyrolysis. For example, Satinover et al. [54] reported the hydrogen production of 5 L H_2/L anode volume-day and 1.5 L H_2/L anode volume-day and power density of 5 and 1.8 A/m^2 at 10 g/L-day of organic loading during the microbial electrolysis of pyrolysis aqueous fraction derived from willow and guayule, respectively. Several compounds present in the pyrolysis oil, like furfural, 5-HMF, and phenolic compounds are commonly inhibitory to the microbial processes. However, during microbial electrolysis the degradation of these inhibitory compounds was observed. For example, Brooks et al. [53] reported the degradation of furfural and phenol as 100% and 92%. However, it is important to note that the concentration of these inhibitory compounds in the studied aqueous fraction of the pyrolysis oil is much lower. The characterization of anode microbial community shows the presence of different families, such as *Geobacteraceae*, *Rhodocyclaceae*, *Comamonadaceae*, *Enterococcaceae*, and *Lachnospiraceae*. According to Lewis et al. [51], presence of the diversified microbial community could be the reason for the degradation of complex composition of the biomass derived pyrolysis oil.

The acid rich fractions are more favorable for microbial electrolysis thus, separating the pyrolysis oil into different fractions and concentrating the acids (i.e. acid rich fraction) could be helpful to improve the overall efficiency of the process. The anode microbial community plays an important role in the conversion of complex composition of pyrolysis oil. Thus, enriching the microbial community, as suggested by Lewis et al. [51], could be helpful in improving the pyrolysis oil conversion efficiency. At the moment, there are very few studies available on the microbial electrolysis of pyrolysis oil fractions. As the composition of the pyrolysis oil depends on the type of biomass and the pyrolysis operating conditions, there is a need to study the microbial electrolysis of pyrolysis oil fractions produced at varied operating conditions to better understand and optimize the overall process.

9.3.4 Anaerobic Digestion (AD) of Pyrolysis Oil

Anaerobic digestion is a biochemical conversion process, where organic matter is converted into biogas (mainly CH_4 and CO_2) through a series of steps. In basic terms, pyrolysis can be integrated with AD by means of different approaches as presented in Fig. 9.4. The pyrolysis oil water soluble fraction (aqueous phase) could be a potential feedstock for AD application when considering its composition (hydroxyl acids, oligomers, water soluble phenols). At the same time, when considering pyrolysis as a downstream processing, the digestate from AD can be pyrolysed to improve its fertiliser properties and possible energy and material recovery. As the microbial degradation of biomass polymers, especially, lignin is limited, the

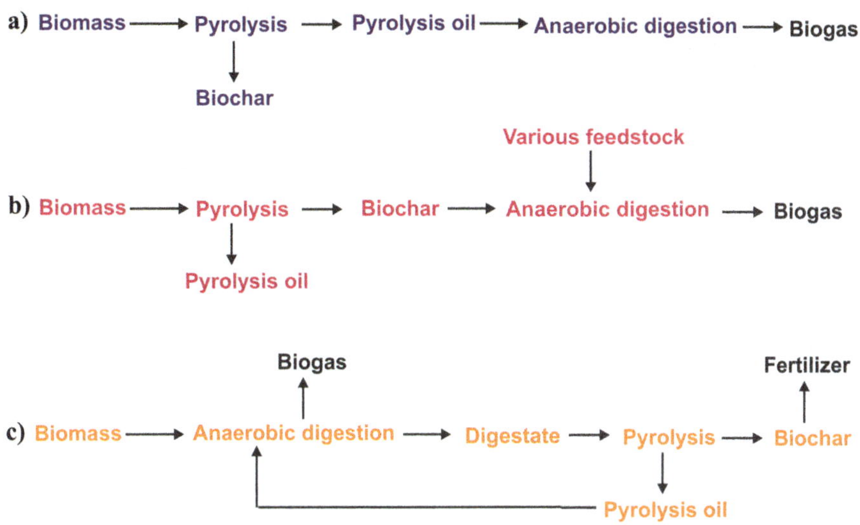

Fig. 9.4 Different approaches of integrating pyrolysis with anaerobic digestion (AD) (**a**) AD of pyrolysis oil (**b**) application of biochar in AD (**c**) pyrolysis of digestate and AD of resulting pyrolysis oil

digestate from the AD of lignocellulosic materials often contains high amount of lignin. In this case, treating the digestate with pyrolysis and recirculating the pyrolysis products back to the AD process could improve the overall conversion efficiency and help achieve flexibility in digestate handling. As an example, the biogas digestate from an agricultural waste feed biogas plant contained 17 wt% of cellulose, 10 wt% of hemicellulose and 31.9 wt% of lignin in the total solid content [55].

There are several inhibitory compounds in the pyrolysis oil that can significantly inhibit the overall AD process. In a fed batch study, Torri and Fabbri [56] observed the increased lag phase and reduced methane yield with each feeding during the AD of corn stalk pellets derived pyrolysis aqueous phase. This could be due to the accumulation of inhibitory compounds and/or the VFA accumulation. The authors observed VFA content of about 19 g COD/L during the first 60 days of the test. Hübner et al. [57] and Yang et al. [58] observed strong correlation between pyrolysis temperature and the methane production rate. For example, when pyrolysis temperature increased from 330 to 530 °C, the methane yield reduced from 220 mL/g COD to 37.5 mL/g COD (using farm scale biogas digestate pyrolysis aqueous fraction, loading rate 12 g/L) [57]. The reason behind this finding could be the increased concentration of lignin derived compounds, mainly water soluble phenols in the pyrolysis oil. At the same time, the concentrations of 5-HMF and furfural, which are derived from biomass carbohydrate fraction, also increase with temperature. Taken all together, these compounds increase the pyrolysis oil toxicity for the AD process with the increase in pyrolysis temperature. However, [56] observed that the addition of bio-char to the AD process significantly increased the methane yield i.e. from 34% theoretical to 60% theoretical.

9.3.5 *Pyrolysis Oil as a Substrate for Algae Cultivation*

Limited studies have also focused on using pyrolysis oil fractions as a carbon source for algal cultivation. Zhao et al. [59] studied the cultivation of microalgae *Chlamydomonas reinhardtii* using alkali treated pyrolysis oil. The authors observed that the acetic acid-rich fraction that was higher than 4 wt% showed itself to be inhibitory to algal growth. Similar fatty acid profiles were observed in algae that were grown on acetic acid-rich pyrolysis oil and in those grown on pure acetic acid substrates. However, the total fatty acid yield was reduced by 50% for pyrolysis oil. In another study, [38] looked at the cultivation of *Chlorella vulgaris* using the pine wood pyrolysis oil aqueous fraction. In contrast to previous study [59], the levoglucosan rich fraction of pyrolysis oil was used as a carbon source. The lipid production and conversion of 1.59 g/L and 0.25 g of lipid/g of glucose, respectively, was observed at 20% (v/v) of pyrolysis oil aqueous fraction (detoxified and hydrolyzed). At the 40% (v/v) of pyrolysis oil aqueous fraction, the complete inhibition of algae growth was observed. The authors concluded that *Chlorella vulgaris* showed high lipid conversion when compared to oleaginous yeast *Rhodosporidium diobovatum*. At the same time, *Chlorella vulgaris* was more sensitive to the pyrolysis oil concentrations.

9.3.6 *Application of Biochar in Microbial Processes*

In the studies of integrating the pyrolysis with microbial process, the primary focus has been given to fractions of pyrolysis oil. However, there are some opportunities for the application of biochar (solid product from biomass pyrolysis) in microbial processes. These applications are (1) using the biochar as an electrode material in microbial fuel cells (2) adding the biochar to AD to enhance the process feasibility.

9.3.6.1 Biochar Application in Microbial Fuel Cell (MFC)

Microbial fuel cells are a bio-electrochemical approach to degrading the organic matter using metabolic activity of exoelectrogenic bacteria and producing electrical energy [60, 61]. In addition to electrical energy, MFCs can be employed to produce biochemicals and also to recover micronutrients and metals from wastewaters [60]. The electrodes of the MFC system play critical role in the large scale applications and economic feasibility as they contribute 20–50% of overall MFC system cost [60]. According to [60], the MFC electrodes should have high surface area, low cost, high conductivity and biocompatibility.

The commonly used carbon based electrode materials are granular activated carbon (GAC), graphite granules (GG), carbon cloth, carbon felt, and carbon paper [60]. These electrodes are mostly manufactured from fossil based materials, such as coal and petroleum. Cost of these electrodes is higher for large scale applications thus, there is a need to identify sustainable cost-effective and technically feasible electrodes. The char produced through biomass pyrolysis could be one such electrode material. Previously, researchers have studied the application of different biomass derived chars as electrodes in MFC system. For example, forestry residual, pine biomass, Kenaf, gaint cane, coconut shells, corn straw, and corn cobs.

In a comparative analysis, [60] observed that the power densities of biochar based MFC system i.e. 532 mW/m^2 for forest residue and 457 mW/m^2 for pine biomass are comparable with the power densities of GAC (674 mW/m^2) and GG (566 mW/m^2). The power densities of activated banana char and Pt/C (30 wt% platinum on carbon) are also comparable at 528 and 695 mW/m^2, respectively [62]. Chen et al. [61] considered the three-dimensional porous structure of kenaf stem to produce an ordered three-dimensional biochar through high temperature pyrolysis (1000 °C and N_2 atmosphere). Such electrode showed higher performance compared with graphite rod in terms of current density i.e. 32.5 A/m^2 for Kenaf biochar and 11.2 A/m^2 for graphite rod. The resistance of the electrode material is an important characteristic and, for an efficient MFC power output, the electrodes must have low resistance [60]. The total resistance values of different biomass derived chars are, milling residue—3 Ω/mm; pine biomass—6 Ω/mm; Kenaf—10 Ω/mm; giant cane—8.9 Ω/mm.

The surface area is also one of the important characteristics and for higher power density the electrodes should possess high surface area [60]. In general, the biochar produced through pyrolysis have low surface area compared with activated carbon. The biochar surface can be increased by increasing the pyrolysis operating temperature and residence time and also by activating the surface through physical and chemical methods. Wang et al. [63] studied the influence of pyrolysis temperature and influence of activating the surface with KOH on the corn straw derived biochar based MFC system. The results showed that the BET surface area increased from 26 to 77 m^2/g when pyrolysis temperature increased from (500 to 900) °C. The same increased to 105 m^2/g for KOH treated biochar. The surface area of KOH treated char is several hundred times higher than carbon felt (0.322 m^2/g) [63]. Following the surface area, power densities increased by 10.7%, 56% and 92% for biochar produced at 500 °C, 900 °C and KOH activated, respectively compared with carbon felt. In another study, [64] studied the influence of pyrolysis temperature on performance of the air cathode MFC, where corn cob derived biochar was used as a cathode. The authors observed that the power density increased from (72.84 to 458) mW/m^2 when the pyrolysis temperature increased from 250 to 650 °C. Similarly, [62] observed the higher power density of 528 mW/m^2 for activated banana biochar compared with non-activated char (483 mW/m^2).

The cost comparative analysis shows that the power production costs with biochar electrode MFC system is very low compared with conventional electrode based MFC system. For example, [60] presented electricity production costs for GAC and GG, pine derived and milled wood waste derived biochar based MFC systems as (402, 392, 17, 35) $/W, respectively. In another costs analysis, [65] reported a huge difference in the electricity production costs between biochar (0.2 $/W) and Vulcan carbon (477 $/W) MFC systems.

Finally, to conclude, biochar based electrodes produce comparable power densities with conventional electrodes. Although the power densities in the laboratory scale experiments look feasible, achieving higher power densities in large scale operations is still under investigation. On the other hand, power densities decline with scale-up of the MFC system because of the reduced conductivity [66]. It is important to note that biochar has less conductivity compared with some of the metal electrodes [67]. The overall efficiency of the biochar based MFC system depends on the properties like active surface area, porosity, conductivity, and biocompatibility of the biochar. To have higher surface area, biochar must be activated through physical and chemical methods. As the thermal conductivity of the biochar increases with increasing pyrolysis temperature, producing the biochar at higher temperature could be beneficial. However, it is important to note that increasing the pyrolysis temperature increases the energy requirement and ultimately may reduce the overall efficiency of the MFC system.

9.3.6.2 Biochar Application in Anaerobic Digestion

The other application of biochar in microbial processes is using it as a stabilizing agent in AD process. Although AD is well established and commercially operated on large scale, it still faces several challenges. These include, substrate inhibition, inhibition from metabolic intermediates, varying pH, nutrient loss. On the other hand, it has been demonstrated that adding the biochar to AD increases the process stability and increases the methane yield. According to Masebinu et al. [68], adding biochar improves the AD process through different mechanisms i.e. immobilizing the microorganism, promoting biofilm growth, increasing the alkalinity, adsorbing the inhibitory compounds, and facilitating direct interspecies electron transfer. The biochar ability to adsorb inhibitory compounds and support biofilm growth depends on the biochar properties, such as specific surface area, porosity, pore volume, and conductivity [68]. These properties are directly linked to the pyrolysis operating temperature. Thus, it is useful to establish the relationship between pyrolysis operating temperature and AD parameters. Previously, Cordiner et al. [69] studied the influence of pyrolysis char produced at (450, 500, 550) °C in co-digestion with activated food waste. The authors observed that, when pyrolysis temperature increased from (450–550) °C, the methane yield increased from (496–547) mL CH_4/g. Shanmugam et al. [70] observed that the biochar produced at temperature higher than 600 °C showed negative effect on methane yield during AD of aqueous fraction of algae liquefaction. For example, the addition of biochar produced at

400 °C and 900 °C resulted in the methane yields of 296 mL/g COD and 37 mL/g COD, respectively. The reason for this significant difference could be the decreased redox active moieties with increased pyrolysis temperature. These moieties facilitate the electron transfer between fermentative bacteria and methanogens [70].

The biochar particle size also showed significant influence on the AD process. When the biochar particle size of 0.5–1 mm, produced through the pyrolysis (800 °C) of fruitwood, was added to AD system (glucose substrate), the lag phase was reduced by 11.4%, 30.3% and 21.6%, and methane yield increased by 86.6%, 21.4% and 5.2% at glucose loading of (4, 6, 8) g/L, respectively [71]. In the same study [71], it was observed that reducing particle size further to 75 μm increased the methane yield by 70% at 6 g/L glucose loading. Lü et al. [72] reported that the methanation lag phase was reduced with the addition of large (2–5) mm and medium (0.5–1) mm sized particles compared with small particles (75–100) μm. They also observed that adding fine biochar particles increased the VFA production. Interestingly, [56] observed that adding biochar to aqueous pyrolysis oil fraction based AD system improved the methane yield by 30% and also improved the methane production rate compared with aqueous fraction only digestion. The aqueous pyrolysis oil fraction contains several inhibitory compounds, such as furfural, 5-HMF, and phenolics. This is a good example to demonstrate that adding biochar to AD system can reduce the substrate induced inhibition. Adding biochar to AD of citrus peel showed an increased methane yield and reduced lag phase compared with citrus peel only AD. The reason could be adsorption of D-limonene, which commonly inhibits the citrus peel AD [73]. Shen et al. [74] observed that adding corn stover biochar to AD system improved the methane production rate and biogas quality to a pipe-line quality (>90% CH_4).

Adding biochar to AD also helps to improve quality of the digestate. For example, [75] observed significant increase in macro and micro nutrients (P, K, Ca Mg and Fe) by 33 times in the digestate when biochar was added to AD process. When digestate is applied to the agricultural field, the biochar supports soil amendment and carbon sequestration.

9.4 Challenges

Although the microbial conversion of pyrolysis oils looks promising, there are still some challenges that need to be addressed before the process can be commercialized. These challenges include, the inhibitory effects of pyrolysis oil on the microorganisms, reduced product yield when compared to the conventional microbial conversion of biomass, economic competitiveness, and environmental feasibility.

9.4.1 Microbial Toxicity of the Pyrolysis Oils

Microbial toxicity of the pyrolysis oils is one of the major challenges that this field of study faces. Pyrolysis oil contains several hundred compounds and most of these compounds are inhibitory to the micro-organisms. Lignin-derived phenolic compounds and oligomers are the major group of inhibitory compounds that are present in pyrolysis oils. In addition to phenolic compounds, furans (furfural and 5-HMF) and organic acids (acetic acid, and formic acid) are also highly inhibitory to the micro-organisms. However, the lignin-derived phenols are ten times more inhibitory when compared to carbohydrate-derived furans and carboxylic acids [76]. The concentration of total furans higher than 0.90 g/L can already highly inhibit the ethanol production [76].

It is worth noting that the inhibitory effects of these compounds depend on the selected microbial strains. For example, organic acids (acetic acid and formic acid) are inhibitory to ethanol-producing *Saccharomyces cerevisiae* while some oleaginous yeasts can use organic acids as a carbon source to produce fatty acid esters. The strains *Pseudomonas putida* and *Acinetobacter ADP1* can metabolise lignin-derived phenolic compounds and produce fatty acid esters. Researchers have developed several methods to detoxify pyrolysis oil. These methods included adsorption, overliming, solvent extraction, and microbial detoxification.

Overliming is one commonly-employed detoxification approach. Previously, $Ca(OH)_2$, NaOH, NH_4OH, and $Ba(OH)_2$ have been used to detoxify pyrolysis oil fractions in the overliming method. Wang et al. [32] studied the detoxification of red oak wood residual-derived pyrolysis fractions using $Ca(OH)_2$, NaOH, and NH_4OH. The results showed that NaOH performed well and removed phenol, vanillin, and guaiacol three times more effectively than $Ca(OH)_2$ and NH_4OH. The authors observed the increased growth of *E. coli* for pyrolytic sugars (at 2 wt % loading), which had been treated with NaOH. Overliming can also reduce concentrations of aldehydes, phenolics, and furans in pyrolytic sugars. For example, overliming treatment using $Ca(OH)_2$ reduced the total phenolic content from 25.6 g/L to 5.1 g/L (a reduction of 80%) [42]. Chi et al. [42] observed a tenfold increase in microbial tolerance following a treatment with $Ca(OH)_2$. Rover et al. [77] also observed similar results for NaOH-treated pyrolytic sugar fractions. For $Ca(OH)_2$, which was treated with pyrolysis oil hydrolysate, the ethanol yield increased from 0 to 0.45 g ethanol/g of glucose at 40% (v/v) [78]. Rover et al. [77] also observed an increased ethanol yield from pyrolytic sugars of 0.9 g/L at 2 wt%, which had been treated with NaOH. However, excess NaOH results in accumulation of sodium salts and could influence the fermentation process [76].

The adsorption of inhibitory compounds from biomass hydrolysates using different adsorbents is one of the more commonly-used techniques and activated carbon is a commonly-used adsorbent when it comes to the treatment of biomass hydrolysates. The other adsorbents - diatomite, bentonite, and zeolite - have also been used to detoxify pyrolysis oils. Treating loblolly pine derived pyrolysis oil with 1% (w/v) activated carbon reduced furfural by a total of 78% [32]. In another study [37], treating the aqueous fraction of Douglas fir pyrolysis oil with activated carbon failed to improve the growth of *Rhodotorula glutinis* (the maximum cell mass was at 0.74 g/L). Yu et al. [79] studied the feasibility of using different adsorbents to remove the inhibitory compounds from pre-treated (neutralisation and overliming) pyrolysis oil. The results showed that the combination of neutralisation and diatomite produced better results with 0.45 g ethanol/g of glucose yield with *Sachromicese cervisiae*. Studies which concentrated on detoxification with ion-exchange resin revealed the removal of more acid fractions when compared with furan and phenolic compounds [77].

Next to adsorption, liquid-liquid extraction is also one of the most commonly-studied detoxification methods. Water can be used to separate the low molecular weight water miscible fraction and high molecular weight fractions in the pyrolysis oil. However, it cannot be an effective method of detoxification. For the purposes of solvent extraction, an immiscible solvent is added to the pyrolysis oil, being mixed at the desired temperature and for the required time. This mixture is subsequently allowed for phase separation. Later, the phases will be separated by gravity. Previously, studies were carried out to understand the applicability of solvents for removal of inhibitory compounds, that have included ethyl acetate, n-butanol, amines, 1-octonal, kerosene, hexane, petroleum ether, chloroform, oleyl alcohol, and oleic acid.

When using the same solvent-to-aqueous ratio (1:1), n-butanol was seen to remove phenolic compounds from the pyrolysis aqueous phase compared with ethyl acetate [80]. Barbary et al. [80] also observed a negligible volume of phenolic components in the case of n-butanol and the aqueous fraction ratio of 3:1. Chan et al. [78] studied the feasibility of using five different amines to reduce microbial toxicity to *Sachromicese cervisiae T2* in ethanol production. The authors observed that tri-n-octylumine, together with co-solvent 1-octonal at 25% (v/v), had better results when compared to the use of other solvents, resulting in an ethanol yield of 0.24 g ethanol/g glucose. Better results were also observed with tri-n-octylumine by [77] in comparison with the use of ionic liquid (1-methyl-3-octylimidazolium tetrafluoroborate), and ion-resin (Dowex 66) in the detoxification of pyrolysis hydrolysates. Ren et al. [81] used hexane, petroleum ether, chloroform, and ethyl acetate to remove the various chemicals from aqueous pyrolysis oil. When compared to other solvents, the use of chloroform showed a higher level of extraction efficiency for furans, phenolics, and ketones. The sequential extraction of chloroform, which had been treated with ethyl acetate removed 62% of the acetic acid. In the same study, the authors observed that hexane showed poor levels of extraction efficiency. For example, the removal efficiency levels of phenols and 5-HMF with hexane was at 1 and 9 wt%, respectively. The extraction efficiency levels for the same compounds when using ethyl acetate was at 100 wt% and 60.8 wt%, respectively. Solvent extraction appears to be a promising approach to carry out detoxification when compared to other available methods. However, the dissolution of solvents into the pyrolysis oil aqueous fraction could result in the inhibition of microbial growth. Sukhbaatar et al. [82] observed a total of 38.43 g/L of n-butanol in the aqueous fraction of pyrolysis oil. An additional evaporation step may be required, followed by solvent extraction, to reduce the solvent content in detoxified pyrolysis oils. According to [14], the important factors that need to be considered in terms of solvent extraction are partition coefficient, immiscibility with the original medium, and the boiling point of the solvent. In addition, the cost of the solvent and applicable green chemistry principles also need to be taken into consideration.

A sequential detoxification approach could be more effective when compared to single step detoxification. Lian et al. [37] studied the sequential detoxification of the pyrolysis oil aqueous fraction in the order of ethyl acetate treatment (to remove phenolics), rotary evaporation (to remove acetol, hydroxyacetaldehyde, and furfural), and finally the activated carbon treatment (to remove furanics, monophenols, and lignin oligomers). The results showed that cell growth increased from 0.74 g/L for the activated carbon-only treated pyrolysis oil aqueous fraction to 3.3 g/L in the sequential detoxified aqueous fraction. The lipid yields also increased, from 0.1 to 7.8 g/L.

Adoptive evaluation of microbial strains to increase the tolerance to the inhibitory compounds could be helpful to improve the microbial conversion of pyrolysis oil fractions. Previously, Wang et al. [32] inoculated the *Saccharomyces pastorianus* on YPD agar media at 10%, 20%, and 30% (V/V) of pyrolysis oil hydrolysates. The inoculation process was repeated for ten times by transferring the grown colonies. The authors observed the colony growth only on 10 (V/V) % and concluded that ten cycles of inoculation may not be sufficient to build the inhibitory resistance of the selected strains. Chan et al. [78] adapted the yeast at 10% and 35% (V/V) pyrolysis oil hydrolysates loading and through 36 inoculations. The adaptation of yeast at 10 (V/V) % of hydrolysate improved the ethanol yield by 39%. The strain *S. cerevisiae* adapted to acid-hydrolysed pyrolysate through 12 inoculations produced 47% higher ethanol yield compared with original strain.

9.4.2 Economic Feasibility

Although pyrolysis oil looks as a promising feedstock for microbial conversion, at the end, the economic feasibility is one of the main deciding factors in the commercialization of integrating the pyrolysis with microbial processes. The cost estimation of microbial conversion of pyrolysis oil can be mainly divided into two group—(1) producing pyrolysis oil and/or extracting the pyrolytic sugars (2) Bioconversion of pyrolysis oil fractions. The overall economic feasibility of the process mainly depends on the cost of pyrolysis oil production because of the increased capital investment. It is also important to note that, as pyrolysis oil contains water and other pyroligneous compounds, only a limited amount of carbon source is available for microbial conversion. Previously, the pyrolysis oil production costs of $1.06/gallon [32] and levoglucosan production cost of $3/kg have been reported [83].

The cost of ethanol production from pyrolysis oil varies from $1.5/gallon to $150/gallon. The huge variation in the production costs is attributed to the variation in the selected process parameters. The pre-processing of pyrolysis sugars and the compounds added to the fermentation medium also influenced the ethanol production cost significantly. For example, with 1% of glucose, 1% of pyrolysis oil aqueous fraction glucose, 2% of peptone and 1% of urea (all w/v %) composition and at a production rate of 80.3 gallon/h, the ethanol production cost was $150/gallon. At the same production rate, but replacing the peptone with corn-steep syrup at 0.6% (w/v), ethanol production costs were reduced to $14/gallon [32]. On the other hand, the fermentation of hydrolyzed levoglucosan increased the ethanol production costs by 26% compared with the direct fermentation of levoglucosan [84]. The reason could be the requirement of additional chemicals, such as sulphuric acid and calcium carbonate for hydrolysis and neutralization [84]. A comparative analysis showed that ethanol production costs of fast pyrolysis integrated fermentation ($1.57/gallon) and acid hydrolysis fermentation process ($1.35/gallon) are comparable to each other [85].

The previous studies show that ethanol production costs for pyrolysis integrated fermentation are in promising range. However, it is not yet fully cost competitive to the ethanol production through conventional fermentation process. There are still several challenges, such as microbial inhibition, reduced loading rate, and reduced yield. At the same time, generating additional revenue from pyrolysis oils heavy fraction could improve the economic feasibility to the microbial conversion of pyrolysis oil. Previously [32], it has been reported that the pyroligneous fraction can be sold at $2.90/gallon considering its heating value as 46 MJ/kg. According to [86], producing chemicals from pyrolysis oil has higher economic feasibility compared with electrical energy production. For example, electrical energy generates $71/ton of pyrolysis oil while fermentable sugars are valued at $128/ton of pyrolysis.

On the other hand, the methane production from pyrolysis oil, is not yet cost competitive compared with natural gas and waste derived biogas [84]. According to Yang et al. [58], only 0.6–6% of energy in the original organic fraction of municipal solid waste can be converted into methane. Thus, the authors considered that additional processing of feedstock before AD through pyrolysis may not be economically feasible because of the increased costs with pyrolysis unit. However, integrating microbial conversion of the pyrolysis oil with combined heat and power (CHP) plants can further improve the economic flexibility through additional revenue from heat and electrical energy. The renewable energy credits available for bio based heat and power may also improve the economic feasibility when integrated with CHP.

9.5 Conclusions and Future Outlook

Pyrolysis can be integrated with different microbial processes, such as fermentation, anaerobic digestion, microbial fuel cells, and algae cultivation. Commonly, the product yield and production rates are lower for pyrolysis integrated microbial processes compared with conventional biochemical conversion processes. The commonly produced microbial products through this process integration are bioethanol, biogas, lipids (fatty acid esters), PHA, bioelectricity, succinic acid, and citric acid. The microbial inhibition is the major challenge with pyrolysis oil. The low specific area compared with activated carbon is an issue with biochar application. The application of biochar in microbial fuel cells significantly reduces the electricity production costs (from 400 $/W to 35 $/W) compared with commonly used carbon based electrodes. The cost competitiveness of integrating the pyrolysis with microbial processes in not yet fully realized. There are several opportunities for future activities, which include, multi-step and co-fermentation with different microbial strains and different substrates, exploring the possibilities of producing high value biochemicals, and detailed techno-economic analysis. Finally, to conclude, technically biomass pyrolysis can be integrated with microbial conversion processes and various microbial products can be produced. However, practically, there are several challenges that need to be addressed before large scale application of this process.

Considering the results from laboratory scale experiments, integrating the pyrolysis with microbial processes looks promising. There are several opportunities for future activities and to improve the processes further. The current studies on reducing the inhibitory effects of pyrolysis oil fractions are mainly focused on conventional approaches, for example adsorption and solvent extraction. These processes are costly and/or not very effective in removal of the inhibitory compounds. However, a two-step fermentation or co-fermentation with different strains could be helpful in reduction of the microbial toxicity of the pyrolysis oils and facilitate the co-production or multiple production. Studying the different combinations of microorganisms to consume different groups of pyrolysis oil compounds simultaneously or sequentially could be interesting for the future. It requires understanding the metabolic activities of different microbial species under synergetic situation. At the same time, genetic engineering of microbial strains could further improve the process feasibility and increase the product yield. At the moment, major focus has been given to the ethanol fermentation from pyrolysis oil sugar fractions. However, there is a need to study the production of various new compounds. For example, several microbial strains can metabolize the acetate-rich substrates and produce different biochemicals. In this context, the acetic acid-rich aqueous fraction of pyrolysis could be interesting. There are very few studies on the economic feasibility and again, the focus has been on ethanol. The detailed cost analysis needs to be carried out for better understanding of the economic feasibility of integrating the pyrolysis with microbial processes.

Acknowledgement Tharaka Rama Krishna C. Doddapaneni gratefully acknowledge the financial support by the European Regional Development Fund and the programme Mobilitas Pluss (Grant No MOBJD405) and base funded project of EMU PM180260TIBT.

References

1. Veum K, Bauknecht D. How to reach the EU renewables target by 2030? An analysis of the governance framework. Energy Policy. 2020;127:299–307. https://doi.org/10.1016/j.enpol.2018.12.013.
2. Pires JCM. COP21: the algae opportunity? Renew Sust Energ Rev. 2017;79:867–77. https://doi.org/10.1016/j.rser.2017.05.197.
3. IEA. Key world energy statistics IEA, Paris. 2019. https://www.iea.org/reports/key-world-energy-statistics-2019.
4. Jiang L, Zheng A, Meng J, Wang X, Zhao Z, Li H. A comparative investigation of fast pyrolysis with enzymatic hydrolysis for fermentable sugars production from cellulose. Bioresour Technol. 2019;274:281–6. https://doi.org/10.1016/j.biortech.2018.11.098.
5. Bridgwater AV. Review of fast pyrolysis of biomass and product upgrading. Biomass Bioenergy. 2011;38:68–94. https://doi.org/10.1016/j.biombioe.2011.01.048.
6. Pinheiro Pires AP, Arauzo J, Fonts I, Domine ME, Fernández Arroyo A, Garcia-Perez ME. Challenges and opportunities for bio-oil refining: a review. Energy Fuel. 2019;33:4683–720. https://doi.org/10.1021/acs.energyfuels.9b00039.
7. Campbell RM, Anderson NM, Daugaard DE, Naughton HT. Financial viability of biofuel and biochar production from forest biomass in the face of market price volatility and uncertainty. Appl Energy. 2018;230:330–43. https://doi.org/10.1016/j.apenergy.2018.08.085.
8. Arnold S, Moss K, Henkel M, Hausmann R. Biotechnological perspectives of pyrolysis oil for a bio-based economy. Trends Biotechnol. 2017;35:925–36. https://doi.org/10.1016/j.tibtech.2017.06.003.
9. Demirbas MF. Biorefineries for biofuel upgrading: a critical review. Appl Energy. 2009;86:S151–61. https://doi.org/10.1016/j.apenergy.2009.04.043.
10. Agbor VB, Cicek N, Sparling R, Berlin A, Levin DB. Biomass pretreatment: fundamentals toward application. Biotechnol Adv. 2011;29:675–85. https://doi.org/10.1016/j.biotechadv.2011.05.005.
11. Raud M, Kikas T, Sippula O, Shurpali NJ. Potentials and challenges in lignocellulosic biofuel production technology. Renew Sust Energ Rev. 2019;111:44–56. https://doi.org/10.1016/j.rser.2019.05.020.
12. Parawira W, Tekere M. Biotechnological strategies to overcome inhibitors in lignocellulose hydrolysates for ethanol production : review. Crit Rev Biotechnol. 2011;31:20–31. https://doi.org/10.3109/07388551003757816.
13. Himmel ME, Ding S, Johnson DK, Adney WS, Nimlos MR, Brady JW. Biomass recalcitrance: engineering plants and enzymes for biofuels production. Science. 2007;315:804–8. https://doi.org/10.1126/science.1137016.
14. Ko JK, Um Y, Park Y, Seo J, Kim KH. Compounds inhibiting the bioconversion of hydrothermally pretreated lignocellulose. Appl Microbiol Biotechnol. 2015;99:4201–12. https://doi.org/10.1007/s00253-015-6595-0.
15. Nor W, Wan R, Hisham MWM, Ambar M, Hin TY. A review on bio-oil production from biomass by using pyrolysis method. Renew Sust Energ Rev. 2012;16:5910–23. https://doi.org/10.1016/j.rser.2012.05.039.
16. Dhyani V, Bhaskar T. A comprehensive review on the pyrolysis of lignocellulosic biomass. Renew Energy. 2018;129:695–716. https://doi.org/10.1016/j.renene.2017.04.035.
17. Oasmaa A, Peacocke C. A guide to physical property characterisation of biomass – derived fast pyrolysis liquids. VTT Publ. 2001;450:65.
18. Jiang L, Fang Z, Zhao Z, Zheng A, Wang X, Li H. Levoglucosan and its hydrolysates via fast pyrolysis of lignocellulose for microbial biofuels: a state-of-the-art review. Renew Sust Energ Rev. 2019;105:215–29. https://doi.org/10.1016/j.rser.2019.01.055.
19. Taherzadeh MJ, Eklund R, Gustafsson L, Niklasson C, Lide G. Characterization and fermentation of dilute-acid hydrolyzates from wood. Ind Eng Chem Res. 1997;11:4659–65. https://doi.org/10.1021/ie9700831.

20. Ibbett R, Gaddipati S, Davies S, Hill S, Tucker G. The mechanisms of hydrothermal deconstruction of lignocellulose: new insights from thermal – analytical and complementary studies. Bioresour Technol. 2011;102:9272–8. https://doi.org/10.1016/j.biortech.2011.06.044.
21. Maduskar S, Maliekkal V, Neurock M, Dauenhauer PJ. On the yield of levoglucosan from cellulose pyrolysis. ACS Sustain Chem Eng. 2018;6:7017–25. https://doi.org/10.1021/acssuschemeng.8b00853.
22. Bennett NM, Helle SS, Duff SJB. Extraction and hydrolysis of levoglucosan from pyrolysis oil. Bioresour Technol. 2009;100:6059–63. https://doi.org/10.1016/j.biortech.2009.06.067.
23. Kwon G, Kim D, Kimura S, Kuga S. Rapid-cooling , continuous-feed pyrolyzer for biomass processing preparation of levoglucosan from cellulose and starch. J Anal Appl Pyrolysis. 2007;80:1–5. https://doi.org/10.1016/j.jaap.2006.12.012.
24. Yu Y, Chua YW, Wu H. Characterization of pyrolytic sugars in bio-oil produced from biomass fast pyrolysis. Energy Fuel. 2016;30:4145–9. https://doi.org/10.1021/acs.energyfuels.6b00464.
25. Li X, Luque-Moreno LC, Oudenhoven SRG, Rehmann L, Kersten SRA, Schuur B. Aromatics extraction from pyrolytic sugars using ionic liquid to enhance sugar fermentability. Bioresour Technol. 2016;216:12–8. https://doi.org/10.1016/j.biortech.2016.05.035.
26. Jönsson LJ, Martín C. Pretreatment of lignocellulose: formation of inhibitory by-products and strategies for minimizing their effects. Bioresour Technol. 2016;199:103–12. https://doi.org/10.1016/j.biortech.2015.10.009.
27. Pattra S, Sangyoka S, Boonmee M, Reungsang A. Bio-hydrogen production from the fermentation of sugarcane bagasse hydrolysate by Clostridium butyricum. Int J Hydrog Energy. 2008;33:5256–65. https://doi.org/10.1016/j.ijhydene.2008.05.008.
28. Ul Z, Yu I, El Z, Hassan B, Dongdong C. Microbial conversion of pyrolytic products to biofuels: a novel and sustainable approach toward second - generation biofuels. J Ind Microbiol Biotechnol. 2015;42:1557–79. https://doi.org/10.1007/s10295-015-1687-5.
29. Kostas ET, Cooper M, Shepherd BJ, Robinson JP. Identification of bio-oil compound utilizing yeasts through phenotypic microarray screening. Waste biomass valor. 2019; https://doi.org/10.1007/s12649-019-00636-7.
30. Ratledge C, Wynn JP. The biochemistry and molecular biology of lipid accumulation in oleaginous microorganisms. Adv Appl Microbiol. 2002;51:1–52. https://doi.org/10.1016/S0065-2164(02)51000-5.
31. Prosen EM, Radlein D, Piskorz J, Scott DS, Legge RL. Microbial utilization of levoglucosan in wood pyrolysate as a carbon and energy source. Biotechnol Bioeng. 1993;42:538–41. https://doi.org/10.1002/bit.260420419.
32. Wang H, Livingston D, Srinivasan R. Detoxification and fermentation of pyrolytic sugar for ethanol production. Appl Biochem Biotechnol. 2012;168:1568–83. https://doi.org/10.1007/s12010-012-9879-1.
33. Chi Z, Zhao X, Daugaard T, Dalluge D, Rover M, Johnston P. Biomass and bioenergy comparison of product distribution , content and fermentability of biomass in a hybrid thermochemical/biological processing platform. Biomass Bioenergy. 2019;120:107–16. https://doi.org/10.1016/j.biombioe.2018.11.006.
34. Yu Z, Zhang H. Ethanol fermentation of acid-hydrolyzed cellulosic pyrolysate with Saccharomyces cerevisiae. Bioresour Technol. 2003;90:95–100. https://doi.org/10.1016/S0960-8524(03)00093-2.
35. Fidalgo RM, Ortigueira J, Mendes B, Lemos PC. Bio-oil upgrading strategies to improve PHA production from selected aerobic mixed cultures. New Biotechnol. 2014;31:297–307. https://doi.org/10.1016/j.nbt.2013.10.009.
36. Lian J, Mckenna R, Rover MR, Nielsen DR, Wen Z, Jarboe LR. Production of biorenewable styrene : utilization of biomass - derived sugars and insights into toxicity. J Ind Microbiol Biotechnol. 2016;43:595–604. https://doi.org/10.1007/s10295-016-1734-x.
37. Lian J, Garcia-Perez M, Chen S. Fermentation of levoglucosan with oleaginous yeasts for lipid production. Bioresour Technol. 2013;133:183–9. https://doi.org/10.1016/j.biortech.2013.01.031.

38. Luque L, Orr VCA, Chen S, Westerhof R, Oudenhoven S, Rossum G. Van, Lipid accumulation from pinewood pyrolysates by Rhodosporidium diobovatum and Chlorella vulgaris for biodiesel production. Bioresour Technol. 2016;214:660–9. https://doi.org/10.1016/j.biortech.2016.05.030.
39. Wang C, Thygesen A, Liu Y, Li Q, Yang M, Dang D. Bio-oil based biorefinery strategy for the production of succinic acid. Biotechnol Biofuels. 2013;6(74):1–10. https://doi.org/10.1186/1754-6834-6-74.
40. Zhuang XL, Zhang HX, Yang JZ, Qi HY. Preparation of levoglucosan by pyrolysis of cellulose and its citric acid fermentation. Bioresour Technol. 2001;79:63–6. https://doi.org/10.1016/S0960-8524(01)00023-2.
41. Islam ZU, Klykov SP, Yu Z, Chang D, Hassan EB, Zhang H. Fermentation of detoxified acid-hydrolyzed pyrolytic anhydrosugars into bioethanol with saccharomyces cerevisiae 2. 399. Appl Biochem Microbiol. 2018;54:58–70. https://doi.org/10.1134/S0003683818010143.
42. Chi Z, Rover M, Jun E, Deaton M, Johnston P, Brown RC. Overliming detoxification of pyrolytic sugar syrup for direct fermentation of levoglucosan to ethanol. Bioresour Technol. 2013;150:220–7. https://doi.org/10.1016/j.biortech.2013.09.138.
43. Guzik MW, Kenny ST, Duane GF, Casey E, Woods T, Babu RP. Conversion of post consumer polyethylene to the biodegradable polymer polyhydroxyalkanoate. Appl Microbiol Biotechnol. 2014;98:4223–32. https://doi.org/10.1007/s00253-013-5489-2.
44. Ward PG, Goff M, Donner M. A two step chemo-biotechnological conversion of polystyrene to a biodegradable thermoplastic. Environ Sci Technol. 2006;40:2433–7. https://doi.org/10.1021/es0517668.
45. Lian J, Garcia-Perez M, Coates R, Wu H, Chen S. Yeast fermentation of carboxylic acids obtained from pyrolytic aqueous phases for lipid production. Bioresour Technol. 2012;118:177–86. https://doi.org/10.1016/j.biortech.2012.05.010.
46. Wei Z, Zeng G, Kosa M. Pyrolysis oil-based lipid production as biodiesel feedstock by Rhodococcus opacus. Appl Biochem Biotechnol. 2015;175:1234–46. https://doi.org/10.1007/s12010-014-1305-4.
47. Papanikolaou S, Aggelis G. Lipids of oleaginous yeasts. Part I: biochemistry of single cell oil production. Eur J Lipid Sci Technol. 2011;113:1031–51. https://doi.org/10.1002/ejlt.201100014.
48. Kim E, Um Y, Bott M, Woo HM. Engineering of Corynebacterium glutamicum for growth and succinate production from levoglucosan , a pyrolytic sugar substrate. FEMS Microbiol Lett. 2015;362:1–6. https://doi.org/10.1093/femsle/fnv161.
49. Dörsam S, Kirchhoff J, Bigalke M, Dahmen N, Syldatk C, Ochsenreither K. Evaluation of pyrolysis oil as carbon source for fungal fermentation. Front Microbiol. 2016;7:1–11. https://doi.org/10.3389/fmicb.2016.02059.
50. Lange J, Müller F, Bernecker K, Dahmen N, Takors R, Blombach B. Valorization of pyrolysis water: a biorefinery side stream , for 1, 2-propanediol production with engineered Corynebacterium glutamicum. Biotechnol Biofuels. 2017;10.277. https://doi.org/10.1186/s13068-017-0969-8.
51. Lewis AJ, Ren S, Ye X, Kim P, Labbe N, Borole AP. Hydrogen production from switchgrass via an integrated pyrolysis – microbial electrolysis process. Bioresour Technol. 2015;195:231–41. https://doi.org/10.1016/j.biortech.2015.06.085.
52. Park LK, Satinover SJ, Yiacoumi S, Mayes RT, Borole AP, Tsouris C. Electrosorption of organic acids from aqueous bio-oil and conversion into hydrogen via microbial electrolysis cells. Renew Energy. 2018;125:21–31. https://doi.org/10.1016/j.renene.2018.02.076.
53. Brooks V, Lewis AJ, Dulin P, Beegle R, Rodriguez M, Borole AP. Hydrogen production from pine-derived catalytic pyrolysis aqueous phase via microbial electrolysis. Biomass Bioenergy. 2018;119:1–9. https://doi.org/10.1016/j.biombioe.2018.08.008.
54. Satinover SJ, Elkasabi Y, Nuñez A, Rodriguez M. Microbial electrolysis using aqueous fractions derived from Tail-Gas Recycle Pyrolysis of willow and guayule. Bioresour Technol. 2019;274:302–12. https://doi.org/10.1016/j.biortech.2018.11.099.

55. Monlau F, Sambusiti C, Antoniou N, Barakat A, Zabaniotou A. A new concept for enhancing energy recovery from agricultural residues by coupling anaerobic digestion and pyrolysis process. Appl Energy. 2015;148:32–8. https://doi.org/10.1016/j.apenergy.2015.03.024.
56. Torri C, Fabbri D. Biochar enables anaerobic digestion of aqueous phase from intermediate pyrolysis of biomass. Bioresour Technol. 2014;172:335–41. https://doi.org/10.1016/j.biortech.2014.09.021.
57. Hübner T, Mumme J. Integration of pyrolysis and anaerobic digestion - use of aqueous liquor from digestate pyrolysis for biogas production. Bioresour Technol. 2015;183:86–92. https://doi.org/10.1016/j.biortech.2015.02.037.
58. Yang Y, Heaven S, Venetsaneas N, Banks CJ, Bridgwater AV. Slow pyrolysis of organic fraction of municipal solid waste (OFMSW): characterisation of products and screening of the aqueous liquid product for anaerobic digestion. Appl Energy. 2018;213:158–68. https://doi.org/10.1016/j.apenergy.2018.01.018.
59. Zhao X, Chi Z, Rover M, Brown R, Jarboe L, Wen Z. Microalgae fermentation of acetic acid-rich pyrolytic bio-oil: reducing bio-oil toxicity by alkali treatment. Environ Prog Sustain Energy. 2013;32:955–61. https://doi.org/10.1002/ep.11813.
60. Huggins T, Wang H, Kearns J, Jenkins P, Jason Z. Biochar as a sustainable electrode material for electricity production in microbial fuel cells. Bioresour Technol. 2014;157:114–9. https://doi.org/10.1016/j.biortech.2014.01.058.
61. Chen S, He G, Hu X, Xie M, Wang S, Zeng D. A three-dimensionally ordered macroporous carbon derived from a natural resource as anode for microbial bioelectrochemical systems. ChemSusChem. 2012;5(6):1059–63. https://doi.org/10.1002/cssc.201100783.
62. Yuan H, Deng L, Qi Y, Kobayashi N, Tang J. Nonactivated and activated biochar derived from bananas as alternative cathode catalyst in microbial fuel cells. Sci World J. 2014;2014:832850. https://doi.org/10.1155/2014/832850.
63. Wang B, Wang Z, Jiang Y, Tan G, Xu N, Xu Y. Enhanced power generation and wastewater treatment in sustainable biochar electrodes based bioelectrochemical system. Bioresour Technol. 2017;241:841–8. https://doi.org/10.1016/j.biortech.2017.05.155.
64. Li M, Zhang H, Xiao T, Wang S, Zhang B. Low-cost biochar derived from corncob as oxygen reduction catalyst in air cathode microbial fuel cells. Electrochim Acta. 2018;283:780–8. https://doi.org/10.1016/j.electacta.2018.07.010.
65. Huggins TM, Pietron JJ, Wang H, Jason Z, Biffinger JC. Graphitic biochar as a cathode electrocatalyst support for microbial fuel cells. Bioresour Technol. 2015;195:147–53. https://doi.org/10.1016/j.biortech.2015.06.012.
66. Wu S, Li H, Zhou X, Liang P, Zhang X, Jiang Y. A novel pilot-scale stacked microbial fuel cell for efficient electricity generation and wastewater treatment. Water Res. 2016;98:396–403. https://doi.org/10.1016/j.watres.2016.04.043.
67. Li S, Cheng C, Thomas A. Carbon-based microbial-fuel-cell electrodes: from conductive supports to active catalysts. Adv Mater. 2017;29:1602547. https://doi.org/10.1002/adma.201602547.
68. Masebinu SO, Akinlabi ET, Muzenda E, Aboyade AO. A review of biochar properties and their roles in mitigating challenges with anaerobic digestion. Renew Sust Energ Rev. 2019;103:291–307. https://doi.org/10.1016/j.rser.2018.12.048.
69. Cordiner S, Manni A, Mulone V, Rocco V, Braglia R, Canini A. Ampelodesmos mauritanicus pyrolysis biochar in anaerobic digestion process: evaluation of the biogas yield. Energy. 2018;161:663–9. https://doi.org/10.1016/j.energy.2018.07.196.
70. Shanmugam SR, Adhikari S, Nam H, Sajib SK. Effect of bio-char on methane generation from glucose and aqueous phase of algae liquefaction using mixed anaerobic cultures. Biomass Bioenergy. 2018;108:479–86. https://doi.org/10.1016/j.biombioe.2017.10.034.
71. Luo C, Lu F. Application of eco-compatible biochar in anaerobic digestion to relieve acid stress and promote the selective colonization of functional microbes. Water Res. 2014;68:710–8. https://doi.org/10.1016/j.watres.2014.10.052.

72. Lü F, Luo C, Shao L, He P. Biochar alleviates combined stress of ammonium and acids by fi rstly enriching Methanosaeta and then Methanosarcina. Water Res. 2016;90:34–43. https://doi.org/10.1016/j.watres.2015.12.029.
73. Fagbohungbe MO, Herbert BMJ, Hurst L, Li H, Usmani SQ, Semple KT. Impact of biochar on the anaerobic digestion of citrus peel waste. Bioresour Technol. 2016;216:142–9. https://doi.org/10.1016/j.biortech.2016.04.106.
74. Shen Y, Linville JL, Urgun-demirtas M, Schoene RP, Snyder SW. Producing pipeline-quality biomethane via anaerobic digestion of sludge amended with corn stover biochar with in-situ CO_2 removal. Appl Energy. 2015;158:300–9. https://doi.org/10.1016/j.apenergy.2015.08.016.
75. Shen Y, Forrester S, Koval J, Urgun-demirtas M. Yearlong semi-continuous operation of thermophilic two-stage anaerobic digesters amended with biochar for enhanced biomethane production. J Clean Prod. 2017;167:863–74. https://doi.org/10.1016/j.jclepro.2017.05.135.
76. Chang D, Yu Z, Islam ZU. Mathematical modeling of the fermentation of acid-hydrolyzed pyrolytic sugars to ethanol by the engineered strain Escherichia coli ACCC 11177. Appl Microbiol Biotechnol. 2015;99:4093–105. https://doi.org/10.1007/s00253-015-6475-7.
77. Rover MR, Johnston PA, Jin T, Smith RG, Brown RC. Production of clean pyrolytic sugars for fermentation. ChemSusChem Full Pap. 2014;7:1662–8. https://doi.org/10.1002/cssc.201301259.
78. Chan JKS, Duff SJB. Methods for mitigation of bio-oil extract toxicity. Bioresour Technol. 2010;101:3755–9. https://doi.org/10.1016/j.biortech.2009.12.054.
79. Yu Z, Zhang H. Pretreatments of cellulose pyrolysate for ethanol production by Saccharomyces cerevisiae , Pichia sp . YZ-1 and Zymomonas mobilis. Biomass Bioenergy. 2003;24:257–62.
80. Barbary E, Abou-yousef H, Steele P. Increasing the efficiency of fast pyrolysis process through sugar yield maximization and separation from aqueous fraction bio-oil. Fuel Process Technol. 2013;110:65–72. https://doi.org/10.1016/j.fuproc.2012.11.003.
81. Ren S, Ye XP, Borole AP. Separation of chemical groups from bio-oil water-extract via sequential organic solvent extraction. J Anal Appl Pyrolysis. 2017;123:30–9. https://doi.org/10.1016/j.jaap.2017.01.004.
82. Sukhbaatar B, Li Q, Wan C, Yu F, Hassan E, Steele P. Inhibitors removal from bio-oil aqueous fraction for increased ethanol production. Bioresour Technol. 2014;161:379–84. https://doi.org/10.1016/j.biortech.2014.03.051.
83. Wang J, Lu Z, Shah A. Techno-economic analysis of levoglucosan production via fast pyrolysis of cotton straw in China. Biofuels Bioprod Biorefin. 2019;13:1085–97. https://doi.org/10.1002/bbb.2004.
84. Claypool JT, Simmons CW. Hybrid thermochemical/biological processing: the economic hurdles and opportunities for biofuel production from bio-oil. Renew Energy. 2016;96:450–7. https://doi.org/10.1016/j.renene.2016.04.095.
85. So KS, Brown RC. Economic analysis of selected lignocellulose-to-ethanol conversion technologies. Appl Biochem Biotechnol. 1999;79:633–40. https://doi.org/10.1385/ABAB:79:1-3:633.
86. Sandvig E, Walling G, Energy A, Brown RC, Pletka R, Radlein D, et al. *Integrated pyrolysis combined cycle biomass power system concept definition final report*. 2003. https://www.osti.gov/servlets/purl/826269-d5wHs7/native/ Accessed 15 Jan 2020.

Chapter 10
Levoglucosan Production by Fast Pyrolysis of Biomass After Dilute Acid Pretreatment

Li-qun Jiang, Xiao-bo Wang, Zeng-li Zhao, and Hai-bin Li

Abstract Fermentable sugars are valuable intermediates that can be produced from lignocellulosic biomass, in which hemicellulose and cellulose are favorable candidates for their production. In this chapter, the adaptability of a comprehensive strategy for lignocellulose saccharification is presented. Hemicellulose of reed pole can be hydrolyzed for xylose production by dilute acid first, and then the retained residue from acid hydrolysis, rich in cellulose, can be pyrolyzed for levoglucosan production. Such a process highlighted in this chapter exhibits several attractive features: (1) hemicellulose can be nearly and completely saccharified, giving a xylose yield as high as 92.2 wt%, (2) due to the alteration of chemical composition and physical structure by dilute acid pretreatment, the levoglucosan yield from acid pretreated biomass can be as high as 45.9 wt% as compared with that of un-pretreated reed pole (4.8 wt%), (3) improvement of levoglucosan formation and the inhibition of acetic acid formation from acid pretreated biomass can improve fermentability of its pyrolysate. Furthermore, it is demonstrated that the effect of demineralization, hydrolysis of hemicellulose and improvement of levoglucosan yield is ordered as $HCl > H_2SO_4 > H_3PO_4$.

Keywords Lignocellulose · Dilute acid pretreatment · Fast pyrolysis · Levoglucosan

10.1 Introduction

The dwindling storage of fossil fuels and the increased attention paid to environmental issues have urged numerous researchers to develop the field of bio-refineries. Lignocellulosic biomass, which is the most abundant natural resource, has a positive

L.-q. Jiang (✉) · X.-b. Wang · Z.-l. Zhao · H.-b. Li
CAS Key Laboratory of Renewable Energy, Guangdong Provincial Key Laboratory of New and Renewable Energy Research and Development, Guangzhou Institute of Energy Conversion, Chinese Academy of Sciences, Guangzhou, China
e-mail: lqjiang@ms.giec.ac.cn

Z. Fang et al. (eds.), *Production of Biofuels and Chemicals with Pyrolysis*, Biofuels and Biorefineries 10, https://doi.org/10.1007/978-981-15-2732-6_10

influence on the renewable energy sector in meeting the growing needs of energy [1]. In general, lignocellulose consists of a massive amount of carbohydrates and predominantly contains various forms of cellulose, hemicellulose, and lignin [2]. Cellulose ($(C_6H_{10}O_5)_n$) is a class of homopolysaccharide consisting of linear chains of β-D-glucan units connected by β-1,4-glycosidic bonds. Naturally, cellulose can be found in crystalline and amorphous forms. The crystalline cellulose is well organized, tightly bound by inter-chain hydrogen bonds, while amorphous cellulose is less pronounced. Hemicellulose is heteropolysaccharide consisting of hexose (D-mannose, D-galactose, D-glucose) and pentose (D-arabinose, D-xylose). Lignin is a heterogeneous phenylpropanoid polymer, where coumarin alcohol, coniferyl alcohol, and sinapyl alcohol are the primary monomers [3]. Both cellulose and hemicellulose are polysaccharides, which can be converted to ample quantities of fermentable sugars, and they are considered as important intermediates in the conversion of lignocellulose, and can be used as the substrate for microbial fermentation to produce chemicals or biofuels [4]. Saccharification of biomass is one of the pivotal challenges in the development of biofuels [5].

Commonly, enzymatic/acid hydrolysis is the most popular route to depolymerize cellulose into glucose [6]. Enzymatic hydrolysis of cellulose is catalyzed by materials derived from several bacteria and fungi. The simultaneous existences of endo-glucanases, exo-glucanases, and β-glucanases are essential in the enzymatic degradation of cellulose. The endo-glucanases first degrade the cellulose chain to generate free chain-ends. Exo-glucanases subsequently target the sugar chains to isolate cellobiose (dimer of glucose) from the free chain-ends. Eventually, β-glucanases cleave glycosidic bonds in cellobiose to yield glucose. To accomplish the depolymerization of cellulose, each enzyme has a well-defined type of action and all three enzymes show synergistic interactions [7]. There are several advantages for the enzymatic hydrolysis, such as mild reaction conditions, the high selectivity of glucose and no elimination of toxic derivatives in hydrolysate [5]. Overall, enzymatic hydrolysis is a promising approach to selectively converting cellulose into glucose. On the other hand, there are many shortcomings of enzymatic hydrolysis which dramatically restrict its economic benefits and industrial applications, such as long reaction times (*ca.* days rather than minutes), high costs of enzymes, loss of enzymatic activity due to the inhibition of released glucose, low concentrations of glucose, and the process being vulnerable to contaminants derived from other biomass components [8].

Acid hydrolysis of lignocellulosic biomass, which could date back to the early 19th industrial century and was commercialized in the early twentieth century, has been extensively investigated [9]. Acids can penetrate lignocellulose, decrystallize cellulose and make it more accessible to reactions, leading to the disruption of the network of hydrogen bonds and breakages of glycosidic bonds. Chemicals used for hydrolysis of lignocellulose include mineral acids, such as HF, HCl, H_2SO_4, H_3PO_4, as well as organic acids, like fumaric, maleic and oxalic acid. To be effectively performed under atmospheric pressure and at room temperature, a high concentration of acid (*e.g.* 72 wt% H_2SO_4) is required. Concentrated acids promote the solubilization of cellulose, and glycosidic bonds are attacked by acid-catalyzed water molecules, followed by hydrolysis of cellulose and hemicellulose to intermediates (*e.g.* oligosaccharides, monosaccharides). The intermediates are further hydrolyzed to sugars and degraded chemicals (*e.g.* aldehydes, furfurals, acids). Low energy is consumed in concentrated acid hydrolysis, while the complexities of recycling acid, the risk of disposal of concentrated acid, severe damage of the reactors caused by the strong corrosiveness of acid restrict its large-scale applications [4]. When dilute acids are used instead, H^+ leads to the protonation of oxygen atoms of cellulose and activates the glycosidic bonds. Dilute acid is usually performed at higher temperatures (>180 °C) and/or pressures (*ca.* 1.3 MPa) for minutes to hours. In dilute acids, amorphous cellulose and hemicellulose are prone to hydrolysis, while crystalline cellulose is less prone to be saccharified [10]. Notably, the final concentration of products (*e.g.* ethanol) is directly proportioned to the initial concentration of the substrate. Thereby, high concentrations of sugars are essential to obtain high ethanol titers to make it easier to distill ethanol from fermentation broth. Nevertheless, the concentration of sugar released from enzymatic/acid hydrolysis is extremely low (approximately 1%) for practical fermentation. It is difficult to overcome the fundamental obstacles and technical limitations of hydrolysis technology [9]. Developing novel efficient and economic strategies for saccharification is warranted.

Despite hydrolysis having attracted much attention, fast pyrolysis is showing a huge potential among researchers aiming towards developing a new road map for lignocellulose saccharification (Fig. 10.1) [1, 11]. Fast pyrolysis, which is a thermochemical conversion method, is typically performed in the absence of oxygen at (450–600) °C at very short residence times (*ca.* seconds), in which lignocellulose thermally decomposes into solid char, non-condensable gas, and pyrolysis oil at the conditions [12]. Pyrolysis oil can contain 75% of the initial energy content of the feed lignocellulose and the bio-oil is typically the desired product [13]. The 1,6-anhydro-β-D-glucopyranose (levoglucosan), derived from cellulose pyrolysis, is the principal component contained in pyrolytic oil [14]. The content of levoglucosan in the pyrolysate is about 80% while the levoglucosan yield obtained from microcrystalline cellulose can reach 70.1 wt% [15]. The feasibility of levoglucosan or pyrolysis oil as a feedstock for eukaryotic and prokaryotic biocatalysts has been demonstrated. The metabolic pathways of levoglucosan have been evaluated for which is understood. Levoglucosan is converted to D-glucose and glucose 6-phosphate in prokaryotic and eukaryotic microorganisms respectively, then metabolized via the glycolytic pathway (Fig. 10.2) [16–18]. Levoglucosan can be fermented to citric acid, itaconic acid, malic acid and lipid by *Aspergillus terreus*, *Aspergillus niger* CBX-209, *Aspergillus oryzae*, *Rhodotorula glutinis* ATCC204091, respectively, giving comparable yields in the glucose fermentation [19–22]. Furthermore, levoglucosan can be converted to glucose via dilute acid hydrolysis under mild conditions (*e.g.* 125 °C) [23–25]. Many researchers have demonstrated that the process via the hydrolysis of pyrolytic oil of lignocellulose and subsequent fermentation is also feasible [24, 26, 27]. Biochemical studies have confirmed that levoglucosan or its hydrolysate can be metabolized as effectively as glucose. Saccharification of lignocellulose via fast pyrolysis may be an alternative to

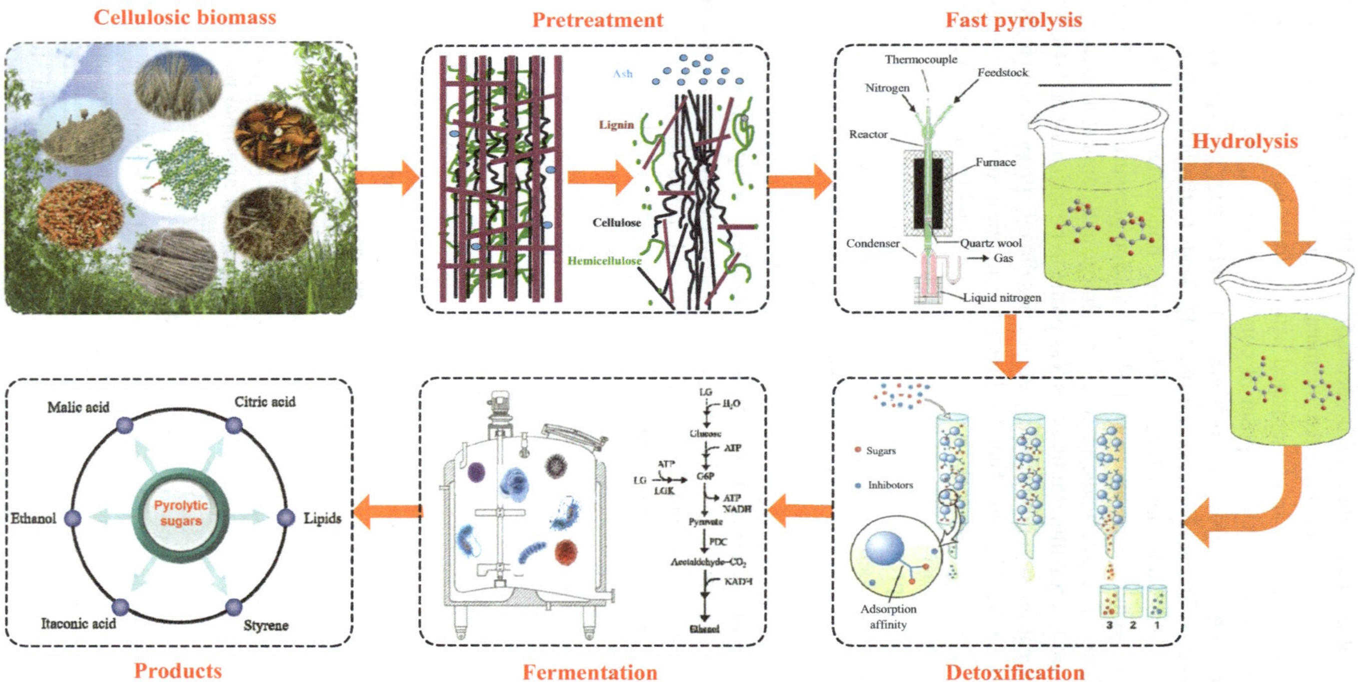

Fig. 10.1 Bio-refining of biomass based on fast pyrolysis (Reprinted with permission from [1], Copyright © 2019 Elsevier)

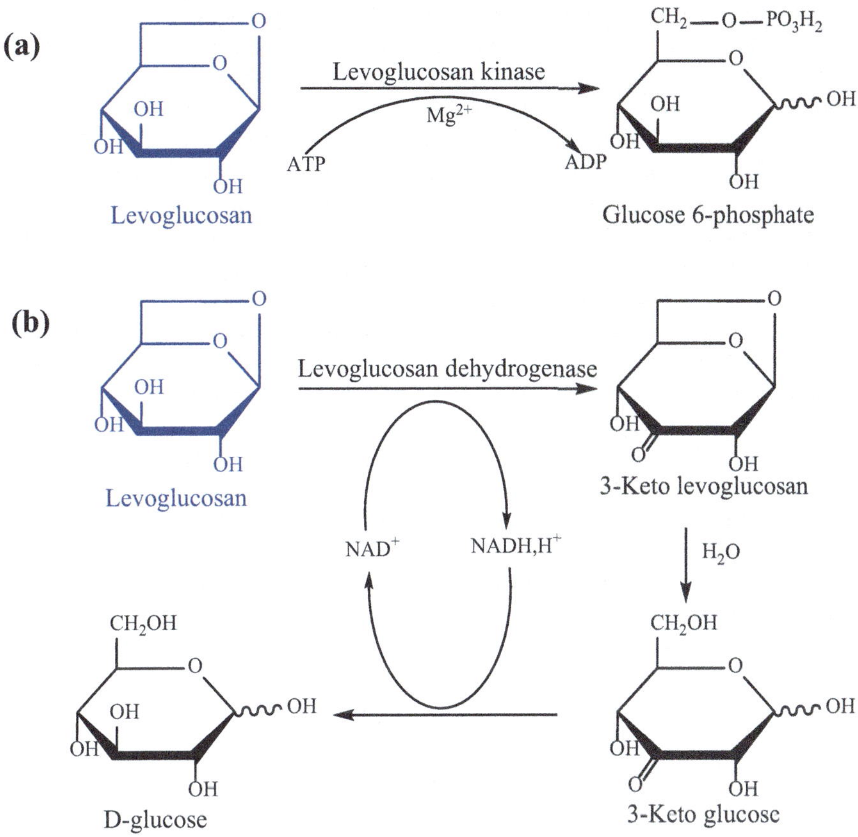

Fig. 10.2 Metabolized pathways for levoglucosan conversion to glucose 6-phosphate and to D-glucose (Adapted with permission from [1], Copyright © 2019 Elsevier. Adapted with permissions from [16, 17], Copyright © 1994 Japan Society for Bioscience and Agrochemistry, Copyright © 1991 Taylor and Francis)

overcome the barriers identified by hydrolysis conversion. There are several predominant advantages of fast pyrolysis compared with enzymatic or acid hydrolysis of lignocellulose. First, the rate of fast pyrolysis process requires seconds rather than hours or days. Then, fast pyrolysis can be implemented without any catalysts such as expensive enzymes or corrosive acids. Additionally, high concentrations of levoglucosan are easily obtained via fast pyrolysis. Most importantly, economic analyses show that saccharification of lignocellulose by fast pyrolysis has lower operating and capital cost, and improves the economics as compared with enzymatic/acid hydrolysis conversion [28–30].

Nevertheless, pyrolytic depolymerization is hindered by its lack of specificity. Generally, lignocellulosic pyrolysate is a complicated mixture that contains a large number of organic compounds, such as sugars, acids, furans, ketones, aldehydes, and phenols [31]. Levoglucosan yields from cellulose vary widely from 5 C% to

80 C% [32]. However, in the fast pyrolysis of lignocellulose, the yield of levoglucosan (<10 wt% based on cellulose) is much lower as compared with that from cellulose. The supramolecular structure and component of lignocellulose have a profound influence on the formation of levoglucosan [33]. Numerous studies have indicated that the formation of levoglucosan is affected by other components in lignocellulose (*e.g.* hemicellulose, lignin, ash), especially the alkaline and alkaline earth metals (AAEMs) [34, 35]. Research on fast pyrolysis saccharification still has a lot of unknowns to be explored and requires the investment of considerable effort to achieve high sugar yields. Nevertheless, it is an efficient route to pretreat lignocellulose into favorable raw material to facilitate levoglucosan formation from cellulose fraction [36]. Bio-pretreatment has been employed to alter the structure and chemical composition by microorganisms (*e.g. Cunninghamella echinulate, Echinodontium taxodii* 2358, *Irpex lacteus* CD2) [37–40]. Bio-pretreatment is performed under mild operations at relatively long times (*e.g.* 25 days) [40]. This pretreatment is environmentally friendly but time-consuming. The improvement in levoglucosan yields is also limited as compared to chemical pretreatments. Hot-water pretreatment, by cooking lignocellulose in water at higher temperatures (180–200 °C) and pressures, is a green hydrothermal reaction. In this process, there exist $[H]^+$ at high concentration and water as an acid-like medium [41]. Due to the removal of part of hemicellulose and lignin, and the removal of ash during the process of hot water washed pretreatment, production of levoglucosan is enhanced, and the quality of bio-oil is improved [42, 43]. Hot-water wash can elevate the yield of levoglucosan to some extent (from 4.9 to 11.7 wt%), but dilute acid pretreatment is more effective (from 4.9 to 23.5 wt%) [43, 44]. Various acid pretreatments (HCl, H_2SO_4, HNO_3 and acetic acid) have been adopted prior to fast pyrolysis, and higher yields of levoglucosan have been obtained [45, 46]. However, acid concentrations higher than a certain threshold (*e.g.* higher concentration than 3M H_2SO_4) caused a reduction in levoglucosan yield due to the degradation of cellulose [47]. Among these pretreatments, dilute acid pretreatment is considered as one of the best approaches to improving the yield of levoglucosan. Despite the high expectations for dilute acid pretreatment, pretreatment conditions suitable for levoglucosan formation should be investigated systematically to maximize levoglucosan production from pretreated lignocellulose. Furthermore, the ideal pretreatment process should not only be able to boost levoglucosan production in fast pyrolysis, but also make full utilization of other components and avoid waste disposal. Herein, with the aim of making full utilization of hemicellulose, further maximizing the levoglucosan production, and minimizing the production of inhibitors in fast pyrolysis, various acid pretreatments have been applied before fast pyrolysis of lignocellulose. The alteration of components and structures of lignocellulose before and after pretreatment and the influence of the alteration on the levoglucosan production are investigated.

10.2 Experimental

10.2.1 Materials

Reed pole derived from Hunan Province was used as the biomass feedstock. The reed pole was ground from (0.11 to 0.18) mm in diameter, dried to constant weight at 378 K in an oven. Acids (including HCl, H_2SO_4 and H_3PO_4) and standard chemicals were purchased from Chuandong Chemical Co., Ltd. (Chongqing) and Sigma (Shanghai).

10.2.2 Dilute Acid Hydrolysis

Reed pole was hydrolyzed by dilute acid in a high-pressure autoclave. Reed pole (5 g) was pretreated with 50 mL, (0.5 and 5)% HCl, H_2SO_4 or H_3PO_4 at 393 K for 1 h. Then, the separation of solid and liquid phases was achieved by filtration. Residual acid in the solid residue was removed with distilled water. Then, the pretreated biomass was dried by lyophilization. The compound contained in the hydrolysate was analyzed by high performance liquid chromatography (HPLC, Waters 2695, Massachusetts, USA) and separated by Bio-Rad Aminex HPX-87H column at 333 K. The content of 5-hydroxymethylfurfural (5-HMF) and furfural was determined using ultraviolet-visible (UV) detector (280 nm), and the content of acetic acid, glucose and xylose was measured using refractive index (RI) detector (323 K). Dilute H_2SO_4 (5 mmol/L) was applied as the mobile phase at a flow rate of 0.6 mL/min. The chemicals were identified and quantified by the external standard method. The yields of xylose and glucose were defined as below:

$$\text{Xylose yield (wt\%)} = \frac{\text{xylose mass in the dilute acid hydrolysate (g)}}{\text{hemicellulose mass of lignocellulsoe (g)}} \times 0.88 \times 100\% \tag{10.1}$$

$$\text{Glucose yield (wt\%)} = \frac{\text{glucose mass in the dilute acid hydrolysate (g)}}{\text{cellulose mass of lignocellulsoe (g)}} \times 0.90 \times 100\% \tag{10.2}$$

where an anhydro correction of 0.88 for xylose and correction of 0.90 for glucose are used to calculate the sugars yields from the corresponding polymers.

10.2.3 Elemental and Componential Analysis

The content of hydrogen (H), carbon (C) and nitrogen (N) in the reed pole was determined by an elemental analyzer (Vario EL Cube, Hanau, Germany). The content of AAEMs, including potassium (K), sodium (Na), calcium (Ca) and magnesium (Mg) was analyzed by inductively coupled plasma-atomic emission spectrometry (ICP-OES) (Optima 8000, PerkinElmer, Massachusetts, USA). Biomass (0.3 g) was dissolved in 4 mL of concentrated acid mixed by HNO_3 and $HClO_4$ (3:1, v/v) before ICP-OES analysis. The composition of carbohydrates was determined by a previous procedure [48]. The un-pretreated and acid pretreated reed pole were hydrolyzed through a two-step H_2SO_4 hydrolysis, consisting of a 72 wt% H_2SO_4 hydrolysis at 303 K for 1 h and a 4 wt% H_2SO_4 hydrolysis at 393 K for 1 h. HPLC was applied to analyze the sugar in the hydrolysate. The content of hemicellulose and cellulose was calculated in light of the content of hexose and pentose.

10.2.4 Thermogravimetric Analysis

Thermogravimetric analysis was carried out in a TGAQ50 analyzer (TA, New Castle, USA) under a nitrogen atmosphere (20 mL/min) to maintain an inert environment. The heating rate was set at 20 K/min and the samples were initially heated from 323 to 378 K and kept for 10 min at 378 K. Then, the samples were heated to 1023 K. The P (integrated pyrolysis coefficient) was defined as follows:

$$P = \frac{DTG_{max}}{T_{\max} \times (T_t - T_i)} \tag{10.3}$$

where DTG_{max} refers to the maximum rate for decomposition, T_i and T_t refer to the initial and final temperature for decomposition respectively, T_{max} refers to the temperature of the maximum rate for decomposition.

The activation energy required for the decomposition of biomass was obtained by numerical calculation with the Distributed Activation Energy Model (DAEM). The typical integral form of DAEM is:

$$\int_0^a \frac{da}{f(a)} = G(a) = \frac{1}{\beta} \int_0^T A \exp\left(\frac{-E}{RT}\right) dT \tag{10.4}$$

where $f(\alpha)$ is the most probable mechanism function, A is the frequency factor, E is the global activation energy (kJ/mol), T refers to the reaction temperature (K), R is the universal gas constant (8.314 J/(mol·K)).

The fundamental equation of DAEM was defined as:

$$a = 1 - \int_{0}^{\infty} \exp\left[-\frac{A_0}{\beta}\psi(E,T)\right] f(E)dE \tag{10.5}$$

where Ψ *(E, T)* is the internal integral:

$$\psi(E,T) = \int_0^T \exp\left(\frac{-E}{RT}\right)dT \xrightarrow{u'=\frac{E'}{R'T'}} \left(\frac{E}{R}\right)\int_u^\infty \frac{\exp(-u')}{u'^2}du' \tag{10.6}$$

The probability distribution *f(E)* was calculated as follows:

$$f(E) = \frac{1}{\sqrt{2\pi\sigma}} e^{-\frac{(E-E_0)^2}{2\sigma^2}} \tag{10.7}$$

10.2.5 Crystallinity Measurement

The crystallinity of biomass was scanned by X-ray diffraction (XRD) in X'Pert PROMPD X-ray diffractometer (PANalytical V.B., Almelo, Netherlands). Biomass was scanned by Cu-Kα radiation (λ = 1.54 Å) at a scanning rate of 0.01° per second at a diffraction angle of 5–45°. The crystallinity index (CrI) was calculated by a peak deconvolution method [49].

10.2.6 Fast Pyrolysis

The process of fast pyrolysis was implemented on a 5200 series CDS pyrolysis probe (CDS analysis, Pennsylvania, USA) connected with Agilent gas chromatograph/ mass spectrometer (GC/MS). During each test, biomass (about 200 μg) was weighed with a microbalance (Xp6152, METTLER TOLEDO, Albstadt, Germany) and loaded into the center of a quartz tube with quartz wool plugged on both sides. Biomass was pyrolyzed at 773 K for 20 s at a heating rate of 10 K ms^{-1}. Helium (20 mL/min) was used as the carrier gas continuously passing through the interface to transport the pyrolysis products from the quartz tube to the GC injection port. The split ratio was 50:1. Helium was used as the mobile phase at a flow rate of 1 mL/min in GC. The temperature of the GC oven equipped with an HP-INNO wax capillary column (Agilent 19,091 N-133) was heated from 323 to 503 K at a heating rate of 10 K/min and kept for a residence time of 0.5 h. The mass meter worked in EI mode (70 eV) and the mass scan ranged from *m/z* of 12–500. The differentiation of compounds was realized by equipping with the NIST mass spectrometry database and quantified by a five-point calibration curve. Each experiment was tested in triplicate. Compound yields were defined as below:

$$\text{Levoglucosan yield (wt\%)} = \frac{\text{levoglucosan mass (g)}}{\text{cellulose mass (g)}} \times 100\% \tag{10.8}$$

$$\text{Acetic acid yield (wt\%)} = \frac{\text{acetic acid mass (g)}}{\text{lignocellulose mass (g)}} \times 100\% \tag{10.9}$$

10.3 Results and Discussion

10.3.1 Elemental Analysis of Biomass

Elemental analysis of reed pole pretreated by different acids at concentrations of 0.5% and 5% is shown in Table 10.1. No evident change of compositions of C (46.4–49.0 wt%), H (5.8–6.1 wt%) and N (0–0.05 wt%) was observed before and after pretreatment, though there was a slight decrease on N. By comparing the compositions of AAEMs detected from pretreated samples with those from un-pretreated samples, it was found that pretreatment by water or acids could remove AAEMs evidently, especially K^+ and Na^+, on which both water and acids had a similar effect. For Ca^{2+} and Mg^{2+}, the contents were decreased after experiencing water-washed, while it was proven that pretreatment by acid had more access to removing Ca^{2+} and Mg^{2+}. Acid pretreatment could remove 84.1–92.0% of AAEMs. After using a higher concentration of acids, AAEMs could be observed to be further removed. Furthermore, abilities to remove these elements were ordered as $HCl > H_2SO_4 > H_3PO_4$, probably due to the differential concentrations of protons released by acid. Higher concentrations of acid are directly related to the further removal of AAEMs from biomass.

The yield of levoglucosan was sensitive to the content of the initial cellulose, which declined due to a small number of impurities, especially inorganic salts [50]. By interacting with oxygen to change the electronic structure of cellulose

Table 10.1 Analysis of elements and components of biomass before and after dilute acid pretreatment

	Organic elements (wt%)			AAEMs (ppm)				Components (wt%)	
Samples	C	H	N	K	Na	Ca	Mg	Cellulose	Hemicellulose
Un-pretreated	46.4	5.8	0.05	2561.1	1055.8	543.2	43.1	53.1	18.9
Water-washed	47.3	6.0	0.02	28.6	252.6	357.1	31.1	57.0	19.1
HCl-0.5%	47.7	6.1	0	20.9	196.6	138.1	8.8	68.5	5.2
HCl-5%	49.0	5.9	0	15.0	185.2	130.7	4.7	78.9	0
H_2SO_4-0.5%	47.7	6.0	0	20.1	207.7	175.5	11.8	66.5	6.3
H_2SO_4-5%	47.6	6.0	0	18.9	194.3	214.8	6.9	78.1	0
H_3PO_4-0.5%	47.6	6.0	0	30.0	235.7	205.4	14.2	62.7	9.5
H_3PO_4-5%	48.7	6.0	0	22.4	200.5	204.8	8.6	75.2	0

and affect the stereochemistry of molecules, AAEMs can favor dehydration, rearrangement reactions, and then fragmentation reaction. Even trace levels of AAEMs naturally found in biomass can shift the pyrolytic pathway, facilitate the formation of lower molecular weight oxygenates while restraining the formation of levoglucosan, by promoting fragmentation rather than depolymerization reaction yielding anhydrosugars [13, 51]. In the research of Patwardhan et al. [51], compared with pure cellulose, when 0.006 mmoles/g NaCl (0.05 wt%) was present in cellulose, the yield of levoglucosan was reduced to 26 wt% from 59 wt%. As the yield of levoglucosan decreased, the yield of low molecular weight compounds increased significantly. As far as the detailed chemical morphology of the pyrolysis products, KCl behaved similarly to NaCl. The participation of 0.004 mmoles KCl/g of cellulose caused a steep decrease in levoglucosan yield (from 59 wt% to 29 wt%). Previous studies had also explored the catalytic role of $CaCl_2$ and $MgCl_2$, which were actually instrumental in the formation of 2-furaldehyde, 5-hydroxymethylfurfural and levoglucosenone, suppressing the formation of levoglucosan [52]. The presence of Ca^{2+} and Mg^{2+} resulted in the preliminary formation of char from cellulose, and the conversion of levoglucosan into furans and other lower molecular weight oxygenates [53]. Nevertheless, compared with the former cases of K^+ and Na^+, the decrease was slightly smaller. Take the yield of levoglucosan into consideration, the order of most mild to strongest influence was $Mg^{2+} < Ca^{2+} < Na^+ < K^+$ [51, 54]. In this study, cellulose could conveniently be decomposed to levoglucosan without the catalytic effect of AAEMs.

10.3.2 Componential Analysis of Dilute Acid Hydrolysate and Solid Residual

The amount of hemicellulose and cellulose in reed pole before and after dilute acid hydrolysis, and the main components in dilute acid hydrolysate are analyzed and listed in Tables 10.1 and 10.2, respectively. The un-pretreated sample consisted of 18.9 wt% hemicellulose and 53.1 wt% cellulose. As the water-soluble fraction was

Table 10.2 Analysis of compound and sugar yield in dilute acid hydrolysate

	Compound concentration in acid hydrolysate (g/L)					Sugar yield (wt%)	
Samples	Acetic acid	Furfural	5-HMF	Xylose	Glucose	Xylose	Glucose
Water-washed	0.1	ND	ND	1.2	0.4	5.6	0.7
HCl-0.5%	2.2	0.2	0.04	15.9	2.7	74.1	4.6
HCl-5%	2.5	3.7	0.04	18.6	4.2	86.6	7.1
H_2SO_4-0.5%	1.0	0.6	0.04	14.7	2.1	68.4	3.6
H_2SO_4-5%	1.7	1.0	0.04	19.8	3.6	92.2	6.1
H_3PO_4-0.5%	0.5	0.02	0.03	11.8	2.3	54.9	3.9
H_3PO_4-5%	2.4	0.1	0.04	19.5	3.8	90.8	6.4

ND not detected

released, the content of hemicellulose and cellulose in water-washed biomass increased. Dilute acid hydrolysis was thought to be a favorable approach for the hydrolysis of hemicellulose and amorphous cellulose, as hemicellulose took up 9.5 wt%, 6.3 wt% and 5.2 wt% in biomass pretreated by 0.5% H_3PO_4, H_2SO_4 and HCl, respectively. Hemicellulose was proven to disappear absolutely at the residue hydrolyzed by 5% acids regardless of the types of acid. As the principal fraction of hemicellulose, xylan was converted to xylose in acid hydrolysis. The increasing concentration of acid enhanced the formation of xylose. The percent of hemicellulose in the acid hydrolysis residue decreased gradually from 18.9 wt% to 0 wt%. The xylose yield increased from 5.6 to 92.2 wt%. At the acidic concentration of 0.5%, the content of xylose in H_3PO_4 hydrolysate was 11.8 g/L, less than that 14.7 g/L pretreated by H_2SO_4 and 15.9 g/L xylose pretreated by HCl. At the acidic concentration of 5.0%, no apparent difference in xylose concentration was observed due to a maximum peak effect achieved at a relatively probable concentration. The content of glucose detected in hydrolysate pretreated differentially was kept in a narrow variation ranging from 2.1 g/L to 4.2 g/L, which was evaluated as a small number compared to xylose, and it could be explained by the harder saccharification of crystalline cellulose than hemicellulose and amorphous cellulose. Acid pretreatment improved the access to removing hemicellulose selectively, with most of the bulk cellulose remaining in the residue. Cellulose took up 75.2 wt%, 78.1 wt% and 78.9 wt% in biomass pretreated by 5% H_3PO_4, H_2SO_4 and HCl, respectively. CrI represented the relative ratio of crystalline cellulose contained in the biomass, which was impacted by the presence of amorphous hemicellulose and lignin. Preferential degradation of amorphous hemicellulose and less ordered cellulose by acid hydrolysis, caused the concentration of crystalline cellulose to increase in CrI of pretreated biomass from 56.5% to 62.1% (Fig. 10.3), whereas hydrolysis went towards increasing the formation of byproducts. The most common degradation products were furfural for pentose and 5-HMF for hexose. With increasing acid concentration, particularly, both furfural and 5-HMF, originating from the further decomposition of carbohydrates, were exponentially generated.

After acid pretreatment, the elimination of hemicellulose and the accumulation of crystalline cellulose could be beneficial to levoglucosan production. There exist some interactions between the components of lignocellulose, dominantly process in pyrolysis. Interactions of cellulose and hemicellulose facilitate the formation of 2,5-diethoxytetrahydrofuran and compounds derived from hemicellulose (*e.g.* furfural and acetic acid), suppressing the pathway leading to the formation of levoglucosan [55, 56]. Liu et al. [57] also observed that hemicellulose promoted the formation of hydroxyacetaldehyde and suppressed the production of levoglucosan. Previous studies have shown that samples with a higher content of cellulose have the potential to produce more levoglucosan [58]. The crystal allomorph and crystallinity of cellulose could influence the yield of levoglucosan, which was mainly originated from crystalline cellulose [59]. Amorphous samples are easier to be decomposed to a liquid phase than crystalline cellulose to form low molecular weight compounds [60]. Higher yields of levoglucosan can be obtained from cellulosic samples with higher crystallinity [59, 61].

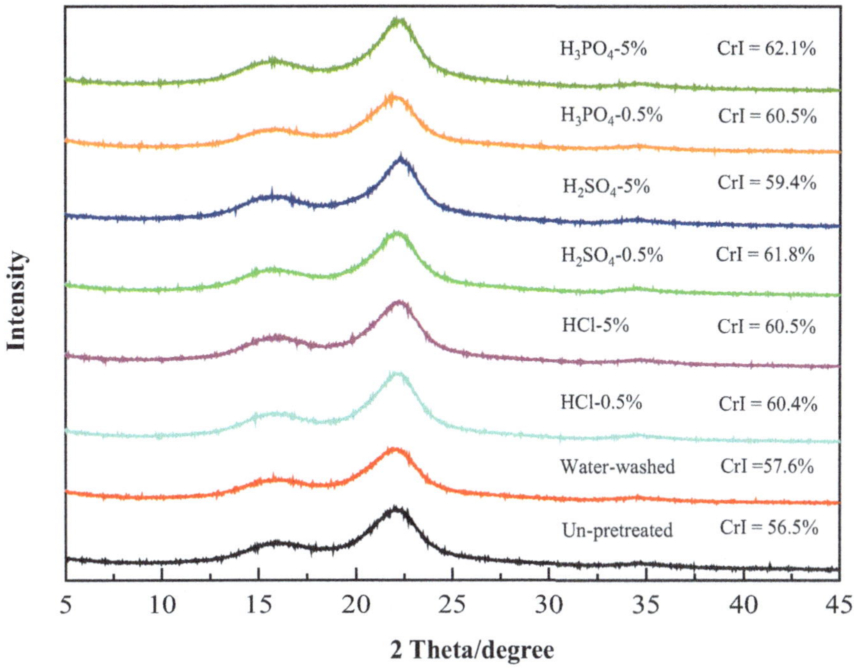

Fig. 10.3 XRD patterns of un-pretreated and pretreated biomass

10.3.3 Thermogravimetric Analysis

The TG and DTG curves of un-pretreated and pretreated biomass are shown in Fig. 10.4. Characteristic parameters of thermal degradation are summarized in Table 10.3. The different relative content of components (hemicellulose, cellulose and lignin) also leads to different pyrolytic profiles of each biomass. Hemicellulose has the strongest reactivity relative to cellulose and lignin, and can be decomposed in a lower temperature range. Cellulose, on the other hand, is the least reactive, requiring higher temperatures to decompose. The thermal decomposition of un-pretreated and water-washed reed pole exhibited two distinct peaks in the DTG curves due to the preferential decomposition of hemicellulose, followed by cellulose. After pretreatment with 5% dilute acid, the first shoulder vanished in the acid pretreated biomass, indicating the removal of hemicellulose. It also confirmed that the profiles of pretreated reed pole had higher T_i, T_{max} and D_{max} compared with those of un-pretreated sample. The proportion of cellulose in biomass pretreated by H_3PO_4 was less than that in samples pretreated by H_2SO_4 or HCl at an acidic concentration of 0.5%. The T_{max} matched the differences exactly on their content of cellulose related to the types of acid used in pretreatment, of which the sample pretreated by H_3PO_4 was 619 K and samples pretreated by H_2SO_4 or HCl was 626.6 K and 627.7 K, respectively. The severity of pyrolysis was expressed by P. Un-pretreated

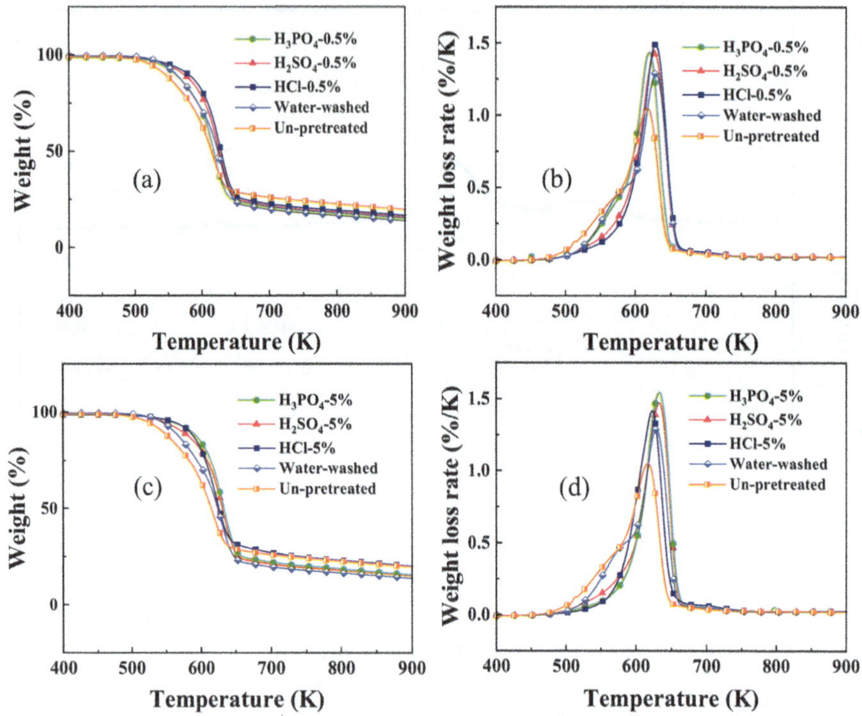

Fig. 10.4 TG (**a**, **c**) and DTG curves (**b**, **d**) of un-pretreated and pretreated biomass

Table 10.3 Pyrolytic parameters of un-pretreated and pretreated biomass showing T_i initial volatilizing temperature, T_t final volatilizing temperature, T_{max} temperature of the maximum mass loss rate, D_{max} maximum mass loss rate, M_r proportion of pyrolytic residue, P comprehensive pyrolysis index

Samples	T_i (K)	T_t (K)	T_{max} (K)	D_{max} (%/K)	M_r (wt%)	$P \times 10^6$ (%/K^2)
Un-pretreated	471.1	698.5	616.6	1.1	11.2	7.8
Water-washed	481.2	676.0	628.0	1.3	5.6	10.6
HCl-0.5%	484.8	674.5	627.7	1.5	6.9	12.6
HCl-5%	511.0	676.0	622.2	1.4	8.8	13.6
H_2SO_4-0.5%	495.8	668.5	626.6	1.4	5.9	12.9
H_2SO_4-5%	477.6	676.9	632.5	1.5	6.0	11.9
H_3PO_4-0.5%	498.3	664.2	619.0	1.4	4.7	13.6
H_3PO_4-5%	497.6	679.3	632.3	1.5	6.3	13.1

biomass had the lowest P value (7.8×10^{-6}%/K^2) compared with the water-washed (10.6×10^{-6}%/K^2) and dilute acid pretreated biomass ($11.9\times10^{-6} - 13.6\times10^{-6}$%/K^2).

Table 10.4 shows that the kinetic parameters for the DAEM model, including $\log_{10}$A at 15, 20 and 25. E is the global activation energy, which represents the

Table 10.4 Distributed activation energy model (DAEM) kinetic parameters for biomass

	$\log_{10}A_0$ (min^{-1}) = 15			$\log_{10}A_0$ (min^{-1}) = 20			$\log_{10}A_0$ (min^{-1}) = 25		
Samples	*E* (kJ/mol)	σ (kJ/mol)	$adjR^2$	*E* (kJ/mol)	σ (kJ/mol)	$adjR^2$	*E* (kJ/mol)	σ (kJ/mol)	$adjR^2$
Un-pretreated	177.6	11.7	0.993	233.9	16.1	0.994	290.4	20.4	0.994
Water-washed	181.2	10.0	0.996	238.6	14.1	0.996	296.3	18.0	0.996
HCl-0.5%	183.5	7.0	0.995	241.7	10.5	0.995	300.1	13.7	0.995
HCl-5%	182.6	7.2	0.993	240.6	10.6	0.993	298.7	13.8	0.993
H_2SO_4-0.5%	182.6	7.8	0.995	240.5	11.4	0.995	298.7	14.9	0.995
H_2SO_4-5%	183.9	8.0	0.994	242.2	11.7	0.994	300.7	15.3	0.994
H_3PO_4-0.5%	180.0	8.6	0.995	237.1	12.4	0.995	294.4	15.9	0.995
H_3PO_4-5%	184.9	6.5	0.994	243.6	9.9	0.994	302.5	13.1	0.994

E is the global activation energy, which represents the average activation energy of all reactions of DAEM
σ is the activation energy deviation, indicating the range of activation energy distribution of DAEM

average activation energy of all reactions. The value of the global activation energy represents the thermal stability of the sample and higher values indicate that the sample is more difficult to decompose. The σ, which is the activation energy deviation, indicates the range of activation energy distribution of a reaction [62]. Smaller activation energy deviations imply more explosive pyrolysis reactions. Each sample could obtain a suitable activation energy density distribution under different pre-exponential factors because of the compensation effect. So, the determination coefficient was above 0.99, which shows that the model agreed well with the experimental curve. Although *f(E)* depended on the change of the pre-exponential factor, the characters and trends of the parameters reflected were similar. The global activation energy E could be arranged in the following order: un-pretreated $<$ water-washed $\approx$ dilute acid pretreated sample. The activation energy deviation σ obeyed the following sequence: un-pretreated $>$ water-washed $>$ dilute acid pretreated sample. Removal of hemicellulose and AAEMs, as well as the accumulation of crystalline cellulose, could enhance the thermal stability of biomass during pyrolysis.

It could be seen from the TG curve that the initial volatilizing temperature for lignocellulose containing a great quantity of hemicellulose and AAEMs was lower than the dilute acid pretreated lignocellulose. Similar to the kinetic analysis results, the activation energy required for the decomposition of raw materials with more hemicellulose and AAEMs was the lowest compared with dilute acid pretreated cellulose. Hemicellulose began to decompose at a lower temperature range while cellulose began to decompose at higher temperature intervals. Previous research demonstrated that the presence of AAEMs could lower the initial temperature of lignocellulose degradation. Where K^+ is considered as an example. The catalytic influence of K^+ has been investigated in previous research [63, 64], where it was shown that K^+ could lower the initial and maximum temperature for decomposition, thus reducing the maximum degradation rate and favoring the formation of char. Since hemicellulose and AAEMs dissolved in the acids during the dilute acid pretreatment of lignocellulose, cellulose was the main component decomposed during the subsequent pyrolysis process, resulting in a smaller activation energy deviation and a more explosive reaction. Due to the removal of AAEMs, the formation of biochar in cellulose pyrolysis was also limited. M_r (proportion of pyrolytic residue) analogically explained the weak removal by H_3PO_4 in hydrolysis, of which the sample pretreated by H_3PO_4 was 4.7 wt% and samples pretreated by H_2SO_4 or HCl was 5.9–6.9 wt%. The crystalline structure of biomass is altered during acid pretreatment, which also plays a significant role in the thermal stability of biomass. The crystalline structure affects both the thermal stability and the required activation energy in the thermal decomposition of cellulose [65]. During pyrolysis, amorphous cellulose decomposes more rapidly by heat and degrades at lower temperatures, and substantial heat is absorbed for cleavage of the hydrogen bond network before decomposition [66]. Higher crystallinity of cellulose acts as a barrier for thermal degradation for its packed structure, and intramolecular hydrogen bonds would play important roles in stabilizing the cellulose and suppressing thermal expansion, which improves the thermal stability of cellulose [67]. Thereby, the acid pretreated biomass with higher crystallinity exhibited a higher endothermic activity.

10.3.4 Fast Pyrolysis

The pyrolysate of lignocellulose is a mixture of various organic compounds, including acids, phenols, furans, ketones, esters, aldehydes and dehydrated sugars [68]. Levoglucosan, which is the target product after pyrolysis, has a sizable proportion in the pyrolysis products. Although a large number of studies have attempted to elaborate on the fast pyrolysis mechanism of the production of levoglucosan from cellulose, there still exists much controversy about the potential generative pathways, mainly including homolytic mechanisms, heterolytic mechanisms and concerted mechanisms [69–71]. As shown in Fig. 10.5, the yield of levoglucosan from the un-pretreated sample was very low. The yield of levoglucosan rose from 4.8 wt% without pretreatment to 37.8 wt% after 0.5% dilute acid pretreatment. Abilities to improve the levoglucosan yield were ordered as HCl > H_2SO_4 > H_3PO_4. Enhancing the concentration of dilute acid can advance the formation of levoglucosan. The highest levoglucosan yield could be obtained from 5% HCl pretreated biomass (45.9 wt%). The promotion as currently envisioned was mainly because of the demineralization, release of hemicellulose, and the concentration of crystalline cellulose after acid pretreatment. Acetic acid was the major product of the pyrolytic degradation of hemicellulose. Due to the release of hemicellulose by dilute acid hydrolysis, the yield of acetic acid decreased from 5.2 wt% to 0.5 wt%. Levoglucosan has the potential to be used as a favorable feedstock for

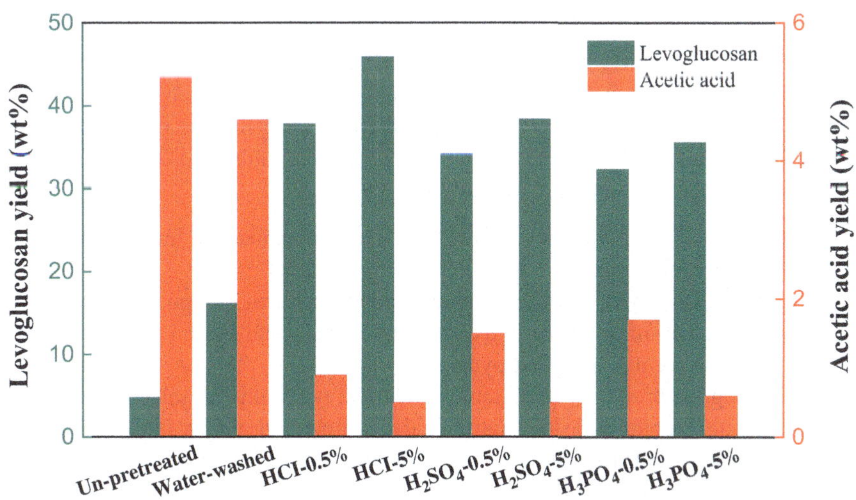

Fig. 10.5 Yields of levoglucosan and acetic acid from biomass

fermentation to produce biofuels. However, a yield less than its theoretical one from lignocellulose, and the existence of toxic compounds to microorganisms used in fermentation limit application of the pyrolysate [1]. In the fermentation of pyrolysate, carboxylic acids are regarded as some of the most toxic inhibitors for growth and metabolization of biocatalysts [72]. In sum, in the current study, the promotion of levoglucosan production and reduction in the amount of inhibitors formed could improve the fermentability of lignocellulosic pyrolysate. The utilization of dilute acid hydrolysate of biomass for the production of biofuel or biochemistry via microorganism fermentation has been widely demonstrated in previous investigations. Therefore, dilute acid pretreatment could not only saccharify hemicellulose efficiently but also eliminate the negative influence of hemicellulose on the pyrolysis of cellulose.

10.4 Conclusions and Future Outlook

Herein, the adaptability of an integrated process for lignocellulosic saccharification is provided. The hemicellulose of the reed pole converts to fermentable sugars via dilute acid hydrolysis first. The remaining solid residue from dilute acid hydrolysis, consisting of abundant cellulose, can be utilized to release levoglucosan via fast pyrolysis. Dilute acid hydrolysis as pretreatment could convert most of hemicellulose of biomass and the yield of xylose could reach values as high as 92.2 wt%. The yield of levoglucosan from acid pretreated biomass increased significantly from acid pretreated biomass (45.9 wt%) compared with that of un-pretreated samples (4.8 wt%) due to the demineralization, the release of hemicellulose and the accumulation of crystalline cellulose through dilute acid pretreatment. The abilities to the removal of AAEMs and hemicellulose, and improvement of levoglucosan yield were ordered as $HCl > H_2SO_4 > H_3PO_4$. Further elevating acid concentration from 0.5% to 5% in pretreatment could promote the formation of levoglucosan. The strategy in this research provides a prospective approach for making full use of hemicellulose and cellulose, and obtaining fermentable sugars for a future bio-refinery. Several issues should be considered for further development:

1. Understanding the fundamental mechanisms responsible for levoglucosan formation from cellulose.
2. Illustrating the physical structure features and chemical compositions of lignocellulose in sub-micrometer and nanometer levels and their relationship with levoglucosan formation in fast pyrolysis.
3. Optimizing the process (*e.g.* suitable feedstock, pretreatment, selective catalysts, pyrolysis temperature, heating rate, residence time and reactor type) to improve levoglucosan yield from biomass and reduce the toxicity of pyrolysate.
4. Developing a multifaceted strategy combining structural, biochemical and metabolic engineering to favor the fermentation of levoglucosan in microorganisms.

Acknowledgements Financial support provided by the National Natural Science Foundation of China (No. 51606204, 51876208), the Project Foundation of Guangdong province and Guangzhou city (No. 2017A020216007 and No. 201707010236), the Open Project of State Key Laboratory of Microbial Technology (No. M2019-10) are gratefully acknowledged.

References

1. Jiang LQ, Fang Z, Zhao ZL, Zheng AQ, Wang XB, Li HB. Levoglucosan and its hydrolysates via fast pyrolysis of lignocellulose for microbial biofuels: a state-of-the-art review. Renew Sust Energ Rev. 2019;105:215–9. https://doi.org/10.1016/j.rser.2019.01.055.
2. Wang SR, Dai GX, Yang HP, Luo ZY. Lignocellulosic biomass pyrolysis mechanism: a state-of-the-art review. Prog Energy Combust Sci. 2017;62:33–86. https://doi.org/10.1016/j.pecs.2017.05.004.
3. Fang Z. Pretreatment techniques for biofuels and biorefineries. Berlin: Springer; 2012. https://doi.org/10.1007/978-3-642-32735-3.
4. Guo F, Fang Z, Xu CC, Smith RL Jr. Solid acid mediated hydrolysis of biomass for producing biofuels. Prog Energy Combust Sci. 2012;38(5):672–90. https://doi.org/10.1016/j.pecs.2012.04.001.
5. Jiang LQ, Zheng AQ, Meng JG, Wang XB, Zhao ZL, Li HB. A comparative investigation of fast pyrolysis with enzymatic hydrolysis for fermentable sugars production from cellulose. Bioresour Technol. 2019;274:281–6. https://doi.org/10.1016/j.biortech.2018.11.098.
6. Tang S, Dong Q, Fang Z, Miao ZD. Complete recovery of cellulose from rice straw pretreated with ethylene glycol and aluminum chloride from enzymatic hydrolysis. Bioresour Technol. 2019;284:98–104. https://doi.org/10.1016/j.biortech.2019.03.100.
7. Coseri S. Cellulose: to depolymerize... or not to? Biotechnol Adv. 2017;35:251–66. https://doi.org/10.1016/j.biotechadv.2017.01.002.
8. Jiang LQ, Zheng AQ, Zhao ZL, He F, Li HB, Wu NN. The comparison of obtaining fermentable sugars from cellulose by enzymatic hydrolysis and fast pyrolysis. Bioresour Technol. 2016;200:8–13. https://doi.org/10.1016/j.biortech.2015.09.096.
9. Binder JB, Raines RT. Fermentable sugars by chemical hydrolysis of biomass. PNAS. 2010;107:4516–21. https://doi.org/10.1073/pnas.0912073107.
10. Jiang LQ, Fang Z, Guo F, Yang LB. Production of 2,3-butanediol from acid hydrolysates of Jatropha hulls with *Klebsiella Oxytoca*. Bioresour Technol. 2012;107:405–10. https://doi.org/10.1016/j.biortech.2011.12.083.
11. Rover MR, Johnston PA, Jin T, Smith RG, Brown RC, Jarboe L. Production of clean pyrolytic sugars for fermentation. ChemSusChem. 2014;7:1662–8. https://doi.org/10.1002/cssc.201301259.
12. Bridgwater AV, Meier D, Radlein D. An overview of fast pyrolysis of biomass. Org Geochem. 1999;30(12):1479–93. https://doi.org/10.1016/S0146-6380(99)00120-5.
13. Carpenter D, Westover TL, Czernik S, Jablonski W. Biomass feedstocks for renewable fuel production: a review of the impacts of feedstock and pretreatment on the yield and product distribution of fast pyrolysis bio-oils and vapors. Green Chem. 2014;16(2):384–406. https://doi.org/10.1039/c3gc41631c.
14. Lindstrom JK, Proano-Aviles J, Johnston PA, Peterson CA, Stansell JS, Brown RC. Competing reactions limit levoglucosan yield during fast pyrolysis of cellulose. Green Chem. 2019;21:178. https://doi.org/10.1039/c8gc03461c.
15. Kwon GJ, Kim DY, Kimura S, Kuga S. Rapid-cooling, continuous-feed pyrolyzer for biomass processing preparation of levoglucosan from cellulose and starch. J Anal Appl Pyrolysis. 2007;80(1):1–5. https://doi.org/10.1016/j.jaap.2006.12.012.

16. Nakahara K, Kitamura Y, Yamagishi Y, Shoun H, Yasui T. Levoglucosan dehydrogenase involved in the assimilation of levoglucosan in *Arthrobacter sp.* I-552. Biosci Biotechnol Biochem. 1994;58(12):2193–6. https://doi.org/10.1271/bbb.58.2193.
17. Kitamura Y, Abe Y, Yasui T. Metabolism of levoglucosan (1,6-anhydro-β-D-glucopyranose) in microorganisms. Agric Biol Chem. 1991;55(2):515–21. https://doi.org/10.1080/00021369.1991.10870617.
18. Yasui T, Kitamura Y, Nakahara K, Abe Y. Metabolism of levoglucosan (1,6-anhydro-β-D-glucopyranose) in bacteria. Agric Biol Chem. 1991;55(7):1927–9. https://doi.org/10.1080/00021369.1991.10870881.
19. Nakagawa M, Sakai Y, Yasui T. Itaconic acid fermentation of levoglucosan. J Ferment Technol. 1984;62(2):201–3.
20. Zhuang XL, Zhang HX. Identification, characterization of levoglucosan kinase, and cloning and expression of levoglucosan kinase cDNA from *Aspergillus niger* CBX-209 in *Escherichia coli.* Protein Expr Purif. 2002;26(1):71–81. https://doi.org/10.1016/S1046-5928(02)00501-6.
21. Lian JN, Garcia-Perez M, Chen SL. Fermentation of levoglucosan with oleaginous yeasts for lipid production. Bioresour Technol. 2013;133:183–9. https://doi.org/10.1016/j.biortech.2013.01.031.
22. Dorsam S, Fesseler J, Gorte O, Hahn T, Zibek S, Syldatk C, Ochsenreither K. Sustainable carbon sources for microbial organic acid production with filamentous fungi. Biotechnol Biofuels. 2017;10(1):242. https://doi.org/10.1186/s13068-017-0930-x.
23. Prosen EM, Radlein D, Piskorz J, Scott DS, Legge RL. Microbial utilization of levoglucosan in wood pyrolysate as a carbon and energy source. Biotechnol Bioeng. 1993;42(4):538–41. https://doi.org/10.1002/bit.260420419.
24. Bennett NM, Helle SS, Duff SJB. Extraction and hydrolysis of levoglucosan from pyrolysis oil. Bioresour Technol. 2009;100(23):6059–63. https://doi.org/10.1016/j.biortech.2009.06.067.
25. Abdilla RM, Rasrendra CB, Heeres HJ. Kinetic studies on the conversion of levoglucosan to glucose in water using Bronsted acids as the catalysts. Ind Eng Chem Res. 2018;57(9):3204–14. https://doi.org/10.1021/acs.iecr.8b00013.
26. Lian JN, Chen SL, Zhou S, Wang ZH, O'Fallon J, Li CZ, Garcia-Perez M. Separation, hydrolysis and fermentation of pyrolytic sugars to produce ethanol and lipids. Bioresour Technol. 2010;101(24):9688–99. https://doi.org/10.1016/j.biortech.2010.07.071.
27. Luque L, Oudenhoven S, Westerhof R, van Rossum G, Berruti F, Kersten S, Rehmann L. Comparison of ethanol production from corn cobs and switchgrass following a pyrolysis-based biorefinery approach. Biotechnol Biofuels. 2016;9(1):242. https://doi.org/10.1186/s13068-016-0661-4.
28. So KS, Brown RC. Economic analysis of selected lignocellulose-to-ethanol conversion technologies. Appl Biochem Biotechnol. 1999;77–79:633–40. https://doi.org/10.1007/978-1-4612-1604-9_57.
29. Anex RP, Aden A, Kazi KF, Fortman J, Swanson RM, Wright MM, Satrio JA, Brown RC, Daugaard DE, Platon A, Kothandaraman G, Hsu DD, Dutta A. Techno-economic comparison of biomass-to-transportation fuels via pyrolysis, gasification, and biochemical pathways. Fuel. 2010;89:S29–35. https://doi.org/10.1016/j.fuel.2010.07.015.
30. Zhang YA, Brown TR, Hu GP, Brown RC. Techno-economic analysis of monosaccharide production via fast pyrolysis of lignocellulose. Bioresour Technol. 2013;127(1):358–65. https://doi.org/10.1016/j.biortech.2012.09.070.
31. Javaid A, Ryan T, Berg G, Pan XM, Vispute T, Bhatia SR, Huber GW, Ford DM. Removal of char particles from fast pyrolysis bio-oil by microfiltration. J Membr Sci. 2010;363(1–2):120–7. https://doi.org/10.1016/j.memsci.2010.07.021.
32. Maduskar S, Maliekkal V, Neurock M, Dauenhauer PJ. On the yield of levoglucosan from cellulose pyrolysis. ACS Sustain Chem Eng. 2018;6(5):7017–25. https://doi.org/10.1021/acssuschemeng.8b00853.

33. Dobele G, Meier D, Faix O, Radtke S, Rossinskaja G, Telysheva G. Volatile products of catalytic flash pyrolysis of celluloses. J Anal Appl Pyrolysis. 2001;58:453–63. https://doi.org/10.1016/S0165-2370(00)00128-5.
34. Zhang J, Choi YS, Yoo CG, Kim TH, Brown RC, Shanks BH. Cellulose-hemicellulose and cellulose-lignin interactions during fast pyrolysis. ACS Sustain Chem Eng. 2015;3:293–301. https://doi.org/10.1021/sc500664.
35. Nallar M, Wong HW. Hydroxyl group stabilization for increased yields of low-molecular-weight products in the copyrolysis of cellulose and thermoplastics. Ind Eng Chem Res. 2019;58:10776–84. https://doi.org/10.1021/acs.iecr.9b01177.
36. Wu YX, Jiang LQ, Lin Y, Qian L, Xu FX, Lang XM, Fan SS, Zhao ZL, Li HB. Novel crude glycerol pretreatment for selective saccharification of sugarcane bagasse via fast pyrolysis. Bioresour Technol. 2019;294:122094. https://doi.org/10.1016/j.biortech.2019.122094.
37. Yang XW, Ma FY, Yu HB, Zhang XY, Chen SL. Effects of biopretreatment of corn stover with white-rot fungus on low-temperature pyrolysis products. Bioresour Technol. 2011;102(3):3498–503. https://doi.org/10.1016/j.biortech.2010.11.021.
38. Liang JJ, Lin YQ, Wu SB, Liu C, Lei M, Zeng C. Enhancing the quality of bio-oil and selectivity of phenols compounds form pyrolysis of anaerobic digested rice straw. Bioresour Technol. 2015;181:220–3. https://doi.org/10.1016/j.biortech.2015.01.056.
39. Yu YQ, Zeng YL, Zuo JE, Ma FY, Yang XW, Zhang XY, Wang YJ. Improving the conversion of biomass in catalytic fast pyrolysis via white-rot fungal pretreatment. Bioresour Technol. 2013;134:198–203. https://doi.org/10.1016/j.biortech.2013.01.167.
40. Wang TP, Ye XN, Yin J, Lu Q, Zheng ZM, Dong CQ. Effects of biopretreatment on pyrolysis behaviors of corn stalk by methanogen. Bioresour Technol. 2014;164:416–9. https://doi.org/10.1016/j.biortech.2014.04.062.
41. Fang Z, Xu C. Near-critical and supercritical water and their applications for biorefineries. Singapore: Springer; 2014.
42. Zhurinsh A, Dobele G, Jurkjane V, Meile K, Volperts A, Plavniece A. Impact of hot water pretreatment temperature on the pyrolysis of birch wood. J Anal Appl Pyrolysis. 2017;124:515–22. https://doi.org/10.1016/j.jaap.2017.01.030.
43. Chandler DS, Resende FLP. Effects of warm water washing on the fast pyrolysis of *Arundo Donax*. Biomass Bioenergy. 2018;113:65–74. https://doi.org/10.1016/j.biombioe.2018.03.008.
44. Wang JQ, Wei Q, Zheng JL, Zhu MQ. Effect of pyrolysis conditions on levoglucosan yield from cotton straw and optimization of levoglucosan extraction from bio-oil. J Anal Appl Pyrolysis. 2016;122:294–303. https://doi.org/10.1016/j.jaap.2016.09.013.
45. Julien S, Chornet E, Overend RP. Influence of acid pretreatment (H_2SO_4, HCl, HNO_3) on reaction selectivity in the vacuum pyrolysis of cellulose. J Anal Appl Pyrolysis. 1993;27(1):25–43. https://doi.org/10.1016/0165-2370(93)80020-Z.
46. David GF, Justo OR, Perez VH, Garcia-Perez M. Thermochemical conversion of sugarcane bagasse by fast pyrolysis: high yield of levoglucosan production. J Anal Appl Pyrolysis. 2018;133:246–53. https://doi.org/10.1016/j.jaap.2018.03.004.
47. Kumagai S, Matsuno R, Grause G, Kameda T, Yoshioka T. Enhancement of bio-oil production via pyrolysis of wood biomass by pretreatment with H_2SO_4. Bioresour Technol. 2015;178:76–82. https://doi.org/10.1016/j.biortech.2014.09.146.
48. Sluiter A, Hames B, Ruiz R, Scarlata C, Sluiter J, Templeton D, Crocker D. Determination of structural carbohydrates and lignin in biomass. Lab Anal Proce. 2008;1617:1–16.
49. Park S, Baker JO, Himmel ME, Parilla PA, Johnson DK. Cellulose crystallinity index: measurement techniques and their impact on interpreting cellulase performance. Biotechnol Biofuels. 2010;3:10. https://doi.org/10.1186/1754-6834-3-10.
50. Kuzhiyil N, Dalluge D, Bai XL, Kim KH, Brown RC. Pyrolytic sugars from cellulosic biomass. ChemSusChem. 2012;5(11):2228–36. https://doi.org/10.1002/cssc.201200341.
51. Patwardhan PR, Satrio JA, Brown RC, Shanks BH. Influence of inorganic salts on the primary pyrolysis products of cellulose. Bioresour Technol. 2010;101(12):4646–55. https://doi.org/10.1016/j.biortech.2010.01.112.

52. Liu Q, Wang SR, Luo ZY, Cen KF. Catalysis mechanism study of potassium salts on cellulose pyrolysis by using TGA-FTIR analysis. J Chem Eng Jpn. 2008;41(12):1133–42. https://doi.org/10.1252/jcej.08we056.
53. Zhu C, Maduskar S, Paulsen AD, Dauenhauer PJ. Alkaline-earth-metal-catalyzed thin-film pyrolysis of cellulose. ChemCatChem. 2016;8(4):818–29. https://doi.org/10.1002/cctc.201501235.
54. Wang K, Zhang J, Shanks BH, Brown RC. The deleterious effect of inorganic salts on hydrocarbon yields from catalytic pyrolysis of lignocellulosic biomass and its mitigation. Appl Energy. 2015;148:115–20. https://doi.org/10.1016/j.apenergy.2015.03.034.
55. Wang SR, Guo XJ, Wang KG, Luo ZY. Influence of the interaction of components on the pyrolysis behavior of biomass. J Anal Appl Pyrolysis. 2011;91(1):183–9. https://doi.org/10.1016/j.jaap.2011.02.006.
56. Wu SL, Sen DK, Hu J, Zhang HY, Xiao R. Cellulose-hemicellulose interactions during fast pyrolysis with different temperatures and mixing methods. Biomass Bioenergy. 2016;95:55–63. https://doi.org/10.1016/j.biombioe.2016.09.015.
57. Liu QA, Zhao ZP, Wang SR, Luo ZY. Interactions of biomass components during pyrolysis: a TG-FTIR study. J Anal Appl Pyrolysis. 2011;90(2):213–8. https://doi.org/10.1016/j.jaap.2010.12.009.
58. Miura M, Kaga H, Yoshida T, Ando K. Microwave pyrolysis of cellulosic materials for the production of anhydrosugars. J Wood Sci. 2001;47(6):502–6. https://doi.org/10.1007/BF00767906.
59. Mukarakate C, Mittal A, Ciesielski PN, Budhi S, Thompson L, Iisa K, Nimlos MR, Donohole BS. Influence of crystal allomorph and crystallinity on the products and behavior of cellulose during fast pyrolysis. ACS Sustain Chem Eng. 2016;4:4662–74. https://doi.org/10.1021/acssuschemeng.6b00812.
60. Wang ZH, McDonald AG, Westerhof RGM, Kersten SRA, Cuba-Torres CM, Ha S, Pecha B, Garcia-Perez M. Effect of cellulose crystallinity on the formation of liquid intermediate and on product distribution during pyrolysis. J Anal Appl Pyrolysis. 2013;100:56–66. https://doi.org/10.1016/j.jaap.2012.11.017.
61. Hosoya T, Sakaki S. Levoglucosan formation from crystalline cellulose: importance of a hydrogen bonding network in the reaction. ChemSusChem. 2013;6(12):2356–68. https://doi.org/10.1002/cssc.201300338.
62. Lin Y, Tian YL, Xia YQ, Fang SW, Liao YF, Yu ZS, Ma XQ. General distributed activation energy model (G-DAEM) on co-pyrolysis kinetics of bagasse and sewage sludge. Bioresour Technol. 2019;273:545–55. https://doi.org/10.1016/j.biortech.2018.11.051.
63. Le Brech Y, Ghislain T, Leclerc S, Bouroukba M, Delmotte L, Brosse N, Snape C, Chaimbault P, Dufour A. Effect of potassium on the mechanisms of biomass pyrolysis studied using complementary analytical techniques. ChemSusChem. 2016;9(8):863–72. https://doi.org/10.1002/cssc.201501560.
64. Zhang LQ, Li SS, Ding HZ, Zhu XF. Two-step pyrolysis of corncob for value-added chemicals and high-quality bio-oil: effects of alkali and alkaline earth metals. Waste Manag. 2019;87:709–18. https://doi.org/10.1016/j.wasman.2019.03.002.
65. Poletto M, Pistor V, Zeni M, Zattera AJ. Crystalline properties and decomposition kinetics of cellulose fibers in wood pulp obtained by two pulping processes. Polym Degrad Stab. 2011;96(4):679–85. https://doi.org/10.1016/j.polymdegradstab.2010.12.007.
66. Patai S, Halper Y. Pyrolytic reaction of carbohydrates.9. Effect of additive of thermal behavior of cellulose samples of different crystallinity. Isr J Chem. 1970;8:655–62. https://doi.org/10.1002/ijch.197000079.
67. Poletto M, Ornaghi HL, Zattera AJ. Native cellulose: structure, characterization and thermal properties. Materials. 2014;7(9):6105–19. https://doi.org/10.3390/ma7096105.
68. Lu Q, Hu B, Zhang ZX, Wu YT, Cui MS, Liu DJ, Dong CQ, Yang YP. Mechanism of cellulose fast pyrolysis: the role of characteristic chain ends and dehydrated units. Combust Flame. 2018;198:267–77. https://doi.org/10.1016/j.combustflame.2018.09.025.

69. Shen D, Gu S. The mechanism for thermal decomposition of cellulose and its main products. Bioresour Technol. 2009;100(24):6496–504. https://doi.org/10.1016/j.biortech.2009.06.095.
70. Ponder GR, Richards GN, Stevenson TT. Influence of linkage position and orientation in pyrolysis of polysaccharides: a study of several glucans. J Anal Appl Pyrolysis. 1992;22 (3):217–29. https://doi.org/10.1016/0165-2370(92)85015-D.
71. Mayes HB, Broadbelt LJ. Unraveling the reactions that unravel cellulose. J Phys Chem A. 2012;116(26):7098–106. https://doi.org/10.1021/jp300405x.
72. Lian JN, McKenna R, Rover MR, Nielsen DR, Wen ZY, Jarboe LR. Production of biorenewable styrene: utilization of biomass-derived sugars and insights into toxicity. J Ind Microbiol Biotechnol. 2016;43(5):595–604. https://doi.org/10.1007/s10295-016-1734-x.

[illegible]

Chapter 11
Production of Phenols by Lignocellulosic Biomass Pyrolysis

Joo-Sik Kim and Ki-Bum Park

Abstract Phenols are a class of aromatic compounds that have at least one benzene ring joined with a hydroxyl group. Phenols are widely used in the manufacture of chemicals, resins, synthetic fibers, pesticides, and dyes and products made from phenols have become an indispensable part of modern life. The simplest form of phenols is phenol, which is industrially prepared on a very large scale from petroleum. Fossil-based chemicals like fossil phenol, however, face depletion in the near future. Hence, the demand for renewable chemicals is increasing, and this will also enhance opportunities of sustainable products. Biomass pyrolysis is an attractive route to produce valuable bio-chemicals. Representative bio-chemicals made via pyrolysis of lignocellulosic biomass are phenols, acetic acid, levoglucosan, and furfural. This chapter deals with the production of renewable phenols, especially phenol, from lignocellulosic biomass pyrolysis. Types of lignocellulosic biomass used in pyrolysis processes, and effect of reaction conditions on the production of phenols are discussed along with applications of phenolic-rich bio-oils.

Keywords Phenolic compounds · Lignocellulosic biomass · Separation · Phenolic resin

11.1 Biomass as Source of Phenols

For the last few decades, concerns about fossil fuel use related to environmental issues, climate change, fluctuation in the price of oil and the depletion of abiotic sources have been greatly discussed. Consequently, the need for alternative sources of fuels and chemicals have encouraged people to explore renewable resources, such as biomass, which have merit of carbon neutrality. Lignocellulosic biomass, plant biomass that is composed of cellulose, hemicellulose, and lignin, is the most

J.-S. Kim (✉) · K.-B. Park
Graduate School of Energy and Environmental System Engineering, University of Seoul, Seoul, Republic of Korea
e-mail: joosik@uos.ac.kr

Z. Fang et al. (eds.), *Production of Biofuels and Chemicals with Pyrolysis*, Biofuels and Biorefineries 10, https://doi.org/10.1007/978-981-15-2732-6_11

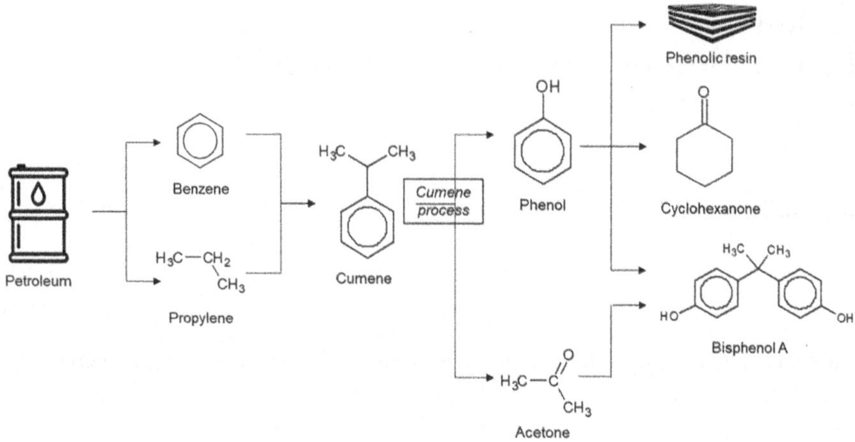

Fig. 11.1 Production and application chain of phenol

abundantly available biomass. However, the complex and rigid structure of lignocellulosic biomass makes it challenging to use as a resource for value-added renewable fuels and chemicals. A representative method for lignocellulosic biomass use since early times has been combustion, in which the principal product is heat. In contrast to combustion, pyrolysis of lignocellulosic biomass produces mainly a liquid product called bio-oil, which can be used as a fuel or a chemical resource. In the early phase of biomass pyrolysis, most research focused on the utilization of bio-oil as fuel in engines or turbines. However, bio-oils produced via pyrolysis cannot be directly adopted in existing energy infrastructures due to their low energy density resulting from their high oxygen content, and their corrosiveness due to acid content. Hence, new attempts have been explored to produce transportation fuels by upgrading bio-oil, with methods such as hydrodeoxygenation or hydrothermal cracking, but those methods have questionable economic viability. As an alternative way to produce fuels, many studies in biomass pyrolysis are aiming to produce biomass-derived valuable chemicals like phenols, acetic acid, and furfural. Of these, phenols (or phenolic compounds), which are a class of chemical compounds consisting of a hydroxyl group bonded directly to an aromatic hydrocarbon group and derived from lignin decomposition, are one of the main compounds in bio-oils. The simplest chemical form of phenols is phenol and it is one of the most important intermediates in the petrochemical industry with applications in production of chemicals like bisphenol-A, phenolic resins, cyclohexanone, and aniline including chemicals in automotive, construction, electronics, paint, and adhesive industries [1]. Figure 11.1 presents production and application examples for phenol.

More than 95% of phenol today is produced by the cumene process, which is an industrial process for synthesizing phenol and acetone from benzene and propylene, both of which originate from fossil fuel [2]. Currently, many researchers are trying to produce biomass-derived phenols, known as renewable phenols, to replace fossil phenols used in synthesizing chemicals and phenolic resins. However, there are still some hurdles to overcome. For example, in the case of the recovery of phenols from bio-oil, low content of target phenols in bio-oil, and the high cost of the multi-step separation are the main issues [1, 3]. In the synthesis of phenolic resins, methoxy substituted phenols in bio-oil have low reactivity in synthesis [4]. Considering that phenols production from biomass through pyrolysis is a promising method, it is worthwhile to discuss the current state of research on biomass pyrolysis that aims to produce phenols, and issues related to renewable phenols. For this purpose, this chapter covers three areas of interest: (1) methods of production for phenolic-rich bio-oils from lignocellulosic biomass using pyrolysis, (2) separation of phenols from bio-oils, and (3) application of phenolic-rich bio-oils.

11.1.1 Biomass Feedstock Selection

Lignocellulosic biomass is the most abundant non-edible terrestrial biomass in the world, and it includes woody and agricultural biomass [5]. As can be expected from its name, lignocellulosic biomass is is mainly composed of holocellulose (cellulose plus hemicellulose) and lignin, with a small amount of ash and extractives. The fraction of each building block is varies for each type of biomass, and that is why biomass selection is important. It is apparent that product distribution in biomass pyrolysis is mostly determined when the feedstock is chosen. Lignocellulosic biomass can be classified into hardwood, softwood, and grasses (herbaceous and agricultural). Hardwood and softwood biomass can be distinguished by simple chemical analyses [6]. Hardwood has a complex cell structure and a characteristic type of cell (vessel element), while softwood has a simpler cell structure and does not have a vessel element [7]. Hardwood comes from flowering plants (angiosperm). Examples of hardwoods are maple, birch, oak, and willow. Softwood comes from seed-producing plants (gymnosperm). Softwood includes pine, spruce, fir, and cypress. Grass biomass is a vascular plant that does not have a woody stem. Agricultural residues, such as rice straw, corn cob, corn stover, miscanthus, and switchgrass are representative grass biomass. Each lignocellulosic biomass group has a content range of cellulose, hemicellulose, and lignin. Hardwood contains 43–58 wt% cellulose, 22–30 wt% hemicellulose, and 14–28 wt% lignin. Softwood contains 33–46 wt% cellulose, 21–25 wt% hemicellulose, and 26–34 wt% lignin. Generally, softwood contains more lignin than hardwood. Grass biomass contains less holocellulose and lignin than woody biomass and in contrast, grass biomass contains more extractives and ash (almost 10–20 wt%) than woody biomass (0–10 wt%) [8]. Some examples of chemical compositions of lignocellulosic biomass are presented in Table 11.1. The values in Table 11.1 have been recalculated by the authors on a dry basis, and the sum of constituents is 100%.

Each constituent of lignocellulosic biomass produces pyrolysis products. Figure 11.2 shows the main pyrolysis products of cellulose, hemicellulose, and lignin.

When cellulose is pyrolyzed, levoglucosan (LG), hydroxymethylfurfural, and furfural are the main products. The main pyrolysis products of hemicellulose are

Table 11.1 Chemical compositions of various lignocelluosic biomass [9–16]

	Cellulose	Hemicellulose	Lignin	Extractive	Ash
Hardwood	*43–58*	*22–30*	*14–28*	*0–5*	*0–2*
Oak	49.9	33.3	14.2	2.2	0.4
Beech	52.1	29.8	16.8	0.9	0.3
Willow	47.8	25.2	18.9	5.1	3.0
Aspen	57.7	22.2	18.2	0.6	1.3
Larch	44.6	24.8	28.5	1.5	0.5
Japanese beech	43.9	28.4	24.0	3.0	0.6
Poplar	53.3	26.8	19.3	0.0	0.5
Walnut	51.9	23.5	24.6	0.0	0.0
Maple	50.4	25.8	23.3	0.0	0.4
Birch	47.0	25.9	22.1	4.7	0.3
Birch bark	15.8	10.5	44.3	27.6	1.8
Softwood	*33–46*	*21–25*	*26–34*	*0–4*	*0–4*
Spruce	46.3	25.5	26.5	0.9	0.9
Gingko	33.5	23.7	30.2	3.5	9.2
Japanese cedar	38.7	23.2	33.8	4.1	0.3
Fir	45.4	21.2	30.3	2.6	0.5
Pine	45.1	24.8	28.4	1.0	0.6
Pine bark	23.5	9.8	32.7	30.0	3.9
Grassy and agricultural residue	*30–48*	*15–32*	*10–31*	*0–18*	*1–15*
Reed	47.3	29.9	10.9	3.2	8.7
Bamboo	42.9	32.7	17.1	5.7	1.6
Miscanthus	53.5	29.1	13.8	1.5	2.1
Banana	69.2	1.1	24.1	0.0	5.7
Sugarcane bagasse	56.3	16.9	23.9	0.0	2.8
Coir pith	26.8	15.5	31.7	14.0	12.0
Saw dust	31.3	21.2	28.7	7.7	11.0
Corn cob	39.3	24.4	17.9	7.9	10.5
Wheat straw	34.9	22.5	21.4	11.9	9.4
Rice straw	32.7	23.4	11.4	17.1	15.4
Switchgrass	32.4	25.5	18.4	17.7	6.1
Hazelnut shell	24.1	24.9	39.6	9.9	1.5
Almond shell	37.9	31.6	27.0	2.3	1.1

LG, furfural, and acetic acid. Phenols (guaiacol, catechol, cresol, and phenol) are the products of lignin pyrolysis [17]. To obtain phenolic-rich oil, hence, it is preferable to select lignin-rich lignocellulosic biomass as a feedstock. As shown in Table 11.1, softwood, barks (birch and pine), and shells (hazelnut and almond) have high lignin content.

The next step after choosing a lignin-rich biomass for production of specific phenols is to determine the characteristics of the lignin structure. Lignin has three basic units: p-hydroxyphenyl (H unit), guaiacyl (G unit), and syringyl (S unit)

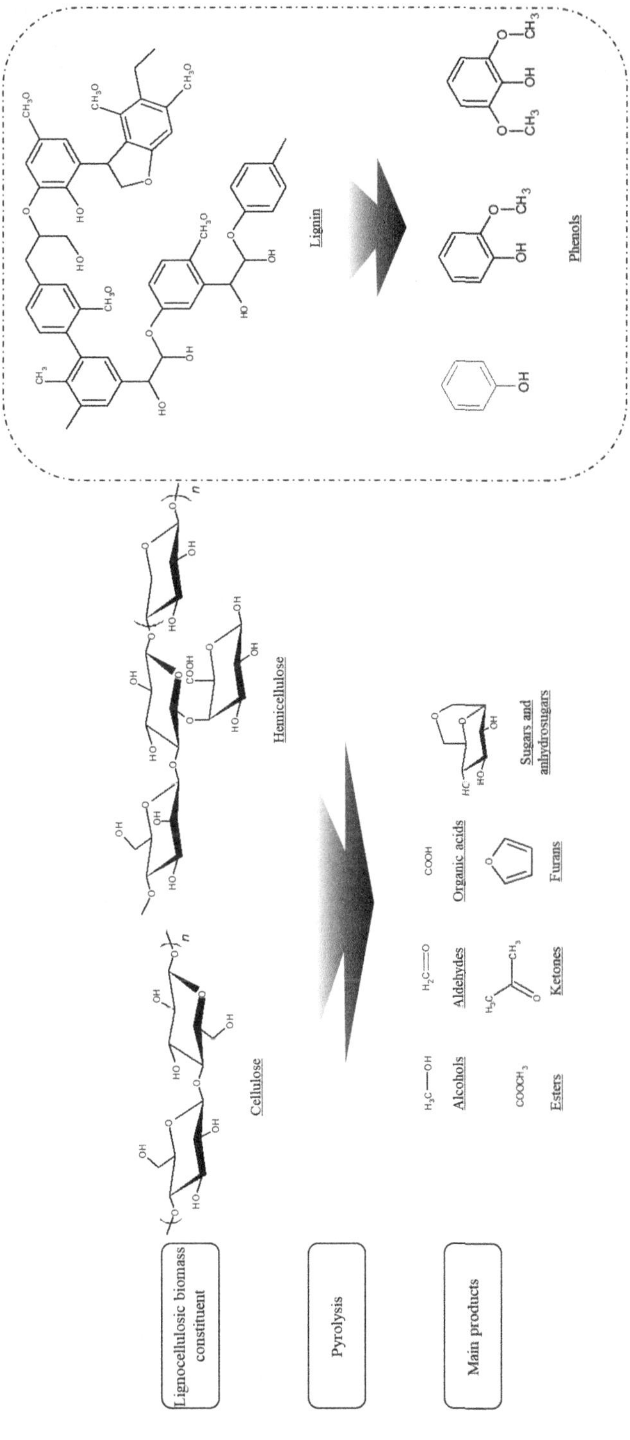

Fig. 11.2 Main pyrolysis products of each biomass constituent

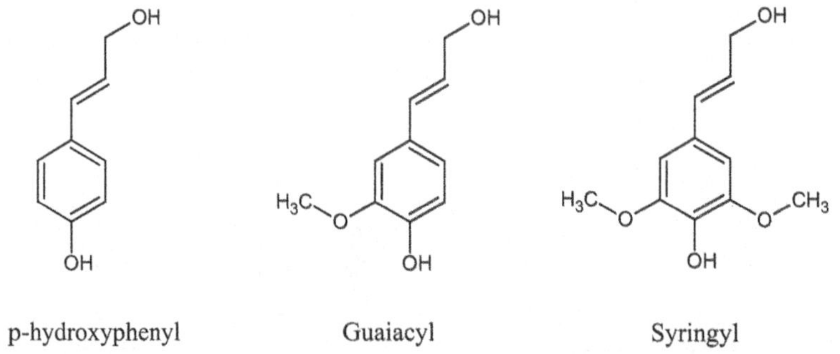

Fig. 11.3 Basic units of lignin

(Fig. 11.3). The three units can be classified by the number of methoxy groups attached to the aromatic ring. When no methoxy group is attached to the aromatic ring, it is denoted as an H unit. Where one or two methoxy groups are attached to the aromatic ring, it is defined as a G or S unit. The ratio of basic units is important in producing a bio-oil with specific phenols, because a biomass having a high content of a specific basic unit would produce a bio-oil rich in that unit. Bio-oils having high content of G and S units have a low reactivity to aldehydes in the synthesis of phenolic resins, due to the occupied methoxy group(s) at the ortho position to the OH group of G and S units to which the aldehydes would be attached [18].

11.2 Production of Phenolic-Rich Bio-oil

Lignin is the most abundant aromatic source in nature [19]. Although many researchers are fascinated by the aromatic orientation of the lignin structure, producing aromatic chemicals, such as BTX (benzene, toluene, and xylene) or phenolics, remains a challenging task [20]. Pyrolysis is a notable thermochemical method that can depolymerize lignocellulosic biomass while preserving features of its aromatic structure. Methods to produce phenolic compounds from lignocellulosic biomass by pyrolysis can be divided into thermal (non-catalytic) and catalytic pyrolysis.

11.2.1 Non-catalytic Pyrolysis

In non-catalytic pyrolysis, decomposition of a material is through heat alone. Decomposed biomass turns into vapor, and a portion of the hot vapor condenses into liquid as it cools, commonly called bio-oil. Uncondensed gases can be used as a heating source for the reaction system. Some solid product remains in the reactor as bio-char. As above, the pyrolysis product streams are bio-char, bio-oil, and uncondensed gas. The product distributions from pyrolysis are highly dependent on the characteristics of the feed material. Reactor design and pyrolysis conditions are important parameters in non-catalytic pyrolysis. Namely, even though the same biomass is chosen, product distribution in non-catalytic pyrolysis is multifarious when different types of reactors and pyrolysis conditions are applied. Hence, to obtain a high phenols yield, an optimal combination of biomass feedstock, reactor design, and pyrolysis conditions must be determined.

In non-catalytic biomass pyrolysis, phenolics yield is generally not high because hundreds of compounds are present in bio-oils [21]. Moreover, monophenols (monomeric phenolic compounds or simple phenols), such as cresol, vanillin, guaiacol, syringol, and phenol, are contained in bio-oil in only small amounts. Oligomeric polyphenols, which contain a varying number of functional groups, are the major phenols in bio-oil. Such oligomeric phenols are in a black, sticky, water-insoluble phase, and they are called pyrolytic lignin (PL). The molecular weight of PL generally ranges from several hundreds to 5000 Da, depending on the type of feedstock and the pyrolysis conditions [22].

To increase phenols yield, especially monophenols, many researchers have tried to determine the degradation mechanism of lignin. Schlosberg et al. [23] hypothesized possible pathways for phenol generation based on a lignin model compound (anisole). They reported that phenol was produced by CH_3-O bond scission, followed by H-abstraction. However, studies on the pyrolysis mechanism of lignin are insufficient, because cellulose, hemicellulose, and lignin are strongly intermeshed and chemically bonded, such bonding networks may affect product distribution. Couhert et al. [24] reported that the pyrolysis product distribution could shift due to interactions between real biomass components that are different from physically mixed components. Hilbers et al. [25] found that phenolics production was enhanced in the presence of cellulose at a low pyrolysis temperature (350 °C). Zhao et al. [26] confirmed those results with TGA and Py-GC/MS. In contrast, Chen et al. [27] reported that hemicellulosic volatiles promoted breaking of the branch chains of lignin, and resulted in further cracking of phenols, which decreased the phenols yield.

11.2.1.1 Effect of Non-catalytic Reaction Conditions

The content of chemicals in bio-oil, especially phenols, vary according to operational conditions, such as pyrolysis temperature, heating rate, and reaction atmosphere. In this section, the effect of each parameter will be discussed.

1. Temperature
   Temperature is one of the most important operational parameters in pyrolysis, and one that significantly influences product distribution and production of special chemicals. Generally, when the temperature increases, the gas yield increases, and char yield decreases, whereas the oil yield increases up to a certain temperature, and then decreases beyond that temperature. For this reason, finding the

optimal temperature for high production of phenolics is critical in non-catalytic pyrolysis. Horne et al. [28] pyrolyzed a mixture of waste wood shavings in a fluidized bed reactor, and reported that the maximum phenolics yield was obtained at 500–550 °C. Hekstra et al. [29] conducted pyrolysis of pinewood with a low mineral content in a fluidized bed reactor and found that the phenolics yield increased with increasing temperature between (400 and 550) °C. Choi et al. [18] conducted pyrolysis experiments of palm kernel shells (PKS) using a fluidized bed reactor, and obtained the maximum content of phenols in bio-oil (24.8 wt%-GC detectable) at 500 °C. Moreover, they reported that the phenol content in bio-oil, which was calculated by the GC external standard method, was also very high (8.1 wt%). Bai et al. [30] determined that during the pyrolysis of lignin, re-oligomerization was stronger than thermal cracking up to 500 °C. Liaw et al. [31] conducted the pyrolysis of woody biomasses using Py-GC/MS, and reported that the phenols yield was maximized at (450–500) °C. Liu et al. [32] found that the selectivity for phenols within each lignin unit in pyrolysis showed a different trend with an increase in temperature. Namely, formation of phenols from the p-hydroxyphenyl unit increased with increasing pyrolysis temperature at (400–800) °C, while phenols from the syringyl unit decreased. Selectivity for phenols from the guaiacyl unit was maximized at 600 °C. In conclusion, the optimal temperature for a high yield of phenols varies with the reactor configuration and other variables, but generally it appears to fall within a range from 400 °C to 550 °C from the viewpoint of phenols yield and energy efficiency.

2. Heating rate

   Heating rate is usually used as a guide for classifying pyrolysis, i.e., flash (>1000 °C/s), fast, and slow pyrolysis (<10 °C/min). Heating rate is a characteristic coupled with reactor configuration. For example, a batch reactor is suitable for slow pyrolysis, which heats the feed material slowly with a long vapor residence time, and a fluidized bed reactor is suitable for fast pyrolysis, which has a short vapor residence time. Safdari et al. [33] conducted the pyrolysis of biomass and found that fast pyrolysis was more favorable for phenol production than slow pyrolysis. The authors reported that the hydroxyl and methyl attachments to the aromatic ring of phenols were removed during fast pyrolysis experiments. Greenhalf et al. [34] conducted pyrolysis of willow using a Py-GC/MS, and they too obtained a higher phenolics yield with fast pyrolysis than with slow pyrolysis. Wang et al. [35] performed the pyrolysis of Jerusalem artichoke stalk with a high heating rate range (100–10,000) °C/s with a Py-GC/MS system and reported that too high a heating rate was unfavorable for the production of phenols. They found that an increase of heating rate improved acids production while reduced phenols production. It is widely accepted that a high heating rate up to a certain level can be applied in phenol production. Further, particle size, which has an influence on the heat transfer rate (or heating rate), is an important influence in phenols production. If a particle is large, a longer time is needed for heat to penetrate (internal heat transfer) to the core of the biomass. It has been reported that when the particle size increased from (0.3 to 3) mm, the PL yield was reduced to half [36].

Table 11.2 Pyrolysis experiments with various reaction atmospheres

Gas or reagent	Reactor configuration	*T* (°C)	Feedstock	Phenols content in bio-oil (wt%)	Phenol content in bio-oil (wt%)	Ref.
Reductive						
N_2	Fluidized bed	400	Corncob	33	3.0	Zhang et al. [37]
CO				28	3.8	
H_2				35	2.5	
No reagent	Fixed bed	700	Syringol		0.3	Zhou et al. [40]
Methanol					1.9	
N_2	Microwave	550	Medicinal herb residue	20.5	1.2	Zhang and Zhang [41]
CO				22.5	3.6	
H_2				19.7	1.5	
Oxidative						
N_2	Fluidized bed	400	Corncob	33	3.0	Zhang et al. [37]
CO_2				27	4.1	
N_2	Microwave	550	Medicinal herb residue	20.5	1.2	Zhang and Zhang [41]
CO_2				16.5	1.8	
N_2	Fluidized bed	290	Radiata pine	1.6	0.03	Butt [42]
$N_2 + O_2$ (20 vol. %)				2.2	0.00	

3. Reaction atmosphere
   Pyrolysis is usually conducted under a nitrogen atmosphere to achieve inert conditions. Sometimes, other gases or reagents are introduced to achieve reductive (CO and hydrogen-donor reagents) or mild oxidative atmospheres (CO_2 and steam) [37–39]. Some experimental data with different reaction atmospheres are presented in Table 11.2.

A reductive atmosphere reinforces the cracking of lignin to enhance phenolics production [40]. Zhang et al. [37] conducted the pyrolysis of corncob in various atmospheres using a fluidized bed reactor. It was reported that phenols content was decreased (33–28) wt% when N_2 atmosphere was substituted by CO, but phenol content was increased (3.0–3.8) wt%. Britt et al. [39] reported that the use of a hydrogen donor augmented the initiation of a free-radical process. Zhou et al. [40] conducted pyrolysis of lignin-model compounds (syringol, guaiacol, and phenol) with hydrogen donors (methanol, acetone, and diethyl ether). Among the hydrogen donors, methanol was found to be the most effective agent for phenols production, accelerating phenols production by promoting reactions with aromatic phenoxy radicals from the O-CH_3 bond homolysis of syringol and inhibiting the dehydroxylation of primary phenols. Further, in oxidative pyrolysis, CO_2 or molecular oxygen proved to promote the depolymerization of lignin by enhancing free radical formation [42]. As shown in Table 11.2, when CO_2 was applied as carrier gas, the phenol content clearly increased, but the total phenols content decreased [37, 41]. When a small amount of oxygen was added to N_2, a small increase (~0.6 wt %) in phenols could be observed.

11.2.1.2 Reactor Design

Along with pyrolysis conditions, reactor design greatly affects the yield of phenols. It can be seen that product distribution is dissimilar for different reactor configurations. Different reactor configurations will allow similar reaction conditions in pyrolysis, but they result in different ways to produce phenols. Many types of pyrolysis reactors have been developed and modified to obtain high yield of special chemicals. Typical reactors for biomass pyrolysis have been fixed bed, fluidized bed, ablative, rotating cone, and auger reactors [5, 21, 43]. Some of those reactors have applied for the production of phenols. For example, Kim et al. [44] carried out pyrolysis of PKS using a fluidized-bed reactor, and obtained a maximum bio-oil yield of 49 wt%. The maximum content of phenol and phenolic compounds in the bio-oil amounted to 22 area% and 70 area%, respectively. Bertero et al. [45] performed pyrolysis of palm residues in a fixed bed reactor. The yields of bio-oil plus tar ranged from 35 wt% to 44 wt%, and the phenols contents in the bio-oils from the two residues were 23–35 wt% and 67–77 wt%, respectively. Conventional pyrolysis reactors, however, showed limited success in the production of phenols due to problems related to heat transfer and heat loss. The following sections cover some promising and distinctive pyrolysis reactor configurations that have been used for chemical production, especially phenols, such as microwave-assisted pyrolysis and stepwise pyrolysis reactors.

1. Microwave-Assisted Pyrolysis Reactor
   Microwave-assisted pyrolysis was one of the intensively investigated technologies of the 2010s. The main characteristic of this reactor is the heating of the feedstock directly from the inside using microwaves. Figure 11.4 shows a comparison between conventional and microwave-assisted pyrolysis.

   In conventional pyrolysis reactors, heat is introduced to feedstock from the outside, and directed to the center of the biomass via conduction, convection, or radiation [47]. Such heating by conventional pyrolysis turns out to be inefficient because high energy is consumed as heat is transferred into the core of the feedstock [46]. However, microwave-assisted pyrolysis can achieve high heating rates, attain high energy efficiency and minimize undesirable secondary reactions due to lower temperatures in the microwave cavity. The other beneficial point of

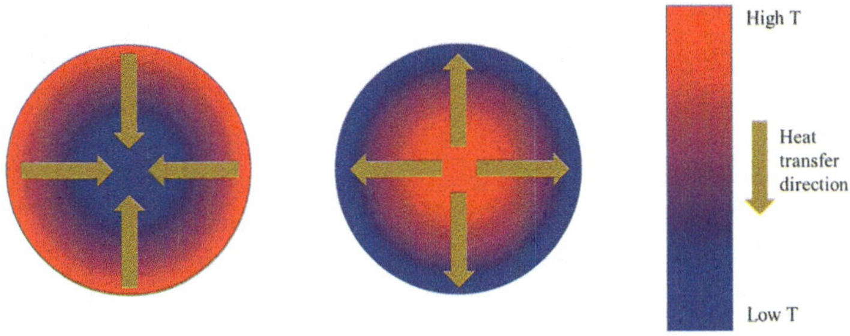

Fig. 11.4 Comparison of heating direction of conventional and microwave-assisted pyrolysis. (Adapted with permission from Morgan et al., Bioresource Technology (2017), 230, 112–121 [46]. Copyright 2017 Elsevier.)

microwave pyrolysis is that it does not require small feedstock sizes, which means that additional energy can be saved in the shredding process. However, the microwave-assisted pyrolysis systems are difficult to scale-up because the penetration limit of microwaves is typically only 1–2 cm [48]. High electricity consumption, high operational cost, and hot spot formation from irregular microwave irradiation are the main obstacles to its commercialization. At any rate, microwave-assisted pyrolysis has proven to be appropriate for phenols production in many works. Aziz et al. [49] conducted pyrolysis of PKS, wood chips, and sago waste (SW) with a microwave-assisted pyrolysis system, and found that the heating rate had a strong influence on the production of phenols. In this work, the bio-oil yield was relatively lower than that from a fluidized bed reactor system, but high phenol contents were obtained with PKS (31 area%) and SW (19 area%). Yerrayya et al. [50] carried out pyrolysis of lignin in a microwave-assisted reactor and investigated the effect of mass ratio of lignin to susceptors and different types of susceptors (activated carbon, charcoal, and graphite) in microwave-assisted pyrolysis. Among the susceptors, activated carbon showed the highest performance in phenol production. The highest content of monophenolic compounds amounted to 64 wt%, and the content of simple phenols (monophenolic compounds without a methoxy group) were up to 40 wt%. Wang et al. [51] conducted pyrolysis of rice straw having a lignin content of 14.5 wt% using a microwave-assisted pyrolysis system and showed that the phenolics and phenol contents were maximized at 40.6 wt% and 8.3 wt%, although the rice had a low lignin content relative to other kinds of biomass.

2. Stepwise Pyrolysis

   Stepwise pyrolysis was developed based on the different thermal degradation behavior of the main biomass components (cellulose, hemicellulose and lignin). Cellulose and hemicellulose are decomposed in the temperature range of (315–400) °C and (220–315) °C, respectively, while lignin is decomposed under a wide temperature range (ambient to 900 °C) [43]. Using the stepwise pyrolysis approach, phenolic-rich bio-oil can be obtained by separating holocellulose-originated products. Murwanashyaka et al. [52] conducted

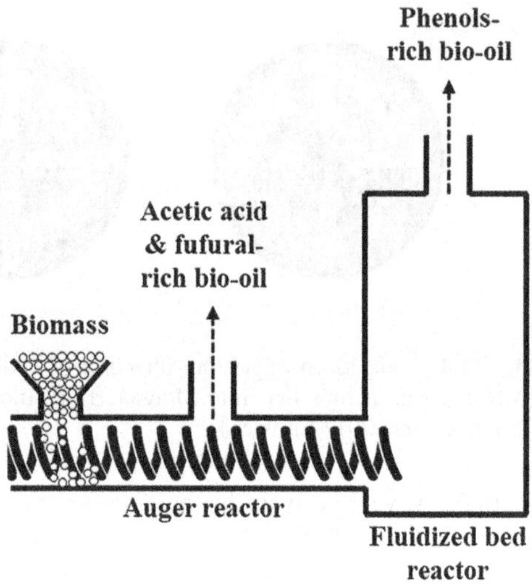

Fig. 11.5 Diagram of a two-stage pyrolysis process

stepwise pyrolysis of birch to study phenols production in certain temperature ranges. The reaction temperature range was divided into five stages: (25–200) °C, (200–275) °C, (275–350) °C, (350–450) °C and (450–550) °C. They found that the active formation of phenols occurred in the temperature range of (275–350) °C, and demethoxylation at (350–450) °C. Zhang et al. [53] carried out the two-step pyrolysis of corncob using a Py-GC/MS, and reported that stepwise pyrolysis enriched valuable chemicals such as phenols in comparison with one-step pyrolysis. Oh et al. [54] conducted pyrolysis of PKS using a continuous two-stage pyrolysis process that employed an auger reactor and a fluidized bed reactor connected in series (Fig. 11.5).

In that work [54], the auger reactor operated in the temperature range of (290–380) °C to separate products derived mainly from hemicellulose and cellulose, and the fluidized bed reactor operated at approximately 520 °C, mainly for lignin degradation. Bio-oil obtained from the auger reactor was abundant in acetic acid and furfural, whereas bio-oil from the fluidized bed reactor contained high levels of acetic acid and phenol. The maximum phenol concentration in the bio-oil was approximately 12 wt%. A further study by the same group reported that a high-temperature auger reactor (typically 300 °C) located in front of the fluidized bed reactor helped to increase the phenol content in the bio-oil obtained in the fluidized bed reactor by activating polymeric molecules inside the auger reactor for the decomposition of the molecules in the fluidized bed reactor.

11.2.2 Catalytic Pyrolysis

Chemical production via non-catalytic pyrolysis typically has low yields of target compounds, because bio-oils derived from lignocellulosic biomass contain several hundreds of chemicals. To enhance the yield and selectivity of specific chemicals, catalysts have been applied to pyrolysis. To obtain the desired products by catalytic pyrolysis, a preliminary understanding of the chemistry of reactions is required. In catalytic pyrolysis, each catalyst induces a featured reaction pathway and specified product(s) in abundance. According to the type of catalyst applied, the distribution, and the kind of active site where the actual interaction occurs between feedstock and catalyst, the product distribution varies. Moreover, it can be shifted even when the same catalyst is used if reaction conditions and reactor configurations are different. For example, Gamliel et al. [55] conducted catalytic fast pyrolysis using a micro-reactor (Py-GC/MS) and a bench-scale reactor (spouted bed) and found that the product distribution was different according to the pyrolysis apparatus used. In the catalytic pyrolysis of lignocellulosic biomass, most research has used batch or fixed bed reactors because they make it easy to control reaction conditions. Two methods of catalytic pyrolysis can be differentiated based on the method of contact between the catalyst and pyrolytic vapors [8]. One is the *in-situ* method, and the other is the *ex-situ* one. In the *in-situ* method, a catalyst is directly introduced to the reactor, and it is then mixed with feedstock during pyrolysis. In the *ex-situ* method, a catalytic reactor is installed after the main pyrolysis reactor to achieve contact between the catalyst and pyrolytic vapor. The *in-situ* method has merit for minimizing secondary reactions (reducing bio-oil yield) between the pyrolytic vapor and catalyst, and for suppressing repolymerization. However, in the *in-situ* catalytic pyrolysis, biomass pyrolysis and the catalytic reaction occur in the same reactor, which makes it difficult for *in-situ* catalytic pyrolysis to occur under optimal reaction conditions. Further, the deactivation of a catalyst by poisoning with biomass ash can occur during *in-situ* catalytic pyrolysis. On the other hand, the *ex-situ* method has merit for controlling each reactor at their respective optimal reaction conditions and minimizing the deactivation of catalysts by char ash. However, the *ex-situ* method is considered less economical than the *in-situ* method. There are many points to consider when using catalysts in the pyrolysis of lignocellulosic biomass, such as catalytic activity and stability, catalyst cost, the reaction mechanism of catalysts, reactor configuration, reaction conditions, and the type of active sites in the catalysts. In this section, the characteristics of catalysts for phenols production is mainly discussed for an overall understanding of phenols production via catalytic pyrolysis.

11.2.2.1 Metallic

Widely adopted metallic catalysts in catalytic pyrolysis are in the form of metal oxides (M_xO_y). Metal oxides can be classified as acidic, base or transition metal oxides. Each group has its own featured characteristics. Acidic metal oxides can promote the dehydration, decarbonylation, and cracking of organic molecules [8]. By such reactions, gas yield increases at the expense of oil yield. Further, more acidic metal oxides, such as alumina, enhance aromatic and polyaromatic production. Base metal oxides promote deoxygenation by ketonization and aldol condensation [56]. Base metal oxides, such as MgO and CaO, can have both acidic and basic roles. Besides metal oxides, other metallic catalysts combined with carbide, or sulfide, phosphide can be used for catalytic pyrolysis. Transition metal oxides show complicated catalytic properties and have versatile reaction mechanisms. Many studies in which transition metal oxides were used indicated that most transition metal oxides, such as Fe_2O_3, NiO, TiO_2, and ZnO, increased gas and char yields, while they decreased bio-oil yield [57]. Of the three types of metallic catalysts, transition metal oxides were found to be appropriate for phenolics production [58]. Nair et al. [59] conducted the catalytic fast pyrolysis of alkali lignin using transition metal oxides (TiO_2, CeO_2, and ZrO_2) and determined that TiO_2 was favorable for phenols production. TiO_2 is thermally excited in the hot atmosphere to generate OH radicals, which induces free radical reactions (demethoxylation and demethylation) to form simple phenols [60]. To enhance catalyst performance, active species can be impregnated onto catalysts. Zhang et al. [61] stated that Fe (III)/CaO showed better performance in transforming heavy phenols into light phenols without a methoxy group than CaO. However, some catalysts could induce further reaction of phenolics to decrease phenolics yield. Cu and Mo based catalysts showed good performance in cleaving the C–O bond to transform phenolics into aromatic hydrocarbons such as benzene, toluene and xylene [62]. Table 11.3 presents characteristics of metallic catalysts that are favorable for formation of phenols.

11.2.2.2 Zeolites

Zeolites are the most extensively applied catalyst in catalytic biomass pyrolysis. Because zeolite has a micro-porous structure, only small molecules can travel into its pores and contact internal acid sites. The acid sites of zeolite depolymerize lignin, and the small pore diameter of zeolite prevents repolymerization and coke formation [71]. When zeolites are used, product distribution is highly affected by its acidity and pore size. For example, high acidity promotes further deoxygenation of phenols to form alkyl aromatics (benzene, toluene, and xylene), and the selectivity of phenol sharply decreases [72]. Pore size is determined by zeolite structure. Ma et al. tested four types of zeolites with similar Si/Al ratio: FAU (~7.4 Å), MOR (~6.7–7 Å), BEA (~6.6 Å), MFI (~5.4–5.6 Å) and determined out that medium pore size zeolite (BEA) was most suitable for phenol production [73]. Research studies have implied that the catalyst to biomass ratio (C/B) and Si/Al ratio of zeolites are important factors in the selectivity of phenols [74–76]. Table 11.4 shows the kinds and characteristics of zeolite catalysts widely used for production of phenols.

Xue et al. [74] reported that bio-oil yield decreased from 50 wt% to 28 wt% when the C/B ratio increased from 0.5 to 2, which was because the addition of more catalyst promoted cracking reactions to transform liquid into gaseous products.

Table 11.3 Examples of metallic catalysts favorable to phenols

Feedstock	Catalyst	Phenol/phenols content in bio-oil (wt%)	Features	Ref.
Poplar wood	Pd-Ce/ TiO_2	~3/37	High cracking activity of oligomeric phenols	Lu et al. [63]
Guaiacol	W_2C/ CNF	46/–	Low ring-hydrogenation activity	Jongerius et al. [64]
Guaiacol	V_2O_5	41/–	Removing an oxygen atom from guaiacol to form water	Filley et al. [65]
Guaiacol	Pd/C	95/–	Lower deoxygenation activity of phenol to benzene	Sun et al. [66]
Guaiacol	Pt-Sn	~70	Deoxygenation of methoxy group	González-Borja and Resasco [67]
Milled wood lignin	Organic Na	30/51	Elimination of alkyl substituents	Jakab et al. [68]
Poplar wood	K_3PO_4/ Fe_3O_4	12/69	Promoting the decomposition of lignin and inhibiting the devolatilization of holocellulose	Zhang et al. [69]
Anisole	Al_2O_3/ MgO	20/–	Formation of phenoxide (PhO^-) at acidic site (Al_2O_3) and hydrogen radical (H^+) at basic site (MgO) to produce phenol	Strassberger et al. [70]

Lazaridis et al. [76] stated that phenols production was enhanced when a high amount of ZSM-5 zeolite having a relatively high Si/Al ratio (approximately 40) and a few strong acid sites were used.

11.2.2.3 Catalyst Deactivation

There is no doubt that catalysts play an important role in biomass pyrolysis. Along with the cost of catalysts, however, catalyst deactivation is a research topic. The deactivation of catalysts is a complicated process mainly derived from the combination of coke deposition, poisoning, and eventual dealumination [5]. Firstly, coking is the most important deactivation mechanism for the catalytic pyrolysis of biomass. Coking generally occurs through the repolymerization and polycondensation of pyrolytic vapor in a reactor. In the case of porous catalysts like zeolites, most coking takes place at external acid sites. In phenols production via catalytic pyrolysis, the coke problem is far more severe because phenols are representative coke intermediates [82]. Secondly, poisoning is generally caused by heteroatoms in the biomass, such as nitrogen, chlorine, and oxygen. Chemicals that contain heteroatoms can cover acid sites or compete with desirable products, resulting in deteriorated catalyst activity. Lastly, dealumination is the loss of active sites in zeolite caused by high temperature and water content [83]. Dealumination is serious because it is an

Table 11.4 Examples of zeolite catalysts favorable to the production of phenols

Feedstock	Catalyst	Phenol/ Phenols content in bio-oil (wt%)	Features	Ref.
Maize straw	ZSM-5@SBA-15	–/54	Primarily cracked in the mesoporous shell and later undergo a series of deoxygenation reaction	Xue et al. [74]
Alkaline lignin	Spent FAU	–/90	Loss of acid site of zeolite inhibit transformation of phenols to BTX	Ma et al. [75]
Kraft lignin	Various pore size ZSM-5	4/30	Promote dealkoxylation of large phenolic compounds and inhibit further dehydration of phenols	Lazaridis et al. [76]
Alkaline lignin	HZSM-5	–/~35	Phenol is tightly bound to the acidic active site without further degradation	Ma et al. [77]
Alkaline lignin	Na/ ZSM-5	–/~32	Porous without acidic site structure stabilize intermediates (such as phenol)	Ma et al. [77]
Beechwood	Co/ ZSM-5	–/39	Decrease of Brønsted sites which are masked by metal ion	Iliopoulou et al. [78]
Beechwood	Co/ ZSM-5	–/41	Decrease of Brønsted sites which are masked by metal ion	Iliopoulou et al. [78]
Rice husk	Pt-Meso-MFI	3/20	Outstanding cracking ability of heavy phenols to light phenols	Jeon et al. [79]
Alkylphenols	ZSM-5 and USY	95/–	Alkylphenols are converted mainly into phenol by efficient dealkylation	Verboekend et al. [80]
Ethylphenols	ZSM-5, beta, etc.	96/–	Brønsted and Lewis acid sites catalyse ethylphenol dealkylation to phenol	Liao et al. [81]

irreversible reaction, which means that recovery by regeneration of zeolite is not possible.

Many researchers have investigated methods to prevent the deactivation of catalysts. Neumann et al. [84] incorporated cerium into a hierarchical HZSM-5 catalyst, and reduced coke formation by reducing the acidity of the external acid sites. Wang et al. [85] conducted a pre-coked treatment of strong external acid sites of HZSM-5, while the internal acid site was preserved, and determined that the stability and activity greatly improved. Ma et al. [86] conducted regeneration of H-USY zeolite by calcination in air at high temperatures and claimed that porosity and acidity were well restored.

11.2.3 Characterization of Bio-oil Components

Bio-oil derived from pyrolysis of lignocellulosic biomass is a complex mixture of chemicals, including water, acids and polar organic compounds. Bio-oils are generally composed of more than 400 different chemical compounds. Organic chemicals in bio-oil can be classified as sugar-based components, alcohols, acids, ketones, ethers, aliphatic, alkyl aromatics, phenols, and furans. In addition to the chemical categories in bio-oil, compounds having a vast range of polarities and molecular weights (up to thousands of Daltons) are present. According to the polarity and molecular weight range, different analytical techniques must be considered for analysis of bio-oil. For low-molecular-weight and weakly polar (or nonpolar) chemicals, gas chromatography (GC) can be adopted. For polar chemicals, use of high-performance liquid chromatography (HPLC) is appropriate. For high molecular weight chemicals, gel permeation chromatography and high-resolution mass chromatography are good options. To determine water content, Karl Fischer titration is widely used. Along with these analytical methods, nuclear magnetic resonance (NMR) aids in unraveling chemical structures and functional groups, while calorimetry helps to estimate the heating value of a bio-oil, and elemental analysis provides molar ratios of elements in bio-oil. For fuel applications of bio-oil, physical properties are measured by several analyzers, such as a viscometer, densitometer, flashpoint tester, and pour point analyzer. As such, a combination of analytical methods are needed to determine bio-oil characteristics, but it can be a time-consuming and costly task. Hyphenated chromatographic and spectrometric technologies, including GC × GC, LC × LC, and GC mass spectroscopy (GC/MS), are being used for efficient analysis of bio-oil [5]. Hyphenated chromatographic systems are useful for bio-oil characterization when peaks of different chemicals overlap. Using columns with dissimilar polarities can separate peaks two-dimensionally [87].

One of the most reliable options for phenols qualification is GC/MS, and for quantification, the GC-flame ionization detector (GC-FID). While GC/MS is usually used as a semi-quantitative method, GC-FID is adopted for accurate quantification. In the quantitative step, a calibration method is essential due to imperfection of the detectors. The internal standard method is appropriate for calibration of overall products. For quantitative analysis of target chemicals, for example phenol, the external standard method is usually applied. A detailed methodology for GC is well presented by Bicchi et al. [88]. The limitations of the GC method are that it is only applicable to low molecular weight compounds. Higher molecular weight compounds, such as PL, are difficult to identify with GC. ^{31}P-NMR is one possible option for estimating the relative amount of phenols in bio-oil since determination by chemical shift can distinguish each functional group [89].

11.3 Separation of Phenols from Bio-oil

Low content of phenols, which is a result of the complexity of bio-oil, limits the direct application of phenols derived from the pyrolysis of lignocellulosic biomass. For effective application of phenols, phenols in bio-oil must be enriched with separation techniques such as solvent extraction, column chromatography, and distillation [48]. In this section, representative separation techniques for phenols will be discussed.

11.3.1 Solvent Extraction

Solvent extraction is a separation technique that isolates target chemicals using selective solvents, leaving other chemicals in other phases. It is the most commonly used separation method for chemicals from bio-oil, and many solvents have been examined by researchers [90]. Some commonly used solvents for phenol extraction are presented in Table 11.5.

In solvent extraction, an understanding of the characteristics of the solvents being used is very important. By using a proper solvent, high recovery of the desired chemicals can be obtained. Wei et al. conducted a liquid-liquid extraction of biomass pyrolysis oil using several solvents (e.g., hexane, petroleum ether, and chloroform) and obtained extracted oil with a high concentration of phenols (85 wt%) [94]. Most works using liquid-liquid extraction have followed a similar procedure. Firstly, bio-oil is dissolved in organic solvent, and phenols are extracted by aqueous alkaline solution. After that, procedures for the recovery of phenols using solvents proceed [1, 91–93]. Mantilla et al. [95] compared extraction methodologies by testing toluene, methyl isobutyl ketone (MIBK), and ethyl acetate as solvents, and concluded that ethyl acetate was the most appropriate solvent for phenols extraction. Other research, however, has reported controversial results. Žilnik and Jazbinše [96]

Table 11.5 Examples of solvents for phenols extraction

Feedstock	Solvent	Procedure	Ref.
Eucalyptus wood	Ethyl acetate	Dissolve bio-oil in ethyl acetate and extract phenols by aqueous alkaline solution in phenolate ion and regenerate phenol with sulfuric acid	Amen-Chen et al. [1]
Alcell lignin	Dichloromethane	Dissolve bio-oil in diethyl ether and extract phenols by NaOH solution and extract phenol with dichloromethane	Thring et al. [91]
Organosolv lignin	Ionic liquid	Dissolve bio-oil in ethyl acetate and extract phenols by NaOH solution and extract phenols with ionic liquid ([chloine][NTf_2])	Cesari et al. [92]
Model bio-oil	Acetic ether	Precipitate bio-oil bi $Ca(OH)_2$ with ammonia solution and filter residue and extract phenols with acetic ether	Wang et al. [93]

conducted solvent extraction with various solvents and aqueous solutions. They concluded that MIBK in combination with NaOH solution was the most appropriate solvent for the extraction of phenols from bio-oil. The difference in the two results seems to arise from different process designs and extraction procedures. Along with water-soluble light-molecular-weight phenols, water-insoluble PL is a valuable product in bio-oil. PL can be applied to the synthesis of phenol novolac resins. Zhang et al. [97] carried out research on the recovery of PL from bio-oil via cold-water precipitation under several mixing methods (magnetic stirring, high-speed homogenization, and ultra-sonication) and determined that the PL samples obtained were similar for every mixing condition. The authors reported that an important factor for the recovery of PL was the water content in bio-oil. When the water content was high, separation of PL was insufficient, and the utilization of both water-soluble fraction and PL was inhibited. Although solvent extraction is an attractive method for the recovery of phenols, it has drawbacks to overcome. For example, a large volume of organic solvent is usually needed in solvent extraction, and hence, a large amount of discharged wastewater containing pollutants is a challenging problem for industrial application [93]. Recently, supercritical fluid extraction (SFE) was developed as an environmentally friendly extraction method. SFE has several merits, such as [48]:

1. Operates at low temperatures, preventing undesirable reactions during extraction;
2. Flexible in modulating solvent power and selectivity;
3. Does not require polluting organic solvents;
4. Gains economic feasibility by not using expensive organic solvents.

The most commonly used supercritical fluid solvent for extraction is carbon dioxide (CO_2), which is relatively inert (reactive with amines, NaOH), inexpensive, readily available, odorless, and safe. Patel et al. [98] conducted the SFE of bio-oil from the pyrolysis of sugarcane bagasse with CO_2 and reported that the phenol concentration in the extract was maximized at 72 wt%, and the total yield of the extracted oil amounted to 15 wt%. Chan et al. [99] conducted the SFE of bio-oil derived from PKS, also with CO_2, reporting that the yield of extract increased with increasing extraction temperature and pressure. In their work, the phenol content in extracts ranged from 7.3 to 8.2 wt%.

11.3.2 Column Chromatography

Column chromatography is a separation method that uses the "like dissolves like" concept. When bio-oil is fed to a column packed with stationary phase, bio-oil components are held on the stationary phase. Then, with the use of an appropriate mobile phase, phenols can be separated from other chemicals. Cao et al. [100] separated phenols using column chromatography with silica as a stationary phase and two mobile phases that included n-hexane for hydrocarbons and ethyl acetate for phenols. The GC $\times$ GC-FID results confirmed that the successful separation of phenols and hydrocarbon fractions was achieved by column chromatography. Hao et al. [101] conducted separation of high-purity syringol and acetosyringone from rice straw-derived bio-oil by column chromatography and obtained a high-purity syringol (91.4%) and acetosyringone (96.2%) with recovery ratios of 73% and 39%, respectively. Further, Zheng et al. [102] carried out isolation of hazardous phthalate esters from bio-oil. Although column chromatography is a generally inexpensive and robust separation method, it has a critical drawback. Namely, it has low throughput. For this reason, column chromatography is usually considered suitable mainly for highly valuable chemicals or for removing hazardous materials.

11.3.3 Distillation

Distillation is one of the oldest separation techniques that is applicable to bio-oil components based on the difference in component boiling points and mean free paths. According to the operating pressure and medium, distillation can be classified as atmospheric, vacuum, steam, or molecular distillation. Atmospheric and vacuum distillations, which separate chemicals at atmospheric and vacuum conditions, are relatively inexpensive, but coke formation at high temperatures, long operational times, and low efficiency are hurdles for these techniques [103]. Elkasabi et al. [104] conducted atmospheric distillation of bio-oil derived from the pyrolysis of eucalyptus and enriched the phenols from 5.3 wt% to 40–45 wt%. In steam distillation, steam is introduced to the distillation apparatus and decreases the boiling points of bio-oil components and the viscosity of bio-oil. Due to low distillation temperatures, secondary thermal reactions (polymerization, degradation, and oxygenation) of thermally sensitive compounds can be suppressed. Murwanashyaka et al. [105] carried out the steam distillation of bio-oil and recovered 88.2% of phenols at a steam-to-oil ratio of 27. Different from other distillation techniques, molecular distillation exploits the difference between the mean free paths of the compounds. In high vacuum conditions, the distance between the evaporation and condensation surfaces is less than or equal to the mean free path. Because of this, molecules can easily evaporate and condense without hindrance [103]. The merits of molecular distillation are its low operating temperatures, short distillation times, high separation efficiency, and viability for thermally sensitive compounds. Wang et al. [106] separated PL into low-molecular-weight PL (LPLH) and high-molecular-weight PL (HPLH) with a series of molecular distillation and methanol-water extraction from sawdust derived bio-oil, and determined that LPLH was suitable for the production of phenolic resins and adhesives because it had abundant active sites for reactions and a low degree of polymerization. Guo et al. [107] conducted a two-step molecular distillation of a bio-oil. The two-step distillation resulted in two distilled fractions (DF-1 and DF-2), and two residual fractions (RF-1 and RF-2). Phenols were enriched in RF-1 and RF-2, the contents of which were greater than 50 wt% and 40 wt%, respectively.

11.3.4 Combination of Separation Techniques

Sometimes, to obtain a high purity of phenols, several staged or stepwise separation processes are needed. Yang et al. [90] conducted separation of phenols using a combination of extraction and column chromatography techniques, and effectively enriched phenols. Li et al. [3] conducted a series of distillations and extractions and obtained an oil with high phenols content (78.8 wt% after distillation, and 93.9 wt% after extraction). The staged condensation of pyrolysis vapor, which is based on the difference of boiling temperatures of bio-oil components, is another possibility for the separation of phenolic compounds from others [108].

11.4 Application of Phenolic-Rich Bio-oil

Bio-oil rich in phenols can usually be used in two ways. One is the use of simple phenols separated from bio-oil with separation techniques such as extraction and distillation. Such simple phenols can be applied in many existing industrial applications, such as fine chemicals, pharmaceuticals, food processing, synthetic flavors, fragrances, germicidal agents, and resin manufacturing. Typical simple phenols that are of interest to industry include guaiacol, cresols, syringol, and phenol. Although several attempts at the recovery of simple phenols from bio-oil have been made, the application of simple phenols from bio-oil in the production of the commercial products mentioned above has not been reported. The second way that bio-oil rich in phenols can be used is the use of PL, which can be obtained simply via cold water precipitation, or the use of whole bio-oils obtained either from the pyrolysis of lignin or lignocellulosic biomass. PL or whole bio-oils have been widely tested in the synthesis of phenolic resins to replace fossil fuel-derived phenols, or for other uses. According to a report [109], the global phenolic resin market was valued at 10.66 billion USD in 2015 and is projected to reach 15.01 billion USD by 2021. The market price of PF-resins is somewhere around (1100–2300) USD/MT [110].

11.4.1 Phenolic Resins Synthesis

Phenolic resins are usually prepared by acid or base-catalyzed reactions between phenol (or substituted phenol) and aldehydes. A representative phenolic resin is phenol-formaldehyde resin, which has two different types: novolacs (thermoplastic) and resoles (thermosetting). Figure 11.6 shows the synthesis methods for phenol-formaldehyde resins.

Novolacs are synthesized under acidic conditions with a molar ratio of formaldehyde to phenol that is less than 1, whereas resoles are synthesized under alkaline conditions with a molar ratio of formaldehyde to phenol greater than 1. In phenol

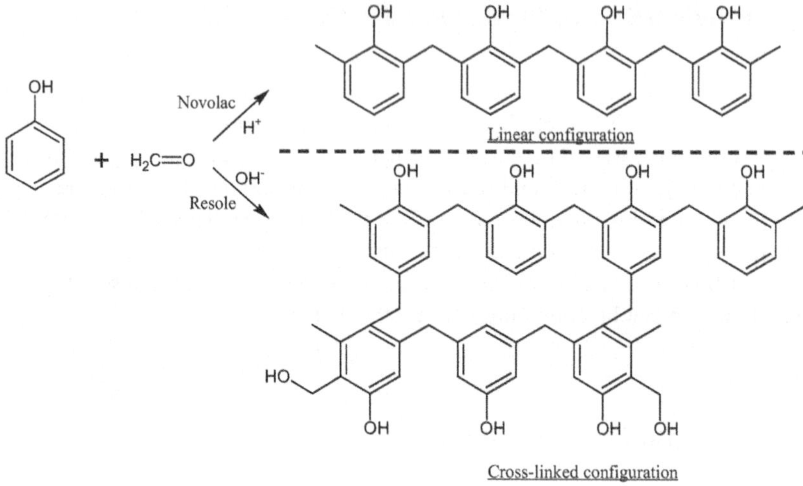

Fig. 11.6 Types of phenol-formaldehyde resins

resins synthesis, reactions occur at the ortho and para positions of the phenol hydroxyl group. Phenolic resins are widely used in many industries due to their excellent properties, such as high rigidity and good corrosion resistance. Hence, they have been widely used as binders for friction materials such as brake linings, and for wood composite products, such as plywood, particle board, and oriented strand board (OSB), because they have exceptional compatibility with cellulose fillers. PL and whole bio-oils containing a high content of phenols have strong potential for use in the preparation of phenolic resins. Particularly, whole bio-oils from the pyrolysis of lignocellulosic biomass have an advantage in the synthesis of phenolic resins in that they contain aldehydes as well as phenols. However, it is very difficult for PL and bio-oils to totally replace fossil phenols because they have low reactivity in the synthesis of phenolic resin. PL and phenols in bio-oil have fewer reactive sites, which means that the ortho and para positions of the phenol hydroxyl group are occupied by other groups, such as methoxy and carbonyls. PL also produces steric hindrance effects in the synthesis of phenolic resin, due to its complex chemical structure. Hence, the major challenge for substituting fossil phenols with PL or bio-oils has been to increase their reactivity in the synthesis of phenolic resin. Since the early 2000s, extensive research on the production and application of wood binders from the pyrolysis of lignocellulosic biomass has been carried out. Amen-Chen et al. [111] conducted the vacuum pyrolysis of bark residues from the pulp, paper, and wood industries, and produced phenolic-rich bio-oils. They prepared resoles with different levels of bio-oil, and then used the synthesized resoles in the preparation of OSB. They found in their study that panels bonded with resins containing 25 wt% and 50 wt% of bio-oils exhibited mechanical properties comparable to those of panels made with a commercial surface resin under the same pressing conditions. Sukhbaatar et al. [112] also explored the possibility of using

PL obtained from the pyrolysis of pinewood using an auger reactor as a binder for OSB. They were able to separate a PL fraction from bio-oil (25 wt% yield on bio-oil weight) and prepare phenol-formaldehyde resins using the PL at (30, 40, 50) wt% phenol replacement levels. The study reported that an incorporation of PL of approximately 40 wt% did not lower the performance of the synthesized phenol-formaldehyde resin. Instead of PL, the entire bio-oil was tested in the synthesis of phenolic resins. Chaouch et al. [113] used bio-oils produced from the pyrolysis of two Canadian whole-tree species in the synthesis of resole-type phenol-formaldehyde resins, and reported that phenol replacement levels up to 50 wt% provided equal or superior reactivity and performance to those of the pure phenol-formaldehyde resin. Choi et al. [18] carried out the pyrolysis of PKS in a fluidized bed reactor, and used a whole bio-oil containing approximately 9 wt% of phenol along with a commercial resole resin in the preparation of plywood panels. The study showed that whole bio-oil could be substituted for up to 25 wt%. of fossil phenol. Cui et al. [114] applied a whole bio-oil obtained from the pyrolysis of larch sawdust in the synthesis of phenol-formaldehyde resin and tested the properties of synthesized resins. The study reported that resin with a bio-oil addition of 20 wt% showed good performance for oxygen index and bending strength, and that the addition of bio-oil had an insignificant impact on the curing characteristics and thermal degradation process of phenol-formaldehyde resin when the amount of bio-oil was relatively low. Mao et al. [115] compared the characteristics and bonding performance of phenol-formaldehyde resins synthesized and blended with bio-oil. For preparation of synthesized phenol-formaldehyde resins, they used bio-oil from the pyrolysis of pinewood in the reaction with phenol and formaldehyde at 10, 25, 50, and 75 wt% phenol substitution levels. For the preparation of blended phenol-formaldehyde resins, they physically blended bio-oil with phenol and formaldehyde resin at addition levels of (4, 13, 23, 38) wt%. The two resins prepared in such a way were then used in the production of plywood. Finally, they concluded in their work that the best bio-oil addition percentage was 13 wt% for blended phenol-formaldehyde resins, while for synthesized phenol-formaldehyde resins, the best phenol substitution percentage was 25 wt%. Further, the possibility of using bio-oil as a part of a polymeric diphenylmethane diisocyanate (pMDI) binder system in the production of flakeboard was also investigated by Mao et al. [116]. They mixed bio-oil obtained from the pyrolysis of pinewood with acetone and pMDI in the preparation of resin and manufactured flakeboard using a hot press. The authors reported that the pMDI resin with a bio-oil content of 25 wt% had bonding properties comparable to those of pure pMDI resin.

11.4.2 Other Uses

In addition to phenolic resins preparation, PL and bio-oils rich in phenols have been tested for other uses. Fini et al. [117] tested PL as a replacement for fossil bitumen (*e.g.* asphalt binder) and reported that the use of bio-binder made from PL improved the petroleum asphalt binder's low-temperature properties while reducing asphalt pavement construction costs. A bio-oil rich in phenols could be used as a wood preservative. Mourant et al. [118] conducted the vacuum pyrolysis of mixed bark residues of balsam fir and white spruce at 450 °C, divided the bio-oil into four fractions, and investigated the anti-fungal properties of each fraction. They concluded in their work that fractions with higher concentrations of 4-ethyl-2-methoxyphenol and 4-propyl-2-methoxyphenol might have a predominant role in fungal inhibition.

11.5 Conclusions and Future Outlook

Renewable phenols are gaining attention because they can be readily produced wherever lignocellulosic biomass is available. A representative technique for producing renewable phenols is fast pyrolysis. To optimize phenols production, a few points must be checked. Firstly, lignin-rich biomass should be screened. Many researchers have recommended softwoods, barks, and shells as proper feedstocks for phenol production. Secondly, appropriate catalysts should be selected and used. The use of catalysts can be a good option when the catalysts have high selectivity for phenols, but problems related to their deactivation and cost should be solved before their application in industry. Non-catalytic pyrolysis of lignocellulosic biomass, which has economic merit but which usually suffers from a low selectivity for phenols, may have good chance to be implemented in industry if proper reactor configurations, lignin-rich biomass, and optimized reaction conditions are used. In a two-stage non-catalytic pyrolysis of palm kernel shells, the phenol content in bio-oil amounted to 12 wt%. After production of phenols-rich bio-oil, efficient and economical separation techniques are needed. Until now, extraction, column chromatography, distillation, and molecular distillation appear to be applicable as separation methods.

If valuable phenols could be separated with high purity from bio-oil, they could effectively replace fossil phenols in many industrial applications. In particular, simple phenols separated from bio-oil can be used in synthesizing materials such as bio-based antioxidants, which would improve the economics of bio-oil through production of value-added products. Although such desirable separations cannot be readily achieved, whole bio-oils with a high phenols content can have other uses, such as in phenolic resins production.

References

1. Amen-Chen C, Pakdel H, Roy C. Separation of phenols from Eucalyptus wood tar. Biomass Bioenergy. 1997;13:25–37. https://doi.org/10.1016/S0961-9534(97)00021-4.

2. Luo G, Lv X, Wang X, Yan S, Gao X, Xu J, Ma H, Jiao Y, Li F, Chen J. Direct hydroxylation of benzene to phenol with molecular oxygen over vanadium oxide nanospheres and study of its mechanism. RCS Advances. 2015;5:94164–70. https://doi.org/10.1039/C5RA17287J.
3. Li J, Wang C, Yang Z. Production and separation of phenols from biomass-derived bio-petroleum. J Anal Appl Pyrolysis. 2010;89:218–24. https://doi.org/10.1016/j.jaap.2010.08.004.
4. Tejado A, Pena C, Labidi J, Echeverria JM, Mondragon I. Physico-chemical characterization of lignins from different sources for use in phenol–formaldehyde resin synthesis. Bioresour Technol. 2007;98:1655–63. https://doi.org/10.1016/j.biortech.2006.05.042.
5. Sharifzadeh M, Sadeqzadeh M, Guo M, Borhani TN, Konda NM, Garcia MC, Wang L, Hallett J, Shah N. The multi-scale challenges of biomass fast pyrolysis and bio-oil upgrading: Review of the state of art and future research directions. Prog Energy Combust Sci. 2019;71:1–80. https://doi.org/10.1016/j.pecs.2018.10.006.
6. Rowell RM, Pettersen R, Tshabalala MA. Cell wall chemistry. In: Rowell RM, editor. Handbook of wood chemistry and wood composites. 2nd ed. Boca Raton: CRC press; 2012. p. 35–76. https://doi.org/10.1201/b12487.
7. Butterfield B. The structure of wood: form and function. In: Walker JCF, editor. Primary Wood Process: Principles and Practice. 2nd ed. Dordrecht: Springer; 2006. p. 1–22. https://doi.org/10.1007/1-4020-4393-7_1.
8. Wang S, Dai G, Yang H, Luo Z. Lignocellulosic biomass pyrolysis mechanism: a state-of-the-art review. Prog Energy Combust Sci. 2017;62:33–86. https://doi.org/10.1016/j.pecs.2017.05.004.
9. Guimarães JL, Frollini E, Da Silva CG, Wypych F, Satyanarayana KG. Characterization of banana, sugarcane bagasse and sponge gourd fibers of Brazil. Ind Crop Prod. 2009;30:407–15. https://doi.org/10.1016/j.indcrop.2009.07.013.
10. Godin B, Lamaudière S, Agneessens R, Schmit T, Goffart J-P, Stilmant D, Gerin PA, Delcarte J. Chemical composition and biofuel potentials of a wide diversity of plant biomasses. Energy Fuel. 2013;27:2588–98. https://doi.org/10.1021/ef3019244.
11. Goenka R, Parthasarathy P, Gupta NK, Biyahut NK, Narayanan S. Kinetic analysis of biomass and comparison of its chemical compositions by thermogravimetry, wet and experimental furnace methods. Waste Biomass Valorization. 2015;6:989–1002. https://doi.org/10.1007/s12649-015-9402-3.
12. Rabemanolontsoa H, Saka S. Comparative study on chemical composition of various biomass species. RSC Adv. 2013;3:3946–56. https://doi.org/10.1039/C3RA22958K.
13. Taherzadeh MJ, Eklund R, Gustafsson L, Niklasson C, Lidén G. Characterization and fermentation of dilute-acid hydrolyzates from wood. Ind Eng Chem Res. 1997;36:4659–65. https://doi.org/10.1021/ie9700831.
14. Zhao X, Zhang L, Liu D. Biomass recalcitrance. Part I: the chemical compositions and physical structures affecting the enzymatic hydrolysis of lignocellulose. Biofuels. Bioprod Biorefining. 2012;6:465–82. https://doi.org/10.1002/bbb.1331.
15. Mattonai M, Licursi D, Antonetti C, Galletti AMR, Ribechini E. Py-GC/MS and HPLC-DAD characterization of hazelnut shell and cuticle: Insights into possible re-evaluation of waste biomass. J Anal Appl Pyrolysis. 2017;127:321–8. https://doi.org/10.1016/j.jaap.2017.07.019.
16. Di Blasi C, Branca C, Galgano A. Biomass screening for the production of furfural via thermal decomposition. Ind Eng Chem Res. 2010;49:2658–71. https://doi.org/10.1021/ie901731u.
17. Collard F-X, Blin J. A review on pyrolysis of biomass constituents: Mechanisms and composition of the products obtained from the conversion of cellulose, hemicelluloses and lignin. Renew Sust Energ Rev. 2014;38:594–608. https://doi.org/10.1016/j.rser.2014.06.013.
18. Choi G-G, Oh S-J, Lee S-J, Kim J-S. Production of bio-based phenolic resin and activated carbon from bio-oil and biochar derived from fast pyrolysis of palm kernel shells. Bioresour Technol. 2015;178:99–107. https://doi.org/10.1016/j.biortech.2014.08.053.
19. Holladay JE, White JF, Bozell JJ, Johnson D. Top value-added chemicals from biomass-Volume II—Results of screening for potential candidates from biorefinery lignin. Pacific

Northwest National Lab.(PNNL), Richland, WA (United States). 2007; https://doi.org/10.2172/921839.
20. Ragauskas AJ, Beckham GT, Biddy MJ, Chandra R, Chen F, Davis MF, et al. Lignin valorization: improving lignin processing in the biorefinery. Science. 2014;344(6185):709–19. https://doi.org/10.1126/science.1246843.
21. Mohan D, Pittman CU Jr, Steele PH. Pyrolysis of wood/biomass for bio-oil: a critical review. Energy Fuel. 2006;20(3):848–89. https://doi.org/10.1021/ef0502397.
22. Effendi A, Gerhauser H, Bridgwater AV. Production of renewable phenolic resins by thermochemical conversion of biomass: a review. Renew Sust Energ Rev. 2008;12:2092–116. https://doi.org/10.1016/j.rser.2007.04.008.
23. Schlosberg RH, Szajowski PF, Dupre GD, Danik JA, Kurs A, Ashe TR, Olmstead WI. Pyrolysis studies of organic oxygenates: 3. High temperature rearrangement of aryl alkyl ethers. Fuel. 1983;62(6):690–4. https://doi.org/10.1016/0016-2361(83)90308-3.
24. Couhert C, Commandre JM, Salvador S. Is it possible to predict gas yields of any biomass after rapid pyrolysis at high temperature from its composition in cellulose, hemicellulose and lignin? Fuel. 2009;88(3):408–17. https://doi.org/10.1016/j.fuel.2008.09.019.
25. Hilbers TJ, Wang Z, Pecha B, Westerhof RJ, Kersten SR, Pelaez-Samaniego MR, Garcia-Perez M. Cellulose-Lignin interactions during slow and fast pyrolysis. J Anal Appl Pyrolysis. 2015;114:197–207. https://doi.org/10.1016/j.jaap.2015.05.020.
26. Zhao S, Liu M, Zhao L, Zhu L. Influence of interactions among three biomass components on the pyrolysis behavior. Ind Eng Chem Res. 2018;57(15):5241–9. https://doi.org/10.1021/acs.iecr.8b00593.
27. Chen Y, Fang Y, Yang H, Xin S, Zhang X, Wang X, Chen H. Effect of volatiles interaction during pyrolysis of cellulose, hemicellulose, and lignin at different temperatures. Fuel. 2019;248:1–7. https://doi.org/10.1016/j.fuel.2019.03.070.
28. Horne PA, Williams PT. Influence of temperature on the products from the flash pyrolysis of biomass. Fuel. 1996;75(9):1051–9. https://doi.org/10.1016/0016-2361(96)00081-6.
29. Hoekstra E, Westerhof RJ, Brilman W, Van Swaaij WP, Kersten SR, Hogendoorn KJ, Windt M. Heterogeneous and homogeneous reactions of pyrolysis vapors from pine wood. AICHE J. 2012;58(9):2830–42. https://doi.org/10.1002/aic.12799.
30. Bai X, Kim KH, Brown RC, Dalluge E, Hutchinson C, Lee YJ, Dalluge D. Formation of phenolic oligomers during fast pyrolysis of lignin. Fuel. 2014;128:170–9. https://doi.org/10.1016/j.fuel.2014.03.013.
31. Liaw SS, Perez VH, Zhou S, Rodriguez-Justo O, Garcia-Perez M. Py-GC/MS studies and principal component analysis to evaluate the impact of feedstock and temperature on the distribution of products during fast pyrolysis. J Anal Appl Pyrolysis. 2014;109:140–51. https://doi.org/10.1016/j.jaap.2014.06.018.
32. Liu C, Hu J, Zhang H, Xiao R. Thermal conversion of lignin to phenols: Relevance between chemical structure and pyrolysis behaviors. Fuel. 2016;182:864–70. https://doi.org/10.1016/j.fuel.2016.05.104.
33. Safdari MS, Amini E, Weise DR, Fletcher TH. Heating rate and temperature effects on pyrolysis products from live wildland fuels. Fuel. 2019;242:295–304. https://doi.org/10.1016/j.fuel.2019.01.040.
34. Greenhalf CE, Nowakowski DJ, Harms AB, Titiloye JO, Bridgwater AV. Sequential pyrolysis of willow SRC at low and high heating rates–Implications for selective pyrolysis. Fuel. 2012;93:692–702. https://doi.org/10.1016/j.fuel.2011.11.050.
35. Wang B, Xu F, Zong P, Zhang J, Tian Y, Qiao Y. Effects of heating rate on fast pyrolysis behavior and product distribution of Jerusalem artichoke stalk by using TG-FTIR and Py-GC/MS. Renew Energ. 2019;132:486–96. https://doi.org/10.1016/j.renene.2018.08.021.
36. Zhou S, Garcia-Perez M, Pecha B, McDonald AG, Westerhof RJ. Effect of particle size on the composition of lignin derived oligomers obtained by fast pyrolysis of beech wood. Fuel. 2014;125:15–9. https://doi.org/10.1016/j.fuel.2014.01.016.

37. Zhang H, Xiao R, Wang D, He G, Shao S, Zhang J, Zhong Z. Biomass fast pyrolysis in a fluidized bed reactor under N2, CO2, CO, CH4 and H2 atmospheres. Bioresour Technol. 2011;102(5):4258–64. https://doi.org/10.1016/j.biortech.2010.12.075.
38. Amutio M, Lopez G, Aguado R, Artetxe M, Bilbao J, Olazar M. Kinetic study of lignocellulosic biomass oxidative pyrolysis. Fuel. 2012;95:305–11. https://doi.org/10.1016/j.fuel.2011.10.008.
39. Britt PF, Buchanan Iii AC, Thomas KB, Lee. Pyrolysis mechanisms of lignin: surface-immobilized model compound investigation of acid-catalyzed and free-radical reaction pathways. J Anal Appl Pyrolysis. 1995;33:1–19. https://doi.org/10.1016/0165-2370(94)00846-S.
40. Zhou J, Jin W, Shen D, Gu S. Formation of aromatic hydrocarbons from co-pyrolysis of lignin-related model compounds with hydrogen-donor reagents. J Anal Appl Pyrolysis. 2018;134:143–9. https://doi.org/10.1016/j.jaap.2018.06.002.
41. Zhang B, Zhang J. Influence of reaction atmosphere (N2, CO, CO2, and H2) on ZSM-5 catalyzed microwave-induced fast pyrolysis of medicinal herb residue for biofuel production. Energy Fuel. 2017;31(9):9627–32. https://doi.org/10.1021/acs.energyfuels.7b02106.
42. Butt DA. Formation of phenols from the low-temperature fast pyrolysis of Radiata pine (Pinus radiata): Part I. Influence of molecular oxygen. J Anal Appl Pyrolysis. 2006;76(1–2):38–47. https://doi.org/10.1016/j.jaap.2005.07.003.
43. Dhyani V, Bhaskar T. A comprehensive review on the pyrolysis of lignocellulosic biomass. Renew Energ. 2018;129:695–716. https://doi.org/10.1016/j.renene.2017.04.035.
44. Kim SJ, Jung SH, Kim JS. Fast pyrolysis of palm kernel shells: influence of operation parameters on the bio-oil yield and the yield of phenol and phenolic compounds. Bioresour Technol. 2010;101:9294–300. https://doi.org/10.1016/j.biortech.2010.06.110.
45. Bertero M, Gorostegui HA, Orrabalis CJ, Guzmán C, Calandri EL, Sedran U. Characterization of the liquid products in the pyrolysis of residual chañar and palm fruit biomasses. Fuel. 2014;116:409–14. https://doi.org/10.1016/j.fuel.2013.08.027.
46. Morgan HM Jr, Bu Q, Liang J, Liu Y, Mao H, Shi A, Lei H, Ruan R. A review of catalytic microwave pyrolysis of lignocellulosic biomass for value-added fuel and chemicals. Bioresour Technol. 2017;230:112–21. https://doi.org/10.1016/j.biortech.2017.01.059.
47. Kim J-S. Production, separation and applications of phenolic-rich bio-oil—a review. Bioresour Technol. 2015;178:90–8. https://doi.org/10.1016/j.biortech.2014.08.121.
48. Bridgwater AV. Review of fast pyrolysis of biomass and product upgrading. Biomass Bioenergy. 2012;38:68–94. https://doi.org/10.1016/j.biombioe.2011.01.048.
49. Aziz SMA, Wahi R, Ngaini Z, Hamdan S. Bio-oils from microwave pyrolysis of agricultural wastes. Fuel Process Technol. 2013;106:744–50. https://doi.org/10.1016/j.fuproc.2012.10.011.
50. Yerrayya A, Suriapparao DV, Natarajan U, Vinu R. Selective production of phenols from lignin via microwave pyrolysis using different carbonaceous susceptors. Bioresour Technol. 2018;270:519–28. https://doi.org/10.1016/j.biortech.2018.09.051.
51. Wang Y, Zeng Z, Tian X, Dai L, Jiang L, Zhang S, Wu Q, Wen P, Fu G, Liu Y, Ruan R. Production of bio-oil from agricultural waste by using a continuous fast microwave pyrolysis system. Bioresour Technol. 2018;269:162–8. https://doi.org/10.1016/j.biortech.2018.08.067.
52. Murwanashyaka JN, Pakdel H, Roy C. Step-wise and one-step vacuum pyrolysis of birch-derived biomass to monitor the evolution of phenols. J Anal Appl Pyrolysis. 2001;60(2):219–31. https://doi.org/10.1016/S0165-2370(00)00206-0.
53. Zhang L, Li S, Li K, Zhu X. Two-step pyrolysis of corncob for value-added chemicals and high quality bio-oil: Effects of pyrolysis temperature and residence time. Energ Convers Manage. 2018;166:260–7. https://doi.org/10.1016/j.enconman.2018.04.002.
54. Oh SJ, Choi GG, Kim JS. Characteristics of bio-oil from the pyrolysis of palm kernel shell in a newly developed two-stage pyrolyzer. Energy. 2016;113:108–15. https://doi.org/10.1016/j.energy.2016.07.044.

55. Gamliel DP, Du S, Bollas GM, Valla JA. Investigation of in situ and ex situ catalytic pyrolysis of miscanthus × giganteus using a PyGC–MS microsystem and comparison with a bench-scale spouted-bed reactor. Bioresour Technol. 2015;191:187–96. https://doi.org/10.1016/j.biortech.2015.04.129.
56. Liu C, Wang H, Karim AM, Sun J, Wang Y. Catalytic fast pyrolysis of lignocellulosic biomass. Chem Soc Rev. 2014;43(22):7594–623. https://doi.org/10.1039/C3CS60414D.
57. Lu Q, Zhang ZF, Dong CQ, Zhu XF. Catalytic upgrading of biomass fast pyrolysis vapors with nano metal oxides: an analytical Py-GC/MS study. Energies. 2010;3(11):1805–20. https://doi.org/10.3390/en3111805.
58. Ma Z, Custodis V, van Bokhoven JA. Selective deoxygenation of lignin during catalytic fast pyrolysis. Cat Sci Technol. 2014;4(3):766–72. https://doi.org/10.1039/C3CY00704A.
59. Nair V, Vinu R. Production of guaiacols via catalytic fast pyrolysis of alkali lignin using titania, zirconia and ceria. J Anal Appl Pyrolysis. 2016;119:31–9. https://doi.org/10.1016/j.jaap.2016.03.020.
60. Dong Z, Yang H, Chen P, Liu Z, Chen Y, Wang L, Wang X, Chen H. Lignin Characterization and Catalytic Pyrolysis for Phenol-Rich Oil with TiO2-Based Catalysts. Energy Fuel. 2019;33(10):9934–41. https://doi.org/10.1021/acs.energyfuels.9b02341.
61. Zhang X, Sun L, Chen L, Xie X, Zhao B, Si H, Meng G. Comparison of catalytic upgrading of biomass fast pyrolysis vapors over CaO and Fe (III)/CaO catalysts. J Anal Appl Pyrolysis. 2014;108:35–40. https://doi.org/10.1016/j.jaap.2014.05.020.
62. Xu L, Zhang Y, Fu Y. Advances in upgrading lignin pyrolysis vapors by ex situ catalytic fast pyrolysis. Energ Technol. 2017;5(1):30–51. https://doi.org/10.1002/ente.201600107.
63. Lu Q, Zhang Y, Tang Z, Li WZ, Zhu XF. Catalytic upgrading of biomass fast pyrolysis vapors with titania and zirconia/titania based catalysts. Fuel. 2010;89(8):2096–103. https://doi.org/10.1016/j.fuel.2010.02.030.
64. Jongerius AL, Gosselink RW, Dijkstra J, Bitter JH, Bruijnincx PC, Weckhuysen BM. Carbon nanofiber supported transition-metal carbide catalysts for the hydrodeoxygenation of guaiacol. Chem Cat Chem. 2013;5(10):2964–72. https://doi.org/10.1002/cctc.201300280.
65. Filley J, Roth C. Vanadium catalyzed guaiacol deoxygenation. J Mol Catal A-Chem. 1999;139(2–3):245–52. https://doi.org/10.1016/S1381-1169(98)00202-7.
66. Sun J, Karim AM, Zhang H, Kovarik L, Li XS, Hensley AJ, McEwen J, Wang Y. Carbon-supported bimetallic Pd–Fe catalysts for vapor-phase hydrodeoxygenation of guaiacol. J Catal. 2013;306:47–57. https://doi.org/10.1016/j.jcat.2013.05.020.
67. González-Borja MÁ, Resasco DE. Anisole and guaiacol hydrodeoxygenation over monolithic Pt–Sn catalysts. Energy Fuel. 2011;25(9):4155–62. https://doi.org/10.1021/ef200728r.
68. Jakab E, Faix O, Till F, Székely T. The effect of cations on the thermal decomposition of lignins. J Anal Appl Pyrolysis. 1993;25:185–94. https://doi.org/10.1016/0165-2370(93)80039-3.
69. Zhang ZB, Lu Q, Ye XN, Li WT, Hu B, Dong CQ. Production of phenolic-rich bio-oil from catalytic fast pyrolysis of biomass using magnetic solid base catalyst. Energ Convers Manage. 2015;106:1309–17. https://doi.org/10.1016/j.enconman.2015.10.063.
70. Strassberger Z, Tanase S, Rothenberg G. Reductive dealkylation of anisole and phenetole: towards practical lignin conversion. Eur J Org Chem. 2011;27:5246–9. https://doi.org/10.1002/ejoc.201101015.
71. Li C, Zhao X, Wang A, Huber GW, Zhang T. Catalytic transformation of lignin for the production of chemicals and fuels. Chem Rev. 2015;115(21):11559–624. https://doi.org/10.1021/acs.chemrev.5b00155.
72. Li X, Su L, Wang Y, Yu Y, Wang C, Li X, Wang Z. Catalytic fast pyrolysis of Kraft lignin with HZSM-5 zeolite for producing aromatic hydrocarbons. Front Env Sci Eng. 2012;6(3):295–303. https://doi.org/10.1007/s11783-012-0410-2.
73. Ma Z, Ghosh A, Asthana N, van Bokhoven J. Optimization of the reaction conditions for catalytic fast pyrolysis of pretreated lignin over zeolite for the production of phenol. Chem Cat Chem. 2017;9(6):954–61. https://doi.org/10.1002/cctc.201601674.

74. Xue X, Liu Y, Wu L, Pan X, Liang J, Sun Y. Catalytic fast pyrolysis of maize straw with a core–shell ZSM-5@ SBA-15 catalyst for producing phenols and hydrocarbons. Bioresour Technol. 2019;289:121691. https://doi.org/10.1016/j.biortech.2019.121691.
75. Ma Z, Ghosh A, Asthana N, Van Bokhoven J. Visualization of structural changes during deactivation and regeneration of fau zeolite for catalytic fast pyrolysis of lignin using nmr and electron microscopy techniques. Chem Cat Chem. 2018;10(19):4431–7. https://doi.org/10.1002/cctc.201800670.
76. Lazaridis PA, Fotopoulos AP, Karakoulia SA, Triantafyllidis KS. Catalytic fast pyrolysis of kraft lignin with conventional, mesoporous and nanosized ZSM-5 zeolite for the production of alkyl-phenols and aromatics. Front Chem. 2018;6:1–21. https://doi.org/10.3389/fchem.2018.00295.
77. Ma Z, Troussard E, van Bokhoven JA. Controlling the selectivity to chemicals from lignin via catalytic fast pyrolysis. Appl Catal A-Gen. 2012;423:130–6. https://doi.org/10.1016/j.apcata.2012.02.027.
78. Iliopoulou EF, Stefanidis SD, Kalogiannis KG, Delimitis A, Lappas AA, Triantafyllidis KS. Catalytic upgrading of biomass pyrolysis vapors using transition metal-modified ZSM-5 zeolite. Appl Catal B-Environ. 2012;127:281–90. https://doi.org/10.1016/j.apcatb.2012.08.030.
79. Jeon MJ, Kim SS, Jeon JK, Park SH, Kim JM, Sohn JM, Lee SH, Park YK. Catalytic pyrolysis of waste rice husk over mesoporous materials. Nanoscale Res Lett. 2012;7(1):1–5. https://doi.org/10.1186/1556-276X-7-18.
80. Verboekend D, Liao Y, Schutyser W, Sels BF. Alkylphenols to phenol and olefins by zeolite catalysis: a pathway to valorize raw and fossilized lignocellulose. Green Chem. 2017;18:297–306. https://doi.org/10.1039/C5GC01868D.
81. Liao Y, d'Halluin M, Makshina E, Verboekend D, Sels BF. Shape selectivity vapor-phase conversion of lignin-derived 4-ethylphenol to phenol and ethylene over acidic aluminosilicates: impact of acid properties and pore constraint. Appl Catal B Environ. 2018;234:117–29. https://doi.org/10.1016/j.apcatb.2018.04.001.
82. Horne PA, Williams PT. The effect of zeolite ZSM-5 catalyst deactivation during the upgrading of biomass-derived pyrolysis vapours. J Anal Appl Pyrolysis. 1995;34(1):65–85. https://doi.org/10.1016/0165-2370(94)00875-2.
83. Gayubo AG, Aguayo AT, Atutxa A, Prieto R, Bilbao J. Role of reaction-medium water on the acidity deterioration of a HZSM-5 zeolite. Ind Eng Chem Res. 2004;43(17):5042–8. https://doi.org/10.1021/ie0306630.
84. Neumann GT, Hicks JC. Novel hierarchical cerium-incorporated MFI zeolite catalysts for the catalytic fast pyrolysis of lignocellulosic biomass. ACS Catal. 2012;2(4):642–6. https://doi.org/10.1021/cs200648q.
85. Wang J, Zhong Z, Ding K, Xue Z. Catalytic fast pyrolysis of mushroom waste to upgraded bio-oil products via pre-coked modified HZSM-5 catalyst. Bioresour Technol. 2016;212:6–10. https://doi.org/10.1016/j.biortech.2016.04.005.
86. Ma Z, van Bokhoven JA. Deactivation and regeneration of H-USY zeolite during lignin catalytic fast pyrolysis. Chem Cat Chem. 2012;4(12):2036–44. https://doi.org/10.1002/cctc.201200401.
87. Murray JA. Qualitative and quantitative approaches in comprehensive two-dimensional gas chromatography. J Chromatogr A. 2012;1261:58–68. https://doi.org/10.1016/j.chroma.2012.05.012.
88. Bicchi C, Liberto E, Matteodo M, Sgorbini B, Mondello L, Zellner BDA, Costa R, Rubiolo P. Quantitative analysis of essential oils: a complex task. Flavour Fragr J. 2008;23(6):382–91. https://doi.org/10.1002/ffj.1905.
89. Fu Q, Argyropoulos DS, Tilotta DC, Lucia LA. Products and functional group distributions in pyrolysis oil of chromated copper arsenate (CCA)-treated wood, as elucidated by gas chromatography and a novel 31P NMR-based method. Ind Eng Chem Res. 2007;46(16):5258–64. https://doi.org/10.1021/ie0702274.

90. Yang HM, Zhao W, Norinaga K, Fang JJ, Wang YG, Zong ZM, Wei XY. Separation of phenols and ketones from bio-oil produced from ethanolysis of wheat stalk. Sep Purif Technol. 2015;152:238–45.
91. Thring RW, Breau J. Hydrocracking of solvolysis lignin in a batch reactor. Fuel. 1996;75 (7):795–800. https://doi.org/10.1016/0016-2361(96)00036-1.
92. Cesari L, Canabady-Rochelle L, Mutelet F. Separation of phenols from lignin pyrolysis oil using ionic liquid. Sep Purif Technol. 2019;209:528–34. https://doi.org/10.1016/j.seppur.2018.07.083.
93. Wang D, Li D, Liu Y, Lv D, Ye Y, Zhu S, Zhang B. Study of a new complex method for extraction of phenolic compounds from bio-oils. Sep Purif Technol. 2014;134:132–8. https://doi.org/10.1016/j.seppur.2014.07.033.
94. Wei Y, Lei H, Wang L, Zhu L, Zhang X, Liu Y, Chen S, Ahring B. Liquid–liquid extraction of biomass pyrolysis bio-oil. Energy Fuel. 2014;28(2):1207–12. https://doi.org/10.1021/ef402490s.
95. Mantilla SV, Manrique AM, Gauthier-Maradei P. Methodology for extraction of phenolic compounds of bio-oil from agricultural biomass wastes. Waste Biomass Valori. 2015;6 (3):371–83. https://doi.org/10.1007/s12649-015-9361-8.
96. Žilnik LF, Jazbinšek A. Recovery of renewable phenolic fraction from pyrolysis oil. Sep Purif Technol. 2012;86:157–70. https://doi.org/10.1016/j.seppur.2011.10.040.
97. Zhang M, Wu H. Pyrolytic lignin from fast pyrolysis bio-oil via cold-water precipitation: Optimal separation conditions and properties. Fuel. 2019;242:580–6. https://doi.org/10.1016/j.fuel.2019.01.092.
98. Patel RN, Bandyopadhyay S, Ganesh A. Extraction of cardanol and phenol from bio-oils obtained through vacuum pyrolysis of biomass using supercritical fluid extraction. Energy. 2011;36(3):1535–42. https://doi.org/10.1016/j.energy.2011.01.009.
99. Chan YH, Yusup S, Quitain AT, Chai YH, Uemura Y, Loh SK. Extraction of palm kernel shell derived pyrolysis oil by supercritical carbon dioxide: Evaluation and modeling of phenol solubility. Biomass Bioenergy. 2018;116:106–12. https://doi.org/10.1016/j.biombioe.2018.06.009.
100. Cao Z, Engelhardt J, Dierks M, Clough MT, Wang GH, Heracleous E, Lappas A, Rindaldi R, Schüth F. Catalysis meets nonthermal separation for the production of (alkyl) phenols and hydrocarbons from pyrolysis oil. Angew Chem Int Ed. 2017;56(9):2334–9. https://doi.org/10.1002/anie.201610405.
101. Hao S, Chen K, Cao L, Zhu X, Luo G, Zhang S, Chen J. Separation of high-purity syringol and acetosyringone from rice straw-derived bio-oil by combining the basification-acidification process and column chromatography. Electrophoresis. 2016;37(19):2522–30. https://doi.org/10.1002/elps.201600126.
102. Zeng F, Liu W, Jiang H, Yu HQ, Zeng RJ, Guo Q. Separation of phthalate esters from bio-oil derived from rice husk by a basification–acidification process and column chromatography. Bioresour Technol. 2011;102(2):1982–7. https://doi.org/10.1016/j.biortech.2010.09.024.
103. Wang S, Gu Y, Liu Q, Yao Y, Guo Z, Luo Z, Cen K. Separation of bio-oil by molecular distillation. Fuel Process Technol. 2009;90(5):738–45. https://doi.org/10.1016/j.fuproc.2009.02.005.
104. Elkasabi Y, Mullen CA, Boateng AA. Distillation and isolation of commodity chemicals from bio-oil made by tail-gas reactive pyrolysis. ACS Sustain Chem Eng. 2014;2(8):2042–52. https://doi.org/10.1021/sc5002879.
105. Murwanashyaka JN, Pakdel H, Roy C. Seperation of syringol from birch wood-derived vacuum pyrolysis oil. Sep Purif Technol. 2001;24(1–2):155–65. https://doi.org/10.1016/S1383-5866(00)00225-2.
106. Wang Y, Wang S, Leng F, Chen J, Zhu L, Luo Z. Separation and characterization of pyrolytic lignins from the heavy fraction of bio-oil by molecular distillation. Sep Purif Technol. 2015;152:123–32. https://doi.org/10.1016/j.seppur.2015.08.011.

107. Guo Z, Wang S, Gu Y, Xu G, Li X, Luo Z. Separation characteristics of biomass pyrolysis oil in molecular distillation. Sep Purif Technol. 2010;76(1):52–7. https://doi.org/10.1016/j.seppur.2010.09.019.
108. Schulzke T, Conrad S, Westermeyer J. Fractionation of flash pyrolysis condensates by staged condensation. Biomass Bioenergy. 2016;95:287–95. https://doi.org/10.1016/j.biombioe.2016.05.022.
109. Phenolic resin market by type (resol resin, novolac resin), application (wood adhesives, molding, insulation), end-use industry (automotive, building & construction, furniture)—Global forecasts to 2021. 2017. Accessed 6 Jan 2020. marketsandmarkets.com.
110. Ľudmila H, Michal J, Andrea Š, Aleš H. Lignin, potential products and their market value. Wood Res. 2015;60(6):973–86.
111. Amen-Chen C, Riedl B, Wang XM, Roy C. Softwood bark pyrolysis oil-PF resols. Part 1. Resin synthesis and OSB mechanical properties. Holzforschung. 2002;56(2):167–75. https://doi.org/10.1515/HF.2002.028.
112. Sukhbaatar B, Steele PH, Ingram LI, Kim MG. Use of lignin separated from bio-oil in oriented strand board binder phenol-formaldehyde resins. Bio Resour. 2009;4(2):789–804.
113. Chaouch M, Diouf PN, Laghdir A, Yin S. Bio-oil from whole-tree feedstock in resol-type phenolic resins. J Appl Polym Sci. 2014;131(6):40014. https://doi.org/10.1002/app.40014.
114. Cui Y, Hou X, Wang W, Chang J. Synthesis and characterization of bio-oil phenol formaldehyde resin used to fabricate phenolic based materials. Materials. 2017;10(6):668. https://doi.org/10.3390/ma10060668.
115. Wan H, Mao A, Xu W, Xi E, Li Q. Evaluation of phenol formaldehyde resins modified/blended with pyrolysis bio-oil for plywood. Forest Prod J. 2018;68(2):113–9. https://doi.org/10.13073/FPJ-D-17-00066.
116. Mao A, Shi SQ, Steele P. Flakeboard bonded with polymeric diphenylmethane diisocyanate/bio-oil adhesive systems. Forest Prod J. 2011;61(3):240–5. https://doi.org/10.13073/0015-7473-61.3.240.
117. Fini EH, Kalberer EW, Shahbazi A, Basti M, You Z, Ozer H, Aurangzeb Q. Chemical characterization of biobinder from swine manure: Sustainable modifier for asphalt binder. J Mater Civil Eng. 2011;23(11):1506–13. https://doi.org/10.1061/(ASCE)MT.1943-5533.0000237.
118. Mourant D, Yang DQ, Lu X, Roy C. Anti-fungal properties of the pyroligneous liquors from the pyrolysis of softwood bark. Wood Fiber Sci. 2007;37(3):542–8.

[illegible]

Part V
Design of Pyrolysis Units and Models

Chapter 12
Syngas Production, Storage, Compression and Use in Gas Turbines

Minjiao Yang, Haiping Yang, Hewen Zhou, Qing Yang, Haibo Zhao, Eid Gul, Mohsin Ali Khan, Øyvind Skreiberg, Liang Wang, He Chao, Pietro Bartocci, Katarzyna Slopiecka, Gianni Bidini, and Francesco Fantozzi

Abstract This chapter analyses syngas production through pyrolysis and gasification, its compression and its use in gas turbines. Syngas compression can be performed during or after thermal treatment processes. Important points are discussed related to syngas ignition, syngas explosion limit at high temperatures and high pressures and syngas combustion kinetics. Kinetic aspects influence ignition and final emissions which are obtained at the completion of the combustion process. The chapter is organized into four subsections, dealing with (1) innovative syngas production plants, (2) syngas compressors and compression process, (3) syngas ignition in both heterogeneous and homogeneous systems and (4) syngas

M. Yang · H. Zhou · Q. Yang · E. Gul
China-EU Institute for Clean and Renewable Energy, Huazhong University of Science and Technology, Wuhan, Hubei, China

H. Yang
China-EU Institute for Clean and Renewable Energy, Huazhong University of Science and Technology, Wuhan, Hubei, China

State Key Laboratory of Coal Combustion, Huazhong University of Science and Technology, Wuhan, Hubei, China

H. Zhao
State Key Laboratory of Coal Combustion, Huazhong University of Science and Technology, Wuhan, Hubei, China

M. A. Khan
University of Haripur Pakistan, Haripur, KPK, Pakistan

Ø. Skreiberg · L. Wang
SINTEF Energy Research, Trondheim, Norway

H. Chao
School of Chemical and Biomedical Engineering, Nanyang Technological University, Singapore, Singapore

P. Bartocci (✉) · K. Slopiecka · G. Bidini · F. Fantozzi
Department of Engineering, University of Perugia, Perugia, Italy
e-mail: bartocci@crbnet.it

Z. Fang et al. (eds.), *Production of Biofuels and Chemicals with Pyrolysis*, Biofuels and Biorefineries 10, https://doi.org/10.1007/978-981-15-2732-6_12

combustion kinetics and experimental methods. Particular attention is given to ignition regions that affect the kinetics, namely systems that operate at temperatures higher than 1000 K can have strong ignition, whereas those operating at lower temperatures have weak ignition.

Keywords Pyrogas · Pyrolysis · Ignition · Syngas · Compression · Gasification

12.1 Innovative Pyrolysis and Gasification Plants

Seven pyrolysis plants are discussed herein: (1) integrated pyrolysis regenerated plant (IPRP), (2) pyrolysis polygeneration plant, (3) Carbofex plant, (4) Carbon Terra plant, (5) Pyreg plant, (6) TCR® plant and (7) 3R Agrocarbon plant. Each of these plants and their developing organizations is described briefly next. The IPRP design was developed by the University of Perugia (Italy) with collaboration of BIONET-Biomass and New Technologies and Biomass Research Centre, University of Perugia and it is now located in the Terni site. The IPRP pilot is based on a rotary kiln pyrolysis reactor. The pyrolysis polygeneration plant was developed by the State Key Laboratoy on Coal Combustion (SKLCC) of Huazhong University of Science and Technology (HUST) and is basd on a moving bed reactor. The Carbofex plant was developed by a Finnish company producing charcoal to be used as fertilizer and filter. The Carbofex reactor is based on an auger reactor. The Carbon Terra plant was developed by a German company, named Carbon Terra which has developed the Schottdorf-Meiler kiln. The Pyreg plant was developed by a German company named Pyreg which has developed a reactor heated at (500–700) °C and in which biomass is moved by screw conveyors. The thermo-catalytic reforming (TCR) plant was developed by a German company (Susteen) who commercialized the TCR® plant for pyrolysis that is coupled to reforming. The 3R Agrocarbon plant was developed by Terra Humana company; the technology is called 3R Agrocarbon. The 3R Agrocarbon plant is based on a rotary kiln reactor.

Together with the seven pyrolysis plants, three gasification plants are presented: (1) Milena (developed at ECN – now TSO – in the Netherlands), (2) Güssing gasifier, developed by a consortium called "Renet Austria" with important participation of Technical University of Vienna and (3) GoBiGas, which was built in Gothenburg (Sweden) with the support of Chalmers university.

12.1.1 IPRP Technology

The integrated pyrolysis regenerated plant (IPRP), is an innovative technology which combines a rotary kiln pyrolyser with gas turbines (GTs) to make full use of the pyrogas produced from biomass and waste thermal conversion.

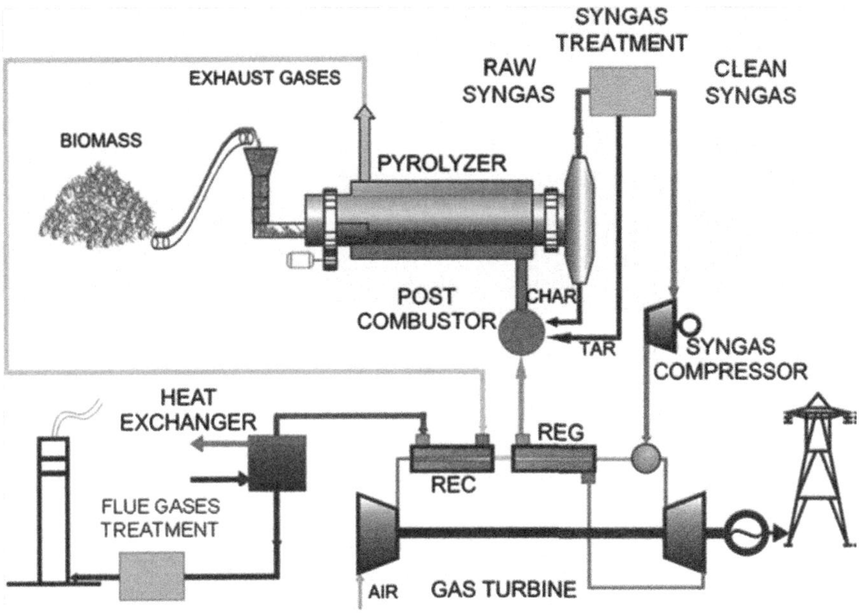

Fig. 12.1 Integrated pyrolysis regenerated plant (IPRP) concept. Reprinted with permission from [1] Copyright © 2013, Elsevier

The IPRP approach was proposed by D'Alessandro et al. [1]. The target of this project was to design a technology which can use the biomass with a higher efficiency. This technology mainly includes a GT fueled with syngas. The exhaust gases produced from the GT can be used to provide the energy required by the pyrolysis reactor. Figure 12.1 shows the exact layout of IPRP technology. The final outputs of the IPRP plant are heat and electricity.

The heat is produced after two heat recovery stages (REG and REC) aiming at the full use of the energy through regeneration of exhaust heat by two-stage air preheating. There is one regenerator (REG) used to recover the thermal energy of exhaust gases out from the turbine, and one recuperator (REC) to recover the thermal energy of exhaust gases out from the pyrolysis reactor. A picture of the plant is shown in Fig. 12.2.

The feedstock is fed through a hopper into a rotary kiln pyrolyser, which has a reaction chamber that thermally degrades the feedstock into syngas, coke and tar in the absence of oxygen. The hermetic seal between the rotary kiln and the oven is achieved by a high temperature resistant graphite ring, while the hermetic seal between the rotary kiln and the inlet and outlet sections is achieved by a soft iron ring. The refractory chamber containing the pyrolysis furnace is equipped with a combustion system at the bottom which continuously feeds the char conveyed by the screw conveyor at the outlet portion of the pyrolysis furnace. Combusting char could provide the heat required for pyrolysis, and the combustion air is provided by a dedicated blower or by the gas turbine exhaust gases. This depends on operational

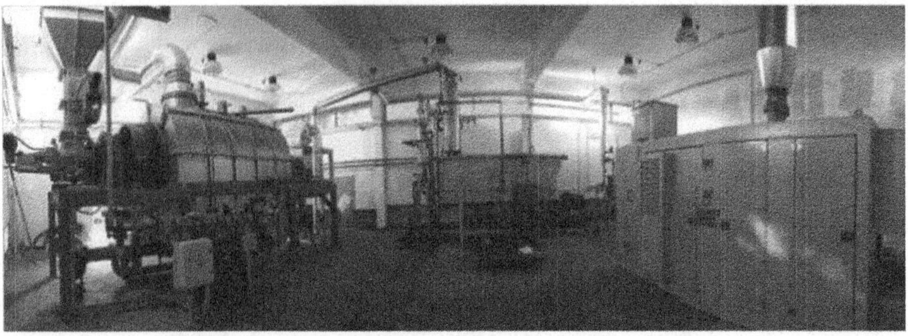

Fig. 12.2 Integrated pyrolysis regenerated plant (IPRP) demonstrative unit. Reprinted with permission from [1] Copyright © 2013, Elsevier

Table 12.1 Parameters of the micro gas turbine used in the IPRP demonstrative unit [1]

Parameter	Value
Electric Power (kW)	80
Electric Efficiency (%)	27
Turbine inlet temperature (°C)	1010
Manometric compression ratio	4
Exhaust gases flow (kg/s)	0.77
Exhaust gases temperature (°C)	270
rpm	68,000

requirements. Dual fuel gas burners exhaust gases (i.e. natural gas and syngas) are discharged directly into the refractory chamber above the char burner to provide the final additional heat for temperature control and startup. The particulates in the syngas coming from the pyrolysis furnace are removed in a cyclone. Syngas finally is cooled to condense tar and water in the wet scrubbing section, which is comprised of a two-stage quencher for temperature reduction. This is made by a variable throat Venturi tube, consisting of a two-stage scrubber with a final demister. Heavy tar and light tar are extracted respectively from the bottom and the top of the tank and can be returned to the dedicated burner in the refractory chamber of the pyrolysis furnace via a hot pipe, although this solution is still under investigation. The syngas is withdrawn from the pyrolysis furnace through the cleaning section by a side channel blower, and the speed of the side blower is adjusted to maintain a slight negative pressure in the rotary kiln, thereby providing the required pressure to the syngas compressor of the micro turbine. The micro turbine is coupled to a radial geometry turbo compressor with an annular combustion chamber suitable for combustion gases (the main parameters of the micro-turbine are shown in Table 12.1).

The compressor has an approximate compression ratio of 4. The gas turbine provides energy to drive compressors and alternators. Electrical energy is generated by a 4-pole permanent magn*et al*ternator that rotates within the oil-cooled stator assembly, which operates as an engine during initial startup, thereby reducing the need for auxiliary starting hardware. The micro turbine is equipped with a REG for preheating the combustion air by using the waste heat generated by the exhaust gas cooling; thereby reducing the exhaust gas temperature to approximately 270 °C.

12.1.2 Polygeneration Plant Based on Moving Bed in HUST

Given the abundance of agricultural residues in China, the biomass energy can play a significant role to satisfy at least part of the energy demand, granting environmental protection [2]. The thermal conversion technologies of biomass mainly involve combustion, gasification and pyrolysis. However, the combustion technology would release large amount of particulate matters with relative low energy efficiency [3]. In addition, biomass gasification has been widely used in China and there has been over 70 gasification systems built. The heating value of syngas produced in applied air-gasification systems is still too lower to be used. Compared with these two technologies, the pyrolysis system converts biomass into three main products (i.e. pyrolysis gas, bio-oil and biochar), in which the heating value of pyrolysis gas is higher because no gasification medium is used in the reactor [4]. In addition, these high-quality products can bring economic benefits to the pyrolysis system. Thus, pyrolysis technology is an important and promising method for Chinese biomass utilization.

Particular interesting results have been obtained in the research and development of poly-generation pyrolysis systems to promote the resource utilization of biomass. HUST has proposed a new poly-generation pyrolysis technology, and developed this technology to the commercial scale (see the demonstration poly-generation pyrolysis plant in Ezhou, Hubei Province, China). In addition, based on this moving-bed poly-pyrolysis technology, they have received the Blue Sky Award from the United Nations Industrial Development Organization in 2014. The layout of the biomass polygeneration system demonstration plant is shown in Fig. 12.3.

In Fig. 12.3, the biomass feedstock (e.g. agricultural residues and forest residues) would be firstly crushed to a particle size (<5 cm) and then sent to a torrefaction oven for 1 hour. The dried biomass pellets are fed from the top to the pyrolysis furnace (designed as moving bed) at high temperature. Both the baking furnace and the pyrolysis furnace are in a closed atmosphere. In the moving bed pyrolysis furnace, the biomass particles move up and down under the action of gravity to complete the pyrolysis, and control the discharge time and quantity of the biochar, so that the biomass particle volatiles are completely released. The high temperature volatiles from the top of the pyrolysis furnace are sequentially passed on to air cooling, water cooling, cleaning and adsorption towers to bring the temperature down to the range (40–60) °C to obtain liquid products (tar and vinegar) and purify the gas product (pyrolysis gas). Throughout the baking and pyrolysis process, heat is provided by the high temperature flue gas of the biomass burner. In order to increase the thermal efficiency of the system, the air required is from the heat exchanger of the air cooling tower. The calcining pyrolysis furnace can process 6 tons of biomass per hour (water content <20 wt%), and the furnace requires about 0.6 tons of biomass

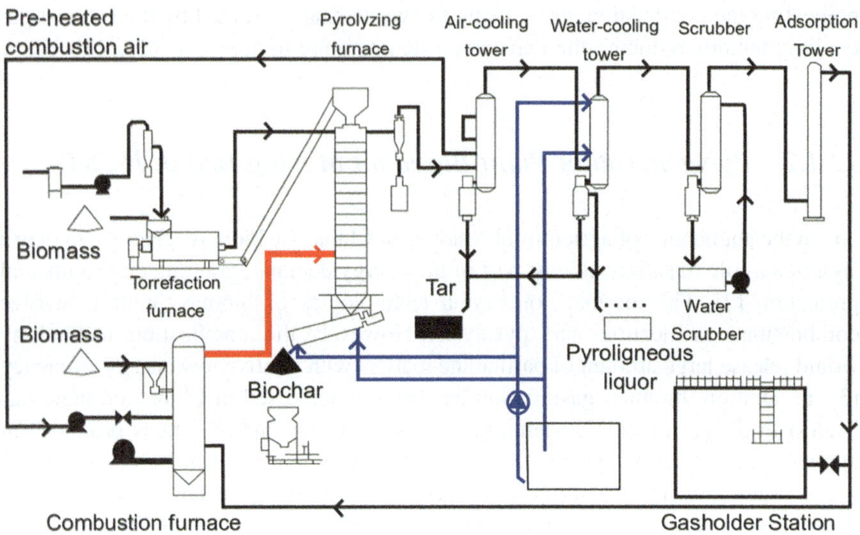

Fig. 12.3 Biomass-based pyrolytic polygeneration demonstration system. Reprinted with permission from [2]. Copyright © 2018, Elsevier

per hour. The temperature of the pyrolysis zone in the operating state of the pyrolysis cogeneration system is 600 °C.

12.1.3 Pyrolysis Plant at Carbofex

Carbofex is a technology developed in Finland. The plant is fed with (400–500) kg/h of dried wood chips and produces (100–140) kg of charcoal and (90–100) L of high quality pyrolysis oils. The syngas is burned to provide the heat required by the pyrolysis process, and to provide heat to a district heating network. When in the summer heat is not required, a significant part of the syngas can be stored, see [5].

12.1.4 Pyrolysis Plant at Carbon Terra

Carbon Terra pyrolysis plant is based on a reactor named Schottdorf kiln. The reactor is fed with 1 MW input of biomass [6, 7]. The power output for the products is composed by: 300 kW pyrolysis gas and 600 kW biochar (i.e. about 600 tons per year). The energy efficiency is between (90 and 95) %. The operating temperature of the process is about (800–900) °C. During the start-up phase, it dry biomass is used. The char outlet is at the bottom of the reactor. Char passes a sieve made by holes with 50 mm diameter. At the exit char is cooled down using water. Biochar contains approximately 60% of the input power. The produced pyrolysis gas pass though the biomass and reach the top of the reactor and then enter a combustion chamber.

12.1.5 Pyrolysis Plant at Pyreg

The plant is fed with sewage sludge. The pyrolysis reactor is heated by exhaust gases obtained from the combustion of pyrolysis gas. The reactor can be fed with about 1000 tons of dry mass per year [8]. The mass flow rate of biochar out of the plant is constant and the process is continuous.

Carbonization efficiency is up to 60%. The energy efficiency of the entire process is comprised between 90% and 95%. The temperature in the reactor is between (400 and 850) °C depending on the fuel moisture. The pyrolysis gas is combusted in a FLOX-burner, which has high efficiency and high combustion temperature (about 1250 °C). The burner is designed to have low NO_x emissions.

12.1.6 Thermo-Catalytic Pyrolysis

The thermo-catalytic reforming (TCR®) reactor is an electrically heated auger type reactor developed and built by Fraunhofer UMSICHT, Sulzbach Rosenberg, Germany. The TCR®-2 reactor setup consists of a hopper (1) connected to an electrically heated horizontal auger reactor (2) which is followed by a vertical reforming unit (3). Then follows a shell and tube type heat exchanger (4) which is used to condense vapors coming from the reforming unit. An external chilling unit (5) maintains the temperature of the condenser at −5 °C. The condensation unit (4) is followed by a gravity settler (6) which collects the liquid products coming from the vapors (7). Provisions have been made to collect condensed liquid products from the vapors condensation unit and separation unit via a common condensate collection system (7). Then follows an ice cooled second heat exchanger (8) and a gas cleaning unit (9) for further purification of incondensable gases. After condensation the gas is filtered with: activated carbons (10), candle filter (11) and silica wool filter (12). The exit pipe of the reactor is connected to a measuring unit composed by: gas flow meter (13), online gas analyzer (14) and a gas calorimeter (15). The plant is shown in Fig. 12.4.

12.1.7 3R Agrocarbon

The 3R Agrocarbon technology is based on a thermochemical (pyrolysis) process. The main input used is represented by food grade cattle and other types of bone grist. The main output is charcoal to be used as a fertilizer, which has a high content of

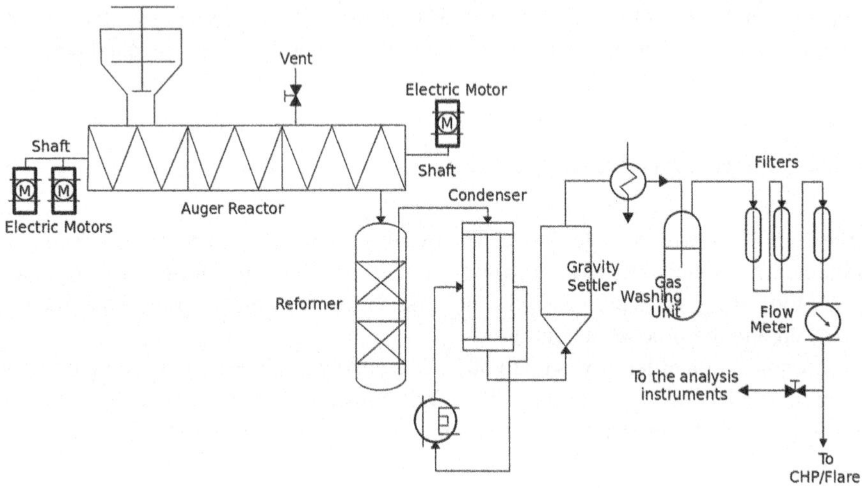

Fig. 12.4 TCR®-2 Laboratory plant. Reprinted with permission from [9]. Copyright © 2018, Elsevier

nutrients (P, K, Ca and N). The capacity of the plant is higher than 12,500 t/y. Technology Readiness Level is very high and can be considered about TRL8/IRL8. In the 3R process the bone grist is heated to as high as 850 °C in the carbonization kiln [10]. Temperature is higher than usual biomass processing temperatures, but it is needed to get high quality products. The main advantage of the technology is represented by feed flexibility. A wide range of different types of organic material streams can be fed to the reactor which is based on a rotary kiln.

12.1.8 Milena Gasifier

The Milena gasifier design uses the coupling of two reactors: a bubbling fluidised bed (BFB) combustor and a riser reactor. The BFB combustor is used to provide heat externally to the riser reactor where gasification take place. For this reason, the Milena gasifier is an allothermal gasifier [11]. The selection of a riser reactor for the gasification process has a positive effect on cold gas efficiency compared to a BFB, because less dilution gas is introduced into the gasifier. Fluidization gas is required to fluidize the bottom part of the riser, not to create the velocity required for vertical transport of the bed material. The amount of required fluidization gas is mainly influenced by reactor area and this is much smaller for a riser than a BFB reactor. The velocity in the riser required for vertical transport of the bed material originates by the fact that the gas produced during the devolatilization of the biomass adds pressure inside the riser and pushes the products of the process out of it.

Compared to other gasifiers, like circulating fluidised bed (CFB) or BFB gasifiers and downdraft gasifiers, the Milena gasifier has a higher cold gas efficiency (CGE). The Milena CGE in fact is about 80% while the other gasifiers can reach usually 70%. The advantage of the gasifier is that a higher efficiency is achieved at lower temperature and also that the heating value of the producer gas is relatively high, about (12–15) MJ/Nm^3 (dry gas basis), with a very low content of nitrogen. The syngas is cleaned with the OLGA system.

12.1.9 Güssing Gasifier

The combined heat and power (CHP) plant at Güssing in Austria is a FICFB (Fast Internal Circulating Fluidised Bed) steam gasifier that converts wood chips to a product gas with a heating value of approximately 12 MJ/Nm^3 (dry basis), see Fig. 12.5.

After passing through a cleaning section (two-stage gas cleaning system), the product gas is used as fuel in an internal combustion engine with a generator producing electricity and heat for the grid. If the engine is not in operation, the product gas can be burned in a boiler, producing only heat. The plant is characterized by a thermal input of 8 MW, electric output of about 2.0 MW, thermal output of about 4.5 MW and an overall efficiency of about 25%. The FICFB gasifier consists of two zones; a gasification zone and a combustion zone. The combustion zone provides heat through the bed material to the gasification zone. Steam is used as gasification agent; this implies a lower tar content, compared to air-blown gasifiers [13]. Olivine sand is used as bed material. The amount of tar in the raw product gas is

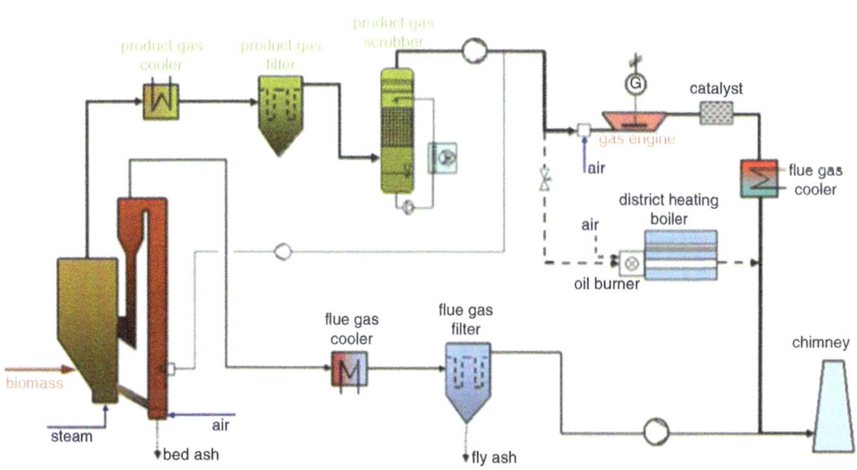

Fig. 12.5 Gasification CHP plant at Güssing. Reprinted with permission from [12]. Copyright © 2013, Elsevier

about (1500–4500) mg/Nm^3 (dry gas basis). Leaving the gasifier, the product gas is cooled down to a temperature of about (160–180) °C. Then it is passed through a fabric filter, removing particles and part of the tar. After the filter comes a scrubber which uses rapeseed methyl ester (RME) as scrubbing liquid. The spent scrubber liquid is recycled in the gasifier combustion zone. The final tar content of the gas is about (10–40) mg/Nm^3 dry gas. The exhaust gas of the engine is catalytically oxidized to reduce CO emissions. In January 2009, the plant had operated for more than 40,000 h since 2002 [14].

12.1.10 GoBiGas Gasifier

The Gothenburg Biomass Gasification (GoBiGas) plant produces 20.5 MW of bio-methane from an input of 32 MW of wood pellets.

The GoBiGas project currently comprises a 32 MW dual fluidized bed (DFB) gasifier (150 dry tons of biomass/day) coupled to a state-of-the-art synthetic natural gas (SNG) synthesis process that produces up to 20 MW of biomethane. The plant has run about 10,000 h [16–18], see Fig. 12.6 for the process steps.

The gasifier in the GoBiGas demonstration plant is one of two, third-generation dual fluidized bed gasifiers which originated from the 8 MW (40 dry tonnes of biomass per day) Güssing CHP plant, which dates back to year 2000. This gasifier was followed by the construction of a CHP plant of the same size in Oberwart, Austria [19–21]. The other third-generation gasifier is a 16 MW (80 dry tons of biomass/day) CHP plant in Senden, Germany. A gasification system similar to the one adopted by the GoBiGas is the one developed and built under the TIGAR trademark by IHI Corporation of Yokohama, Japan. This has been later scaled up to a 15 MW (30 dry tons of biomass/day) demonstration unit in Kujan, Indonesia and brought into operation in 2015 [22, 23]. In general, the GoBiGas process can be divided into three conversion steps Heat Generation (1); Gasification (2); and

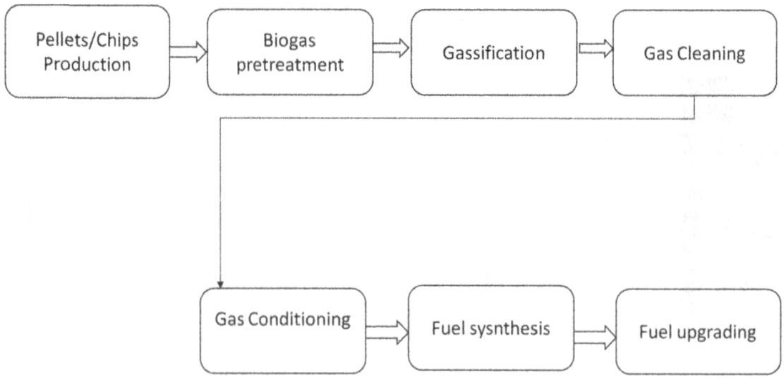

Fig. 12.6 GoBiGas process steps. Reprinted with permission from [15]. Copyright © 2016, Elsevier

Synthesis (3). In addition, there are 2 units for compression (4) and BTX removal (5).

12.2 Production of Pressurized Syngas During Pyrolysis

To produce compressed gas there are two strategies: (i) compression after thermal treatment or (ii) compression during the thermal treatment. The pyrolysis process is influenced by many parameters such as final temperature, heating rate, dimensions of the biomass particles, residence time and pressure. The influence of pressure on pyrolysis product yields and composition has been not fully considered in many studies [24]. In the study of Mahinpey et al. the effect of pressure on the pyrolysis of wheat straw in a tubular reactor is considered and reported [25]. A heating rate of 12 °C/min is considered. Experiments have shown that reactor pressure can influence significantly both yields and quality of the obtained pyrolysis products. However, the work of [25] considered a narrow range of pyrolysis pressures, that is, (0.689–2.758) bar. Mahinpey et al. inferred that 1.379 bar is a favorable pressure for the pyrolysis process of wheat straw in a tubular reactor, to obtain syngas.

Other researchers [26] have worked with an entrained flow pyrolysis reactor fed with pulverized wheat straw and working at pressures of (10 and 20) bar and temperatures of (700–1000) °C. Results have shown that the product yields in this case are not greatly influenced by the operating pressure.

Whitty et al. focused their study on the effect of pyrolysis pressure on char [27] and noted that with an increase in pressure, the particle size of the char obtained from kraft liquors decreased. These results are not as in Mahinpey et al. [25], who used wheat straw as a feedstock.

While there are many studies on the influence of pressure on biomass pyrolysis products, even more studies are available on coal pressurized pyrolysis. Roberts et al. [28] found that.

1. yields of coal pyrolysis products (liquid and gaseous) decreased with an increase in pressure, even though the rate of increase was not easy to predict;
2. large differences were found in the morphologies of the chars produced at different pressures;
3. char porosities were greatly influenced by pressure inside the reactor along with char morphologies [29].

The results obtained for coal cannot be easily translated to biomass because of physical and chemical characteristics, E.g. the bonds between carbon atoms in coal and in biomass substrates are very different. The major components of biomass are cellulose, hemicellulose and lignin. While cellulose and hemicellulose are characterized by glycosidic linkages between the sugar molecules, lignin is characterized by ether linkages between the aromatic components and between the aromatic functionalities and the phenylpropane components. The energy needed to break these bonds of (380–420) kJ/mol is lower than that needed to break the polycyclic aromatic hydrocarbon structures which are characteristic of some coals. These bonds in fact require about 1000 kJ/mol of energy to be cleaved [30]. These large differences in bond energies that are encountered in the pyrolysis of biomass, compared with that of coal can be studied by analyzing the char, its morphology and active sites. In this way also the effect of pressure can be better understood [31]. In the work of Cetin et al. [31] pressurized pyrolysis tests were used with a wire mesh reactor working at high heating rates (500 °C/s), which can reach an internal pressure of about 100 bar.

A typical way to study char morphology is with scanning electron microscopy (SEM) in which charcoal structure can be analyzed instead of or as supplement to Brunauer-Emmett-Teller (BET) porosity analysis.

Many works in the literature have reported that pressure has an important effect on the composition of pyrolysis products. Pressure is an important effect on the porosity of charcoal. In fact, with an increase in pressure, volatiles escape from the pores slower and so will have a longer residence time inside the pores, producing secondary char and reducing porosity which is a phenomenon that can be detected using NMR spectroscopy and indicates that chars formed at high pressures have an aromatic structure that tend to fuse and collapse in presence of high pressure. Chars formed under high pressure conditions have usually higher carbon content, due to secondary char formation. Figure 12.7a, b show SEM images of samples of char produced from Radiata pine pyrolysis at different pressures [31]. Char samples reported in Fig. 12.7 have been produced using a wire mesh reactor operating under a nitrogen atmosphere. As shown in Fig. 12.7a, char particles have been generated at pressures of (5, 10, 20) bar. It can be seen by comparing Fig. 12.7a-c that generally the cross section image of charcoal obtained at high pressure is characterized by particles that have larger cavities and thinner cell walls, compared with chars that have been generated at low or atmospheric pressure. If pressure is applied, bigger particles are obtained, which are characterized by a perforated surface structure. So with an increase in temperature, the porosity of charcoal enlarges, leaving space to large voids inside the particles and making cell walls thinner. This means that microporosity is reduced while macroporosity increases with an increase in pressure.

Similar results to those shown in Fig. 12.7 have been obtained with eucalyptus wood [31]. These tests have been performed at 20 bar pressure at different heating rates in the same wire mesh reactor. It has been noted that when high heating rates are used the char particle melts down and loses completely its structure. Regarding pyrolysis gas composition, Porada [32] shows analysis of the kinetics of gaseous products evolved during coal pressurized pyrolysis, namely, methane, ethane, ethene, propane, propene and hydrogen at (0.1, 2.5, 5.0, 10) MPa. The experiments were performed by heating a 1 g sample of coal with particle size comprised between (0.8 and 1) mm. The heating process was performed in a quartz tube reactor in argon atmosphere, starting from ambient temperature and reaching up to 1200 K at a heating rate of 3 K/min using electric heaters. Pyrolysis gases were cooled down to enable tar and water to condense. The noncondensable gases at the exit of the condenser were analyzed with the use of gas chromatography: an HP 5890-A type

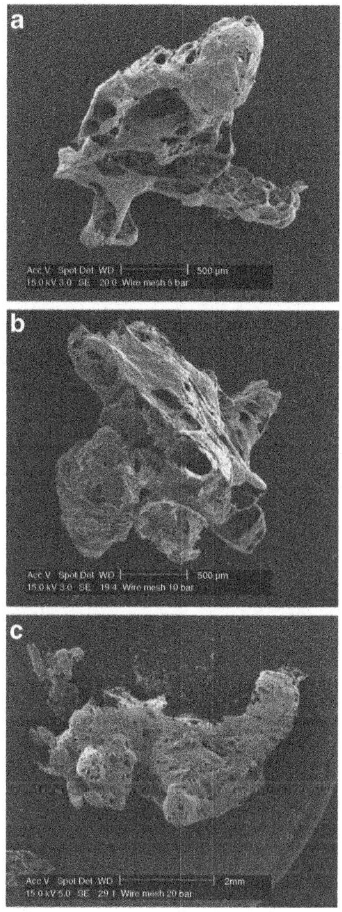

Fig. 12.7 SEM analysis of sawdust pine chars produced under pressure (**a**) 5 bar, (**b**) 10 bar, (**c**) 20 bar. Reprinted with permission from [31]. Copyright © 2005, Elsevier

gas chromatograph (GC) equipped with FID for analysis of C_1–C_3 hydrocarbons and another GC equipped with TCD to measure H_2, CO and CO_2. Analysis of the pyrolysis gases have shown that an increase in pyrolysis pressure gives an increase in methane yield and a reduction of hydrogen yield where the C_2 and C_3–hydrocarbons are given by single reactions. With an increase in pyrolysis pressure, an important decrease in the yield of CO and CO_2 occurs.

12.3 Syngas Compression

12.3.1 Compressors Classification and Selection

As reported by Blackmer [33] for compressor selection, compression ratio is a key parameter as defined by:

$$R = P_d / P_s \tag{12.1}$$

where P_s represents the absolute suction pressure and P_d represents the absolute discharge pressure. Usually the identification of the compression ratio is the first step in compressor selection, which involves globally the following steps:

1. Calculation of compression ratio;
2. Selection of a single-stage or a two-stage compressor;
3. Calculation of discharge temperature;
4. Determination of volumetric efficiency;
5. Determination of required piston displacement;
6. Selection of compressor model;
7. Determinaton the minimum rpm required of selected compressor;
8. Selection of actual rpm;
9. Calculation of actual piston displacement;
10. Calculation of power required;
11. Selection of appropriate options.

Details on how to address each step are proposed in ref. [33]. Another important aspect to take into account in compressor selection is which type of compressor fulfills better the process requirements. Generally the goal of a compressor is to increase the static or inlet pressure of the gas and deliver it at the specified discharge pressure and flow rate. This can be done by different types of compressors. The two basic categories of compressors are: dynamic (centrifugal and axial) and positive displacement (reciprocating and rotary types). Table 12.2 shows a summary of typical operating characteristics of compressors. Dynamic compressors are based on the principle that they provide velocity to a gas which passes through impellers or blades. The gas then exits the blades and enters into a stationary volume were its velocity is transformed into pressure. In centrifugal compressors the acceleration of the gas is obtained through the action of one or more rotating impellers, while in axial compressors we find both rotating and stationary. The shaft rotates a drum inside a casing, which is stationary. Between the drum and the casing rows, airfoils are placed, which are either connected to the drum or the casing.

In positive displacement compressors the compressing action is due to a mechanical part which reduces the volume of a chamber. As shown in Table 12.2, an important parameter of the compressor is the efficiency. Different types of efficiencies can be considered, as explained in the next section.

12.3.2 Compression Efficiency

For a compressor the efficiency can be defined as the ratio of output work (head) to input work (shaft power) of a system [35]. The difference between the two works is due to friction and results in a higher temperature at discharge. Assuming negligible

Table 12.2 Typical operating conditions of compressors [34]

Type	Inlet capacity (m^3/h)	Maximum discharge pressure (bar)	Adiabatic efficiency (%)	Operating speed (rpm)	Maximum power (MW)	Application
Dynamic compressors						
Centrifugal	170-850,000	690	70-87	1800 – 50,000	38	Process gas & air
Integrally geared centrifugal compressors	500 – 500,000	350	80	7000 – 50,000	60	Process gas & air
Axial	50,000 – 850,000	17	87-90 +	1500 – 10,000	75	Mainly air
Positive displacement compressors						
Reciprocating (piston)	20 – 34,000	4150	80-90	200 - 900	15	Air & process gas
Diaphragm	0 - 250	1400	60 - 70	300 - 500	1.5	Corrosive & hazardous process gas
Rotary screw (wet)	100 – 12,000	24	65-70	1500 – 3600	1.5	Air, refrigeration & process gas
Rotary screw (dry)	200 – 100,000	1 - 50	55 - 70	1000 – 20,000	6	Air & dirty process gas
Rotary lobe	25 – 50,000	0-3 – 1.7	55 - 65	300 – 4000	0.4	Pneumatic conveying, process gas & vacuum
Sliding vane	15 – 5000	10	40 – 70	400 – 1800	0.35	Vacuum service & corrosive process gas
Liquid ring	10 – 17,000	5.5 – 10.5	25 - 50	200 – 3600	0.3	Vacuum service & corrosive process gas

heat transfer, minimal velocity effect and an isentropic process, the work can be calculated as follows:

$$W = c_p(T_2 - T_1) \tag{12.2}$$

which is derived directly from the general energy equation given by the first law of thermodynamics. In Eq. (12.2), W is the work, c_p is the specific heat of the compressed material and T_1 and T_2 are respectively the inlet and the outlet temperature. So, based on this definition, the isoentropic (or adiabatic) efficiency results from Eq. (12.3).

$$\eta_{\mathrm{ad}} = c_p(T_{2\mathrm{ad}} - T_1)/c_p(T_2 - T_1) \tag{12.3}$$

where $T_{2\mathrm{ad}}$ represents the isentropic compression temperature and η_{ad} represents the isentropic efficiency. For an adiabatic process for a gas, the following equation applies:

$$\frac{T_1}{T_2} = \left(\frac{P_1}{P_2}\right)^{(k-1)/k} \tag{12.4}$$

where k is the ratio between specific heats and it is a constant, P_1 is the pressure at the inlet and P_2 is the pressure at the outlet. Through mathematical calculation, the denominator of Eq. (12.3) can be written as:

$$T_2 - T_1 = T_1\left(\frac{T_2}{T_1} - 1\right) \tag{12.5}$$

Substitution of Eq. (12.4) into (12.5) gives:

$$T_2 - T_1 = T_1\left[(r_p)^{(k-1)/k} - 1\right] \tag{12.6}$$

where r_p is the ratio between P_2 and P_1. If Eq. (12.6) is substituted into Eq. (12.3), the equation for isentropic efficiency changes into:

$$\eta_{\mathrm{ad}} = \frac{T_1\left[(P_2/P_1)^{(k-1)/k} - 1\right]}{T_2 - T_1} \tag{12.7}$$

The adiabatic or isentropic efficiency is useful to determine the power of the compressor and its performance. Nevertheless, the polytropic efficiency gives a more accurate efficiency for the compression of an ideal gas. For real gases instead it has to be considered that k is not always constant.

Once compression efficiency has been calculated, overall efficiency has to be taken into account [36], defined as the ratio between the adiabatic power (W_{ad}) to the shaft power (W_{sh}), Eq. (12.8):

$$\eta_{\text{overall,ad}} = \frac{\dot{W}_{\text{ad}}}{\dot{W}_{\text{sh}}} \tag{12.8}$$

If a compressor system consists of a compressor, a motor, a controller and other devices, the efficiency of the system has to take into account also other losses (*e.g.* electrical), that gives:

$$\eta_{\text{sys}} = \bar{\eta}_{\text{overall,ad}} \cdot \eta_{\text{motor}} \cdot \eta_{\text{controller}} \cdot \eta_{\text{auxiliary}} + \ldots . \tag{12.9}$$

12.3.3 Syngas Compression in IGCC Plants

Gaseous fuels derived from thermochemical treatment of biomass, coal and waste are commonly called "syngas", "pyrogas" or "producer gas". Usually the word "syngas" is used when the gas final use is the synthesis of useful final products that can be a biochemical or a fuel. Dry syngas is usually composed of the following gaseous compounds: H_2, CO, CO_2, CH_4, N_2, C_xH_y, traces of tar and water vapor. This gas has, in many cases, a low calorific value and can be burnt in turbines and microturbines. If coal is gasified to produce syngas and then combusted to produce electricity, this results in a more clean process than combusting coal directly as a solid fuel. Syngas can be used also as a raw material, to produce hydrogen and methanol. Gasification is the first process encountered in Integrated Gasification Combined Cycle (IGCC) power plants, which can emit lower quantities of SO_x, NO_x, particulate matter and heavy metals, and have high electrical efficiency [36], compared to combusting the solid fuel directly.

Although IGCC plants have many advantages, they are hampered by economic and technical barriers because it is difficult to operate these plants under fast changing loads. A possible solution to this problem is represented by the compression of the produced syngas and its storage. This technology can grant more flexibility to the plant and widen its field of operation. [37]. If the syngas can be successfully stored, this can help to use it in the power plant only during peak-load demand and so having higher prices per unit of electricity produced from the distribution system operator [38]. So syngas compression can permit operation of a plant when the price of electricity is more convenient and this increases the IGCC plant profitability (see Fig. 12.8). The compression process is simple but at the same time needs large volumes for storage if the compressing pressure is low, (10–20) bar.

Both volumetric and centrifugal compressors can be used for syngas compression. Volumetric compressors can be used with different ranges of mass flow and

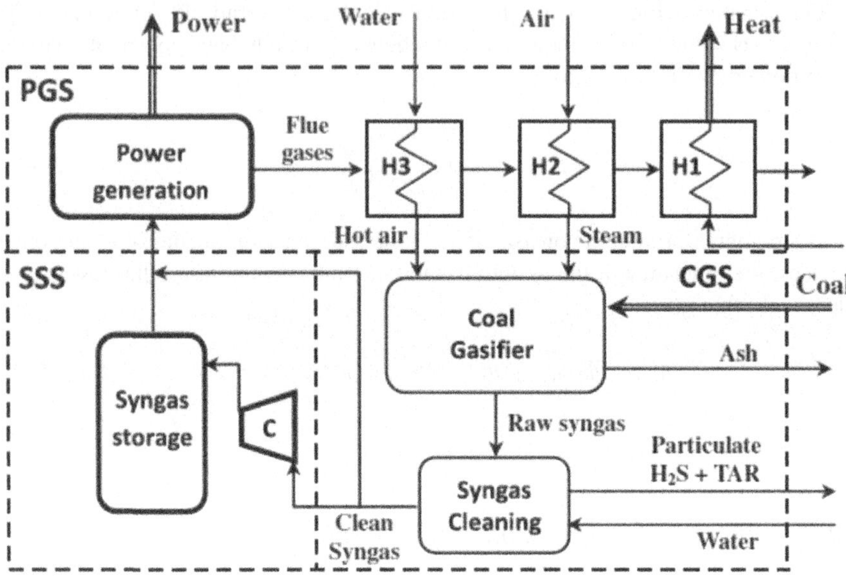

Fig. 12.8 Coal gasification plant with syngas storage. Reprinted with permission from [38]. Copyright © 2012, Elsevier

different compression ratios. Centrifugal compressors can be operated with higher efficiencies [39]. The two parameters to be considered when designing a syngas storage facility are: power demand and syngas storage capacity. These are directly influenced by the peak energy demand and the conversion efficiency of the IGCC plant [40]. An optimum volume has to be found, given that with the decrease of the storage volume the energy consumption for compression increases (Fig. 12.9).

Once the syngas has been produced and stored, to transform it into energy, it has to be burnt in gas turbines, internal combustion engines, or combined cycles that couple a gas turbine with a steam turbine recovering the residual heat contained in the exhaust combustion gases, which have expanded in the gas turbine. Many researchers have studied the energy and economic performance of different prime movers (see Cau et al. [38]). In this study, different prime movers (i.e gas turbines and internal combustion engines) are analyzed when integrated into a small and medium size coal gasification power plant, in which a syngas storage section is present. The results of the comparison are shown in Table 12.3. The data shown in the Table 12.3 are the result of plant optimization realized in the software Aspen Plus™. Plant performance and load modulation capacity are linked deeply with two parameters: syngas storage ratio and gasifier capacity ratio. The syngas storage ratio is the ratio between the syngas produced in the gasifier section and the syngas stored in the syngas storage section. The gasifier capacity ratio is defined as the ratio between the coal energy input of the IGCC and the coal energy input of the base load coal gasification plant. The power ratio defined in the Table 12.3 is the ratio between the base load power and the peak load power. The peak load energy ratio is

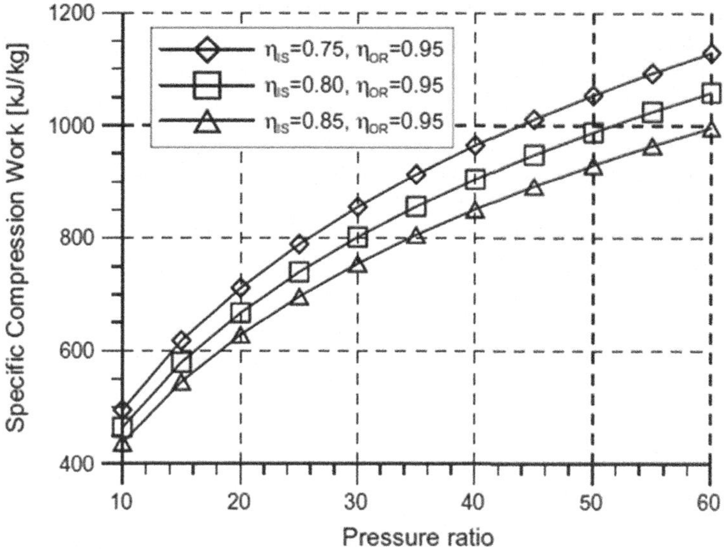

Fig. 12.9 Energy consumption for syngas compression in function of pressure ratio and isentropic efficiency. System efficiency (η_{SYS}) takes into account the mechanical and electrical energy losses and is assumed to be 95% Reprinted with permission from [38]. Copyright © 2012, Elsevier

Table 12.3 Main performance indicators of Integrated Gasification Combined Cycle power plants with gas turbines (GT) or internal combustion engines (ICE) [38]

	IGCC-GT	IGCC-ICE
Power ratio	0.6	0.6
Peak-load energy ratio	0.1333	0.1333
Gasifier capacity ratio	1.241	1.201
Syngas storage ratio	0.065	0.075
Overall efficiency (%)	20.6	32.3
Coal energy input (kW)	10980.0	7028.0
Syngas mass flow rate (kg/s)	1.541	0.986
Stored syngas (t/day)	8.65	6.38
Storage volume (m^3)	466.7	344.0
Compressor power (kW)	81.6	60.1
CGS power consumption (kW)	140.1	89.7
PGS power output (kW)	3473.4	3423.0
Base-load power output (kW)	2000.0	2000.0
Peak-load power output (kW)	3333.3	3333.3

the ratio between peak-load and base-load. The Table 12.3 shows that the IGCC configuration using a GT has significantly lower efficiency. This can be explained by the fact that at off-design conditions the efficiency of the gas turbine decreases significantly. At design conditions the turbine in fact has an electrical efficiency of 27.1% and a thermal efficiency of 59%. The thermal efficiency is calculated assuming cooling down the exit gases from the turbine from a temperature of 594 °C to a temperature of 120 °C, but what happens in practice is that the turbine works at design conditions only at peak load. At off-design conditions the performance of the GT results is influenced by the load rate. So what actually happens is that in off-design conditions the performance of the GT decreases significantly, contrary to an internal combustion engine.

The performance of the internal combustion engine is evaluated based on literature data for capacities ranging between (1–5) MWe. The thermal efficiency of the ICE is determined based on heat recovery from exhaust gases, lubricating oil and engine cooling mediums. Generally, the ICE has a higher efficiency at part load compared to the gas turbine.

Once the syngas is compressed before it can be used in gas turbines. Here the issue is how to maintain an efficiency comparable to the one which can be achieved with natural gas, using syngas and without changing turbine materials and design too much. As it is shown in the study of He et al. [41], gas turbines (GT) are designed to be fed with natural gas; if they have to be fed with gases with a lower heating value, the input flow has to be increased, to obtain a Wobbe index which is similar to that of natural gas. The Wobbe index is defined as the ratio between the higher heating value of the gas and the specific gravity. On one hand, if syngas is to be fed to a gas turbine, both the inlet flow and the inlet pressure have to increase; on the other hand, if the pressure ratio is too high, there is the risk of encountering instability at the compressor side and too high temperature on the turbine blades.

To respond to the problems caused by the increase of the syngas mass flow other measures can be adopted, such as: air bleeding and fuel dilution [42, 43]. Air bleeding is compressed air extracted from the compressor, before it arrives in the combustion chamber. Usually bleed air can be used to cool down the gas turbine blades, and to move pneumatic actuators, for example. In this study [42] it is assumed that air bleed is used in an air separation unit (ASU), the nitrogen obtained from the ASU is fed to the combustion chamber to dilute the fuel. An example of plant layout, where the bleed air is fed to an ASU (which is also fed with the air supplied by the auxiliary air compressor) is shown in Fig. 12.10. The performance of the considered plant can be evaluated based on two fundamental parameters: air bleed ratio and integration degree.

$$\text{Air bleed ratio} = \text{Air to ASU from GT}/\text{Air at GT inlet} \tag{12.10}$$

$$\text{Integration degree} = \text{Air to ASU from GT}/\text{Total air to ASU} \tag{12.11}$$

Both air bleeding and integration decrease the quantity of exhaust gases flowing through the turbine expander, therefore they reduce the adaptations required in case of an important increase in the volumetric flow through the expander. Many papers have shown that for IGCC plants the recommended integration ratio ranges between (25–30) %, which provides the best balance between the maximum plant electricity production and the efficiency and it does not compromise the reliability and availability of the plant [44, 45]. Generally speaking, in a conventional IGCC plant, the amount of air required at the ASU is about (20–25)% of the air that flows into the

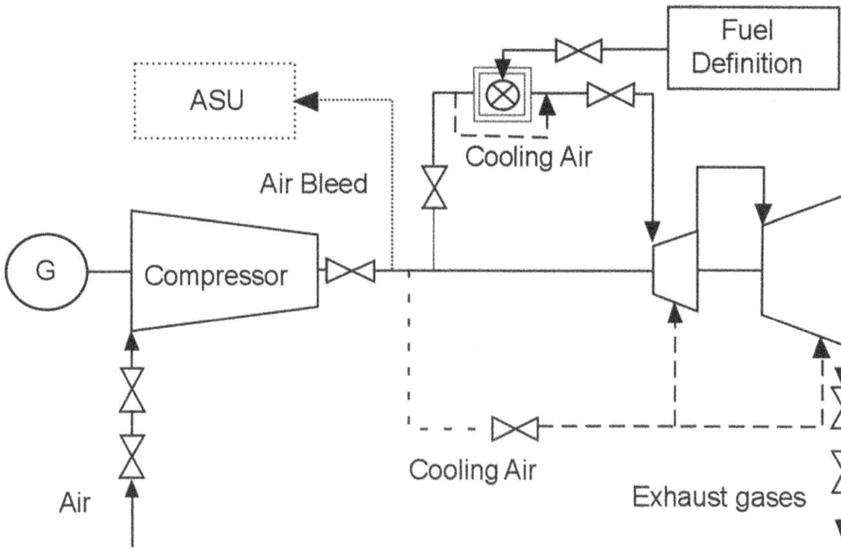

Fig. 12.10 Gas turbine with air cooling arrangement. Reprinted with permission from [41]. Copyright © 2012, Elsevier

GT. The air required by the ASU is directly determined by the oxygen flow required by the gasification reactor. So, an integration degree of (25–30)% determines an air bleed ratio of about (5–7)%.

The injecting of compressed nitrogen coming from the ASU into the fuel gas before combustion is considered a fuel dilution technique; which has numerous advantages, among them the reduction of NO_x emissions and the increase in power production [46].

A possible layout of a gas turbine with air bleed used as a cooling medium is shown in Fig. 12.10. The plant shown in the figure is basically composed by a compressor, a combustion chamber and a turbine expander. The compressor provides two output flows of bleed air which is used as a coolant: one before the turbine inlet and the other behind the first expansion vane. The air entering at the turbine inlet is assumed to do mechanical work, while the air entering after the first vane produces work only in the second rotor [47].

The work of He et al. 2010 [41] calculates the turbine operation window in the case of the plant shown in Fig. 12.10. The window shows the power production from the turbine at different flame temperatures. The change in flame temperature is calculated referring to the operating parameters of the turbine, referring to a E-class GE gas turbine power plant, with the following characteristics: 166.5 MW (power output), 17.6 compressor pressure ratio, 1100 °C turbine inlet temperature, 524 °C turbine exhaust temperature, net efficiency equal to 35.7%.

The operation window of a plant fed with CO-rich syngas, is shown in Fig. 12.11. Figure 12.11a shows the operation window without air bleed and Fig. 12.11b shows the operation window with 5% air bleed.

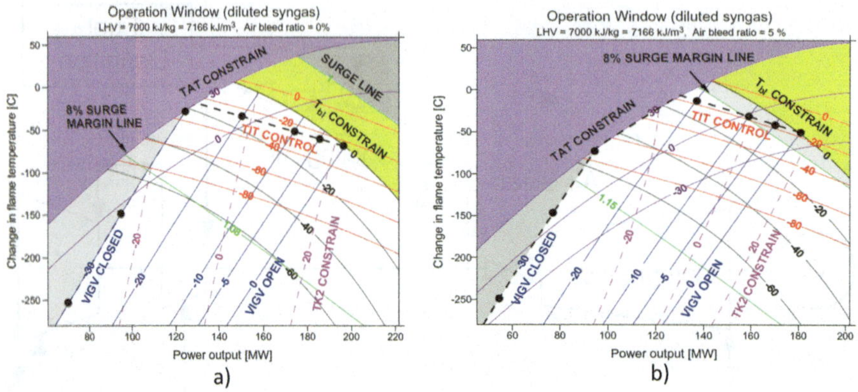

Fig. 12.11 (**a**) Operation window of a diluted CO-rich syngas fired gas turbine without air bleed; (**b**) Operation window of a diluted CO-rich syngas fired gas turbine with 5% air bleed. Reprinted with permission from [41]. Copyright © 2012, Elsevier. Line definitions. (red line): Turbine inlet temperature (TIT); (black line): Blade temperature (Tbl); (pink dashed lines): compressor outlet temperature (TK2) is represented by pink dashed lines; (purple lines): turbine exhaust temperature (TAT)

The operational field of the gas turbine at full load is comprised between the 'VIGV = 0' and the "Tbl = 0" contour lines. Regarding the Variable Inlet Guide Vane (VIGV), when it is equal to zero it represents the position at the design point, while when it is equal to 30 it refers to a fully closed position where the compressor inlet mass is reduced by 30%.

Compared to the case where the GT is fed with natural gas, given that the syngas has a lower LHV and assuming that the air flow in the compressor is constant, this implies an increase in the fuel mass flow rate to maintain the same flame temperature. Increasing the syngas mass flow rate will cause also the increase of the hot gas mass flow and so the heat output at turbine exit.

Secondly, due to the fact that the volumetric flow of the gas turbine (also known as swallowing capacity) can be changed in a very limited way, the increase in the exhaust gases mass flow can be done only by increasing the compression ratio. If the compression ratio increases then the temperature of the cooling flow extracted from the compressor also increases. In Fig. 12.11a, it can be seen that if the gas turbine is operated with the design Tbl, the flame temperature will be reduced significantly. This implies also a reduction of 22 °C for the TIT. It can be seen also that the operating point of the gas turbine approaches the surge line. This is mainly due to the significant increase in the compression ratio. Another important change is that the Turbine Exhaust Temperature (TAT) decreases of about 20 °C, because of the increased turbine expansion ratio. This has as a consequence also an increased power production of about 16.6%, compared to the design value.

Figure 12.11b shows the results for the scenario using 5% of bleed air extracted from the compressor to produce oxygen in the ASU, this layout is quite common in

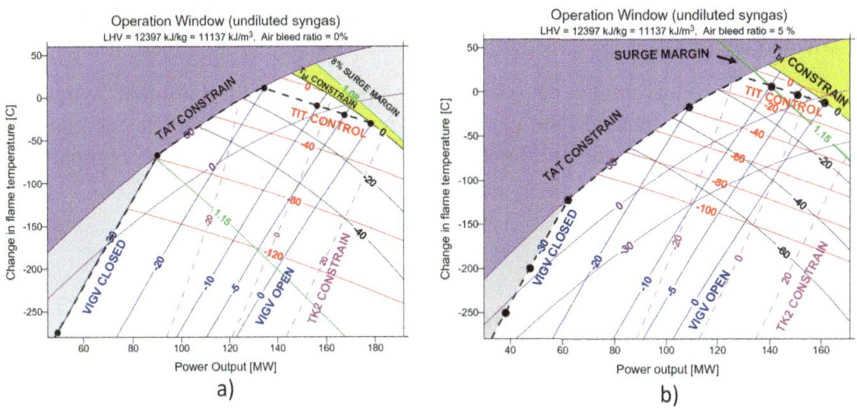

Fig. 12.12 (**a**) Operation window of an undiluted CO-rich syngas fired gas turbine without air bleed; (**b**) Operation window of an undiluted CO-rich syngas fired gas turbine with 5% air bleed. Reprinted with permission from [41]. Copyright © 2012, Elsevier. Line definitions. (red line): Turbine inlet temperature (TIT); (black line): Blade temperature (Tbl); (pink dashed lines): compressor outlet temperature (TK2) is represented by pink dashed lines; (purple lines): turbine exhaust temperature (TAT)

an IGCC plant with partial integration. This new layout is better performing and has many advantages compared to the previous configuration, among them:

- an increase in the compressor surge margin;
- in this case a big reduction of the flame temperature is not needed;
- the decrease in the Turbine Inlet Temperature (TIT) is also smaller.

When the gas turbine is operated at full capacity, the power output increases by 8% from the design-point value.

If the plant is fed with undiluted CO-rich syngas we will have the operational maps shown in Fig. 12.12a, b. Compared with the case of dilution, non-dilution has a larger operation window. The following positive effects can be noted:

- first of all, turbine blade cooling is remarkably improved;
- the reduction in the TIT, which is necessary to prevent overheat of the turbine blades is less intense. In fact a smaller reduction (about 5 °C) in the TIT is required at full load for a gas turbine with 5% air bleed, compared to the 10 °C which are needed for a gas turbine without air bleed.
- the compressor surge margin is further enlarged. In fact this parameter is about 8% for a gas turbine without air bleed and about 15% for a gas turbine with 5%air bleed.
- the plant can be controlled based on a constant turbine exhaust temperature (TAT) control mode in the load range comprised between 51% and 77% for a gas turbine without air bleed and from 39% to 78% for one with air bleed. This allows to have a relatively higher temperature of the exhaust gases in a wider load range. This

can increase the amount of heat which is recovered by the HRSG (heat recovery steam generator).

The main disadvantage caused by the operation with undiluted syngas is represented by the fact that the gas turbine has limited possibility to increase the power output given that the undiluted fuel has a smaller mass flow rate compared to the diluted gas.

12.3.4 Syngas Ignition Behavior during Compression

Micro-turbines and gas turbines have been tested using different biofuels in many works reported in Literature [48]. Among them: bio-ethanol, bio-methanol, straight vegetable oils, bio-diesel, biogas, hydrogen and pyrolysis oils, syngas, DME.

Despite some difficulties in using syngas in gas turbines, many experiments have been performed in various plants all over the world, especially in the United States [49–51]. Syngas is mainly composed of mixtures containing different concentrations of carbon monoxide (CO) and hydrogen (H_2) [50]. The variation in syngas composition can complicate the turbine design and operation. Hydrogen can increase the temperature in the combustion chamber, leading to important nitrogen oxide (NOx) emissions. The actual dry low-NOx gas turbine technologies cannot easily support high mass flow rates and fuel concentrations between 15% and 40% [49, 51]. Thus, firing diluted syngas, typically using nitrogen (N_2) or steam in the dilution, can be a good approach [52]. Gas turbines operated in lean premixed combustion conditions can overcome the problem of high emissions and low efficiency. The new concept for syngas turbines is focused on such premixed systems. For premixed operation, there could be concerns over pre-ignition, flashback and safety and performance problems.

Some important publications on the use of H_2 and CO_2 mixtures on GTs are represented by [52–55]. There are also many studies on developing revised H_2 and CO reaction mechanisms which will be explained in Sect. 12.4.1 [56–58]. However, lack of data at elevated pressures and temperatures and use of non-diluted fuel/air mixtures make reaction kinetics of syngas blends widely unexplored at conditions relevant to gas turbine applications.

Walton et al. [59] gave an important database of combustion kinetic benchmarks for syngas combustors at typical conditions of temperature and pressure. The study has been done to understand properly the ignition behavior of the simulated syngas mixtures as a function of combustion conditions and syngas composition. To analyze high temperature and high pressure combustion phenomena the use of rapid compression facilities (RCF) can be beneficial. These are test benches which are used also for internal combustion engines, because their piston compresses the test-gas mixture in the chamber, in a similar way to what happens in an internal combustion (IC) engine, leading to a very rapid increase in pressure and temperature.

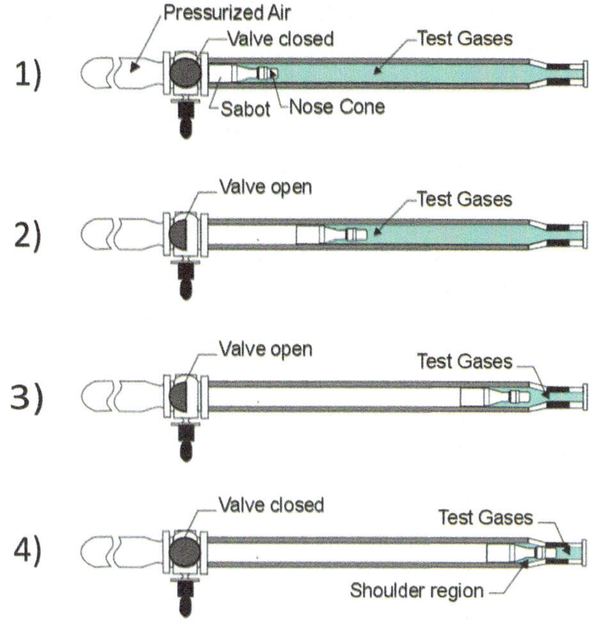

Fig. 12.13 Depiction of the Michigan University rapid compression facility in operation. Reprinted with permission from [60]. Copyright © 2004, Elsevier

These facilities can be used also to analyze the behavior of gas turbines combustion chambers.

Different tests have been realized at the rapid compression facility (RCF) of the Michigan University (see Fig. 12.13) [60].

The UM-RCF is mainly composed of the following:

- a driven section, that is where the syngas is contained and compressed;
- the test manifold, where the compressed syngas is stored at uniform high pressure and high temperature conditions;
- the driver section, that is where compressed air is contained and used to drive the sabot;
- the hydraulic control valve assembly; which liberates the compressed air;
- the sabot (free piston), which is driven by compressed air.

Pressurized gas is used by the UM-RCF to move the sabot down the bore of the driven section. The test syngas is loaded in front of the sabot, into the driven section and then it is compressed when the sabot traverses the driven section length.

The phases in which a compression test is divided are:

1. loading of the gas in the driver section. Driver section is loaded with pressurized air.
2. The hydraulic valve is opened to allow compressed air to exit the driver section and to push the sabot along the length of the driven section.
3. the sabot moves through the driven section compressing the syngas ahead of the sabot.

4. the sabot arrives to the end of its movement and traps the syngas in the test manifold.

In detail it can be said that before starting each test a diffusion pump makes the void in the driven section. To begin the experiment the hydraulic control valve and a sheet of polyester film (0.05 mm thick, Mylar®) are used to separate the driver and driven sections. When the driven section is charged with the gas mixture, the hydraulic control valve opens and the high-pressure gas, coming from the driver section, breaks the polyester film, enters the driven section and accelerates the sabot. In the driven section, the test gas mixture compression is done in front of the sabot. When, the sabot nose cone seats by an annular fit situated in the test manifold wall, the compressed test gas is sealed within the test manifold. The compression process can be considered isentropic.

The University of Michigan (UM) used the rapid compression facility (RCF) to perform ignition experiments to analyze the chemical kinetics related to the syngas combustor operation. The UM-RCF is an effective experimental apparatus that can be employed [60] to create constant high-temperature (T = 500–3000 K) and high pressure (P = 0.5–60 atm) conditions. The test facility recreates an environment which is typical of GTs working conditions. The test bench can be used to analyze the ignition delay time (τ_{ign}) of the syngas. This parameter is the most important kinetic characteristic of the combustion process and it is strongly influenced by the syngas chemical composition and also by the temperature and the pressure conditions.

The τ_{ign} value obtained from the experiments is of fundamental importance to indicate the extent of the reaction kinetics and to verify detailed, skeletal and reduced kinetic schemes [56, 58, 61].

With the RCF test bench different kinds of syngas with different H_2:CO ratios (ranging from 0.25 to 4.0) can be tested. Also different syngas dilution rates can be tested by mixing N_2 with syngas; so that the behavior of the syngas can be analyzed at high pressures and lean or stoichiometric conditions. These experiments are of paramount importance to design modern gas turbines. In the study of Walton et al. 2007 [59], to simulate the ignition behavior of syngas mixtures consisting of CO and H_2, different parameters have been analyzed, such as: the H_2:CO ratio, the equivalence ratio, the oxygen concentration, the pressure and the temperature. The pressures derivative time-history and the pressure time-history for both pure H_2 and pure CO ignition experiments of the ignition are shown in Fig. 12.14; those data can be compared with the data reported in [62].

The first peak in the pressure curve shows the end of the compression phase. The pressure remains practically steady for a while, after the gases are sealed in the test section. This stabilization phase is followed by a fast rise in pressure, which coincide with the ignition of the syngas (set as $t = 0$ s in Fig. 12.14). After the time of ignition the pressure data show an increase in the pressure, which is slow at first and then is followed by a rapid increase. The rapid increase in pressure is due to the presence of reaction fronts before the volumetric ignition. The τ_{ign} is given from every test as the time between Pmax at the end of compression and the maximum value of dP/dt

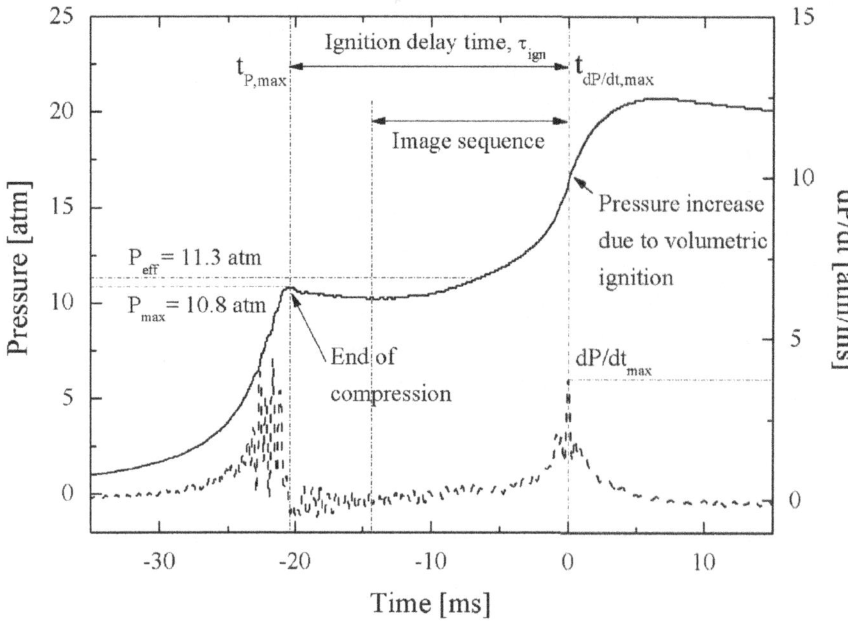

Fig. 12.14 Typical experimental results for pressure and pressure derivative time-histories for experimental conditions of $T_{eff} = 1004$ K, $P_{eff} = 11.3$ atm, H_2:CO ratio = 0.4, inert:O_2 ratio = 3.76, syngas fuel = 20% H_2, 80% CO. Reprinted with permission from [59]. Copyright © 2007, Elsevier

(Fig. 12.14). Repeating the tests in different conditions and with different syngas composition and measuring the ignition delay time for all the cases, a database of ignition times can be produced. Through regression analysis the ignition delay time has been calculated by Walton et al. 2007 [59], as shown in Eq. (12.12).

$$\tau_{\text{ign}} = 3.7 * 10^{-6} P^{-0.5} \phi^{-0.4} \chi_{O_2} * \exp\left(12,500/\overline{R}_{[\text{cal/mol/K}]} T\right) \tag{12.12}$$

where,

- P is the pressure;
- ϕ is the equivalence ratio;
- χ_{O_2} is oxygen mole fraction;
- $\overline{R}_{[\text{cal/mol/K}]}$ is the universal gas constant.

In the ignition tests previously reported in literature and performed on hydrogen and hydrocarbons, when shock tube and rapid compression facilities have been used, similar phenomena of creation of localized reaction fronts and propagation to the entire combustion chamber have been noticed [63–67]. This phenomenon is depending on syngas composition. The reaction front propagation is absent in the

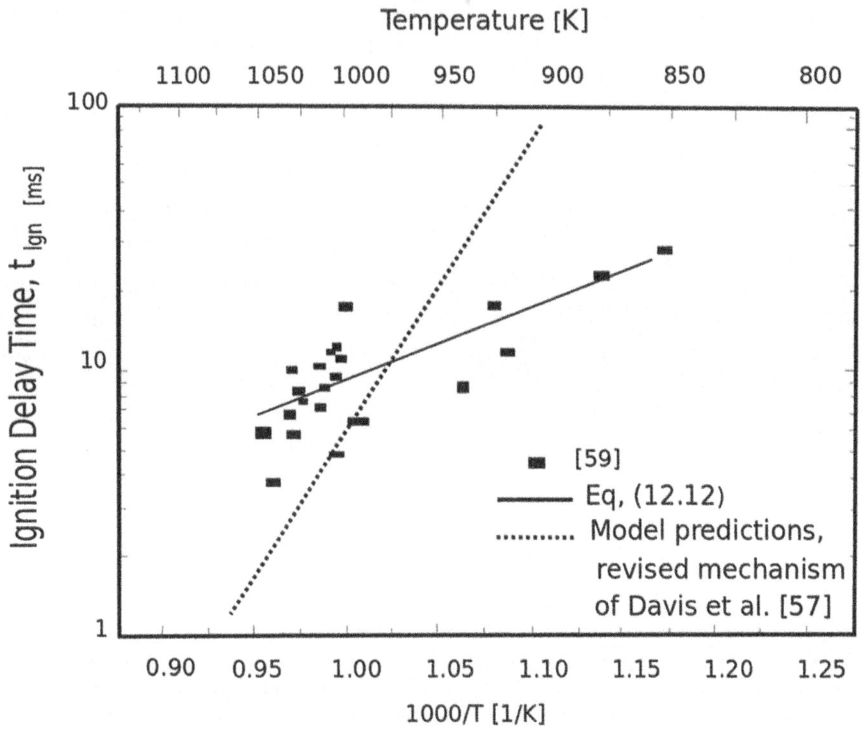

Fig. 12.15 Comparison of experimentally measured H_2 and CO ignition delay time with model predictions as a function of inverse temperature. The experimental data have been normalized to $P = 15$ atm, $\phi = 0.4$, $\chi_{O2} = 18\%$ using Eq. (12.12). Reprinted with permission from [59]. Copyright © 2007, Elsevier

case of lean mixtures (with fuel mole fractions below a critical limit), above a limit of fuel concentration ignition fast propagation is observed (Fig. 12.15).

Ignition delay time can be also described by the kinetic mechanism developed by Wang and co-workers [57, 68]. This derives from the mechanism previously developed by Davis et al. [57], which is based on the work of Mueller et al. [56].

The scheme of [56] was re-evaluated by Davis et al. [57], updating it with reaction chemistry and thermodynamic data to obtain results which were more correspondent to the experimental conditions.

Figure 12.15 shows the complete dataset of experiments performed by [50]. The results of ignition delay time are plotted against 1000/T. The results are also interpolated with Eq. (12.12). This kind of interpolation is also compared with the model of Davis et al. [57]. It can be seen that the interpolation had a better performance than the model of ref. [59].

12.3.5 Syngas Homogeneous and Heterogeneous Ignition

In combustion processes we can have premixed flames and diffusion flames. Diffusion flames in particular can be classified as homogeneous and heterogeneous. In homogeneous flames the characteristics are determined by the flow of gases and by interdiffusive phenomena; while in heterogeneous flames the characteristics are determined by the exchange of heat and material with the surface of the solid. An important study authored by Chaos and Dryer [69] has evaluated the recent research on chemical kinetics of syngas combustion. The study presents a critical comparison of the differences between kinetic model predictions and experimental measurements in high-pressure ignition tests.

Auto-ignition can be of two types: strong and weak [70]. Weak ignition is defined as non-uniform ignition of the mixture; while strong ignition is defined as a phenomenon which starts with a wave that propagates through the whole combustible mixture.

Voevodsky and Soloukhin were the first to study this phenomenon [71]. They arrived at the conclusion that the weak ignition can be explained by a kinetic phenomenon which is delimited by the second explosion limit. Thereby, the chemical kinetic process in determining the behavior of ignition is essential.

Meyer and Oppenheim [72] linked the weak ignition to the post shock temperature in the shock tube reactor. Assuming stoichiometric H_2/O_2 mixtures, the following equation for the ignition delay time can be inferred:

$$\left.\frac{\partial \tau_{\text{ign}}}{\partial T_5}\right|_{p_5} = -2(\mu\text{s/K}) \tag{12.13}$$

where T_5 and p_5 are temperature and pressure in the shock region. The interesting conclusion of Mewer and Oppenheim [72] is that weak ignition is a phenomenon linked with the dynamics of the gases and it is induced by perturbations to the flow field.

To understand better the results of [71, 72], Fig. 12.16 presents the data of the study of [71, 72] on weak and strong ignition of different syngases (obtained from mixtures of H_2 and CO in different concentrations). From Fig. 12.16 it can be seen that the sensitivity criterion developed by Meyer and Oppenheim (reported in dashed lines) performs better in predicting the weak ignition boundary than the second explosion limit, especially in the low-pressure region.

In the work of Mansfield and Wooldridge [73] an experimental campaign was conducted to analyze the ignition delay of syngas, differentiating between strong and weak ignition.

From Fig. 12.17 it can be seen that the strong ignition is located usually at temperatures which are higher than 1000 K. The strong ignition limit is not existent anymore at high pressures, where it has been assessed only strong ignition (i.e homogeneous ignition).

Figure 12.17 shows two explosion limits:

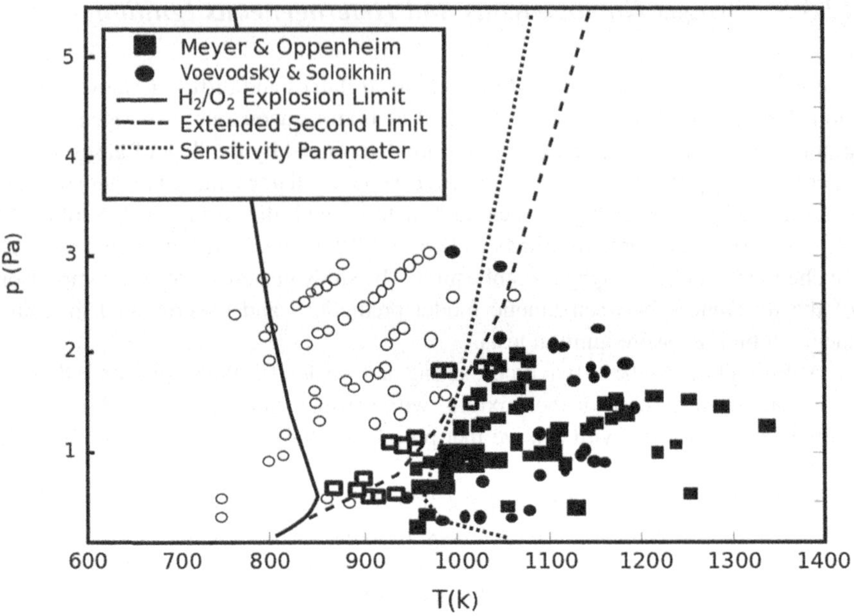

Fig. 12.16 Weak and strong ignition limits as a function of temperature and pressure for a H_2/O_2 stoichiometric mixture. Reprinted with permission from [70]. Copyright © 2014, Elsevier

- the second explosion limit and
- the extended explosion limit.

The second explosion limit represents the competition between the dominant chain-branching combustion pathway and the dominant chain-terminating pathway. The extended second explosion limit is defined as the state in which no net radicals (H, O, OH, HO_2) are produced. In this case also HO_2 chemical reactions are included, which are only significant at pressures higher than 1 atmosphere.

12.4 Combustion of Pressurized Syngas in Gas Turbines

12.4.1 Modeling Aspects

Syngas is highly variable in composition. Given that major gaseous components are always the same, i.e. H_2, CO, CO_2, CH_4, N_2 and eventually O_2; the percentage content of those gases can be very different, depending on thermochemical process conditions (final temperature, heating rate) and on the use of reagents, like steam. The variable composition of syngas influences substantially the combustion behavior both in steady state and transient regimens [74].

Generally syngas composition has an important influence on:

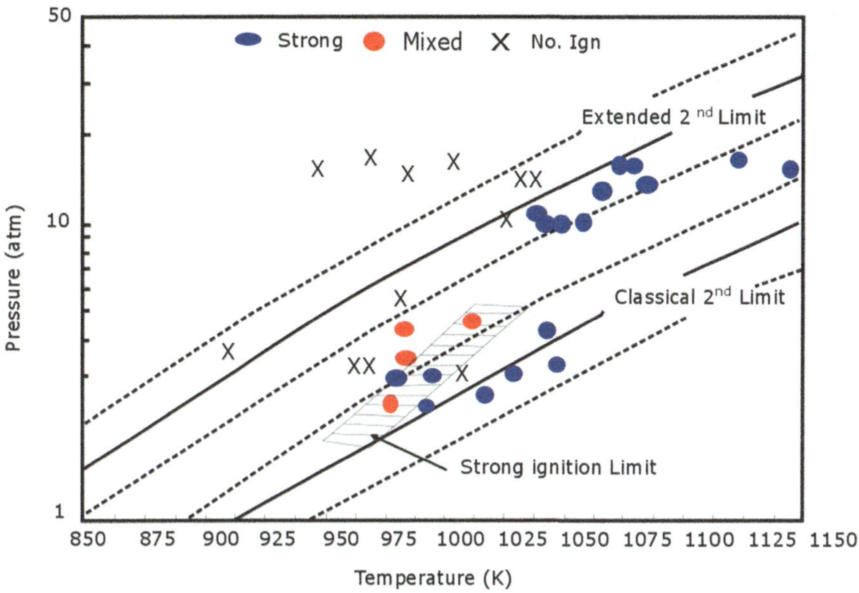

Fig. 12.17 Ignition behavior as a function of thermodynamic state for mixtures with equivalence ratio equal to 0.1. The strong ignition limit is shown as a hashed area. H_2/O_2 explosion limits are shown as solid lines with upper and lower bounds shown as dashed lines, representing uncertainty in the rate coefficient of reactions Reprinted with permission from [73]. Copyright © 2014, Elsevier

- combustion instabilities (different instabilities can happen during combustion, for example: acoustic instabilities, shock instabilities, fluid-dynamic instabilities, chemical-kinetic instabilities, diffusive-thermal instabilities, hydrodynamic instabilities, feed-system interactions, exhaust-system interactions etc.)
- blowoff (flame blowoff can be indicated also as static instability, this happens usually due to a change in the composition of the syngas-oxidizer mixture or in the flow environment of the flame);
- blowout (when excessive flow of oxidizer quenches the flame);
- flashback (upstream propagation of the flame back to the burner);
- pollutant emissions (CO, CxHy, PM, NOx, PAH, etc.);
- flame structure (which can be mainly analyzed based on the shape and on the colors of the flame).

Combustion instability is an important topic for gas turbines, which are usually designed to reach very low emissions levels and so are often operated in conditions which are quite close to the blowout limits. Flashback is also an important instability generated by high flame speed fuels which contain high hydrogen levels.

Given the wide variety of composition of syngas it has to be taken into account that a system designed for a low hydrogen containing syngas cannot be easily adapted to work with a high hydrogen content syngas.

Table 12.4 $CO/H_2/O_2$ mechanisms [74]

Number	Mechanism	Species	Reactions
1	NUIG-NGM-2010 [80]	15 species (obtained from an original number of 293)	41 reactions (obtained from an original number of 1593)
2	Kéromnès-2013 [81]	15 species (obtained from an original number of 17)	49 reactions
3	Davis-2005 [48]	14 species	38 reactions
4	Li-2007 [82]	15 species (obtained from an origi nal number of 21)	45 reactions (obtained from an original number of 93)
5	USC-II-2007 [83]	14 species (obtained from an original number of 111)	48 reactions (obtained from an original number of 784)
6	SanDiego-2014 [84]	15 species (obtained from an original number of 50)	37 reactions (obtained from an original number of 244)
7	CRECK-2012 [85]	14 species	34 reactions
8	Li-2015 [86]	14 species	37 reactions
9	Starik-2009 [87]	16 species	44 reactions
10	GRI3.0-1999 [88]	15 species (obtained from an original number of 53)	48 reactions (obtained from an original number of 325)
11	Rasmussen-2008 [89]	15 species (obtained from an original number of 24)	59 reactions (obtained from an original number of 105)
12	SaxenaWilliams-2006 [78]	14 species	30 reactions
13	Sun-2007 [90]	15 species	48 reactions
14	Ahmed-2007 [91]	14 species (obtained from an original number of 246)	37 reactions (obtained from an original number of 1284)
15	Zsély-2005 [49]	13 species	44 reactions
16	Dagaut-2003 [92]	13 species (obtained from an original number of 132)	34 reactions (obtained from an original number of 922)

To reduce combustion instabilities much modeling work has been performed, based on kinetic studies. These have been developed to characterize the combustion behavior of H_2/CO mixtures which can be used to model the behavior of syngas [48, 75–78]. Studies have been performed using both approaches: experimental and modeling. Modeling can also play an important role in emissions reduction. Burning hydrogen rich syngas higher combustion temperatures are obtained and this can cause an increase in NO_x emissions. For this reason the current approach is to dilute syngas using N_2 or steam [58]. So to reduce emissions a possible solution can be a lean premixed turbine operation, here combustion modeling can be used to check safety issues. For this purpose, very reliable kinetic models are needed. The development of kinetic mechanisms of syngas combustion is a key step toward the optimization of syngas fed gas turbines [75].

Several works are available in the literature on syngas combustion modeling, as presented in Table 12.4.

An interesting comparison of the different kinetic schemes that have been developed for syngas combustion in gas turbines is proposed in the work of Olm et al. [79]. Sixteen combustion mechanisms for syngas are compared, based on the performances on 5 parameters:

- ignition delay;
- flame velocities;
- species concentrations;
- overall results, all diluents except He;
- overall results, all diluents including He.

To evaluate the performance on the 5 parameters an error function is considered, which is calculated in the following way:

$$E_i = \frac{1}{N_i} \sum_{j=1}^{N_i} \left(\frac{Y_{ij}^{\text{sim}} - Y_{ij}^{\text{exp}}}{\sigma\left(Y_{ij}^{exp}\right)} \right)^2 \tag{12.14}$$

where,

- N is the number of available data.
- Y_{ij}^{sim} is the j-th simulated data;
- Y_{ij}^{exp} is the j-th experimental data;
- $\sigma(Y_{ij}^{\text{exp}})$ is the standard deviation of the j-th experimental data.

The total error is the sum of the errors obtained from Eq. (12.14):

$$E = \frac{1}{N_i} \sum_{j=1}^{N} E_i \tag{12.15}$$

The mechanisms are assigned a number based on their performance on the 5 chosen parameters (number 1 is the best, number 16 is the worst), they are shown in Table 12.4.

Here the model used to simulate CO/H_2 mixtures by the CRECK modeling group is shown. This model is the first step in the implementation of the CRECK-2012 model and contains a large number of the reactions. We chose to present this model given the experience on biomass modeling of the CRECK research group. This model has been realized by adding the hydrogen oxidation mechanism [76] to the oxidation of CO. In this way the model is mainly based on the oxidation of H_2/CO mixtures, where CO and H_2 are the main syngas components.

The model has been verified by comparing the results with experimental data obtained mainly under high-pressure conditions. It is thought to be used to model syngas combustion in gas turbines in a wide range of conditions. The kinetic model includes:

- recent and accurate thermodynamic and kinetic estimates;

– the interaction of the combustion system with pollutant species (NO_x) [93].

The scheme published in [74], which is based on 32 elementary reactions is proposed in Table 12.5. The part on methane combustion is taken from Ranzi et al. [94].

The first reaction consists of the oxidation of a single hydrogen atom, which is a chain branching reaction, where O is a diradical; so in this step from one H radical 3 radical products are formed. The diradical can react with a hydrogen molecule to form two radicals (see branching reaction number 2). The two branching reactions can be responsible of explosion phenomena.

$$H + O_2 = OH + O \tag{12.16}$$

$$O + H_2 = OH + H \tag{12.17}$$

The third-order reaction, $H + O_2 + M = HO_2 + M$, is important as a chain terminating reaction in combustion. The reaction competes with the branching reaction, $H + O_2 = OH + O$, at temperatures less than ~900 K and, therefore, has a substantial effect in the later stages of combustion in both flames and practical combustors [100, 101].

$$H + O_2 + [M] = HO_2 + [M] \tag{12.18}$$

Equation (12.18) can involve also a H radical and two molecules of oxygen ($H + O_2 + O_2 = HO_2 + O_2$).

The reaction $OH + HO_2 = H_2O + O_2$ plays an important role in combustion chemistry. It is a major HO_2 termination path in lean combustion [102, 103], and it is responsible for the depletion of both OH and HO_2 radicals in burnt gases [104, 105].

$$OH + HO_2 = H_2O + O_2 \tag{12.19}$$

The H_2/CO_2 sub-mechanism consists of 20 reversible reactions, as reported in [76]. The extension to syngas combustion requires the addition of only three new species (CO, CO_2 and HCO) and 12 new reactions.

As reported in the work of Davis [48], the reaction $CO + OH = CO_2 + H$ is the critical step in syngas oxidation and combustion and special attention should be focused on its rate constant. The importance of this reaction in syngas combustion is linked with the significant heat released during CO to CO_2 conversion. This reaction has been experimentally studied by Wooldridge et al. [106] who suggested a new expression to define the rate coefficient, based on experimental measures performed with infrared absorption of CO_2 and the UV laser absorption of OH. This rate is confirmed also by other measures reported in the literature [107, 108].

Davis et al. [48] added two modified Arrhenius expressions to describe more accurately the high temperature data of Wooldridge et al. [109] and of Golden et al. [110]. Similarly, a combination of two expressions was used by Joshi and Wang [108] while three expressions were combined by Sun et al. [111].

Table 12.5 $CO/H_2/O_2$ mechanism with rate coefficients in the form k = A · T n · exp(−Ea/RT), A units: mol/l/s/K; E_a units: cal/mol [74]

	Reaction	A	n	E^a	Source
1	H+ O_2 = OH + O	2.21E+11	0	16,650	[76]
2	O+ H_2 = OH + H	4.33E+10	0	10,000	[76]
3	H+ O_2 + [M] = HO_2 + [M]	4.65E+09	−0.8	0	[76]
Low-pressure limit:		7.00E + 11	0.4	0	
Troe parameters: 0.5, 1E − 30 1E + 30					
Enhanced third-body efficiencies: H_2O = 18.0, H_2 = 2.5, N_2 = 1.26, O_2 = 0, Ar = 0.8, He = 0.8, CO = 1.2, CO_2 = 2.4					
	H + O_2 + O_2 = HO_2 + O2	8.90E+0.8	0	−2822	[76]
4	OH+ HO2 = H2O + O2	5.00E+10	0	1000	[76]
5	H+ HO_2 = OH + OH	2.50E+11	0	1900	[76]
6	O+ HO_2 = O_2 + OH	3.25E+10	0	0	[76]
7	OH + OH=O + H_2O	7.36E+09	0	1100	[76]
8	H_2 + [M] = H + H+[M]	2.23E+11	0	96,081	[76]
Enhanced third-body efficiencies: H_2O = 12.0, H_2 = 2.5, CO = 1.9, CO_2 = 3.8, Ar = 0.5, He = 0.5					
9	O_2 + [M] = O + O+[M]	1.55E+11	0	115,120	[76]
Enhanced third-body efficiencies: H_2O = 12.0, H_2 = 2.5, CO = 1.9, CO_2 = 3.8, Ar = 0.2, He = 0.2					
10	H + OH+[M] + H_2O+[M]	4.502E+16	−2	0	[76]
Enhanced third-body efficiencies: H_2O = 16.0, H_2 = 2.0, CO_2 = 1.9					
11	H + HO_2 = H_2 + O_2	2.50E+10	0	700	[76]
12	HO_2 + HO_2 = H_2O_2 + O_2	2.11E+09	0	0	[76]
13	OH + OH+[M] = H_2O_2 + [M]	7.40E+10	−0.37	0	[76]
Low-pressure limit:		2.30E + 12	−0.9	−1700	
Troe parameters: 0.7346, 94.00, 1756, 5182					
Enhanced third-body efficiencies: H_2O = 6, H_2 = 2, CO = 1.5, CO_2 = 2.0, CH_4 = 2.0, C_2H_6 = 3.0, Ar = 0.7, He = 0.7					
14	O + OH+[M] = HO_2 + [M]	1.00E+10	0	0	[76]
15	H + H_2O=H_2 + OH	4.00E+07	1	19,000	[76]
16	H_2O_2 + H=H_2O + OH	2.41E+10	0	3970	[76]
17	H_2O_2 + H=H_2 + HO_2	6.03E+10	0	7950	[76]
18	HO_2 + H_2O → H_2O_2 + OH	5.39E+05	2	28,780	[76]
19	OH + H_2O_2 → H_2O + HO_2	3.20 + E+05	2	−4170	[76]
20	O + H_2O_2 → OH + HO_2	1.08E+06	2	−1657	[95]
21	CO + O+[M] = CO2 + [M]	9.64E+07	0	3800	[48]
Low-pressure limit:		2.07E + 20	−3.34	7610	
Enhanced third-body efficiencies: H_2O = 12.0, H_2 = 2.0, CO = 1.5, CO_2 = 2.0, Ar = 0.5					
22	CO + OH=CO_2 + H	9.60 + 08	0.14	7352	
23	CO + H_2O=CO_2 + OH	3.01E+10	0	23,000	[47]
24	CO + H_2O=CO_2 + H_2	2.00E+08	0	38,000	[95]
25	O_2 + CO=CO_2 + O	2.53E+09	0	47,000	[96]
26	HCO+[M] = CO + H+[M]	1.20E+14	−1	17,000	[97]

(continued)

Table 12.5 (continued)

	Reaction	A	n	E[a]	Source
Enhanced third-body efficiencies: $H_2O = 5.0$, $H_2 = 1.9$, $CO = 1.9$, $CO_2 = 3.0$					
27	$HCO + O{=}CO_2 + H$	3.00E+10	0	0	[96]
28	$HCO + H{=}H_2 + CO$	1.00E+11	0	0	[98]
29	$HCO + OH{=}H_2O + CO$	5.00E+10	0	0	[99]
30	$HCO + HO_2 = H_2O_2 + CO$	4.00E+08	0	0	[94]
31	$O_2 + HCO{=}H_2O + CO$	1.00E+09	0	0	[94]
32	$HCO + HO_2{=}{>}H + OH + CO_2$	3.00E+10	0	0	[99]

This is due to the complexity of the CO + OH reactions and to the difficulty in describing them using a single Arrhenius expression which should perform well over a wide range of temperatures. The CO + OH reactions can proceed in both ways: passing and not passing through the chemically activated complex HOCO [108]. Consequently, the temperature dependence of the reaction can be explained better using two rate expressions and a two-channel reaction.

CO to CO_2 conversion can also proceed through the following recombination reaction:

$$CO + O + M = CO_2 + M \tag{12.20}$$

This happens mainly at high pressures and/or in anhydrous systems, as reported by [74]. Figure 12.18 shows that in literature we can find very different values of rate constants [47, 96, 112–119]. In the model presented in Table 12.5 we can see that the reaction number 21 is modelled with the following Arrhenius expression:

$$k = 2.07 \times 1020 \times T - 3.34 \times \exp\left(-7160/\mathrm{RT}\right) \tag{12.21}$$

Which in case of high-pressure is combined with the parameters of Kondratiev [120] as suggested by Allen et al. [121]. The value of the kinetic rate at high-pressure value is about 2–3 times faster than the one suggested by Troe [122]. Dealing with the reaction, Eq. (12.22),

$$CO + HO_2 = CO_2 + OH \tag{12.22}$$

the equation was modified, according to what was reported in Mueller et al. [47], to take into account syngas reactivity at low temperature and high pressure.

Nitrogen chemistry and kinetics has been analyzed in many papers [76, 123, 124]. The kinetic model presented in Table 12.5 can be combined with a NO_x kinetic scheme in which the following reaction was updated to improve the model predictions:

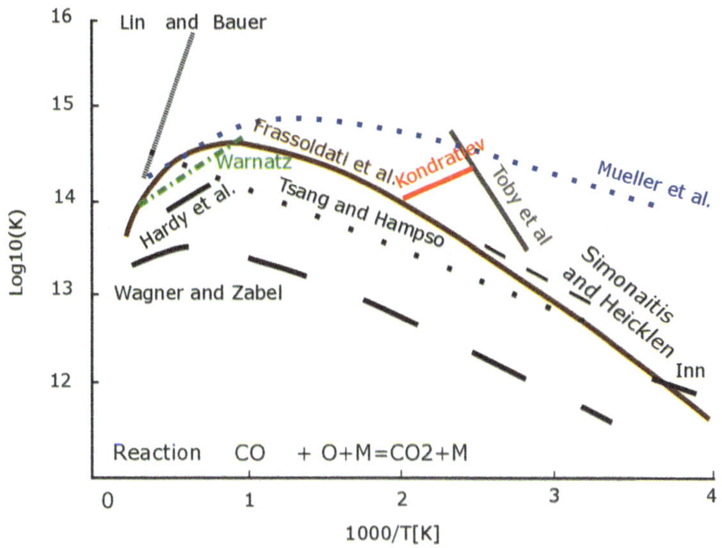

Fig. 12.18 Reaction CO + O + M = CO_2 + M. Comparison of the low-pressure rate constants suggested by [47, 74, 96, 112–119] (K units: mol cm^3 s). Reprinted with permission from [74]. Copyright © 2007, Elsevier

$$HONO + OH = NO_2 + H_2O \tag{12.23}$$

Another interesting scheme, which is not comprised in Table 12.5 is the one proposed by Boivin *et al* [125]. This is a four step mechanism which builds upon a three step mechanism for H_2 combustion in air [126]. Also in this case CO combustion is added to an already existing scheme describing H_2 combustion. The final scheme is quite simple and composed by 16 elementary steps.

12.4.2 Experimental Pressurized Syngas Combustion Test Benches

Burke et al. [127] used a dual-chambered, pressure-release type high-pressure combustion apparatus (see Fig. 12.19) to assess the influence of temperature and pressure on mass burning rates of syngas with different compositions.

In recent studies, the facility shown in Fig. 12.19 was managed in the "closed configuration", as discussed in [128], such that the combustion chamber can be considered as a constant-volume cylindrical bomb. Experimental and numerical studies have been done to investigate the dependencies between the pressure and flame temperature of mass burning rates for H_2/CO/O_2/diluents blends. Mass burning rates and flame speeds were extracted from externally propagating flames for equivalence ratios from 0.85 to 2.5, flame temperatures of (1500–1800) K, pressures

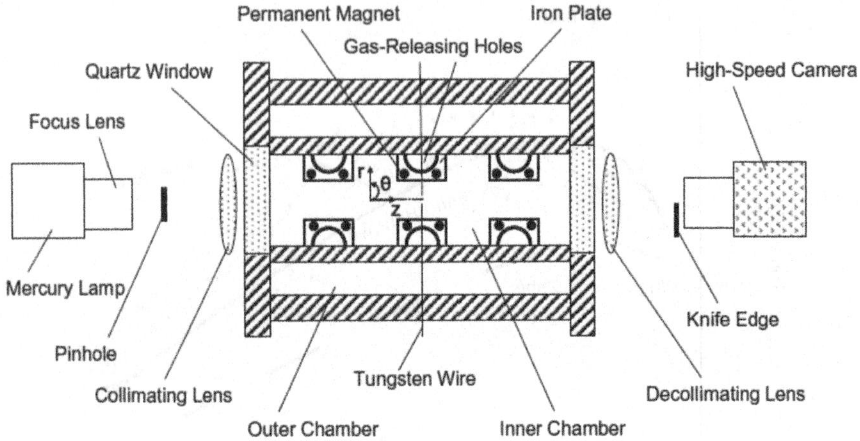

Fig. 12.19 Dual chamber high pressure combustion apparatus. Reprinted with permission from [127]. Copyright © 2010, Elsevier

from (1 to 25) bar, CO fuel fractions from 0 to 0.9 and dilution concentrations of He, Ar and CO_2 up to respectively 0.8, 0.6 and 0.4. The main conclusions are explained in the following paragraphs:

At low pressure, the mass burning rate increases with pressure, while at high pressure conditions there is an inversely proportional relationship between the mass burning rate and the pressure for both lean and rich conditions. A negative dependence of the mass burning rate on pressure is observed at lower pressure and lower temperature. With the increase of pressure the dependence on temperature of the burning rate increases. Compared to pure H_2 at the same pressure and temperature conditions, a 0.5 mass fraction of carbon monoxide addition to the fuel doesn't have an important effect on the pressure dependence. Dilution with CO_2 enhances the pressure and temperature dependence.

Currently there is no kinetic model which can predict the dependence of the burning rate on pressure across all the pressure and temperature conditions.

Considering experiments which applied the typical conditions of a gas turbine combustion chamber. In order to keep the same flame temperature, the dilution can be increased together with the preheating of the mixture; this will give the same burning rate-pressure dependence with respect to the non-preheated test. Burning rate-pressure dependence and sensitivity results are reduced in the case of preheating the mixture with fixed dilution rates. Referring to the previous case, if dilution is not maintained constant, but decreased, then we will have also an increase in the flame temperature and a weaker dependence of burning rate from pressure. The strength of the burning rate-pressure dependence of H_2/O_2/diluent flames is influenced by diluent according to the following order: He $<$ N_2 $<$ CO_2 $<$ H_2O.

When the pressure is increased, the extended second explosion limit, which delimits straight-chain kinetics from chain-branching kinetics, translates to higher temperatures. As a result, the part of the flame zone in which intense branching happens is restricted to a narrower, higher temperature window as the pressure is increased—resulting in a shift of the peak for all radical reactions to higher temperatures. According to the sensitivity analysis, the pressure dependence for rich conditions is controlled by two competing chemical reactions:

- $H + O_2$.
- and $H + HO_2$.

which largely govern the fate of H. The same reactions are important for lean conditions together with:

- $OH + H_2$.
- and $OH + HO_2$.

which are mainly responsible of the OH radical fate.

The sensitivity of mass burning rates to elementary rate constants increases greatly with pressure. The big sensitivities at elevated pressures raise the effects of uncertainties in rate constants for elementary reactions. Consequently, the inconsistencies in the burning rate predictions by different models may be due to uncertainties associated with a number of reactions, radical recombination reactions; such as HO_2 + radical reactions, radical recombination reactions and even well-studied reactions (like hydrogen oxidation reactions). The collision efficiencies, as well as the treatment of fall-off for the reaction: $H + O_2(+M) = HO_2(+M)$ can influence in an important way the predictions at high pressures, whereas for “fall-off” a reaction has an increasing rate with an increase in temperature.

It becomes clear that one or more elementary reactions could be significant in high-pressure H_2 flames but are not considered by most of the current H_2 combustion models (e.g. $O + OH + M = HO_2 + M$).

In the case of increasing pressure and decreasing flame temperature, the recombination reactions are thought to become more important. For this reason the flux through branching and recombination channels may become approximately equal in the most reactive portion of the flame at low flame temperatures and high pressures conditions. Hence, the sensitivity of the predictions to the rate parameters for those reactions increase dramatically. The high sensitivities to input parameters gives important implications. For example it will be very difficult to model exactly the flame at high-pressure, and uncertainty in a rate parameter will result in a larger uncertainty in the flame speed/burning rate.

Another experimental apparatus is shown in Fig. 12.20; this has been realized in the ‘Centro Combustione Ambiente’ Gioa del Colle, through the collaboration of Ansaldo and Turbec, it is based on a pressurized combustion chamber which simulated the combustion chamber of a micro-gasturbine. Tests have been performed to understand the combustor performances when fed with compressed air from bottles and a mixture of syngas and steam; where syngas is provided by bottles.

Figure 12.20 shows a clarified view of the experimental installation used to fire the pressurized combustion on NG/syngas or a mixture of both. The flows of

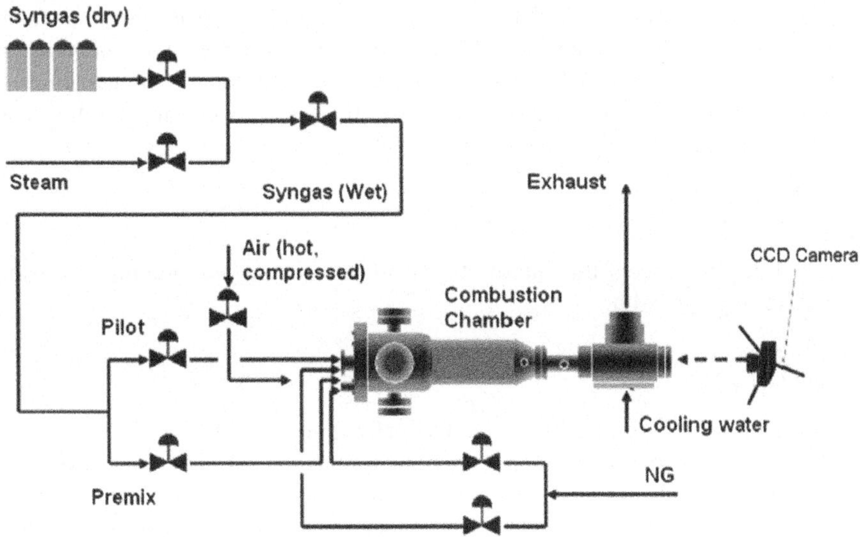

Fig. 12.20 Simplified overview of the experimental setup used to test different fuel types in the pressurized combustor. Reprinted with permission from [129]. Copyright © 2010, Elsevier

compressed carbon monoxide (20.5 vol.%) and hydrogen (79.5 vol.%) are regulated via electronic valves. The steam is generated externally and premixed to create a wet syngas. The temperature of the combustion mixture is monitored using 24 K type thermocouples. Pressure sensors are used to monitor the pressure ratio of the air when it is compressed externally to the combustion chamber. Downstream of the combustion chamber, there is a diaphragm that regulates the pressure and represents the turbine backpressure. Electric heaters preheat the air up to a theoretical maximum of 450 °C. To maintain the same pressure drop used for methane the injection nozzles were slightly to optimize the injection of hydrogen-rich synthesis gas.

The combustion gases are cooled by a water-cooling circuit respectively after passing the adjustable diaphragm and before being analyzed to determine their composition and evacuated through the exhaust pipe.

After the syngas combustion tests in a microturbine combustion chamber, it can be inferred that:

- the values obtained during measurements of wall-temperature for the combustion of synthesis gas did not exceed those obtained with natural gas. So there should be no problem regarding temperature limits of the materials due to the use of wet hydrogen-rich gas.
- during partial load and full load operation of the syngas combustion chamber, no observations have been made which could indicate a declining flame anchoring position. According to the temperature profiles, it can be noted that the flame does not come close to the combustor rim, so there are not flashback problems.

- the production of NO_x and CO measured over combustion remain notably low. It can be noted that the combustion regime was stable producing very low CO emissions during the operation with syngas. So no incomplete combustion was observed. The small values for the NOx emissions are due to the positive effect of the steam content on flame temperature. So thermal NOx have been minimized. Prompt NO_x was not expected to have significant contributions. Correspondingly, the total NO_x emissions generated during syngas combustion were found to be very low, under (<5 ppm).
- no auto-ignition, flashback instabilities or blow-off were noticed.

12.5 Conclusions and Future Outlook

Syngas is the product of pyrolysis and gasification processes which can be used both for CHP and for the production of chemicals. Syngas compression is a useful process which can boost its utilization in many cases: for the chemical conversion of the gas; for power production in IGCC plants; for heat production for cooking and heating needs. The compression has to be performed in safe and effective ways. This chapter shows that:

- it is important to analyze the explosion limit of the syngas, the ignition delay and the behavior of the syngas in high pressure and high temperature environments.
- this can be done following two approaches: experimental and modeling.
- experiments can be performed in Rapid Compression Facility (RCM) test bench and also shock tube reactors.
- kinetic models can be verified on experimental data and used to simulate syngas behavior and the combustion performances of gases with different composition which are compressed and ignited in different conditions of temperature and pressure.
- experimental tests have shown that after ignition the pressure data show an increase, which is slow at first and then is followed by a rapid increase.
- further studies have shown that auto-ignition can be of two types: strong and weak. Weak ignition is defined as non-uniform ignition of the mixture; while strong ignition is defined as a phenomenon which starts with a wave that propagates through all the combustible mixture. The weak ignition can be explained with a kinetic phenomenon which is delimited by the second explosion limit.
- weak ignition is a phenomenon linked with the dynamics of the gases and it is induced by perturbations to the flow-field. It has to be taken into consideration also that the strong ignition is occurring usually at temperatures which are higher than 1000 K. The strong ignition limit is not existent anymore at high pressures, where it has been assessed only strong ignition (i.e homogeneous ignition).
- when the pressure is increased, the extended second explosion limit, which delimits straight-chain kinetics from chain-branching kinetics, translates to higher

temperatures. As a result, the part of the flame zone in which intense branching happens is restricted to a narrower, higher temperature window as the pressure is increased – resulting in a shift of the peak for all radical reactions to higher temperatures.

- according to sensitivity analysis, the pressure dependence for rich conditions is controlled by two competing chemical reactions: $H + O_2$ and $H + HO_2$; which largely govern the fate of H. The same reactions are important for lean conditions together with: $OH + H_2$ and $OH + HO_2$; which are mainly responsible of the OH radical fate.

Dealing with future outlook and future areas of research, if the syngas has to be compressed and stored in closed vessels, the condensation of tar will be very relevant in the compressed syngas systems, and this is a new topic which deserves to be better studied to protect the components of the compressor and ensure the long life of the process. For this aspect, modeling of tar dew point will be of paramount importance in the design and optimization. Systems will be needed to:

- extract the condensed tar from the compressor;
- avoid tar entering the compressor itself;
- model tar dew point and tar condensation.

References

1. D'Alessandro B, D'Amico M, Desideri U, Francesco F. The IPRP (integrated pyrolysis regenerated plant) technology: from concept to demonstration. Appl Energy. 2013;101 (1):423–31. https://doi.org/10.1016/j.apenergy.2012.04.036.
2. Zhang X, Che Q, Cui X, Wei Z, Zhang X, Chen Y, Wang X, Chen H. Application of biomass pyrolytic polygeneration by a moving bed: characteristics of products and energy efficiency analysis. Bioresour Technol. 2018;254:130–8. https://doi.org/10.1016/j.biortech.2018.01.083.
3. Minchener AJ. Coal gasification for advanced power generation. Fuel. 2005;84(17):2222–35. https://doi.org/10.1016/j.fuel.2005.08.035.
4. Beaudin M, Zareipour H, Schellenberglabe A, Rosehart W. Energy storage for mitigating the variability of renewable electricity sources: an updated review. Energy Sustain Dev. 2010;14 (4):302–14. https://doi.org/10.1016/j.esd.2010.09.007.
5. Tukiainen S. Carbofex EBC-certified biochar technology, Carbofex. 2020. https://www.carbofex.fi/. Accessed 18 Jan 2020.
6. Carbon Terra. Das unternehmen. 2020. https://www.carbon-terra.eu/en/schottdorf-meiler/functionality. Accessed 18 Jan 2020.
7. Gustafsson M. 2013Pyrolysis for heat production. Master thesis. 2013. http://www.diva-portal.org/smash/get/diva2:655188/FULLTEXT02.pdf. Accessed 18 Jan 2020.
8. EliquoStulz. PYREG, Minimise the cost of your sludge disposal. 2020. https://www.eliquostulz.com/en/pyreg.html. Accessed 18 Jan 2020.
9. Ahmad E, Jäger N, Apfelbacher A, Daschner R, Hornung A, Pant KK. Integrated thermo-catalytic reforming of residual sugarcane bagasse in a laboratory scale reactor. Fuel Process Technol. 2018;171:277–86. https://doi.org/10.1016/j.fuproc.2017.11.020.
10. 3R Agrocarbon. The 3R zero emission pyrolysis technology generates new resources and added value recovered BIO-PHOSPHATE products. 2020. https://www.3ragrocarbon.com/sites/default/files/attachments/3r_fact_sheet.pdf. Accessed 18 Jan 2020

11. Rietveld G, van der Drift A, Grootjes AJ, van der Meijden CM, Vreugdenhil BJ. Commercialization of the ECN MILENA gasification technology, ECN report ECN-M--14-061. 2014. https://publications.tno.nl/publication/34631595/N4tX73/m14061.pdf. Accessed 18 Jan 2020.
12. Ahrenfeldt J, Thomsen TP, Henriksen U, Clausen LR. Biomass gasification cogeneration – a review of state of the art technology and near future perspectives. Appl Therm Eng. 2013;50 (2):1407–17. https://doi.org/10.1016/j.applthermaleng.2011.12.040.
13. Hofbauer H, Rauch R, Bosch K, Koch R, Aichernig C. Biomass CHP plant güssing - a success story. 2002. https://www.semanticscholar.org/paper/Biomass-CHP-Plant-G%C3%BCssing-A-Success-Story-Hermann-Reinhard/0122da66ae2088ad3f046a150f251a4a345b8038. Accessed 18 Jan 2020.
14. Biollaz S. The SNG technology platform in güssing, a status report of bio-SNG project. In: Poster presented at European Biofuels Technology Platform: Second Stakeholder Plenary Meeting (SPM2), 22nd of January 2009, Brussels. 2009. http://www.etipbioenergy.eu/images/Poster_BioSNG_PSI.pdf. Accessed 18 Jan 2020.
15. Alamia A, Magnusson I, Johnsson F, Thunman H. Well-to-wheel analysis of bio-methane via gasification, in heavy duty engines within the transport sector of the European Union. Appl Energy. 2016;170:445–54. https://doi.org/10.1016/j.apenergy.2016.02.001.
16. Hedenskog M. The GoBiGas project: bio-methane from forest residues – from vision to reality. Presentation at SVEBIO2015. 2015.
17. Alamia A, Larsson A, Breitholtz C, Thunman H. Performance of large-scale biomass gasifiers in a biorefinery, a state-of-the-art reference. Int J Energy Res. 2017;41:2001–19. https://doi.org/10.1002/er.3758.
18. Larsson A, Hedenskog M, Thunman H. Monitoring the bed material activation in the GoBiGas-Gasifier. In: Nordic Flame days. 2015. https://www.researchgate.net/publication/283578378_Monitoring_the_Bed_Material_Activation_in_the_GoBiGas-Gasifier. Accessed 18 Jan 2020.
19. Rauch R, Hofbauer H, Bosch K, Siefert I, Aichernig C, Voigtlaender K et al. Steam gasification of biomass at CHP plant guessing-status of the demonstration plant. In: World biomass conference; biomass for energy industry and climate protection, pp 1687–1690. 2004.
20. Pfeifer C, Koppatz S, Hofbauer H. Steam gasification of various feedstocks at a dual fluidised bed gasifier: impacts of operation conditions and bed materials. Biomass Convers Biorefin. 2011;1:39–53. https://doi.org/10.1007/s13399-011-0007-1.
21. Kotik J. Über den Einsatz von Kraft-Wärme-Kopplungsanlagen auf basis der wirbelschichtdampfvergasung fester biomasse am beispiel des biomassekraftwerks oberwart. Vienna: Vienna University of Technology; 2010.. (in German)
22. Suda T, Liu Z, Takafuji M, Hamada K, Tani H. Gasification of lignite coal and biomass using twin IHI Gasifier (TIGAR®). 2012. https://www.ihi.co.jp/var/ezwebin_site/storage/original/application/fae45aac0eb82bef2ca20cf8cc2cb0f0.pdf. Accessed 18 Jan 2020.
23. Paethanom A. Twin IHI Gasifier (TIGAR®) – current status of Indonesian demonstration project and its business plan. In gasification and syngas technologies conference. Vancouver, BC. 2016. https://www.globalsyngas.org/members/2016-conference-presentations/. Accessed 18 Jan 2020.
24. Melligan F, Auccaise R, Novotny EH, Leahy JJ, Hayes MHB, Kwapinski W. Pressurised pyrolysis of Miscanthus using a fixed bed reactor. Bioresour Technol. 2011;102(3):3466–70. https://doi.org/10.1016/j.biortech.2010.10.129.
25. Mahinpey N, Murugan P, Mani T, Raina R. Analysis of bio-oil, biogas, and biochar from pressurized pyrolysis of wheat straw using a tubular reactor. Energy Fuel. 2009;23:2736–42. https://doi.org/10.1021/ef8010959.
26. Fjellerup J, Gjernes E, Hansen LK. Pyrolysis and combustion of pulverized wheat straw in a pressurized entrained flow reactor. Energy Fuel. 1996;10:649–51. https://doi.org/10.1021/ef950204e.

27. Whitty K, Backman R, Hupa M. Influence of pressure on pyrolysis of black liquor: 1. Swelling. Bioresour Technol. 2008;99:663–70. https://doi.org/10.1016/j.biortech.2006.11.065.
28. Roberts DG, Harris DJ, Wall TF. On the effects of high pressure and heating rate during coal pyrolysis on char gasification reactivity. Energy Fuel. 2003;17:887–95. https://doi.org/10.1021/ef020199w.
29. Matsuoka K, Akiho H, Xu WC, Gupta R, Wall TF, Tomita A. The physical character of coal char formed during rapid pyrolysis at high pressure. Fuel. 2005;84:63–9. https://doi.org/10.1016/j.fuel.2004.07.006.
30. Vuthaluru HB. Investigations into the pyrolytic behaviour of coal/biomass blends using thermogravimetric analysis. Bioresour Technol. 2004;92:187–95. https://doi.org/10.1016/j.biortech.2003.08.008.
31. Cetin E, Gupta R, Moghtaderi B. Effect of pyrolysis pressure and heating rate on radiata pine char structure and apparent gasification reactivity. Fuel. 2005;84:1328–34. https://doi.org/10.1016/j.fuel.2004.07.016.
32. Porada S. The influence of elevated pressure on the kinetics of evolution of selected gaseous products during coal pyrolysis. Fuel. 2004;83(7–8):1071–8. https://doi.org/10.1016/j.fuel.2003.11.004.
33. Blackmer®. Compressors, 04/99 CB-207. 2020. https://www.gasequipment.com/catalogs/cryogenic/pdf/Blackmer/Compressors/Comp%20Selection%20and%20Sizing.pdf. Accessed on 17 Jan 2020.
34. Process Industry Practices Machinery. Compressor selection guidelines. 2013. https://pip.org/docs/default-source/practices-documents/reec001bf1ca80395a262f789edff00008ddc6a.pdf?sfvrsn=d3beca9e_0. Accessed 17 Jan 2020.
35. Gresh MT. Compressor performance: aerodynamics for the user. Amsterdam: Elsevier; 2001.
36. Rezvani S, McIlveen-Wright D, Huang Y, Dave A, Deb Mondol J, Hewitt N. Comparative analysis of energy storage options in connection with coal fired integrated gasification combined cycles for an optimized part load operation. Energy Convers Manag. 2012;101:154–60. https://doi.org/10.1016/j.fuel.2011.07.034.
37. Amos WA. Costs of storing and transporting hydrogen. National Renewable Energy Lab, Golden, CO. 1999. https://www.nrel.gov/docs/fy99osti/25106.pdf. Accessed 17 Jan 2020.
38. Cau G, Cocco D, Serra F. Energy and cost analysis of small-size integrated coal gasification and syngas storage power plants. Energy Convers Manag. 2012;56:121–9. https://doi.org/10.1016/j.enconman.2011.11.025.
39. Richards GA, McMillian MM, Gemmen RS, Rogers WA, Cully SR. Issues for low-emission, fuel-flexible power systems. Prog Energy Combust Sci. 2001;27(2):141–69. https://doi.org/10.1016/S0360-1285(00)00019-8.
40. Lieuwen T, McDonell V, Santavicca D, Sattelmayer T. Burner development and operability issues associated with steady flowing syngas fired combustors. Combust Sci Technol. 2008;180(6):1169–92. https://doi.org/10.1080/00102200801963375.
41. He F, Li Z, Liu P, Ma L, Pistikopoulos EN. Operation window and part-load performance study of a syngas fired gas turbine. Appl Energy. 2012;89(1):133–41. https://doi.org/10.1016/j.apenergy.2010.11.044.
42. Frey HC, Zhu Y. Improved system integration for integrated gasification combined cycle (IGCC) systems. Environ Sci Technol. 2006;40:1693–9. https://doi.org/10.1021/es0515598.
43. Smith AR, Klosek J. A review of air separation technologies and their integration with energy conversion processes. Fuel Process Technol. 2001;70(2):115–34. https://doi.org/10.1016/S0378-3820(01)00131-X.
44. Geosits RF, Schmoe Lee A. IGCC – the challenges of integration. In: Proceedings of GT2005 ASME turbo expo 2005: power for land, sea, and air, Reno, NV. 2005. https://www.edockets.state.mn.us/EFiling/edockets/searchDocuments.do?method=showPoup&documentId={1AB68706-28A0-4706-B9A8-5C412EAF4BDE}&documentTitle=281454. Accessed 17 Jan 2020.

45. Jaeger H. Plant design net rated 644 MW and 38% HHV on low rank coal, gas turbine world; March–April 2006. 2006.
46. Lee C, Lee SJ, Yun Y. Effect of air separation unit integration on integrated gasification combined cycle performance and NOx emission characteristics. Korean J Chem Eng. 2007;24 (2):368–73. https://doi.org/10.1007/s11814-007-5047-7.
47. NATO. Performance prediction and simulation of gas turbine engine operation for aircraft, marine, vehicular, and power generation. Technical report RTO-TRAVT-036. Research and Technology Organisation, North Atlantic Treaty Organisation. 2007. https://apps.dtic.mil/docs/citations/ADA466188. Accessed 17 Jan 2020.
48. Gupta KK, Rehman A, Sarviya RM. Bio-fuels for the gas turbine: a review. Renew Sustain Energy Rev. 2010;14(9):2946–55. https://doi.org/10.1016/j.rser.2010.07.025. Accessed 17 Jan 2020
49. Todd DM, Battista RA. Demonstrated applicability of hydrogen fuel for gas turbines. In: Proceedings of gasification for the future, Noordwijk, Nederland. 2000.
50. Shilling N, Jones RM. The response of gas turbines to a CO2 constrained environment. In: Gasification technology conference report, GE Power Systems.
51. Chiesa P, Lozza G, Mazzocchi L. Using hydrogen as gas turbine fuel. Trans ASME J Eng Gas Turb Power. 2005;127(1):73–80. https://doi.org/10.1115/1.1787513.
52. Gardiner WC Jr, McFarland M, Morinaga K, Takeyama T, Walker BF. Initiation rate for shock-heated hydrogen-oxygen-carbon monoxide-argon mixtures as determined by OH induction time measurements. J Phys Chem. 1971;75:1504–9. https://doi.org/10.1021/j100680a022.
53. Dean AM, Steiner DL, Wang EE. A shock tube study of the $H_2/O_2/CO/Ar$ and $H_2/N_2O/CO/Ar$ systems: measurement of the rate constant for H + N2O = N2 + OH combust. Flame. 1978;32:73–83. https://doi.org/10.1016/0010-2180(78)90081-0.
54. Fotache CG, Tan Y, Sung CJ, Law CK. Ignition of $CO/H_2/N_2$ versus heated air in counterflow: experimental and modeling results. Combust Flame. 2000;120:417–26. https://doi.org/10.1016/S0010-2180(99)00098-X.
55. Wierzba I, Kilchyk V. Flammability limits of hydrogen–carbon monoxide mixtures at moderately elevated temperatures. Int J Hydrogen Energy. 2001;26:639–43. https://doi.org/10.1016/S0360-3199(00)00114-2.
56. Mueller MA, Yetter RA, Dryer FL. Flow reactor studies and kinetic modeling of the $H_2/O_2/$NOX and $CO/H_2O/O_2/$NOX reactions. Int J Chem Kinet. 1999;31:705–24. https://doi.org/10.1002/(SICI)1097-4601(1999)31:10<705::AID-JCK4>3.0.CO;2-%23.
57. Davis SG, Joshi AV, Wang H, Egolfopoulos F. An optimized kinetic model of H2/CO combustion. Proc Combust Inst. 2005;30:1283–92. https://doi.org/10.1016/j.proci.2004.08.252.
58. Zsely IG, Zador J, Turanyi T. Uncertainty analysis of updated hydrogenand carbon monoxide oxidation mechanisms. Proc Combust Inst. 2005;30:1273–81. https://doi.org/10.1016/j.proci.2004.08.172.
59. Walton SM, He X, Zigler BT, Wooldridge MS. An experimental investigation of the ignition properties of hydrogen and carbon monoxide mixtures for syngas turbine applications. Proc Combust Inst. 2007;31(2):3147–54. https://doi.org/10.1016/j.proci.2006.08.059.
60. Donovan MT, He X, Zigler BT, Palmer TR, Wooldridge MS, Atreya A. Demonstration of a free-piston rapid compression facility for the study of high temperature combustion phenomena. Combust Flame. 2004;137(3):351–65. https://doi.org/10.1016/j.combustflame.2004.02.006.
61. He X, Donovan MT, Zigler BT, Palmer TR, Walton SM, Wooldridge MS, Atreya A. Combust. Flame. 2005;142:266–75. https://doi.org/10.1016/j.combustflame.2005.02.014.
62. Lee D, Hochgreb S. Hydrogen autoignition at pressures above the second explosion limit (0.6–4.0 MPa). Int J Chem Kinet. 1998;30:385–406. https://doi.org/10.1002/(SICI)1097-4601(1998)30:6<385::AID-KIN1>3.0.CO;2-O.

63. Elsworth JE, Haskell WW, Read IA. Non-uniform ignition processes in rapid-compression machines combust. Flame. 1969;13(4):437–8. https://doi.org/10.1016/0010-2180(69)90115-1.
64. Vermeer DJ, Meyer JW, Oppenheim AK. Auto-ignition of hydrocarbons behind reflected shock waves combust. Flame. 1972;18(3):327–36. https://doi.org/10.1016/S0010-2180(72)80183-4.
65. Fieweger K, Blumenthal R, Adomeit G. Self-ignition of S.I. engine model fuels: a shock tube investigation at high pressure combust. Flame. 1997;109(4):599–619. https://doi.org/10.1016/S0010-2180(97)00049-7.
66. Walton SM, He X, Zigler BT, Wooldridge MS, Atreya A. Demonstration of distinct ignition regimes using high-speed digital imaging of Iso-octane mixtures. In: Proc. fourth joint meeting of the US Sections of the Combust. Inst. https://ci.confex.com/ci/2005/techprogram/P1498.HTM
67. Walton SM, He X, Zigler BT, Wooldridge MS, Atreya A. An experimental investigation of iso-octane ignition phenomena. Combust Flame. 2007;150(3):246–62. https://doi.org/10.1016/j.combustflame.2006.07.016.
68. Wang H. Private communication. 2005.
69. Chaos M, Dryer FL. Syngas combustion kinetics and applications. Combust Sci Technol. 2008;180(6):1053–96. https://doi.org/10.1080/00102200801963011.
70. Grogan KP, Ihme M. Weak and strong ignition of hydrogen/oxygen mixtures in shock-tube systems. Proc Combust Inst. 2015;35(2):2181–9. https://doi.org/10.1016/j.proci.2014.07.074.
71. Voevodsky VV, Soloukhin RI. On the mechanism and explosion limits of hydrogen-oxygen chain self-ignition in shock waves. Proc Combust Inst. 1965;10(1):279–83. https://doi.org/10.1016/S0082-0784(65)80173-4.
72. Meyer J, Oppenheim A. On the shock-induced ignition of explosive gases. Symp Combust. 1971;13(1):1153–64. https://doi.org/10.1016/S0082-0784(71)80112-1.
73. Mansfield AB, Wooldridge MS. High-pressure low-temperature ignition behavior of syngas mixtures. Combust Flame. 2014;161(9):2242–51. https://doi.org/10.1016/j.combustflame.2014.03.001.
74. Frassoldati A, Faravelli T, Ranzi E. The ignition, combustion and flame structure of carbon monoxide/hydrogen mixtures. Note 1: detailed kinetic modeling of syngas combustion also in presence of nitrogen compounds. Int J Hydrog Energy. 2007;32(15):3471–85. https://doi.org/10.1016/j.ijhydene.2007.01.011.
75. Mittal G, Sung CJ, Yetter RA. Autoignition of H_2/CO at elevated pressures in a rapid compression machine. Int J Chem Kinet. 2006;38:516. https://doi.org/10.1002/kin.20180.
76. Frassoldati A, Faravelli T, Ranzi E. A wide range modeling study of NOx formation and nitrogen chemistry in hydrogen combustion. Int J Hydrogen Energy. 2006;31(15):2310–28. https://doi.org/10.1016/j.ijhydene.2006.02.014.
77. Kalitan DM, Petersen EL. Ignition and oxidation of lean CO/H2 fuel blends in air 41st AIAA/ASME/SAE/ASEE joint propulsion conference and exhibit, 10–13 July 2005, Tucson, Arizona (USA), paper 2005-3767. 2005. https://doi.org/10.2514/1.28123.
78. Saxena P, Williams FA. Testing a small detailed chemical-kinetic mechanism for the combustion of hydrogen and carbon monoxide. Combust Flame. 2006;145:316–23. https://doi.org/10.1016/j.combustflame.2005.10.004.
79. Olm C, Zsély IG, Varga T, Curran HJ, Turányi T. Comparison of the performance of several recent syngas combustion mechanisms. Combust Flame. 2015;162:1793–812. https://doi.org/10.1016/j.combustflame.2014.12.001.
80. Healy D, Kalitan DM, Aul CJ, Petersen EL, Bourque G, Curran HJ. Oxidation of C1−C5 alkane quinternary natural gas mixtures at high pressures. Energy Fuel. 2010;24:1521–8. https://doi.org/10.1021/ef9011005.
81. Kéromnès A, Metcalfe WK, Heufer KA, Donohoe N, Das AK, Sung CJ, Herzler J, Naumann C, Griebel P, Mathieu O, Krejci MC, Petersen EL, Pitz WJ, Curran HJ. An experimental and detailed chemical kinetic modeling study of hydrogen and syngas mixture

oxidation at elevated pressures. Combust Flame. 2013;160:995–1011. https://doi.org/10.1016/j.combustflame.2013.01.001.

82. Li J, Zhao Z, Kazakov A, Chaos M, Dryer FL, Scire JJJ. A comprehensive kinetic mechanism for CO, CH2 O, and CH 3 OH combustion. Int J Chem Kinet. 2007;39(3):109–36. https://doi.org/10.1002/kin.20218.
83. Wang H, You X, Joshi AV, Davis SG, Laskin A, Egolfopoulos F, Law CK. USC Mech version II. High-temperature combustion reaction model of H2/CO/C1-C4 compounds. http://ignis.usc.edu/USC_Mech_II.htm/. Accessed 17 Jan 2020.
84. Mechanical and Aerospace Engineering (Combustion Research), University of California at San Diego: Chemical-Kinetic Mechanisms for Combustion Applications, San Diego Mechanism, version 2014-02-17. http://combustion.ucsd.edu. Accessed 17 Jan 2020.
85. CRECK modeling Group Hydrogen/CO mechanism version 1212. http://creckmodeling.chem.polimi.it/kinetic.html/. Accessed 17 Jan 2020.
86. Li X, You X, Wu F, Law CK. Uncertainty analysis of the kinetic model prediction for high-pressure H_2/CO combustion. Proc Combust Inst. 2015;35(1):617–24. https://doi.org/10.1016/j.proci.2014.07.047.
87. Starik AM, Titova NS, Sharipov AS, Kozlov VE. Syngas oxidation mechanism. Combust. Explos. Shock Waves. 2010;46:491–506. https://doi.org/10.1007/s10573-010-0065-x.
88. Smith GP, Golden DM, Frenklach M, Moriary NW, Eiteneer B, Goldenberg M, Bowman CT, Hanson RK, Song S, Gardiner WC, Lissianski VV, Qin Z. GRI-Mech 3.0. http://combustion.berkeley.edu/gri-mech/version30/text30.html. Accessed 15 Nov 2019.
89. Rasmussen CL, Hansen J, Marshall P, Glarborg P. Experimental measurements and kinetic modeling of CO/H_2/O_2/NOx conversion at high pressure. Int J Chem Kinet. 2008;40:454–80. https://doi.org/10.1002/kin.20327.
90. Sun H, Yang SI, Jomaas G, Law CK. High-pressure laminar flame speeds and kinetic modeling of carbon monoxide/hydrogen combustion. Proc Combust Inst. 2007;31(1):439–46. https://doi.org/10.1016/j.proci.2006.07.193.
91. Ahmed SS, Mauß F, Moréac G, Zeuch T. A comprehensive and compact n-heptane oxidation model derived using chemical lumping. Phys Chem Chem Phys. 2007;9:1107–26. https://doi.org/10.1039/B614712G.
92. Dagaut P, Lecomte F, Mieritz J, Glarborg P. Experimental and kinetic modeling study of the effect of NO and SO_2 on the oxidation of CO-H_2 mixtures. Int J Chem Kinet. 2003;35(11):564–75. https://doi.org/10.1002/kin.10154.
93. Wu KT, Lee HT, Juch CI, Wan HP, Shim HS, Adams BR, Chen SL. Study of syngas co-firing and reburning in a coal fired boiler. Fuel. 2004;83(14–15):1991–2000. https://doi.org/10.1016/j.fuel.2004.03.015.
94. Ranzi E, Sogaro A, Gaffuri P, Pennati G, Faravelli T. A wide range modeling study of methane oxidation. Combust Sci Technol. 1994;96(4–6):279–325. https://doi.org/10.1080/00102209408935359.
95. Nist best fit. 2006. http://kinetics.nist.gov/index.php. Accessed 15 Nov 2019.
96. Tsang W, Hampson RF. Chemical kinetic data base for combustion chemistry. Part I. Methane and related compounds. J Phys Chem Ref Data. 1986;15(3):1087–279. https://doi.org/10.1063/1.555759.
97. Timonen RS, Ratajczak E, Gutman D. The addition and dissociation reaction atomic hydrogen + carbon monoxide. dblharw. oxomethyl. 2. Experimental studies and comparison with theory. J Phys Chem. 1987;91:5325. https://doi.org/10.1021/j100304a037.
98. Jachimowski CJ. Chemical kinetic reaction mechanism for the combustion of propane. Combust Flame. 1984;55(2):213–24. https://doi.org/10.1016/0010-2180(84)90029-4.
99. Gardiner WCJ, editor. Combustion chemistry. New York: Springer; 1984. https://doi.org/10.1007/978-1-4684-0186-8.
100. Michael JV, Su MC, Sutherland JW, Carroll JJ, Wagner AF. Rate constants for H + O2 + M f HO2 + M in seven Bath gases. J Phys Chem A. 2002;106:5297–313. https://doi.org/10.1021/jp020229w.

101. Westbrook CK, Dryer F. Chemical kinetic modeling of hydrocarbon combustion. Prog Energy Combust Sci. 1984;10(1):1–57. https://doi.org/10.1016/0360-1285(84)90118-7.
102. Peeters J, Mahnew G. Reaction mechanisms and rate constants ofelementary steps in methane-oxygen flames. Symp Combust. 1973;1973(14):133–46. https://doi.org/10.1016/S0082-0784(73)80015-3.
103. Petersen EL, Davidson DF, Hanson RK. Kinetics modeling of shock-induced ignition in low-dilution CH4/O2 mixtures at high pressures and intermediate temperatures. Combust Flame. 1999;117:272–90. https://doi.org/10.1016/S0010-2180(98)00111-4.
104. Goodings JM, Hayhurst ANJ. Heat release and radical recombination in premixed fuel-lean flames of H2+ O2+ N2. Rate constants for H + OH + M → H2O + M and HO2+ OH → H2O + O2. Chem Soc Faraday Trans 2. 1988;84:745–62. https://doi.org/10.1039/F29888400745.
105. Hong Z, Vasu SS, Davidson DF, Hanson RK. Experimental study of the rate of OH + HO2 f H2O + O2 at high temperatures using the reverse reaction. J Phys Chem A. 2010;114:5520–5. https://doi.org/10.1021/jp100739t.
106. Wooldridge MS, Hanson RK, Bowman CT. A shock tube study of CO + OH → CO2 + H and HNCO + OH → products via simultaneous laser absorption measurements of OH and CO_2. Int J Chem Kinet. 1996;28:361–72. https://doi.org/10.1002/(SICI)1097-4601(1996)28:5<361::AID-KIN5>3.0.CO;2-T.
107. Zhao Z, Li J, Kazakov A, Dryer FL. Temperature-dependent feature sensitivity analysis for combustion modeling. Int J Chem Kinet. 2005;37:282. https://doi.org/10.1002/kin.20080.
108. Joshi AV, Wang H. Master equation modeling of wide range temperature and pressure dependence of CO + OH → products. Int J Chem Kinet. 2006;38:57. https://doi.org/10.1002/kin.20137.
109. Wooldridge MS, Hanson RK, Bowman CT. A shock tube study of the CO+OH → CO2 +H reaction. Proc Combust Inst. 1994;25:741–8. https://doi.org/10.1016/S0082-0784(06)80706-X.
110. Golden DM, Smith GP, McEwen AB, Yu CL, Eitneer B, Frenklach M, Vaghjiani GL, Ravishankara AR, Tully FP. OH (OD) + CO: measurements and an optimized RRKM fit. J Phys Chem A. 1998;102(44):8598–606. https://doi.org/10.1021/jp982110m.
111. Sun HY, Yang SI, Jomaas G, Law CK. High-pressure laminar flame speeds and kinetic modeling of carbon monoxide/hydrogen combustion. Proc Combust Inst. 2006;31(1):439–46. https://doi.org/10.1016/j.proci.2006.07.193.
112. Lin MC, Bauer SH. Bimolecular reaction of N2O with CO and the recombination of O and CO as studied in a single-pulse shock tube. J Chem Phys. 1969;50:3377. https://doi.org/10.1063/1.1671561.
113. Warnatz J. Rate coefficients in the C/H/O system. In: Gardiner Jr WC, editor. Combustion chemistry. New York: Springer; 1984. p. 197–360. https://doi.org/10.1007/978-1-4684-0186-8.
114. Hardy JW, Gardiner WC Jr, Burcat A. Recombination of carbon monoxide and oxygen atoms. Int J Chem Kinet. 1974;10:503–17. https://doi.org/10.1002/kin.550100508.
115. Wagner HG, Zabel F, Bunsenges B. Neuere Untersuchungen zum thermischen Zerfall von CO2. Teil II. Phys Chem. 1974;72:705. https://doi.org/10.1002/bbpc.19740780717.
116. Inn ECY. Rate of recombination of oxygen atoms and CO at temperatures below ambient. J Chem Phys. 1974;61:1589. https://doi.org/10.1063/1.1682139.
117. Simonaitis R, Heicklen J. Kinetics and mechanism of the reaction of O (3P) with carbon monoxide. J Chem Phys. 1972;56:2004. https://doi.org/10.1063/1.1677490.
118. Toby S, Sheth S, Toby FS. The chemistry of combustion processes. In: Sloane TM, editor. ACS symposium series, vol. 249. Washington: DC pp; 1984. p. 267–76. https://doi.org/10.1021/bk-1983-0249.ch016.
119. Kondratiev VN. On the rate of CO + O recombination. React Kinet Catal Lett. 1974;1:7–13. https://doi.org/10.1007/BF02075114.
120. Kondratiev VN. Proc Combust Inst. 1959;7:41.

121. Allen MT, Yetter RA, Dryer FL. High pressure studies of moist carbon monoxide/nitrous oxide kinetics. Combust Flame. 1997;109:449–70. https://doi.org/10.1016/S0010-2180(96)00181-2.
122. Troe J. Thermal dissociation and recombination of polyatomic molecules. Proc Combust Inst. 1975;15(1):667–80. https://doi.org/10.1016/S0082-0784(75)80337-7.
123. Faravelli T, Frassoldati A, Ranzi E. Kinetic modeling of the interactions between NO and hydrocarbons in the oxidation of hydrocarbons at low temperatures. Combust Flame. 132:188–207. https://doi.org/10.1016/S0010-2180(02)00437-6.
124. Frassoldati A, Faravelli T, Ranzi E. Kinetic modeling of the interactions between NO and hydrocarbons at high temperature. Combust Flame. 2003;135:97–112. https://doi.org/10.1016/S0010-2180(03)00152-4.
125. Boivin P, Jiménez C, Sánchez AL, Williams FA. A four-step reduced mechanism for syngas combustion. Combust Flame. 2011;158(6):1059–63. https://doi.org/10.1016/j.combustflame.2010.10.023.
126. Boivin P, Jiménez C, Sánchez AL, Williams FA. An explicit reduced mechanism for H2–air combustion. Proc Combust Inst. 2011;33(1):517–23. https://doi.org/10.1016/j.proci.2010.05.002.
127. Burke MP, Chaos M, Dryer FL, Ju Y. Negative pressure dependence of mass burning rates of H2/CO/O2/diluent flames at low flame temperatures. Combust Flame. 2010;157(4):618–31. https://doi.org/10.1016/j.combustflame.2009.08.009.
128. Burke MP, Chen Z, Ju Y, Dryer FL. Effect of cylindrical confinement on the determination of laminar flame speeds using outwardly propagating flames. Combust Flame. 2009;156(4):771–9. https://doi.org/10.1016/j.combustflame.2009.01.013.
129. Delattin F, Di Lorenzo G, Rizzo S, Bram S, De Ruyck J. Combustion of syngas in a pressurized microturbine-like combustor: experimental results. Appl Energy. 2010;87(4):1441–52. https://doi.org/10.1016/j.apenergy.2009.08.046.

Chapter 13
Review on Modelling Approaches Based on Computational Fluid Dynamics for Biomass Pyrolysis Systems

Przemysław Maziarka, Frederik Ronsse, and Andrés Anca-Couce

Abstract Modelling is a complex task combining elements of knowledge in the field of computer science, mathematics and natural sciences (fluid dynamics, mass and heat transfer, chemistry). In order to correctly model the process of biomass thermal degradation, in-depth knowledge of multi-scale unit processes is necessary. A biomass conversion model can be divided into three main submodels depending on the scale of the unit processes: the molecular model, single particle model and reactor model. Molecular models describe the chemical changes in the biomass constituents. Single-particle models correspond to the description of the biomass structure and its influence on the thermo-physical behaviour and the subsequent reactions of the compounds released during decomposition of a single biomass particle. The largest scale submodel and at the same time, the most difficult to describe is the reactor model, which describes the behaviour of a vast number of particles, the flow of the reactor gases as well as the interaction between them and the reactor. This chapter contains a basic explanation about which models are currently available and how they work from a practical point of view.

Keywords Biomass · Conversion · Pyrolysis · Modelling · CFD · Multiscale

13.1 Introduction

One of the most important processes of primary biomass conversion into carbonaceous materials is pyrolysis. It can be defined as the thermal conversion of biomass in an atmosphere with no oxygen to prevent its burnout. The "idea" of this process is

P. Maziarka · F. Ronsse (✉)
Department Green Chemistry and Technology, Ghent University, Ghent, Belgium
e-mail: Frederik.Ronsse@UGent.be

A. Anca-Couce
Graz University of Technology, Graz, Austria
e-mail: anca-couce@tugraz.at

Z. Fang et al. (eds.), *Production of Biofuels and Chemicals with Pyrolysis*, Biofuels and Biorefineries 10, https://doi.org/10.1007/978-981-15-2732-6_13

not a new concept and has been known since ancient times [1]. As one can presume, these traditional technologies are based on very basic solutions, like kilns or burning pits, which are simple in use, but their efficiency and process control are relatively poor. In the past, the knowledge about the conversion process itself was not profound and did not allow for significant improvements in the technology. In the last four decades, due to social pressure favouring renewables and though research initiatives, the knowledge gaps started to fill, and new, more efficient solutions started to appear. Unfortunately, despite the increasing pressure for replacing fossil fuels, the alternative materials produced using novel renewable technologies are in many cases not sufficiently engineered, or their price is uncompetitive on the current market. For this purpose new and more sophisticated methods of research as well as new technological ideas, including modelling, are being developed to meet both economic and engineering ends of the problem.

13.2 Biomass Conversion: The Modeller's Approach

13.2.1 General Overview of Simulation and Its Uncertainties

Substantial improvements in computer science in the last 30 years eased and spread access to a robust tool—numerical modelling. Simulations conducted on numerical models have allowed to significantly improve the pace of research and development in the biomass processing field.

Some commonly used terms need to be defined and clarified before the topic of computational modelling can be dealt with. A "model" is the mathematically described (by algorithms and equations) representation of a system existing in real life, and a "simulation" is an act of performing a test on a model. The term "numerical" means that the mathematical model will be translated through informatics into a numerical language, known by a numerical tool (more straightforward, a computer) to perform the computations [2, 3]. Models can be various, depending on the field where they are used, but in natural sciences and engineering, the most commonly used ones are numerical models.

A simplified scheme of a simulation study with the linkage between the experiments, theory, and model is shown in Fig. 13.1. As can be seen in this figure, the simulation has to be validated to obtain proof of its usefulness. Models based on experimental data are reliable only in a specified range of experimental values and only for this range results are valid. In general, it is always better to set the foundation of the model on fully established theories, which have a broader range of validity.

It needs to be kept in mind that models are only a representation of a real system, and in most cases, they include simplifications and approximations. Moreover, the model background lies often in experimental data, which could be burdened with errors. Therefore, simulation results in most cases show discrepancies from "true/real" results, caused by unknown deviations of the model elements. These deviations

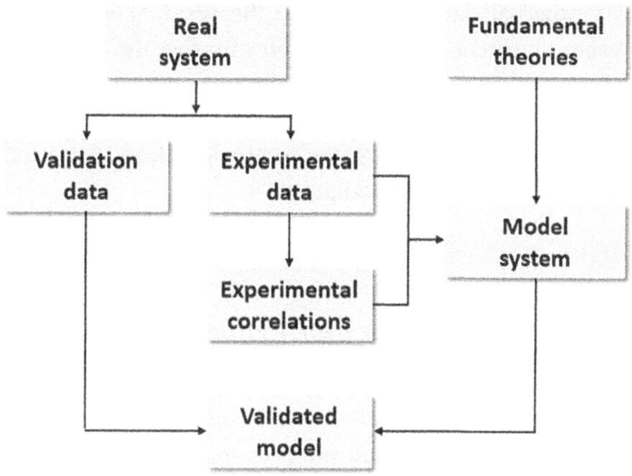

Fig. 13.1 Simplified scheme of a simulation study

are known as "uncertainties". To be able to bring the model result's closer to reality, the uncertainties need to be found, quantified and clarified. The sources of uncertainty can be divided into [4]:

- **Parameter uncertainty**—related to the parameters used in the model, which cannot be experimentally measured (too hard or too expensive) and have to be assumed in the model
- **Model inadequacy**—lack of full knowledge about the theory behind the modelled system or influence of the simplifying assumptions
- **Residual variability**—simulation output differs from experimentally obtained results through random fluctuations of parameters in a real situation (low repeatability of the real system)
- **Parametric variability**—the modelled system is not sufficiently described/measured, and input values have to be assumed
- **Observation error (experimental uncertainty)**—stemming from deviation in values due to the variability of experimental measurements
- **Interpolation uncertainty**—related to the assumption of the parameter trend in the range of experimental results between two consecutively measured data points
- **Code uncertainty (numerical uncertainty)**—the strongest uncertainty related to numerical procedures, caused by the inability to exactly solve the problem (technical boundaries) and the use of approximations while solving, e.g., in solving partial differential equations by a finite element solution method

A clear indication of the individual share of each uncertainty on the total uncertainty is not simple if at all possible, because of their strong interdependencies. For example, application of thermo-physical data from literature can influence parametric variability and residual variability. The initially implemented experimental correlations in the model and the simplification of a real system introduce model inadequacy, and the model's validation with its consecutive adjustment to experimental data can increase the residual variability and the observation error. Proper clarification of errors can improve the modeller's awareness about possible flaws within the model. Modellers are advised to keep a critical and very careful approach due to the possible implementation of unknown (unexpected) errors. The aforementioned errors, after implementation, are usually difficult to identify and time-consuming to remediate.

13.2.2 Simulation and Profit

Simulations on a properly constructed model provide valuable information about the system behaviour, which often cannot be obtained through experimental measurements. Such knowledge can give a significant boost for the development of innovative solutions and helps to identify the critical points within the system (bottlenecks). In general, the use of modelling studies brings four main advantages [2]:

- Allows for conduction of proof-of-concept (PoC) at the very beginning of the project (low sunk cost in case of failure)
- Allows for a performance of numerous tests with a low unit cost
- Increases the knowledge about dependencies in a real system
- Accumulates the obtained datasets and simplifies their treatment and sharing (big data processing)

All of the mentioned advantages can have a crucial impact on the economic feasibility of new technological solutions. As it is shown in Fig. 13.2, the application of simulations can reduce the overall cost of new solution implementations and reduce the risk of the project's unprofitability, which in the development of new technologies is a strong benefit.

Models are more flexible than real processes, so changes in modelled systems and their influence can be quickly verified. The model allows for solving technical problems in the early stage, which is the lowest cost extensive option. Modelling

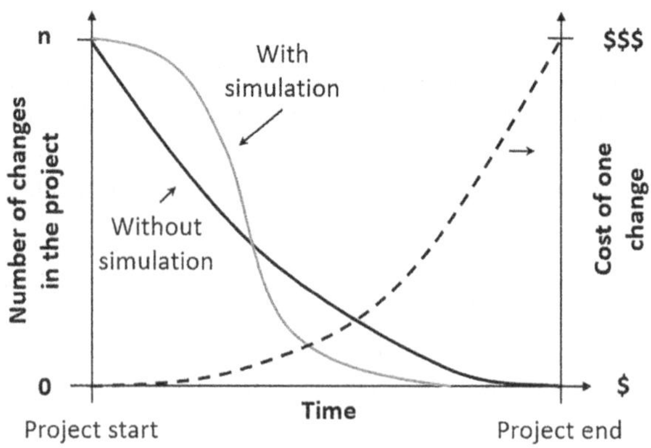

Fig. 13.2 Changes to the new idea implementation costs, through the project time (adapted with permission from [2] Copyright © 1990, Taylor and Francis Group, LLC, a division of Informa plc.)

can also expand the knowledge about the investigated process. If the model is detailed and mimics the real system well, there is a possibility to investigate and validate new correlations and theories through large and detailed databases of the process history.

13.2.3 *Theoretical Framework of a Comprehensive Model for Pyrolytic Biomass Conversion*

As it is illustrated in Fig. 13.3, a comprehensive/multi-scale model for biomass conversion can be divided into few submodels according to the scale in which the crucial processes take place. Besides combined implementation, each submodel can be studied separately, experimentally or through simulation, leading to expanding the knowledge of certain biomass conversion phenomena.

The smallest considered scale in a comprehensive model is the molecular model. It describes the chemical reactions of organic compounds and catalytic effects of inorganic compounds which take place during biomass conversion. Chemical reactions themselves are not necessarily bound to spatial dimensions, so the implementation of geometry (i.e. biomass particle) can be omitted. The amount of data which is used for this model scale allows for simulations without the need for robust numerical solvers.

A submodel covering a larger size is the single-particle model. It describes the behaviour of one individual biomass particle for which temperature, species concentration and pressure gradients during the process play a crucial role. A single-particle model needs to contain a description of the heat and mass transport phenomena and fluid dynamics. The model may cover changes in particle size, shape

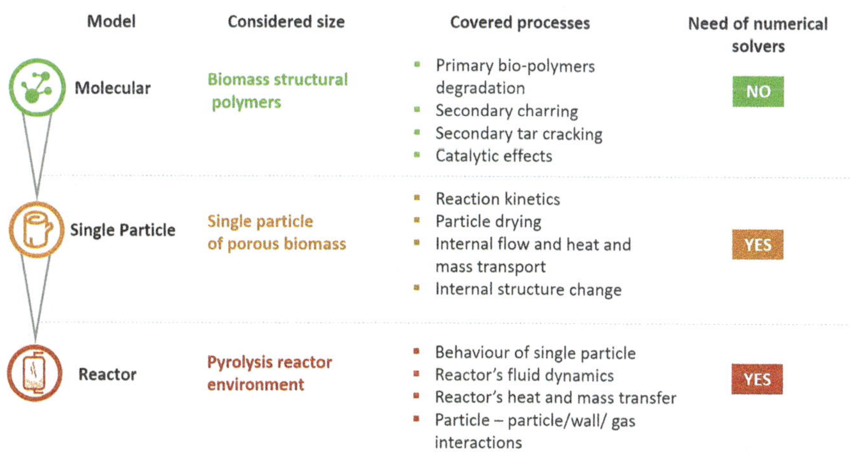

Fig. 13.3 Framework of a comprehensive biomass conversion model (adapted with permission from [5] Copyright © 2016, Elsevier)

and structure (porosity) as well as bio-polymers chemical reactions and water evaporation processes. The particle properties, intra-particle processes and boundary conditions have a strong influence on the final products yield and composition [5]. Therefore, the intra-particle phenomena, as well as their chemistry, has to be described in a very detailed manner. In the model, the gases and liquids are treated as fluids and biomass as a stagnant solid. The Eulerian description (see later) is sufficient to cope with such physical behaviour for both phases. The single particle model is strongly dependant on the geometry, so the use of a numerical solver is necessary to perform simulations at this stage.

The last submodel of a comprehensive biomass conversion model is the reactor model. It covers the description of every relevant process in a reactor for biomass thermochemical conversion. The behaviour of each biomass particle in most cases should be, if possible, described separately with a single-particle submodel. Besides the particles' conversion, the model also consists of flow and thermal behaviour of gases, particles movement (collisions with each other and walls) and thermo-physical interactions between gas and solid phases. Therefore in the reactor model, the Eulerian description of fluids needs to be combined with biomass particles movement described with a Lagrangian approach (more complex and precise, simultaneously harder and more computationally extensive option), or with an Eulerian approach (this simplification is not always possible and valid but less complex and less computationally burdening). The quantity of equations and the amount of data needed to be processed in the reactor submodel is the largest among all submodels of a comprehensive biomass conversion model. To perform simulations in an efficient manner, the model requires appropriately large computational power resources, adequate to the chosen sub-models and their complexity.

13.3 Molecular Model

13.3.1 Brief Overview of Biomass Composition

Before the description of chemical reactions that occur in biomass during thermochemical conversion, a brief explanation of biomass composition should be made. There are several biomass sources such as wood and woody biomass, herbaceous and agricultural residues, starchy crops, oil crops, aquatic biomass and, animal and human biomass wastes. The most commonly employed biomasses for energetic purposes, such as woody biomass, herbaceous biomass or agricultural residues, have a lignocellulosic structure. In lignocellulosic biomass, organic matter is mainly made from 3 main structural biopolymers: cellulose, hemicellulose, lignin and other, minor compounds which are organics named extractives and inorganics called mineral matter. The concentration of each substance varies with biomass type, and even within the same species, they are distributed in different ways among the plant organs (e.g. leaves, stem, bark, roots in wood) [6]. Detailed characterisation of the structure of bio-polymers and their thermal degradation has been extensively investigated and can be found in numerous literature reports [7–22].

13.3.2 Single Component and Competitive Schemes

Historically, the description of the pyrolysis reaction started with the introduction of simple biomass thermal degradation models. Those models are largely based on mass-loss data obtained in thermo-gravimetric (TG) experiments and up to this day are very common among researchers due to their simplicity. The core of said models is the biomass degradation kinetic, in which biomass is treated as a bulk material. Those models only take into consideration the primary biomass degradation reactions. Models based on TG show strong fluctuations between publications in obtained kinetic values. Differences can be caused by using feedstocks with different bio-composition, size, and morphology as well as by the applied methodology and calculation procedures [5]. In order to systematise TG measurements, the International Confederation for Thermal Analysis and Calorimetry (ICTAC) presented guidelines for an experimental procedure for kinetic investigations, including researches related to biomass degradation [23]. Discrepancies between the kinetic data among publications can also be caused by inappropriate assumptions regarding the kinetic mechanism. In most cases, TG models consider only the primary biomass degradation and they do not take into account the low-temperature tar-char interactions (<500 °C). Additionally, the secondary charring reactions in most TG-based models are not distinguished nor considered. Those reactions are usually lumped together with the primary degradation reactions, which leads to a shift in the value of primary kinetic parameters and as such, discrepancies in values between sources. A detailed overview of the experimental approach of a mass-loss based biomass degradation study can be found in a recent and comprehensive review by Anca-Couce [5].

Introduction of the single-component competitive models led to an improvement of TG models accuracy. Those models, besides prediction of mass loss, aim to predict also the three main products of biomass pyrolysis: char, tar, and gas—without distinction on their detailed composition. Single-component competitive models are covering only primary biomass degradation reactions, which have an influence on the prediction accuracy of product's yields [24]. Further development of the single-component competitive models was made by the introduction of cracking reactions of high molecular mass vapours (tars) at temperatures higher than 500 °C [25]. The most often used kinetic scheme is the one proposed by Shafizadeh and Chin [26].

When a higher prediction accuracy is required, the degradation of individual biomass components has to be considered in the kinetic scheme. Such schemes are named the multi-component parallel schemes, and they cover the degradation of the main biomass components (e.g. cellulose) and their intermediary products [27]. In literature extensions and improvements of the original Shafizadeh and Chin's competitive scheme can be found, e.g. via the addition of intermediate compounds or considering the three main biomass constituents. Nevertheless, the expanded models show only moderate improvement regarding the accuracy in model prediction [28, 29]. For more detailed outcomes, kinetic schemes need to cover the description of the thermal degradation of all bio-components, combined with a description of the consecutive degradation of the primary pyrolysis products.

13.3.3 Detailed Reaction Schemes: Ranzi Scheme

A more detailed description of biomass degradation in a kinetic scheme was first introduced by Ranzi et al. [30], and was further improved by him and co-workers [31–35]. The most recent extension of the model was published by Debiagi et al. [36], which improves the accuracy of the prediction of char yield. In general, the Ranzi model combines all findings related to the thermal decomposition of each major component of biomass: cellulose, hemicellulose (2 types), and lignin [11, 16, 37]. In the scheme, the overall lignin is divided into 3 artificial types of lignin: LIG-H, LIG-O, and LIG-C (hydrogen-, oxygen- and carbon-rich, respectively). Another innovation of the Ranzi model is a description of char, which distinguishes "pure" char and the volatiles "trapped" within a char metaplastic phase. Thermally unstable "traps" degrade according to the applied kinetic, releasing captured volatiles. Such a description allows for the introduction of the char devolatilisation into the kinetic scheme. The Ranzi model does not cover all possible evolved species in pyrolysis, but reduces their amount to 20 representative volatile compounds, being the most abundant in non- and condensable vapours. The Ranzi scheme allowed for the derivation of a complex reaction scheme, combining separate mechanisms into a consolidated form. The latest version of the composition of vapours, kinetic parameters, and the reaction heats can be found in the work of Ranzi et al. [34, 35].

The Ranzi model is a milestone in the description of pyrolysis kinetics, but there are a few areas in which improvements or extensions can be made. The kinetic scheme was developed for a description of fast pyrolysis, so it does not cover the secondary charring reactions. Moreover, it does not consider the catalytic influence of the of mineral matter (mainly AAEM's) contained in biomass, which leads to overprediction of the sugars and underprediction of the non-condensable gases and char. Also, the pyrolytic mechanism of the evolution of phenolic compounds is not contained in the base scheme, causing an underprediction of BTXs at higher temperatures [5, 19]. Accuracy improvement can be made by the implementation of secondary cracking reactions of the primary pyrolysis products in the gas phase. For example, it can be done by the implementation of the POLIMI kinetic mechanism, developed by the CRECK modelling group, recently revised by Ranzi et al. [35]. The POLIMI kinetic mechanism is a complex, radical, kinetic scheme, whose application improves the accuracy of prediction, but is also time-consuming to implement and increases the computational burden significantly.

13.3.4 Detailed Reaction Schemes: Ranzi—Anca-Couce Model

As was mentioned in the previous section, the Ranzi model was intended for the prediction of products from fast pyrolysis, so it shows some limitations in terms of describing biomass conversion in less severe thermal regimes. Lower thermal gradients or extended gas-solid reactions, e.g. in pyrolysis of larger samples, can lead to losses in prediction accuracy in case of application of the Ranzi model. An extension of secondary charring reactions to the Ranzi scheme, named as RAC (Ranzi—Anca-Couce) scheme was introduced by Anca-Couce et al. [19]. Their adaptation aimed to incorporate the secondary charring phenomena with the possibility of their adjustment to the severity of the conversion regime. A full description of the model with its kinetic parameters and reactions heat values can be found in the works of Anca-Couce et al. [19, 37].

The RAC model introduces an adjustable parameters "x" which defines the share of the alternative degradation, named "charring" or "secondary charring" in the overall degradation process. The adjustable parameter also partially takes into account the influence of inorganics which have a role in promoting "charring" reactions. As the main factors which increase the extent of charring, the adjustable parameter value can be modified to account for [5]:

- Decrease in the pyrolysis temperature,
- Decrease in the heating rate,
- Increase in volatiles retention time in the particle (larger particle or slower gas movements),
- Increase of the pressure in the reactor,
- High concentration of the mineral matter, especially AAEMs.

The extent of secondary charring can be different for each bio-component, so the value of the "x" parameter should be assigned separately. Unfortunately, lack of quantitative correlations between the pyrolysis conditions, biomass composition and amount of secondary charring reactions cause the need for the iterative fitting of the "x" parameter to the experimental results. A common approach is to set the adjustable parameter for all bio-components a priori, based on the available experimental data and then slightly adjust to the experimental result [5, 38]. It is worth to mention that the amount of secondary charring reactions have as well a noticeable influence on the heat of the reaction, as it was observed by Rath et al. [39].

The RAC scheme also does not cover all areas which the Ranzi scheme lacks, e.g., a detailed description of AAEM's influence or insights into polycyclic aromatic compounds formation. The base RAC scheme does not take into account the secondary gas-phase tar cracking kinetics. As well as the Ranzi scheme, it can be extended with the POLIMI kinetic mechanism. Another possible option is the simple one-step kinetics firstly introduced by Blondeau and Jeanmart [40], and consecutively improved by Mellin et al. [41] and most recently by Anca-Couce et al.

Table. 13.1 Brief summary of the comparison of kinetic models

	Detailed mass loss prediction	Detailed product composition
Single component competitive scheme	No	Limited
Multi-component parallel scheme	Yes	No
Detailed schemes (Ranzi, RAC)	Yes	Yes

[42]. Application of the one-stage kinetic cracking scheme is relatively simple, and it improves the accuracy of predictions of the vapour composition.

Constant work and recent findings on the subject gives promise of improvement and further extension of the pyrolysis reaction schemes, which would allow for better understanding of biomass pyrolysis and the ability to predict its outcome with higher accuracy [17, 43, 44]. In Table 13.1 is shown a brief summary of the comparison of kinetic models. As it can be anticipated, the more detailed the model, the better the accuracy of the predictions that can be attained. From the practical point of view, the application of a detailed model needs a lot more initial information about the processed feedstock. It also increases the complexity of the model, which leads to a higher computational burden. Therefore, the complexity of the calculation has to be chosen with caution, in relation to the desired precision of the model outcome.

13.4 Single-Particle Model

As was mentioned previously in Sect. 13.2.3, the single-particle model focuses on the influence of the composition of a particle and its thermo-physical properties on the particle's behaviour during pyrolysis. The biomass particle, due to its structure, cannot be treated as an impermeable solid object, so the description of a porous structure needs to be implemented. In practical pyrolysis applications, the biomass is rarely fed to the process in a completely dry state. Therefore, besides the description of the pyrolytic behaviour, the drying process and description of water movement within the particle have to be included in single particle models.

Due to the geometrical dependence as well as the complexity of the phenomena occurring in this stage, robust numerical solvers have to be applied. Having in mind that the Eulerian approach is able to handle the description of the processes, suitable numerical tools have to be applied, such as the Computational Fluid Dynamics (CFD).

13.4.1 Modelling Conversion Based on CFD

Prior to the mathematical description of the thermo-physical phenomena occurring in the single particle, a brief explanation of CFD will be provided here. It should give the reader a basic insight in the Eulerian approach, which is applied in single particle models as well as in the modelling of gas flow at the reactor scale.

Computational Fluid Dynamics (CFD) is the analysis of systems involving fluid flow, heat transfer and associated phenomena (e.g., chemical reactions) by using computer-based simulation [45]. In general, CFD can be treated as the integration of the following fields: natural sciences (physics and chemistry), mathematics and computer science [46].

The model behaviour is based on governing equations—in which physical phenomena like transport phenomena are mathematically described through differential equations (e.g. Navier-Stokes equation). To solve the governing equations, high-level computer programs and software packages convert them with the use of computer programming languages to numerous, simple commands that can be understood by a computing machine.

CFD for its computation needs dimensional geometrical domains. As the first step of the model's construction, the initially specified geometry ("domain") needs to be subdivided into a finite number of smaller, non-overlapping subdomains called "cells". The process of dividing a domain into subdomains is called "meshing", and it results in a grid of cells ("mesh"), that occupies the whole geometry. The cell can be defined as a representative element or a representative volume, depending on the division method ("finite element" or "finite volume", respectively). Geometry division techniques are already included in most commercially available CFD software packages. Each cell in the domain has a "node", which holds information about this certain area in the geometry. Information stored in the node changes according to the applied physical phenomena and chemical reactions.

The fluid dynamics principle employed in CFD means that it treats the flow of matter (fluid) as a continuum (Eulerian approach). In the Eulerian description of fluid dynamics, points in the geometry do not change their position with respect to the fluid motion [47, 48]. The only change that occurs is the change of the values of parameters stored at specific, fixed points (nodes). Therefore, it allows only for a description of changes taking place in nodes in the investigated geometry. As a consequence, the approach makes no distinction of single molecules or particles, so their time-based investigation is not possible.

The accuracy and precision of a CFD simulation are determined by the number of cells contained in the grid ("mesh coarseness"). An increase in the number of cells improves a simulation accuracy, until the moment when a simulation becomes grid-independent. In other words, there exists a number of cells above which the addition of new cells no longer influences the simulation quality. The simulation is called a grid-independent simulation when further mesh densification does not lead to an improvement in solution accuracy [45]. Grid independent simulations have a major advantage, which is the smallest numerical error is achieved with the most coarse mesh (least computational burden).

A detailed explanation of the CFD solution procedure is complex and goes beyond the purpose of this chapter. Nevertheless, a brief introduction to the matter will be provided. The CFD framework consists of three main elements [46]:

- **Pre-processor**—is a part of a CFD code that is responsible for the creation of an investigated geometry and its consecutive meshing. The mesh obtained in the pre-processor is a foundation for implementation of governing equations.
- **Solver**—through implemented solution methods, the solver simulates the changes of the variables in the nodes according to the applied governing equations and boundary conditions. The solver processes information regarding the applied physics and chemistry located on the nodes of the grid. Therefore, the solver is responsible for performing the simulation.
- **Post-processor**—is responsible for the visualisation of the simulation results. Most post-processors allow for quick creation of 1D, 2D or 3D plots and representation of variables of interest on the applied geometry.

The CFD solution scheme which can be found in [45] provides a general scheme, which is valid for any model based on the Eulerian approach. The specification of parameter values in the governing equations depend on the characteristics of the process which one needs to solve. Moreover, the reliability of a simulation's results is linked directly to data and auxiliary correlations, so to their compliance with the modelled system and range of application. Therefore further subsections will be focused on the reliable description of the phenomena occurring in the single particle models as well as the validity of the thermo-physical parameters applied in modelling of biomass pyrolysis.

13.4.2 Definitions of Phases in a Particle's Structure

Biomass feedstock which has not been dried previously, and is typically used for conversion, consists in most cases of four different phases: liquid water, bound water, solid and gas. The bound water is distinguished from liquid water due to its significant difference in behaviour. Each of the mentioned phases needs to be identified and described separately.

A detailed theoretical description of each phase was first made by Whitaker [49], in which a boundary surface between each phase has to be differentiated and known during the whole process. Wood has a very complex geometric structure, which strongly changes during pyrolysis, so identification of boundary surfaces at every point in time is a very difficult and complex task. Also, the amount of computation for such a sophisticated model would be very high.

The efficient description of phases has been investigated by Perre and his co-workers [50, 51], and on this basis, an elegant description of the system was presented in the work of Grønli [52]. In their approach, all of the phases are treated as a continuum for which conservation laws must be satisfied. The description assumes averaging of variables and parameters over a finite volume, which can simultaneously contain all phases. This results in a set of conservation equations for every phase, valid within the applied geometry.

For further model description, it will be helpful to define the spatial average over the geometry's total volume for any given variable (φ) valid for every phase. The spatial average is defined as:

$$<\varphi>=\frac{1}{V}\int\limits_{V}\varphi\, dV \tag{13.1}$$

The spatial average for one of the phases (γ) is defined as:

$$<\varphi>^{\gamma}=\frac{1}{V_{\gamma}}\int\limits_{V_{\gamma}}\varphi\, dV \tag{13.2}$$

where $<\varphi>^{\gamma}$ is the variable 's averaged value in the phase $\gamma$and V_{γ} is the volume of the phase in the representative volume V. The volume fraction occupied by the phase γ is defined as:

$$\varepsilon_{\gamma}=\frac{V_{\gamma}}{V} \tag{13.3}$$

A relation between the averaged value in phase γ and a spatial average is described as:

$$<\varphi>=\varepsilon_{\gamma}<\varphi>^{\gamma} \tag{13.4}$$

In other words, $<\varphi>^{\gamma}$ is an intrinsic or true value of the variable and $<\varphi>$ is an averaged value in the representative volume. For example, if $<\rho_S>^S$ would be defined as the true density of the solid phase, then $<\rho_S>$ will be defined as the density of contained solids in a representative volume of the porous particle structure (i.e. bulk density). The notation with the <,> brackets is based on the authors believe that it is clearer, and of course, it is not mandatory.

Since the particle is made in most cases out of four phases, the representative volume can be treated as a sum of volumes of each phase:

$$V=V_S+V_L+V_B+V_G \tag{13.5}$$

where subscripts S, L, B, and G represent solid, liquid water, bound water and gas, respectively. Sum of volume fractions occupied by each phase sums into one, so:

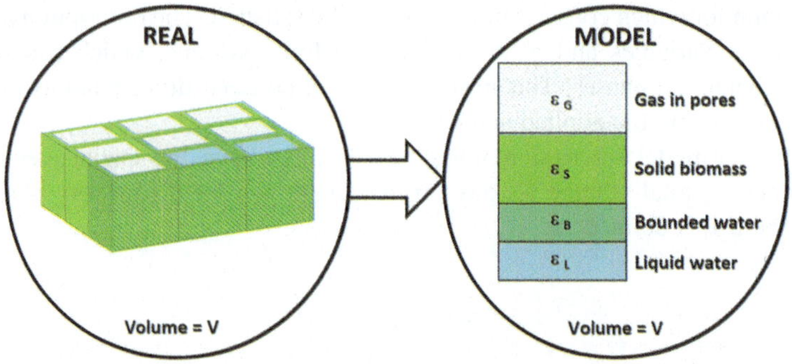

Fig. 13.4 Visual representation of the conversion of a real system (woody biomass) into a model system according to Whitaker's theory

$$\varepsilon_G = 1 - (\varepsilon_S + \varepsilon_L + \varepsilon_B) = 1 - \left(\frac{< \rho_S >}{< \rho_S >^S} + \frac{< \rho_L >}{< \rho_L >^L} + \frac{< \rho_B >}{< \rho_B >^B} \right) \tag{13.6}$$

which means that knowing the intrinsic and average density of a solid, and both types of water, a volume fraction occupied by the gas can be calculated. Visual representation of a real system in the Whitaker theory is shown in Fig. 13.4.

13.4.3 Governing Equations

In this section, an explanation of the conservation laws will be provided. Nonetheless, the theoretical derivation of the formulas will be omitted. In here, the fundamental description of the mathematical description of the governing equations is applied. Therefore, the negative signs in the equations originated purely from mathematical derivations, and they are reflecting the actual values of parameters (positive or negative). All equations mentioned in this subsection are valid only within the applied particle geometry, and they do not describe the interactions of the particle with its external environment. Reading this subsection is worth to keep in mind that all conservation equations are referring to a single, finite and representative volume.

For clarity purposes, the one component kinetic scheme will be used for explaining the principles. All kinetic schemes described in this section are treated as first-order Arrhenius kinetics with the pre-exponential parameter set as constant or temperature dependent. Additionally, from now on, wood will be treated as the exemplary lignocellulosic biomass type in the model description.

13.4.3.1 Mass Conservation Equations: Solids

At any given time of a pyrolysis reaction, the solid is represented by a mix of unconverted biomass and biochar, so it can be stated that:

$$< \rho_S >=< \rho_{BM} > + < \rho_{BC} > \quad (13.7)$$

where $<\rho_S>$, $<\rho_{BM}>$ and $<\rho_{BC}>$ are the volume-averaged densities of solid, biomass and biochar respectively. Mass conservation equation of biomass is defined as:

$$\frac{\partial}{\partial t} < \rho_{BM} >= \dot{\omega}_{BM} \quad (13.8)$$

where $\dot{\omega}_{BM}$ is the mass change rate of biomass caused by degradation and devolatilisation reactions. Although the degradation reactions lead to a reduction in mass, a negative sign is not used in Eq. (13.8). Similarly, the mass conservation equation of biochar is defined as:

$$\frac{\partial}{\partial t} < \rho_{BC} >= \dot{\omega}_{BC} \quad (13.9)$$

In most general form, the mass conservation equation is defined as:

$$\frac{\partial}{\partial t} < \rho_S >= \dot{\omega}_S \quad (13.10)$$

where $\dot{\omega}_S$ is the total mass change of a solid obtained from a sum of the biomass degradation and char formation.

13.4.3.2 Mass Conservation Equations: Single Component in the Gas Mixture

The equation for mass conservation of the ith component in a gas mixture is defined as:

$$\frac{\partial}{\partial t}(\varepsilon_G < \rho_i >^G) + \nabla < u_i \rho_i >= \dot{\omega}_i \quad (13.11)$$

where $<\rho_i>^G$ is the density of the ith component in the gaseous phase, $<u_i\rho_i>$ is i^{th} component's transport term and $\dot{\omega}_i$ is the mass change rate caused due to formation/degradation reactions of the ith gas component. Transport of the gas is driven by two phenomena: convection and diffusion. Therefore the transport term can be described as:

$$< u_i\rho_i >= u_G < \rho_i >^G - < \rho_G >^G D_{eff} \nabla \left(\frac{< \rho_i >^G}{< \rho_G >^G} \right) \quad (13.12)$$

where u_G is the superficial gas velocity, $<\rho_G>^G$ is the total density of the gas mixture, D_{eff} is the effective gas diffusion coefficient. The low permeability of biomass structures (small pores) leads to relatively low Reynolds numbers (<10) for the gas movement inside a particle. Therefore the viscous resistance force is much larger than the inertial one, which simplifies the description of flow from Darcy and Forchheimer's description to a pure Darcy's description [53]:

$$u_G = \frac{K_{G,eff}}{\mu_G} \nabla (< P_G >^G) \quad (13.13)$$

where $K_{G,eff}$ is the effective gas permeability, μ_G is the gas dynamic viscosity and $<P_G>^G$ is the pressure in the gas mixture.

13.4.3.3 Mass Conservation Equations: Liquid Water

Mass conservation equation for liquid water is defined as:

$$\frac{\partial}{\partial t} < \rho_L > + \nabla < u_L\rho_L >= \dot{\omega}_L \quad (13.14)$$

where $<\rho_L>$ is the volume-averaged liquid water density, $<u_L\rho_L>$ is its transport term and $\dot{\omega}_L$ is a mass change rate caused by evaporation or re-condensation. It is assumed that liquid water migrates through the structure entirely due to a pressure change (convectively), so its transport term is expressed as:

$$< u_L\rho_L >= u_L < \rho_L > \quad (13.15)$$

where u_L is a superficial velocity of the liquid water. Similar to the gas mixture, Darcy's law is also avalid to obtain the superficial liquid velocity:

$$u_L = \frac{K_{L,eff}}{\mu_L} \nabla (< P_L >^L) \quad (13.16)$$

where $K_{L,eff}$ is the effective liquid water permeability, μ_L is the liquid water dynamic viscosity and $<P_L>^L$ is the pressure in the liquid water.

13.4.3.4 Mass Conservation Equations: Bound Water

Mass conservation equation of bound water is defined as:

$$\frac{\partial}{\partial t} < \rho_B > + \nabla < u_B \rho_B > = \dot{\omega}_B \tag{13.17}$$

where $<\rho_B>$ is the volume-averaged bound water density, $<u_B\rho_B>$ is the bound water's transport term and $\dot{\omega}_B$ is the mass change rate caused by water's unbinding. In opposition to the liquid water, it is assumed that the bound water migrates entirely by diffusion, so its transport term is:

$$< u_B \rho_B > = - < \rho_S > D_B \nabla \left(\frac{< \rho_B >}{< \rho_S >} \right) \tag{13.18}$$

where D_B is the bound water's diffusion coefficient.

13.4.3.5 Energy Conservation Equation

The energy conservation equation is based on the assumption that the Péclet number for heat transfer is sufficiently large, so a local thermal equilibrium is obtained by all phases [53]. Therefore the equation is defined as:

$$\begin{aligned} &\frac{\partial T}{\partial t} (< \rho_S > C_{P,S} + < \rho_L > C_{P,L} + < \rho_B > C_{P,B} + \varepsilon_G < \rho_G >^G C_{P,G}) \\ &\quad + \nabla T \left(< u_L \rho_L > C_{P,L} + < u_B \rho_B > C_{P,B} + \varepsilon_G \sum_{i=1}^{N} < u_i \rho_i > C_{P,i} \right) \\ &\quad = \nabla (\lambda_{eff} \nabla T) + Q \end{aligned} \tag{13.19}$$

where C_P is the heat capacity/specific heat and subscripts S, L, B and i indicate solid, liquid water, bound water, and the ith component of the gas mixture, respectively, λ_{eff} is the effective thermal conductivity and Q is the total heat produced by the occurring reactions, and it is defined as:

$$Q = \sum_{i}^{N} H_i \dot{\omega}_i + H_L \dot{\omega}_L + H_B \dot{\omega}_B + H_S \dot{\omega}_S \tag{13.20}$$

where H is the overall heat of the reaction. In the most general case, the transport terms are implemented in the conservative form, so the energy conservation equation takes into account the heat transfer through conductive, convective and diffusion transport [52, 54, 55]. Some authors apply simplifications in defining the transport, by omitting the heat transported through diffusion, assuming that the amount of heat exchanged through this phenomenon is negligible [28, 56–58]. Taking abovementioned simplification into account, the energy conservation equation takes the form:

$$\frac{\partial T}{\partial t}(<\rho_S> C_{P,S}+ <\rho_L> C_{P,L}+ <\rho_B> C_{P,B} + \varepsilon_G <\rho_G>^G C_{P,G})$$
$$+ \nabla T(u_L <\rho_L> C_{P,L} + u_B <\rho_B> C_{P,B} + u_G \varepsilon_G <\rho_G>^G C_{P,G})$$
$$= \nabla(\lambda_{eff} \nabla T) + Q \quad (13.21)$$

13.4.3.6 Reactions

The mass change rate of every reaction in the kinetic scheme is defined as:

$$\dot{\omega}_j = k_j <\rho_j> = k_j \varepsilon_\gamma <\rho_j>^\gamma \quad (13.22)$$

where $\dot{\omega}_j$ is the mass change rate of the jth species (e.g., biomass, tar, gas), k_j is a reaction rate of the jth species, $<\rho_j>$ is the averaged volume density of the jth species and $<\rho_j>^\gamma$ is the intrinsic density of the jth species in phase γ. Water can be an exception to this definition. Depending on the applied drying/evaporation model (equilibrium, heat sink, kinetic model) the mass change rate for the liquid and bound water will take a form suitable for the chosen model.

13.4.4 *Evaporation of Water*

Moisture evaporation is one of the most energy-intensive phenomena occurring during pyrolysis of wet biomass particles. Therefore, its appropriate description has much importance. Three common ways of implementing biomass drying can be used in practice: the kinetic model, heat sink model and equilibrium model.

13.4.4.1 Kinetic Model

The kinetic model represents the simplest way of describing evaporation. It was first introduced by Chan et al. [59], and then, due to its simplicity, it has been widely applied by other authors [60–63]. The kinetic model assumes a first-order Arrhenius reaction of the liquid water phase turning into vapour. In work by Haberle et al. [64] a summary of the commonly used parameters for this model can be found.

The kinetic model is very convenient, but it treats a physical phenomenon via a chemical description, so it does not reflect the process well in real terms. In practice, in the kinetic model, water evaporation starts before water obtains its boiling temperature (100 °C at 1 atm), and the temperature during evaporation does not stay constant during the whole process. Therefore, such a model may be suitable for specific cases, but it is not advised for general application.

13.4.4.2 Heat Sink Model

The heat sink model (thermal drying model, heat flux model) [57, 64, 65] assumes that water evaporation in a representative volume occurs only at the boiling temperature, and the temperature stays constant until all water is evaporated. To maintain a constant temperature, the evaporation reaction needs to consume all the energy transferred to the representative volume. Thus all the energy delivered to the volume is absorbed (sunk) by the evaporation reaction. Mathematically the model is formulated as:

$$\dot{\omega}_e = \begin{cases} \frac{j_{Heat}}{H_e} & T \geq T_e \; and < \rho_L >> 0 \\ \\ 0 & otherwise \end{cases} \tag{13.23}$$

where $\dot{\omega}_e$ is the evaporation rate, T_e is the water boiling temperature, H_e is the heat of water evaporation and j_{Heat} is the heat flux towards to the representative volume. With the assumption that heat is not transferred by water, the heat flux is defined as:

$$j_{Heat} = \nabla(\varepsilon_G u_G < \rho_G >^G C_{p,G} - \lambda_{eff} \nabla T) \tag{13.24}$$

The heat sink model of Lu et al. [65] assumes that the boiling temperature of water is fixed at 373 K. Nevertheless, strong local evaporation can cause noticeable changes in pressure which shifts the boiling temperature. The pressure effect on the boiling temperature can be modelled as [64]:

$$T_e = T_{e,0} \log\left(\frac{< P_G >^G}{P_0}\right) + T_0 \tag{13.25}$$

where $<P_G>^G$ is the actual gas pressure, P_0 is atmospheric pressure (1 atm), $T_{e,\,0}$ is an empirical constant (32.7 K) and T_0 is the water boiling temperature at atmospheric pressure (373 K).

The heat sink model describes the evaporation phenomena more accurately than the kinetic model, and it suits very well the models of large particles, which are subjected to a high temperature and a high heating rate. Nevertheless, it also has its flaws. The model assumes an infinitely thin moving volume where evaporation takes place, so it is not valid in case if the thickness of the drying volume is not negligible in comparison to the size of the domain [5]. Another disadvantage of the model is the application of a step function (Eq. (13.23)), which is hard to handle by a numerical solver and results in numerical instability [57, 66]. The step function was investigated by Haberle et al. [64], who advised using an evaporation fraction factor (f_{evap})

as the multiplier of the heat flux. The purpose of this limiting factor is to reduce the amount of the heat sunk by the evaporation reaction. In that way, the drying is distributed over neighbouring nodes, leading to the smoothing of the step and reduction of numerical instability. The disadvantage of such an approach is the forced broadening of the thickness of the drying volume.

13.4.4.3 Equilibrium Model

The equilibrium model assumes that an equilibrium between liquid water and water vapour exists inside the particle's pores. The water vapour's partial pressure at any given time tends to be equal to the saturation vapour pressure (when the biomass moisture content is above the fibre suration point, or FSP) or saturation vapour pressure reduced by the relative humidity factor (moisture content below the FSP). For a whole range of moisture concentrations, it can be stated that:

$$< P_v^{eq} >^G = \begin{cases} P_{sat}(T) & (MC > MC_{FSP}) \\ P_{sat}(T)\ \kappa(MC_B, T) & (MC \leq MC_{FSP}) \end{cases} \tag{13.26}$$

where $< P_v^{eq} >^G$ is the equilibrium's partial pressure of water vapour, $P_{sat}(T)$ is the saturation pressure in function of the temperature, $\kappa(MC_B, T)$ is the relative humidity factor calculated from the wood isotherm. This parameter depends on the bound water content and the temperature. The saturation pressure in function of temperature can be obtained from Raznjevic's [67] experimental correlation:

$$P_{sat} = \exp\left(24.21 - \frac{467.35}{T}\right) \tag{13.27}$$

The equation for the wood's relative humidity can be obtained based on data from the Encyclopedia of Wood [68], which was obtained by Grønli [52]:

$$\kappa(MC_B, T) = 1 - \left(1 - \frac{MC_B}{MC_{FSP}}\right)^{6.453 \cdot 10^{-3}\ T} \tag{13.28}$$

From the equilibrium partial pressure, the vapour density can be obtained through:

$$< \rho_v^{eq} >^G = \frac{< P_v^{eq} >^G M_{H_2O}}{RT} \tag{13.29}$$

where M_{H_2O} is the molecular mass of water. Taking into account all above, the final equation for water evaporation rate can be defined as:

$$\dot{\omega}_e = \frac{\varepsilon_G(< \rho_v^{eq} >^G - < \rho_v >^G)}{t_{eq}} \tag{13.30}$$

where $< \rho_v^{eq} >^G$ is the equilibrium vapour density, $<\rho_v>^G$ is the water vapour density at a given time and t_{eq} is the time it takes to reach equilibrium between the actual vapour density and theoretically assumed saturation vapour density ("equilibration time"). Jahili et al. [54] stated that the equilibration time has to be appropriately short in relation to the pore diameter of wood and proposed a constant value of 10^{-5} s. Lu et al. [65] proposed a correlation of the equilibration time based on particle specific surface area and pore diameter, expressed as:

$$t_{eq} = S_{SSA} \frac{3.66\, D_{eff,H_2O}}{d_{pore}} \tag{13.31}$$

where S_{SSA} is the specific surface area of a porous particle, D_{eff,H_2O} is the effective diffusivity of water, calculated according to the work of Olek et al. [69] and d_{pore} is the average pore diameter. In their work, Lu et al. applied values obtained experimentally from N_2 adsorption [65].

The equilibrium model was designed initially for the modelling of slow, low-temperature drying. Nevertheless, it was also applied in the modelling of fast, high-temperature drying, but only with moderate success [57, 64, 65, 70, 71]. In the literature, hybrid evaporation models can also be found. Those models combine different models for liquid and bound water evaporation [63, 64].

13.4.4.4 The Heat of Water Evaporation

The most convenient way to implement the heat of evaporation is by using a constant value. For models without differentiation between liquid and bound water or models with liquid water only, the heat of evaporation can be assumed to be equal to 2440 kJ/kg (at 20 °C) [64, 65] or as 2257 kJ/kg (at 100 °C) [57]. A more appropriate way to implement the heat of evaporation can be done by using a temperature-dependent heat of evaporation correlation, e.g. the equation suggested by Ranzjevic [67]:

$$H_L = 3179 - 2.5\, T \tag{13.32}$$

where H_L is the heat of water evaporation. In models where both liquid and bound water are distinguished, a more complex approach for describing the heat of evaporation is needed. Such a model should include an additional term to account for the energy required for unbinding of the bound water prior to its evaporation. As such, the heat of evaporation for a whole range of moisture contents (liquid and bounded water) can be defined as:

$$H_e = \begin{cases} H_L & \text{if } MC \geq MC_{FSP} \\ H_L + H_B & \text{if } MC < MC_{FSP} \end{cases} \tag{13.33}$$

where H_e is the total evaporation heat of water and H_B is the the energy needed to unbind the water. The latter can be calculated using the equation proposed by Stanish [72]:

$$H_B = 0.4\, H_L \left(1 - \frac{MC_B}{MC_{FSP}}\right)^2 \tag{13.34}$$

13.4.5 Shape Specification and Coordinate Systems

The most common coordinate system for fluid dynamics is the Cartesian coordinate system. In cases where the particle anisotropy in a direction other than Cartesian's the implementation of another coordinate system can be beneficial. A wood particle does not have large property differences in the radial and tangential direction. Therefore in case of a wood particle, despite the particle's anisotropy, the Cartesian system can be applied without significant error. Table 13.2 shows the changes in description between coordinate systems for particles of different shapes: block (Cartesian), cylinders (Cylindrical) and spheres (Polar).

Table. 13.2 Coordinate systems for CFD systems

Coordinate system	D	$\nabla < up>$
Cartesian (x, y, z)	1	$\frac{\partial}{\partial x} < up >$
	2	$\frac{\partial}{\partial x} < up > + \frac{\partial}{\partial y} < up >$
	3	$\frac{\partial}{\partial x} < up > + \frac{\partial}{\partial y} < up > + \frac{\partial}{\partial z} < up >$
Cylindrical (r, θ, z)	1	$\frac{1}{r}\frac{\partial}{\partial r}(r < up >)$
	2	$\frac{1}{r}\frac{\partial}{\partial r}(r < up >) + \frac{\partial}{\partial z} < up >$
	3	$\frac{1}{r}\frac{\partial}{\partial r}(r < up >) + \frac{\partial}{\partial z} < up > + \frac{1}{r}\frac{\partial}{\partial \theta}(< up >)$
Polar (r, θ, φ)	1	$\frac{1}{r^2}\frac{\partial}{\partial r}(r^2 < up >)$
	2	$\frac{1}{r^2}\frac{\partial}{\partial r}(r^2 < up >) + \frac{1}{rsin(\theta)}\frac{\partial}{\partial \theta}(\sin(\theta) < up >)$
	3	$\frac{1}{r^2}\frac{\partial}{\partial r}(r^2 < up >) + \frac{1}{rsin(\theta)}\frac{\partial}{\partial \theta}(\sin(\theta) < up >) + \frac{1}{rsin(\theta)}\frac{\partial}{\partial \varphi}(< up >)$

D number of dimensions

13.5 Thermal and Physical Properties of Lignocellulosic Biomass

13.5.1 Density

13.5.1.1 Density of Biomass

The composition and the structure of biomass differ significantly not only with plant species but also within individual specimens of the same species. Moreover, the climate, the availability of nutrients, solar radiation and genetic changes have an influence on the plant growth, hence its structure and composition. Also, different plant organs differ in structure and composition. This leads to significant differences in biomass densities among others. Analysis of apparent density (oven dry) data of 167 measurements of the *Pinaceae* family from the Global Wood Density Database shows a significant heterogeneity within one family of a single plant (n =167, average = 435 kg/m^3, st. dev. = 65 kg/m^3).

Measurement of the solid's apparent density can be conducted by a simple measurement of weight over mass. This is not a very accurate method, especially for finely ground biomass or char samples, due to the free spaces between the grains of a solid. A more sophisticated method for measuring the apparent density is mercury porosimetry, in which Hg displaces gas around the grain. At atmospheric pressure, mercury is not able to penetrate pores whose size is below 15 μm. Therefore, the result of the measurement by mercury porosimetry is only slightly overestimated [52]. Due to the high toxicity of mercury, recently more interest is devoted to measurement methods with micro-granular suspensions. Their role is similar to mercury and relies on displacement of the gas from spaces between the grains. Some sources call the density measured with micro-granular suspensions as "envelope" density [73], in order to distinguish it from bulk density, but stay with the name "apparent" [74].

The true (skeletal, intrinsic) density is measured by helium pycnometry. The method uses helium as the pore displacement gas because it can penetrate pores with a diameter larger than 40 nm [52]. If the analysed material does not have closed pores, helium pycnometry allows for very accurate true density measurements. As is shown in the work by Brewer et al. [75], some pores in the biochar structure are not penetrable by helium, without prior grinding of the material.

Knowing both true and apparent densities and in case that samples were measured with zero moisture (dry state), the volume fraction occupied by gas, can be calculated using:

$$\varepsilon_G = 1 - \varepsilon_S \tag{13.35}$$

The orientation of the cut plane of a sample during true density measurement influences the result due to the anisotropy within the wood cell walls. Table 13.3 shows a summary of the apparent and true densities together with resulting porosity

Table. 13.3 Apparent and true densities together with resulting porosity of selected biomasses

Species (common name)	Type	Apparent density (kg/m^3)	True density (kg/m^3)	Porosity	Ref.
Birch	HW	580	1450	0.600	[52]
Spruce	SW	470	1390	0.662	
Bilinga	TW	603	1458	0.586	[74]
Beech	HW	781	1472	0.469	
Boxwood	HW	940	1506	0.376	
Danta	TW	698	1480	0.528	
Afzelia	TW	826	1501	0.450	
Yew	SW	626	1481	0.577	
Maple	HW	483	1512	0.681	
Spruce	SW	401	1524	0.737	
Idigbo	TW	616	1501	0.590	
Birch	HW	594	1502	0.605	
Larch	SW	588	1481	0.603	
Mansonia	TW	625	1466	0.574	
Merbau	TW	902	1518	0.406	
Gaboon	TW	426	1473	0.711	
Ramin	TW	608	1505	0.596	
Black locust	HW	726	1509	0.519	
Oak	HW	706	1528	0.538	
Pine	SW	451	1489	0.697	
White alder	HW	538	1492	0.639	
White lauan	TW	627	1474	0.575	
Spruce (2 mm)	SW	420	1470 (L)	0.714	[76]
			1290 (T)	0.674	
Spruce (6 mm)	SW	420	1380 (L)	0.696	
			1310 (T)	0.679	
Maple (2 mm)	HW	520	1510 (L)	0.656	
			1430 (T)	0.636	
Maple (6 mm)	HW	520	1430 (L)	0.636	
			1400 (T)	0.629	
Ash wood (2 mm)	SW	660	1360 (L)	0.515	
			1350 (T)	0.511	
Ash wood (6 mm)	SW	660	1320 (L)	0.500	
			1330 (T)	0.504	
Mesquite wood	SW	n.a.	1204	-	[75]
Miscanthus	GR	n.a.	1322	-	

SW softwood, *HW* hardwood, *TW* tropical wood, *GR* grass, *L* longitudinal, *T* transverse

for selected biomasses. If not specified, the sample anisotropy was not taken into account in the measurement.

13.5.1.2 Density of Char

The char's density and porosity depend on the initial composition and structure of biomass, as well as on the conditions of a pyrolysis process. The production temperature has a significant effect on the char's true density, as opposed to the heating rate, which seems to not have a relevant influence [75, 77]. In Table 13.4 data of the true and apparent (if available) density as well as the porosity of chars obtained from different biomasses distinguished by pyrolysis conditions is summarised. The theoretical maximum of the true density of a char is 2250 kg/m^3, which refers to the true density of graphite [78], but in practice, the maximum that can be obtained is within the range between 2000 kg/m^3 and 2100 kg/m^3.

13.5.1.3 Densities of Bound and Liquid Water

Bound water is water that exists in the biomass structure, and which is partially incorporated into the cell wall. In literature an explanation of the interaction between bound water and the cell structure as well as information about the storage locations of bounded water can be found [79]. In general, the cell wall of biomass, due to its chemical structure, is hydrophilic in its nature, and it has the ability to interact with water molecules through hydrogen bonding. Through this mechanism, water is able to stick to the wall and occupy empty spaces in its structure [80].

The cells wall of biomass has only a finite ability to bind water. To describe the amount of water that can be bound to a wall, the term fibre saturation point (FSP) was introduced first by Tiemann in 1906 [79]. It is defined as the moisture content below which only bound water exists in a biomass structure. Above the fibre saturation point, cell walls cannot bind more water, so both bound and liquid water can exist. In literature, the two most commonly applied values of the base FSP have been reported: 30% proposed by Stamm in 1971 [81] and 40% proposed by Skaar in 1988 [82]. Measurements show that above the FSP, the density of the bound water is close to 1110 kg/m^3 and with moisture content close to zero its value rises up to 1300 kg/m^3 [83]. The bound water's density increases at lower moisture content, according to the cell wall binding strength per amount of available water molecules [80]. In order to avoid over-complexity of the problem, authors typically use a constant value of 1000 kg/m^3 for the true density of the bound water [52, 54, 57, 64, 65].

The true density of the liquid water depends on the temperature, due to its thermal expansion. In the pyrolysis conditions, the water does not significantly exceed 100 °C, so the simplification that the true density of water has a constant value of 1000 kg/m^3 does not induce strong inaccuracies in the model.

Table 13.4 True and apparent (if available) densities with resulting porosity of chars obtained from different biomasses

Species	Final pyro. temp. (°C)	Heating rate (°C/min)	Apparent density (kg/m^3)	True density (kg/m^3)	Porosity	Ref.
Birch	600	5.0	390	1570	0.752	[52]
Spruce	600	5.0	390	1540	0.747	
Mesquite wood	300	5.0	603	1340	0.550	[75]
	350		532	1382	0.615	
	400		523	1384	0.622	
	450		476	1433	0.668	
	500		492	1520	0.676	
	600		447	1634	0.726	
	700		509	1735	0.707	
Miscanthus	350	5.0	262	1392	0.812	
	400		282	1438	0.804	
	450		274	1466	0.813	
	550		286	1611	0.822	
	600		293	1722	0.830	
	700		271	1965	0.862	
Miscanthus	350	23.3	284	1357	0.791	
	360	24.0	307	1368	0.776	
	370	24.7	271	1380	0.804	
	400	26.7	270	1402	0.807	
	425	28.3	295	1432	0.794	
	450	30.0	253	1432	0.823	
Pitch pine	450	0.5	n.a.	1360	–	[77]
		3.3	n.a.	1370	–	
		10.8	n.a.	1370	–	
		16.7	n.a.	1390	–	
	525	0.5	n.a.	1400	–	
		3.3	n.a.	1400	–	
		10.8	n.a.	1410	–	
		16.7	n.a.	1420	–	
	750	0.5	n.a.	1740	–	
		3.3	n.a.	1740	–	
		10.8	n.a.	1720	–	
		16.7	n.a.	1760	–	
	1000	0.5	n.a.	1970	–	
		3.3	n.a.	1980	–	
		5.8	n.a.	2000	–	
		8.3	n.a.	2010	–	
		10.8	n.a.	2000	–	
		12.5	n.a.	2010	–	
		16.7	n.a.	2010	–	

13.5.1.4 Density and Pressure of Gases and Vapours

Temperatures and intrinsic pressures during pyrolysis allow for the assumption that gases and vapours can be treated as ideal gases, so:

$$< P_i >^G = \frac{< \rho_i >^G RT}{M_i} \tag{13.36}$$

where $< P_i >^G$ and M_i are the partial pressure and molar mass of ith component in the gas mixture, respectively. The total gas density can be calculated from:

$$< \rho_G >^G = \sum_i^N < \rho_i >^G \tag{13.37}$$

The molecular mass of the gas mixture is defined as:

$$M_G = \left(\sum_i^N \frac{< \rho_i >^G}{< \rho_G >^G M_i} \right)^{-1} \tag{13.38}$$

where M_G is the mean molar mass of the gas mixture. The total gas pressure can be calculated as:

$$< P_G >^G = \frac{< \rho_G >^G RT}{M_G} \tag{13.39}$$

where $< P_G >^G$ is the total pressure. In case of the application of a simple, single-component model, permanent gases and tars are often treated not as a product mixture, but as single representative species of the mixture. For example in the work of Grønli [52], tars are represented by benzene with a molecular mass of 110 g/mol and gases are represented by a 1:1 mixture of carbon monoxide and carbon dioxide with a molecular mass of 38 g/mol.

13.5.2 Moisture Content and Saturation

The amount of water in biomass is described by the moisture content (MC), and calculated as:

$$MC = \frac{mass\ of\ water}{mass\ of\ biomass\ (db.)} \tag{13.40}$$

The water in biomass can exist in two phases, so:

$$MC = MC_L + MC_B \tag{13.41}$$

where MC_L is the moisture related to the liquid water and MC_B is the moisture related to the bound water. To calculate both moisture contents, the value of the fibre saturation point (function of the temperate) has to be obtained, for example, with the equation proposed by Siau [84]:

$$MC_{FSP}(T) = \left(M_{FSP}^{o} + 0.298\right) - 0.001\ T \tag{13.42}$$

where MC_{FSP} is the fibre saturation point at a certain temperature, and M_{FSP}^{o} is the base fibre saturation point (value between 0.3 or 0.4). Knowing that only above the fibre saturation point both types of water can be found in biomass, it can be stated that:

$$MC_B = \min\left(MC_{FSP}, MC\right) \tag{13.43}$$

$$MC_L = \max\left(MC - MC_{FSP}, 0\right) \tag{13.44}$$

With the assumption that the water content in the gas phase is negligible, the apparent density of bound and liquid water can be calculated respectively:

$$MC_B = \frac{<\rho_B>}{<\rho_S>} \tag{13.45}$$

$$MC_L = \frac{<\rho_L>}{<\rho_S>} \tag{13.46}$$

where $<\rho_S>$ is the solid's apparent density in the dry state. Having the value of the true and apparent density for both water types, the volume fraction occupied by these phases can be calculated.

Saturation of a particle quantifies to what extent the space within pores is occupied by water. This value should not be confused with the MC_{FSP}. Saturation is defined as:

$$S = \frac{liquid\ volume}{pore\ volume} \tag{13.47}$$

where pore volume is a particle's empty (filled with gas) volume which theoretically can be occupied by the liquid water. When equal representative volumes are considered:

$$S = \frac{MC_L}{MC_{sat} - M_{FSP}} \tag{13.48}$$

where MC_{sat} is the maximum moisture content which can be retained by a biomass structure:

$$MC_{sat} = MC_{FSP} + MC_{sat,L} \tag{13.49}$$

where $MC_{sat,L}$ is the maximum liquid water content which can be retained by a biomass structure. Assuming that during the maximum saturation state all pores of biomass are filled with water, and that liquid and bound water have the same density, MC_{sat} can be obtained from the equation:

$$MC_{sat} = <\rho_L>^L \left(\frac{1}{<\rho_S>} - \frac{1}{<\rho_S>^S} \right) \tag{13.50}$$

In the literate devoted to wood drying, a parameter "irreducible water content of structure" (S_{irr}) can be found. It refers to the water bound so strongly to a cell wall structure that it is not removed during a conventional drying processes (up to 120 °C). In the model of a pyrolysis process of biomass, it is not advisable to implement such parameter for two reasons. First, the energy flux added to water is much higher than in conventional drying due to higher temperatures. Theoretically, it should allow for complete unbinding of water. Second, even if the energy flux would be insufficient during the pyrolysis, the structure of biomass changes and cell walls lose their binding ability (hydrophilicity).

13.5.3 Capillary Pressure

For models in which the transportation term for the liquid water is included in the mass conservation equation, the capillary pressure needs to be defined. Capillary pressure in the lumens of wood is defined as:

$$<P_L>^L = <P_G>^G + P_C \tag{13.51}$$

where P_C is a capillary pressure and $<P_L>^L$ is pressure in of the liquid water. In literature different correlations for the capillary pressure can be found. An extensive comparison can be found in the work of Jalili et al. [54]. Here are shown only two, most commonly used empirical correlations, one by Spolek and Plumb [85]:

$$P_C = \frac{8.4 \cdot 10^4}{S^{0.63}} \tag{13.52}$$

where S is the saturation. The second, by Perre and Degiovanni [86]:

$$P_C = \frac{1.364 \cdot 10^5 \, \sigma(T)}{\left(MC_L + 1.2 \cdot 10^{-4}\right)^{0.63}} \tag{13.53}$$

where $\sigma(T)$ is the temperature-related coefficient, defined as:

$$\sigma(T) = (128 - 0.185\,T) \cdot 10^{-3} \tag{13.54}$$

Both above mentioned empirical correlations were established for softwood. Therefore they should be applied only for modelling those biomasses due to significant differences in pore size, pore shape and surface wettability with other wood types.

13.5.4 Permeability

The permeability has a major influence on the fluid movement through a porous structure. The permeability determines the superficial velocity and pressure formation of gases and transport of liquid water in a porous biomass structure.

13.5.4.1 Intrinsic Permeability of Biomass

The proper assumption regarding biomass permeability is not an easy task. As it was pointed out by Grønli [52], the value of the intrinsic gas permeability of wood shows high variability and strongly depends on:

- type of wood: hardwood or softwood
- position in the plant from which the wood sample was taken: heartwood (older part) or sapwood (younger part)
- cut plane direction (related to sample anisotropy): longitudinal, tangential or radial

Table 13.5 contains experimental data of the intrinsic gas permeability of selected biomasses. As it can be noticed, sapwoods show higher intrinsic gas permeability than heartwoods. Regarding the cut plane direction, the permeability in the longitudinal direction is much higher than in the radial or tangential direction, for which values are comparable. Taking this into account, the assumption that radial and tangential permeability are equal does not lead to a significant loss in model accuracy. In publications related to modelling, the implemented values of the intrinsic gas permeability sometimes differ significantly from those experimentally obtained. For example, some authors adjust the permeability values according to the simulation's result, or, as it was done by Di Blasi [71], the author adapted permeability to obtain the same pressure as in the experimental data from Lee et al. [87].

Analysis of the intrinsic gas permeability with differentiation on the cut plane direction, for c.a. 100 different wood samples was made by Smith and Lee in 1958 [84]. Results of their study are presented in Fig. 13.5. Values of the longitudinal permeability used by modellers are in general within the range of experimental data, but for the radial permeability, values are usually overstated by at least one order of magnitude [50, 71, 90–93]. From experimental data, it can be stated that the valid

Table 13.5 Intrinsic gas permeability for selected biomass in different directions

Species	P	Permeability (m^2) L	R	T	Ratio L/R	R/T	L/T	Ref.
Pine	*h*	2.98×10^{-11}	2.07×10^{-15}	3.65×10^{-16}	14,381	5.68	81,621	[88]
Pine	*h*	1.86×10^{-12}	3.55×10^{-16}	7.80×10^{-17}	5222	4.56	23,797	
Fir	*s*	8.88×10^{-13}	7.90×10^{-17}	1.28×10^{-17}	11,250	6.15	69,230	
Fir	*h*	4.44×10^{-14}	1.51×10^{-17}	1.68×10^{-18}	2941	9.00	26,470	
Douglas-fir	*h*	1.78×10^{-14}	5.43×10^{-19}	1.48×10^{-18}	32,727	0.37	12,000	
Redwood	*s*	1.40×10^{-11}	3.95×10^{-16}	1.23×10^{-14}	35,500	0.03	1136	
Redwood	*h*	5.38×10^{-12}	3.95×10^{-16}	5.92×10^{-16}	13,625	0.67	9083	
Red cedar	*s*	1.63×10^{-12}	1.97×10^{-16}	1.97×10^{-15}	8250	0.10	825	
Red cedar	*h*	1.04×10^{-12}	1.38×10^{-15}	1.97×10^{-15}	750	0.70	525	
Spruce	*s*	1.90×10^{-16}	2.90×10^{-18}	*n.a.*	65.52	*n.a.*	*n.a.*	[89]
Maritime pine	*s*	1.47×10^{-17}	8.60×10^{-16}	*n.a.*	0.02	*n.a.*	*n.a.*	
Scots pine	*s*	7.10×10^{-16}	4.20×10^{-17}	*n.a.*	16.90	*n.a.*	*n.a.*	

P place in the wood, *s* sapwood, *h* heartwood, *L* longitudinal, *T* tangential, *R* radial, *n.a.* not available

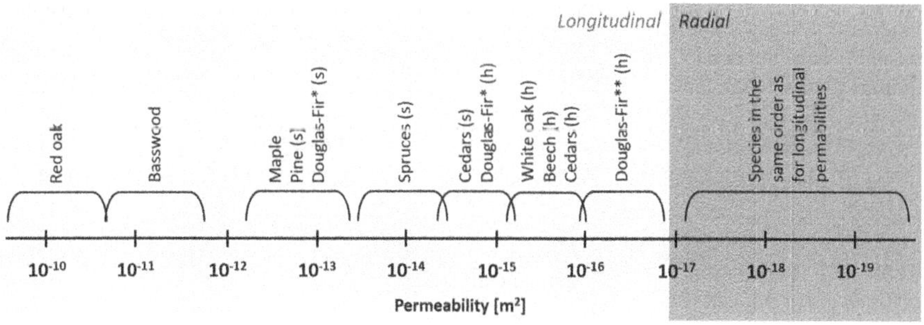

Fig. 13.5 Intrinsic gas permeability range for woods, based on the data from Smith and Lee [84] (*s* sapwood, *h* heartwood,∗ sample from the coast, ∗∗ sample from mountains)

range for the longitudinal intrinsic gas permeability is between 10^{-11} m^2 and 10^{-17} m^2 and for the radial between 10^{-15} m^2 and 10^{-19} m^2.

13.5.4.2 Intrinsic Permeability of Char

The thermal decomposition of biomass increases the internal volume of the structure. Therefore, chars formed in pyrolysis show higher permeability than the initial biomass due to an increase of the size of the channels (pore size) and development of new pores and cracks in the cell walls. Experimentally measured permeabilities of char are rarely found in the literature. Hence, most works related to the modelling of biomass pyrolysis estimate its value. Usually, the permeability of a char in the longitudinal direction is estimated to be about 1–2 orders of magnitude higher, and in the radial and tangential direction from 1 to 4–5 orders of magnitude higher than a value of the initial biomass. In Table 13.6 data of the intrinsic permeability of a pinewood char is presented. Unfortunately, the data source did not provide information regarding the direction other than the longitudinal.

13.5.4.3 Intrinsic Permeability of Liquid Water

Table 13.7 shows a summary of the relationship between the intrinsic permeability of a gas and liquid water in biomass. According to the literature, the liquid permeability should be in the range of ±1 order of magnitude different than that of the gas permeability. It is worth to mention that during pyrolysis at any given time, the liquid water does not co-exist with the char.

Table 13.6 Pinewood char's longitudinal intrinsic gas permeability as a function of pyrolysis temperature [94]

Temperature (°C)	Permeability (m^2)	Raw/char
20	5.42×10^{-13}	–
200	9.27×10^{-13}	1.71
250	1.20×10^{-12}	2.22
300	2.68×10^{-12}	4.94
350	5.74×10^{-12}	10.58

Table 13.7 Relationship between gas and liquid intrinsic permeability in biomass

Empirical correlation	Ref.
$K_L = 10K_G$	[72]
$K_L = 5K_G$	[95]
$K_L = K_G$	[96]
$K_L = 0.1K_G$	[86]

Table 13.8 Most commonly used correlations for relative gas and liquid permeabilities [96]

Direction	Relative permeability	
	Gas ($K_{G,rel}$)	Liquid ($K_{L,rel}$)
Longitudinal	$1 + (4S - 5)S^5$	S^8
Tangential	$1 + (2S - 3)S^2$	S^3

13.5.4.4 Intrinsic, Relative and Effective Permeability

The intrinsic permeability at any time of the reaction is defined as:

$$K_{ph} = X_{BM}K_{ph,BM} + X_{BC}K_{ph,BC} \quad (13.55)$$

where K_{ph} is the intrinsic permeability of a phase and X_{BM} and X_{BC} are the mass ratio of the unreacted biomass and biochar in the solid matrix, respectively. The subscript *ph* refers to a particular phase (gas or liquid).

The relative permeability reflects the difference between a material effective permeability in a wet state and the intrinsic permeability in a dry state. The correlation of moisture content and the permeability is expressed by the saturation. The most commonly used correlation is the one developed by Perre et al. [97] and is shown in Table 13.8. It is based on experimental data retrieved on softwood. In literature, other correlations between saturation and relative permeability are also available [54].

The effective permeability consists of two parts: a first related to the solid porous structure (intrinsic permeability) and a second related to the effect of saturation of pores on the fluid movement (relative permeability). Effective permeability can be calculated as:

$$K_{ph,eff} = K_{ph} \cdot K_{ph,rel} \quad (13.56)$$

where $K_{ph,eff}$ is the effective permeability of a phase, K_{ph} is the intrinsic permeability of a phase, and $K_{ph,rel}$ is the relative permeability of a phase.

13.5.5 *Diffusion*

13.5.5.1 Bound Water Diffusion

The migration of bound water arises only from diffusion through cell walls of biomass. Mathematically, such transport can be described using Fick's law [98]. During pyrolysis and at any given time, bound water does not co-exist with biochar.

By fitting the experimental data of bound water diffusivity in a transverse direction, the following correlations based on the Arrhenius expression were proposed:

Perre and Degiovanni [86]:

$$D_{B,T} = \exp\left(-9.9 + 9.8\, MC_B - \frac{4300}{T}\right) \tag{13.57}$$

Perre and Turner [98]:

$$D_{B,T} = \exp\left(-12.818 + 10.895\, MC_B - \frac{4300}{T}\right) \tag{13.58}$$

Stamm [99] stated that the following dependency exists between diffusion of bound water in different directions:

$$D_{B,T} \cong \frac{1}{3} D_{B,L} \cong \frac{2}{3} D_{B,R} \tag{13.59}$$

where subscripts T, L, and R denote the transverse, longitudinal and radial direction respectively. More complex dependency between bound water diffusion and direction can be found in the works of Pierre and Turner [98, 100].

13.5.5.2 Gas Binary Diffusion

The gas-vapour mixture, which exists in the pores during pyrolysis consists of a variety of compounds in different concentrations and its composition changes as the process progresses. Mathematical description of such a process is not straightforward.

Application of binary diffusion description is valid only for systems where only two major components interact with each other, and there are no other components or their influence on a mixture is negligible. Also, binary diffusion is based on the assumption that one compound has to be indicated as an inert during the whole process. Such a situation is far different from the one that takes place in the pores during the pyrolysis process of biomass. Therefore, the application of the binary diffusion description can lead to significant inaccuracies in prediction. Hence, other more complex ways of describing diffusion have to be applied. A satisfactory procedure which is always valid for a multi-component system is the Maxwell-Stefan equations system, so in theory, its application would be the most valid option [101].

Diffusion is the dominating transport phenomenon only in systems where large pressure gradients do not exist. An increase in the pressure gradient leads to a reduction of the diffusion's share in the overall transport of gases, as convection becomes the dominating phenomenon of transport [52]. During pyrolysis of dry biomass, especially at high temperatures and with a high heating rate, the pressure gradients are significant, which indicates that the diffusion does not play a major role in gas transport. It leads to the conclusion that implementation of the binary diffusion model, which will be rather inaccurate, but fairly simple in implementation and easy in computation should not add a significant inaccuracy to the prediction of fast pyrolysis. In general, it is always advised to try to avoid the application of a robust, global description, which can be overcomplex and simultaneously not lead to visible improvement in modelling accuracy.

On the other hand, for a pyrolysis process of wet biomass, so combined with particle's drying, the diffusion of water vapour can be significant. Especially for pyrolysis of a large particle that is exposed to moderate thermal conditions, where evolved pressure gradients can be insufficient to shift the convection into the dominant transport process. For such situations, an assumption that diffusion is negligible will not be valid. During drying, an inert (most often nitrogen)—water vapour system will appear, which can be described satisfactorily by binary diffusion. Often in practice, the binary diffusion of an inert-water vapour system is treated as an air-water vapour system instead of nitrogen-water vapour system due to the marginal difference in gas properties and higher availability of data for the air-water vapour system.

The air-water vapour binary diffusion coefficient ($D_{A/V}$), in function of the temperature and the pressure inside a particle, can be calculated with the equation proposed by Siau [84]:

$$D_{A/V} = 2.23 \cdot 10^{-5} \frac{T^{1.81}}{< P_G >^G} \tag{13.60}$$

Alternatively, it can be calculated with a more often used equation, proposed by Grønli [52]:

$$D_{A/V} = 1.192 \cdot 10^{-4} \frac{T^{1.75}}{< P_G >^G} \tag{13.61}$$

Correlations above can be used not only for the water vapour but also for other compounds in the pyrolysis gas mixture without introduction of a significant error. If higher accuracy is needed, a discrete description of the binary diffusion coefficient for each component of a system can be calculated with the Chapman-Enskog equation, based on the kinetic gas theory, or with the equation proposed by Poling et al. [102]:

$$D_{inert/i} = 1.43 \cdot 10^{-7} \frac{T^{1.75}}{P\, M_{inert/i}{}^{1/2} \left[\left(\Sigma_v^{inert}\right)^{1/3} + \left(\Sigma_v^{i}\right)^{1/3} \right]^2} \tag{13.62}$$

where $D_{inert/i}$ is the binary diffusion coefficient between an *inert* and an ith component, Σ_v is the sum of the atomic diffusion volumes (from Poling et al. [102]) and $M_{inert/i}$ is the mean molecular mass ratio between an *inert* and an ith compound.

The diffusion phenomena are omitted in certain publications related to modelling of pyrolysis of dry biomass [28, 42, 56, 103]. Authors who modelled the pyrolysis of wet biomass have treated the diffusion coefficients as constant values (range from

10^{-6} m²/s to 10^{-5} m²/s) for all gas species in order not to overcomplicate the model [59, 91, 92, 104]. Such approaches are not fully invalid with respect to the minor role of diffusion in the overall transport of gases in specific cases.

13.5.5.3 Effective Gas Diffusion Coefficient

Besides the gas mixture composition, the structure of the porous material in which the diffusion process takes place has an influence on the diffusion coefficient. The effective gas diffusion coefficient can be defined as:

$$D_{eff,\ inert/i} = \theta\, D_{inert/i} \tag{13.63}$$

where $D_{eff,\ inert/i}$ is the effective inert—i^{th} component diffusion coefficient, $D_{inert/i}$ is the inert—i^{th} component diffusion coefficient and θ is the structure resistance factor (tortuosity factor). The structure resistance factor is an artificial parameter describing the restriction of diffusion in narrow pores, which can be linked to the porosity. The correlation of the structure resistance factor to porosity is obtained by fitting a function to the experimental data. A summary of the correlations available in literature is shown in Table 13.9.

13.5.6 Heat Capacities

13.5.6.1 Heat Capacity of Biomass

In the literature devoted to drying of biomass, empirical correlations can be found which combine the influence of temperature and moisture content (liquid and bound water) on the specific heat of biomass. Since there are no theoretical reasons to combine the effects of both parameters into one correlation, the specific heat of biomass and water will be treated separately.

Biomass starts its degradation in the temperature range from 200 °C to 250 °C. Therefore the range of temperature for which specific heat of biomass has to be described is more narrow than for gas and vapour compounds. One of the most commonly used correlations is the one obtained experimentally by Grønli [52] for spruce wood and is valid in the range from 80 °C to 230 °C:

Table 13.9 Proposed correlations for the structure resistance factor

Theta (θ)	Ref.
$\varepsilon_G^{3/2}$	Bruggeman [105]
$\varepsilon_G^{4/3}$	Millington and Quirk [106]
$0.05\varepsilon_G^{2}$	Stannish [107]
ε_G^{6}	Bonneau and Puiggali [108]
$\frac{\varepsilon_G^{6}}{1.37}$	Fernandez and Howell [109]

$$C_{P,BM} = 1500 + T \tag{13.64}$$

where $C_{P,\ BM}$ is the specific heat of biomass. Dupont et al. [110] conducted an analysis of the specific heat of 19 different biomasses in the temperature range from 40 °C to 80 °C. The result for every biomass shows a linear change of the specific heat with temperature in the investigated range. Taking into account Grønli's correlation, it can be assumed that this trend will be kept until the temperature at which biomass starts to thermally decompose. Averaged for all biomasses used in the study of Dupont et al., the correlation between the specific heat and the temperature has the form:

$$C_{P,BM} = 1032.8 + 3.783\ T \tag{13.65}$$

It is proven that the specific heat of biomass is a function of temperature, but in some older publications, it can be found that the parameter as a constant value [87, 91, 92]. Recent work of Gorensek et al. [111] deserves attention in where the authors, starting from fundamentals of thermodynamics, calculated missing heat capacities of artificial, initial components and their transitional forms from the Ranzi scheme. Thereby, they allowed for the implementation of biomass into the model as a mixture of individual bio-components.

13.5.6.2 Heat Capacity of Char

The most well-known correlation between the specific heat of char and the temperature is the one provided by Raznjevic [67], valid in the range from 0 °C to 1000 °C:

$$C_{P,BC} = 1430 + 0.355\ T + 6.85 \cdot 10^{-4}\ T^2 \tag{13.66}$$

where $C_{P,\ BC}$ is the specific heat of biochar. In literature, also other correlations for specific heat capacity can be found, e.g. one proposed by Larfeldt et al. [93], valid in the range from 0 °C to 800 °C:

$$C_{P,BC} = 420 + 2.09\ T - \frac{7.32 \cdot 10^7}{T^2} \tag{13.67}$$

The specific heat for solids at any given time of the reaction is defined as:

$$C_{P,S} = X_{BM} C_{P,BC} + X_{BC} C_{P,BM} \tag{13.68}$$

where $C_{P,S}$ is the specific heat of the solid.

13.5.6.3 Heat Capacity of Bound and Liquid Water

Liquid water heat capacity ($C_{P,L}$) at the atmospheric pressure does not change significantly within the range from 20 °C to 100 °C. Therefore the value of its heat capacity can be assumed as a constant value of 4.20 kJ/(kg K), which is an averaged value within the mentioned temperature range. The specific heat of the bound water ($C_{P,B}$) is assumed to be slightly higher than the liquid water. Hunt et al. [112] proposed a value of 4.66 kJ/(kg K), but this is a rough estimated value, not measured analytically. For the sake of simplicity, the value of $C_{P,B}$ can be treated as equal to $C_{P,L}$ without introducing significant error.

13.5.6.4 Heat Capacity of Gases and Vapours

The specific heat correlation of compounds in the gas mixture applied in a model depends on the complexity of the kinetic scheme. For all low-molecular compounds and most of the high-molecular compounds data can be obtained from the NIST Chemistry WebBook [113] and Gorensek et al. [111]. In case of missing heat capacity data for a specific compound, the authors suggest to find the data record of a compound with similar mass, chemical structure, and chemical properties and treat it as a representative. If more accuracy is needed, the use of thermodynamically based approaches provided by Gordon and McBride [114] is advised.

For the single component reaction scheme, only four representative compounds have to be described: air, water vapour, gas (1:1 mixture of CO and CO_2) and tar (benzene). For the mentioned compounds, Grønli's correlations [52] can be used:

$$C_{P,Air} = 950 + 0.188\,T \tag{13.69}$$

$$C_{P,v} = 1670 + 0.64\,T \tag{13.70}$$

$$C_{P,Tar} = -100 + 4.4\,T - 1.57 \cdot 10^{-3}\,T^2 \tag{13.71}$$

$$C_{P,Gas} = 770 + 0.629\,T - 1.91 \cdot 10^{-4}\,T^2 \tag{13.72}$$

where $C_{P,}$ is the specific heat and subscript *Air, v, Tar* and *Gas* denotes air, water vapour, tars and gases, respectively. The specific heat for the gas-vapour mix at any time in the process can be obtained from an equation:

$$C_{P,G} = \frac{\sum_{i}^{N} C_{P,i} < \rho_i >^G}{< \rho_G >^G} \tag{13.73}$$

where $C_{P,G}$ is the specific heat of the gas-vapour mix and $C_{P,i}$ is the specific heat for the ith component of the gas mixture.

13.5.7 Dynamic Viscosities of Fluids

13.5.7.1 Dynamic Viscosity of Gases-Vapour Mixture

According to the definition, viscosity is a property of a fluid which indicates its resistance to flow (i.e. continual deformation). The viscosity of fluids depends strongly on temperature and pressure. In the atmospheric pyrolysis, a pressure change during the process is not significant in relation to viscosity, so the pressure influence on fluid viscosity can be omitted. The temperature between the start and the end of the pyrolysis usually exceeds a few hundred degrees, so its influence on the viscosity is significant. Therefore, the temperature dependence of the viscosity should be to implemented into a model.

Similar to heat capacity, the correlations of the viscosity of compounds in the gas mixture applied in a model depend on the complexity of the kinetic scheme. Data for permanent gases and light organic compounds can be found in the NIST database [113]. Heavy organic compounds, for which data is lacking, can be replaced by other, similar compounds and treat them as representatives. The missing data can also be calculated, according to the procedure provided by Poling et al. [102]. For the single component kinetic scheme, the correlations valid in the range from 0 °C to 1000 °C, for air, water vapour, tars and gases, provided by Grønli [52] can be applied:

$$\mu_{G,Air} = 9.12 \cdot 10^{-6} + 3.27 \cdot 10^{-8}\, T \tag{13.74}$$

$$\mu_{G,v} = -1.47 \cdot 10^{-6} + 3.78 \cdot 10^{-8}\, T \tag{13.75}$$

$$\mu_{G,Tar} = -3.73 \cdot 10^{-7} + 2.62 \cdot 10^{-8}\, T \tag{13.76}$$

$$\mu_{G,Gas} = 7.85 \cdot 10^{-6} + 3.18 \cdot 10^{-8}\, T \tag{13.77}$$

where μ_G is the dynamic viscosity of gaseous matter and subscript *Air*, *v*, *Tar* and *Gas* denote air, water vapour, tars and non-condensable gases, respectively. To calculate the viscosity of a gas mix at any given time, the Graham model can be used:

$$\mu_G = \frac{\sum_i^N \mu_{G,i} < \rho_i >^G}{< \rho_G >^G} \tag{13.78}$$

where μ_G is the viscosity of the gas mix and $\mu_{G,\,i}$ is the viscosity of the ith component of the mixture. Above mentioned Eq. (13.78) is appropriate for rough calculations, and it is fully valid only when the molar masses of the mixture components are relatively similar [115]. For a more accurate calculation it is advised to use the Wilkie model with the Herning and Zipperer approximation:

$$\mu_G = \frac{\sum_i^N \mu_{G,i} < \rho_i >^G \sqrt{M_i}}{\sum_i^N < \rho_i >^G \sqrt{M_i}} \tag{13.79}$$

where M_i is the molar mass of the ith component in the mixture. In most of the publications related to modelling, the subject of viscosity is treated with neglect. Most of the authors apply the assumption that the viscosity of gases and vapours is invariant to either the gas mix composition and the temperature and its value is constant, equal to 3×10^{-5} Pa s.

13.5.7.2 Dynamic Viscosity of Liquid Water

As it was mentioned in Sect. 13.4.3.3, only the liquid water has the ability to move actively through convection. The viscosity of liquid water as a function of temperature can be calculated with the equation proposed by Grønli [52]:

$$\mu_L = 1.40 \cdot 10^{-2} - 7.30 \cdot 10^{-5}\, T + 9.73 \cdot 10^{-8}\, T^2 \tag{13.80}$$

where μ_L is the liquid water viscosity. Alternatively the correlation proposed by de Paiva Souza et al. [116] can be used:

$$\log(\mu_L) = -13.73 + \frac{1828}{T} + 1.97 \cdot 10^{-2}\, T - 1.97 \cdot 10^{-5}\, T^2 \tag{13.81}$$

13.5.8 *Thermal Conductivity*

13.5.8.1 Thermal Conductivity of Biomass

For particles in the thermally thick regime, thermal conductivity and radiative thermal conductivity have a major influence on the thermal behaviour of the biomass sample. Therefore their appropriate implementation into the model is crucial in terms of the model accuracy.

In Table 13.10 is shown a summary of thermal conductivity data of different biomasses. The thermal conductivity of biomass depends on the bio-composition and structure of the cell wall as well as on the direction of the cut plane (direction of fibres). A rough analysis of the data indicates that the thermal conductivity of hardwoods in the longitudinal direction is c.a. 1.6 times higher than the thermal conductivity in the radial direction. The difference for softwoods is much higher and the ratio of longitudinal to radial thermal conductivity has a value of 2.7. On average,

Table 13.10 Data of thermal conductivity of different biomasses

Biomass species	Type	Temp. (°C)	Density (d.b.) (kg/m^3)	$\lambda_{BM, L}$ (W/(m K))	$\lambda_{BM, R}$ (W/(m K))	Ref
Fir	S	20	370	0.305	0.112	[117]
Fir	S	20	430	0.387	0.118	
Spruce	S	20	385	0.422	0.087	
Pine	S	20	414	0.450	0.105	
Pine	S	20	438	0.246	0.111	
Pine	S	20	440	0.358	0.313	
Fir	S	20	540	0.350	0.140	[118]
Pine	S	60	450	0.260	0.110	
Pine	S	20	450	0.259	0.098	[119]
Fir	S	20	540	0.340	0.138	
Oak	H	15	710	0.361	0.209	
Spruce	S	20	414	0.279	0.128	[67]
Maple	H	30	710	0.419	0.158	
Beech	H	20	700	0.349	0.209	
Birch	H	21	680/680	0.323	0.214	[120]
			567/473	0.293	0.196	
			543/443	0.291	0.177	
		100	680/680	0.370	0.250	
			567/473	0.309	0.244	
			543/443	0.318	0.207	

S softwood, *H* hardwood, in case two values are mentioned for thermal conductivity, they represent longitudinal and radial thermal conductivity respectively

the difference in thermal conductivity in the longitudinal direction between both wood types is relatively low. The difference between both wood types is more visible for the radial thermal conductivity, where hardwoods show *c.a.* 1.5 times higher thermal conductivity than for softwoods.

13.5.8.2 Thermal Conductivity of Char

The thermal conductivity of char depends strongly on the initial thermal conductivity of the parent biomass, as well as on the pyrolysis process conditions. In Table 13.11 summarised data of char thermal conductivity originating from different biomasses are shown, at different pyrolysis temperatures. In general, an increase in the pyrolysis temperature results in a decrease in the char thermal conductivity. Data indicate that the thermal conductivity in the longitudinal direction is much less sensitive to the pyrolysis temperature than the one in the radial direction (relative change of 1.3 for the longitudinal direction and 2.4 for the radial direction). For chars originating from softwood and pyrolysed at 470 °C, the longitudinal thermal conductivity is on average five times higher than the radial thermal conductivity. It is suspected that

Table 13.11 Data of thermal conductivity of char originated from different biomasses

Biomass species	Type	Temp. (°C)	Temp. of pyro. (°C)	Density (d.b.) (kg/m^3)	$\lambda_{BC,L}$ (W/(m K))	$\lambda_{BC,R}$ (W/(m K))	Ref
Fir	S	50	270	340	0.338	0.112	[117]
			450	264	0.255	0.034	
Fir			270	331	0.325	0.087	
			450	255	0.223	0.032	
Spruce			270	337	0.344	0.105	
			450	249	0.186	0.052	
Pine			270	330	0.265	0.118	
			450	248	0.247	0.049	
Pine			270	360	0.198	0.111	
			450	251	0.188	0.046	
Pine			270	364	0.180	0.131	
			450	269	0.216	0.072	
Maple	H	–	450	200	0.105	0.071	[87]
Miscanthus	GR	–	500	–	0.152		[121]
Switchgrass	GR	–	500	–	0.153		

S softwood, *H* hardwood, *GR* grass

such a large change in the radial direction is related by breaking the continuity of the cell wall's structure caused by the bio-polymers degradation.

The thermal conductivity of solids in a given direction (D = L, R, T) at any given time of the reaction is defined as:

$$\lambda_{S,D} = X_{BM}\lambda_{BM,D} + X_{BC}\lambda_{BC,D} \tag{13.82}$$

where $\lambda_{BM,D}$ and $\lambda_{BM,D}$ denote the thermal conductivity in a given direction for biomass and biochar, respectively.

13.5.8.3 Thermal Conductivity of Liquid and Bound Water

The thermal conductivity of liquid water as a function of temperature can be obtained through the correlation of data from the NIST database [113]:

$$\lambda_L = 0.7695 + 7.5 \cdot 10^{-3}\,T - 1 \cdot 10^{-5}\,T^2 \tag{13.83}$$

In literature, constant values of thermal conductivity of liquid water, i.e. 0.658 W/(m K) [52] can be found. Due to a lack of experimental data regarding the thermal conductivity of bound water, it has to be assumed that its thermal conductivity value is similar to that of liquid water.

13.5.8.4 Thermal Conductivity: Gas Mixture

The thermal conductivity of most of the permanent gases, light and heavy organic compounds can be found in tables [67, 122, 123] or in the NIST database [113]. Heavy organic compounds, for which data is lacking, can be replaced by other, similar compounds and treat them as representatives. The missing data can also be calculated, according to the procedure provided by Poling et al. [80]. For the single component kinetic scheme, correlations between the thermal conductivity and the temperature for air, water vapour and permanent gases are based on data from NIST [113], and they are valid in range from 0 °C to 1000 °C. The correlation for tar (benzene) can be obtained from the work of Zaitseva et al. [124], and it is valid in the range from 320 °C to 660 °C.

$$\lambda_{G,Air} = 9.3 \cdot 10^{-3} + 6 \cdot 10^{-5}\, T \tag{13.84}$$

$$\lambda_{G,v} = -8.1 \cdot 10^{-3} + 1 \cdot 10^{-3}\, T \tag{13.85}$$

$$\lambda_{G,Tar} = -5.07 \cdot 10^{-1} + 1.1 \cdot 10^{-3}\, T \tag{13.86}$$

$$\lambda_{G,Gas} = 1.01 \cdot 10^{-2} + 4 \cdot 10^{-5}\, T \tag{13.87}$$

Analogous to the viscosity, the thermal conductivity of a gas mixture at any time in the pyrolysis process can be calculated with the equation:

$$\lambda_G = \frac{\sum_i^N \lambda_{G,i} < \rho_i >^G}{< \rho_G >^G} \tag{13.88}$$

where $\lambda_{G,i}$ is the thermal conductivity of the ith component in the gas mix. For more accurate calculations, the Wassilijewa's equation with the Herning and Zipperer approximation can be used:

$$\lambda_G = \frac{\sum_i^N k_{G,i} < \rho_i >^G \sqrt{M_i}}{\sum_i^N < \rho_i >^G \sqrt{M_i}} \tag{13.89}$$

where M_i is the molar mass of the ith component in the mixture. Many authors use simplifications and implement the thermal conductivity of the gas mix as a constant value in the range from 0.025 W/(m K) to 0.026 W/(m K) [52, 56, 91, 103].

13.5.8.5 Radiative Heat Transfer

When the pyrolysis temperature exceeds 600 °C, the share of the heat transferred through radiation within the particle starts to become significant. In such cases, implementation of the radiative heat transfer into the model is necessary. Radiative thermal conductivity within a particle can be defined as:

$$\lambda_{rad} = A\, \ell_n\, \sigma\, T^3 \tag{13.90}$$

where λ_{rad} is the radiative thermal conductivity, ℓ_n is the photon's mean free path, σ is the Stefan-Boltzman constant and A is the coefficient of the radiative model. In Table 13.12 are presented the most commonly used correlations for radiative thermal conductivity, others can be found in work of Grønli [52].

Where ω is the surface emissivity and d_{pore} is the average diameter of the pores, calculated as:

$$d_{pore} = X_{BM}\, d_{pore,BM} + X_{BC}\, d_{pore,BC} \tag{13.91}$$

where d_{pore} is the average pore's size and subscripts BM and BC denote the biomass and the biochar, respectively. The average pore size of biomass or biochar in the equation above is obtained from the whole range of pores existing in the structure (micro-, meso- and macropores). Therefore, its value should be obtained by helium pycnometry.

Regarding the influence of the pore size on the radiative thermal conductivity, the work of the Janse et al. [104] is worth to mention. They proposed a division of the radiative thermal conductivity in the macropore radiative conductivity and the micropore radiative conductivity. Such an approach seems intuitively reasonable and in theory, it should be more accurate. Nonetheless, the lack of reliable data regarding the pore size distribution of the biochar and its evolution throughout pyrolysis does not allow to obtain solid proof. In the literate applications of Janse et al.'s radiative heat transfer model with an averaged pore size [57, 65] can be found.

13.5.8.6 Effective Thermal Conductivity

The effective thermal conductivity depends on the following factors: anisotropy of the structure, porosity and pore size distribution, bio-composition of the cell wall and water content. In literature examples of correlations for the thermal conductivity of

Table 13.12 Models of radiative thermal conductivity

A	ℓ_n	Ref.
4	$\frac{\varepsilon_G \omega d_{pore}}{(1-\varepsilon_G)}$	Pantoon and Ritman [125]
13.5	$\frac{d_{pore}}{\varepsilon_G \omega}$	Chan et al. [59]
1	$\frac{d_{pore}}{\omega}$	Di Blasi [92]

wet biomass particles obtained empirically can be found. In most cases, they were applied for the description of a drying process, not pyrolysis combined with drying [50, 72, 84, 86, 126]. In order to describe the thermal conductivity of biomass, the following general equation is often used:

$$\lambda_{eff,D} = \lambda_{cond,D} + \lambda_{rad} \tag{13.92}$$

where $\lambda_{eff,D}$ is the effective thermal conductivity, $\lambda_{cond,D}$ is the thermal conductivity, λ_{rad} is the internal radiative thermal conductivity and the subscript D denotes the direction of the conduction. The thermal conductivity can also be treated as a function based on conduction through the solid matter with respect to the heat transfer direction ($\lambda_{S,D}$), the conduction through the liquid and bound water (λ_L, λ_B) and the conduction through gas filling the pores (λ_G). The last three terms are not directionally dependent.

$$\lambda_{cond,D} = f(\lambda_{S,D} + \lambda_L + \lambda_B + \lambda_G) \tag{13.93}$$

The share of each thermal conductivity component is proportional to its volume fraction, so the effective thermal conductivity within a particle can be defined as:

$$\lambda_{eff,D} = (\varepsilon_S\, \lambda_{S,D} + \varepsilon_L\, \lambda_L + \varepsilon_B\, \lambda_B + \varepsilon_G\, \lambda_G) + \lambda_{rad} \tag{13.94}$$

More detailed approaches on the modelling of the thermal conductivity can be found in work of Suleiman et al. [120], Thunman and Leckner [127], Blondeau and Jeanmart [40] and Gentile et al. [53]. All mentioned approaches are based on the comprehensive thermal conduction model developed by Kollmann and Côte [128].

13.5.9 Surface Emissivity

Radiative heat emissivity from natural surfaces is usually modelled as a "grey body". According to the definition and the Stefan-Boltzmann law, a "grey body" is an intermediate material between a perfect absorber of light (ideal "black body") and a perfect reflector of light (ideal "white body"). The value of the emissivity of the "grey body" depends on the surface's temperature, colour and roughness. For biomass, emissivity (ω_{BM}) is usually assumed to be in the range between 0.7 and 0.85 [64, 65] and for biochar (ω_{BC}) in the range between 0.9 and 0.95 [52, 59]. The surface emissivity (ω) at any time of the process can be defined as:

$$\omega = X_{BM}\omega_{BM} + X_{BC}\omega_{BC} \tag{13.95}$$

13.5.10 Particle Shrinking

Drying and thermal degradation of a biomass particle have an influence not only on its porosity and thermo-physical properties but also on its overall geometry and shape. To model a particle's change in geometry and shape, usually, one out of three methods of shrinking process description is applied: uniform shrinkage, shrinking shell, or shrinking cylinder.

An analysis of shrinkage during pyrolysis shows that final shrinkage in the longitudinal direction is lower than in the radial direction. Additionally, for small particles, the shrinking takes place mostly at the end of the conversion. During the conversion under a high heating rate, strong mechanical tensions occur within a particle, which leads to particle cracking and in some cases, even to the fragmentation of a particle. Besides the reduction in particle shape by shrinking, the expansion via the swelling can take place. The expansion can be observed usually at the beginning of the conversion, especially for large particles [129]. The details regarding cracking and swelling are not incorporated in mentioned models of shrinking.

A detailed description of mentioned shrinking models can be found in the work of Bryden et al. [61] and Bellais [130]. The most commonly used shrinking model is the uniform shrinking model, so its basis will be briefly described here. Shrinking in a selected direction can be defined as [61]:

$$f_D = \frac{current\ dimension}{original\ dimension} = \frac{L_D}{L_{D0}} \tag{13.96}$$

where f_D is the shrinkage factor in the D direction, L_D is the dimension after shrinkage in the D direction and L_{D0} is the original dimension in the D direction. The uniform shrinking model assumes, that particle size change is directly proportional to the mass loss, so it can be stated that:

$$f_D = \left[(\overline{\eta} - 1)p_D + \left(\overline{\mathrm{M}} - 1\right)m_D\right] \tag{13.97}$$

where $\overline{\eta}$ is the conversion extent of pyrolysis, averaged over the particle's geometry, p_D is the parameter of final shrinkage due to pyrolysis in direction D, $\overline{M}$ is the progress of the particle's drying, averaged over the particle's geometry, m_D is the parameter of final shrinkage due to drying in the D direction. The drying influence on the shrinkage in any direction is not large (6–7%) [61], so its omission does not introduce significant accuracy loss in modelling [131]. Applying the aforementioned simplification, the following equation is obtained:

$$f_D = [(\overline{\eta} - 1)p_D] \tag{13.98}$$

The averaged conversion extent of the pyrolysis reaction is defined as:

Table 13.13 Exemplary values of final shrinkage parameter in pyrolysis, and obtained experimentally

$p_L = \alpha$	$p_R = \beta$	$p_T = \gamma$	Ref.
0.34	0.50	1	[93]
0.30	0.40	1	Low HR [132]
0.30	0.05–0.20	1	High HR [132]

HR heating rate

Table 13.14 Parameters of Davidson et al. correlation [133]

D	a_D	b_D	c_D
L	−4.7	1.08×10^{-2}	-5.86×10^{-6}
R	4.4	-8.56×10^{-3}	4.55×10^{-6}
T	−1.0	3.95×10^{-3}	-2.62×10^{-6}

D direction, *L* longitudinal, *R* radial, *T* tangential

$$\bar{\eta} = \frac{current\ mass\ of\ biomass}{inital\ mass\ of\ biomass} = \frac{< \rho_{BM} >}{< \rho_{BM0} >} \tag{13.99}$$

where $<\rho_{BM0}>$ is the initial apparent density of biomass. A more in-depth description of changes to the conservation equations due to implementation of the uniform shrinking model can be found in the work of Bryden et al. [61] and Anca-Couce et al. [56].

Exemplary final shrinkage parameters that can be found in the literature are summarised in Table 13.13. The symbols α, β and γ denote the final shrinkage in the longitudinal (p_L), radial (p_R) and transversal (p_T) direction, respectively. The data in the table indicates that the final particle shrinkage depends not only on the pyrolysis temperature but also on the heating rate.

An extensive analysis of particle shrinkage was performed by Davidson et al. [133]. It resulted in a correlation between the highest temperature in pyrolysis and the final shrinkage parameter in direction *D*. The correlation is valid for temperatures from 350 °C to 700 °C.

$$p_D = {a_D}^2 \left[a_D + b_D T + c_D T^2\right] \tag{13.100}$$

where a_D, b_D and c_D are correlation parameters obtained by fitting to experimental data. Their values are shown in Table 13.14.

13.6 Boundary Conditions

13.6.1 Boundary Conditions Equations

Boundary conditions are the drivers of the modelled process through the description of phenomena occurring on the geometrical surface of a particle. In other words, boundary conditions define the behaviour of the nodes located on the geometry edge. In the most general case, they are defined as [134]:

- For pressure:

$$<P_G>^G|_{x=x_p} = P_\infty \tag{13.101}$$

- For heat transfer:

$$\nabla(\lambda_{eff}\nabla T)|_{x=x_p} = h_T\left(T_{flow,\infty} - T|_{x=x_p}\right) + \sigma\omega\left(T^4_{wall} - T^4|_{x=x_p}\right) \tag{13.102}$$

- For mass transfer:

$$D_{eff}\nabla(<\rho_i>^G)|_{x=x_p} = h_m\left[\rho_{i,\infty} - <\rho_i>^G|_{x=x_p}\right] \tag{13.103}$$

where $|_{x=x_p}$ denotes the position (x_p point of the surface, "x" can be adapted according to the appropriate coordinate system), P_∞ is the pressure of the environment (ambient), h is the convective transfer coefficient, subscript T and m denote heat and mass respectively, $T_{flow,\infty}$ is the temperature of the flowing fluid at a considerable distance from the particle's surface, T_{wall} is the temperature of the reactor wall, σ is Stefan-Boltzmann coefficient, ω is surface emissivity and $\rho_{i,\infty}$ is the density of the ith compound at a considerable distance from the particle's surface. Even though the radiative heat transfer at a temperature below 600 °C does not have a large share in the total heat exchange [135], its implementation is not complex and can result in improvements in model accuracy.

13.6.2 Dimensionless Numbers and Transfer Coefficients

Convective heat and mass transfer coefficients can be obtained from correlations of the dimensionless numbers.

- The convective heat transfer coefficient from the Nusselt number:

$$Nu = \frac{h_T L}{\lambda_{eff}} \tag{13.104}$$

- The convective mass transfer from the Sherwood number:

$$Sh = \frac{h_m L}{D_{eff}} \tag{13.105}$$

where L is the characteristic dimension of a particle, h_T is the convective heat transfer coefficient, h_m is the convective mass transfer coefficient, λ_{eff} is the effective thermal conductivity and D_{eff} is the effective diffusion coefficient. For laminar flow, the dimensionless numbers can be obtained from flow-shape correlations presented in Table 13.15.

Where Pr is Prandtl's number, Sc is Schmidt's number, μ is the dynamic viscosity of the gas mix, and subscript ∞ and S denote the free stream and the surface (on the fluid's side) respectively. Above mentioned correlations are valid only for particles immersed in a single-phase flow. It is advised to use other correlations to calculate the convective heat and mass transfer coefficient of particles immersed in two-phase flows (e.g. gas-solid systems in a fluidised bed), [92, 137–140].

Rapid evaporation or ignition of evolved pyrolysis gases can cause a temporary disturbance in the convective transfer. In order to account for it in a model, Stefan's correlation can be used to calculate the mass and heat convective transfer coefficients with an extensive outflow from surfaces [57, 65, 92, 101]. Correlations are defined as:

- Heat transfer:

$$h_{Ts} = \frac{A_S u_G \varepsilon_G < \rho_G >^G C_{P,G}}{\exp\left(\frac{A_S u_G \varepsilon_G < \rho_G >^G C_{P,G}}{h_T}\right) - 1} \tag{13.106}$$

- Mass transfer:

$$h_{ms} = \frac{A_S u_G}{\exp\left(\frac{A_S u_G}{h_m}\right) - 1} \tag{13.107}$$

where h_{T_s} is the convective heat transfer coefficient with surface outflow, h_{ms} is the convective mass transfer coefficient with surface outflow and A_S is the external surface area of a particle.

In literature, exemplary values of the convective heat transfer coefficient can be found: flat plate—5 W/(m^2 K) [52], sphere—20 W/(m^2 K) [28], shapes with different size from 8.4 W/(m^2 K) to 20 W/(m^2 K) [141], particles in a fixed bed—50 W/(m^2 K) [142] or particles in a fluidised bed—400 W/(m^2 K) [143]. Not as many examples for the convective mass transfer coefficient can be found: flat plate—0.03 m/s [52]. It is possible that many authors consider that the mass transfer from a particle is not hindered nor enhanced. Therefore the convective mass transfer coefficient is equal to the superficial gas velocity on the surface.

Table 13.15 Dimensionless number correlations as a function of shape [134, 136]

Particle shape	Convective heat transfer	Convective mass transfer
Flat plate (Kerith and Black eq.)	$Nu = 0.644Re^{0.5}Pr^{0.33}$	$Sh = 0.644Re^{0.5}Sc^{0.33}$
Cylindrical (Churchill and Bernstein eq.)	$Nu = 0.3 + \frac{0.62Re^{0.5}Pr^{0.33}}{\left[1+\left(\frac{0.4}{Pr}\right)^{0.66}\right]^{0.25}}\left[1+\left(\frac{Re}{2.82\cdot 10^{5}}\right)^{0.625}\right]^{0.8}$	$Sh = 0.3 + \frac{0.62Re^{0.5}Sc^{0.33}}{\left[1+\left(\frac{0.4}{Sc}\right)^{0.66}\right]^{0.25}}\left[1+\left(\frac{Re}{2.82\cdot 10^{5}}\right)^{0.625}\right]^{0.8}$
Spherical (1st Ranz-Marshall eq.), (2nd Whitaker eq.)	$Nu = 2 + [0.6Re^{0.5}Pr^{0.33}]$ $Nu = 2 + \left[0.4Re^{0.5} + 0.06Re^{0.66}\right]Pr^{0.4}\left(\frac{\mu_{\infty}}{\mu_s}\right)^{0.25}$	$Sh = 2 + [0.6Re^{0.5}Sc^{0.33}]$ $Sh = 2 + \left[0.4Re^{0.5} + 0.06Re^{0.66}\right]Sc^{0.4}\left(\frac{\mu_{\infty}}{\mu_s}\right)^{0.25}$

13.7 Reactor Model and Multiscale

The gases and the solid phase (processed biomass) in a given reactor have significant differences in physical, chemical and thermal behaviour. Therefore, in this section, the description of each phase separately needs to be considered as well as interactions between the phases.

13.7.1 Lagrangian method: Particle Movement Description

In reactor systems, the movement of every single particle is independent. The method which allows for describing the behaviour of each individual particle is the Lagrangian approach, which is based on Newton's second law of motion [56, 144]. In the Lagrangian framework, each particle is modelled with its own body (subdomain), which moves independently in an applied geometry according to the forces affecting the particle. This framework allows for investigating the time-position relation of each particle (e.g. trajectory).

The framework of the Lagrangian method also allows for the implementation of mechanical interactions between particles and between phases. Consecutively, it opens the possibility for implementation and investigation of particle-particle and particle-wall interactions. The visualisation of the basic difference between the Eulerian and the Lagrangian approach is shown in Fig 13.6.

13.7.2 Methods of Two-Phase Flow Description

The Eulerian approach is sufficient to describe a single-phase flow and all significant unit processes occurring in it. Unfortunately, such an approach may not be sufficient to describe two-phase flows (e.g. gas-solid systems) appropriately.

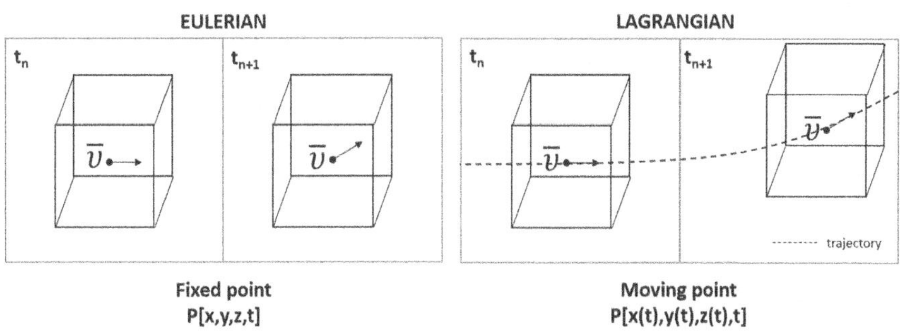

Fig. 13.6 Simplified visualisation of the difference between Eulerian and Lagrangian approach

A comprehensive and complete description of the behaviour of two-phase flow is provided by a combination of Computational Fluid Dynamics (CFD) with the Discrete Element Method (DEM) resulting in the so-called Eulerian-Lagrangian approach (CFD-DEM method). The first part of the name indicates that the gas phase description (fluid with continuum properties) is done according to the Eulerian method. In the Eulerian approach, fluid properties are stored in grid nodes of the applied geometry. The fluid movement does not interfere with the grid arrangement. The second part of the name indicates that the description of the solid phase (particles) is done with the Lagrangian approach. In this approach, solid particles are not linked to the grid used for modelling fluid dynamics, and the subdomains of particles can move freely through the applied geometry. Nevertheless, both phases are interconnected, so, e.g. the movement of the particles causes changes in the fluid phase, and the flow of the fluid can alter the movement of particles.

With an increase in the number of investigated particles as well as with an increase of the complexity of the single-particle behaviour, the quantity of data that needs to be handled by the solver grows exorbitantly. Therefore, a proper description of the investigated system with the full Eulerian-Lagrangian method, besides an in-depth knowledge about its fundamentals, needs robust numerical software tools and powerful computing hardware.

The Discrete Phase Model (DPM) is a hybrid method, and it is based on a partial simplification of the DEM method. The DPM method still has its base in the Lagrangian description and takes into accountsthe particle's movement resulting from forces like: gravity, drag force, pressure force and Magnus force, but the particle-particle collisions are neglected. Additionally, the Discrete Phase Model method omits the fluid volume's displacement by particles, so the volume of a fluid phase remains constant. Recommendations with respect to choosing the simplification from DEM to DPM, are not clear in the literature. The cause of this can be linked to the difference in types of reactors that were modelled with the use of the simplification. The most general recommendation is to apply the simplification in cases when the solid phase is strongly dispersed, and its volume fraction is less than 5 vol. % [145].

The Dense Discrete Phase Model method (DDPM), an improved version of the DPM is an another, more recently developed hybrid method. The DDPM method is capable of handling higher volume fractions of the solid phase, and it has an in-built particle-particle collision sub-model through a collision component taken from the DEM method. One of the drawbacks of the DDPM method is that the flow around particles is not taken into account during the simulation, so the dynamic behaviour of particles still can differ from reality. Both hybrid approaches (simplifications) lead to a significant reduction in the computational burden in comparison to the full DEM description method. Also, the influence of applying these simplifications on the accuracy loss strongly depends on the modelled scenario (reactor type, number of particles and their size).

In peculiar cases, when the size of the particles is sufficiently small and the particles are strongly dispersed among the fluid phase (suspension), there is the possibility of a strong simplification with the assumption that the particles suspended in a fluid are "dissolved" in it. Therefore they can be treated as part of the fluid, and they can behave as such (quasi-continuous solid phase). The method of describing a two-phase system where both phases are treated as a continuum is called the Eulerian-Eulerian approach. It indicates that both phases, fluid and solid, are described by the Eulerian approach, so the model does not distinguish each particle in the solid phase. Therefore in this approach, it is impossible to investigate the single particle movement. The Eulerian-Eulerian approach is the least computationally burdening method of simulating two-phase flow. Moreover, the simplification is very convenient in terms of mathematical description. Expressions used for describing the movement, thermal and chemical behaviour of the solid phase have the same construction as those used for the fluid phase description. There is a strong restriction regarding the application of this simplification. The Eulerian-Eulerian approach for particles with relatively large size introduces a significant deviation from reality in the model. In such cases, the result of the simulation is burdened with a considerable inadequacy, so its accuracy is low.

13.7.3 Particle Conversion Regimes and Two-Phase Flow Models

The fluid phase in a reactor is always described with the Eulerian approach. In comparison to the description of the solid phase, it makes the fluid phase a less challenging part of the reactor model. The description of the fluid phase has to contain, among others: fluid motion within the reactor geometry, changes of fluid phase volume due to particle movement and rotary elements (if any), the heat exchange between the fluid and reactor's walls and particles. The description of the fluid phase in the reactor has to cope also with the chemical behaviour of the compounds (e.g., secondary tar cracking) that are contained in it.

As it has been mentioned, the possibility of simplifying the description of the two-phase flow into the a quasi-one phase fluid flow (Eulerian-Eulerian) is only valid when particles immersed in the fluid are sufficiently small. To assess if this simplification is valid, values of two non-dimensional numbers have to be checked: the thermal Biot number (Bi) and the Pyrolysis number (Py), the latter is also called reversed thermal Thiele modulus [56]. Those numbers indicate to which thermal regime the investigated particles belong. Each of the regimes indicate which thermal phenomena (chemical reactions, intra-particle or extra-particle heat exchange) have the strongest influence on the rate of the particle's conversion [24, 91, 146–148]. Particles can be assigned to one of the four following thermal regimes: pure kinetic, thermally thin, thermal wave and thermally thick.

The simplification through the Euler-Euler approach is the most valid for particles in the pure kinetic regime, whose size usually is smaller than 1 mm in any direction [5]. Conversion of particles in the thermal thin regime is also driven mostly by the reaction kinetics, but also external heat transfer starts to play a significant role. Due to a relatively small size, those particles do not show high thermal or internal pressure gradients during conversion. The application of this simplification for particles in the thin thermal regime is not advised, but it would not introduce a critical error to the model. For this regime, the dilution of the solid phase also has to be taken into account. In case of a highly concentrated solid phase, the Eulerian-Eulerian approach is not valid, so more sophisticated description methods (DPM or DDPM) have to be applied to obtain more accurate and reliable description.

The conversion of particles assigned to the thermal wave regime is mostly driven by the internal and external heat transfer. Additionally, a significant temperature and pressure gradient is formed during the conversion. For particles in the thermal wave regime, the particle's location during the process starts to play a major role in its conversion. Therefore, applying the Euler-Euler simplification is not valid for particles in this regime, and thus they have to be described with a Lagrangian approach. It is expected that, in the thermal wave regime the conversion of the particle takes place in a thin surface front, so the assumption that the conversion front thickness strives to 0, is not a large departure from reality. Such an approach opens a possibility of a partial simplification of describing the conversion process. The simplification can be made by implementation of the unreacted shrinking core model or the layer model [5, 24, 149].

The internal heat transfer has the largest share in the control of the conversion of particles in the thermally thick regime. To this regime are assigned the largest particles, which show the highest temperature and pressure gradients during conversion. There is no stiff border, from which point the particles have to be assigned to the thermally thick regime. In literature, it can be found that the particle is considered to be in the mentioned regime if the Bi number is higher than 40 or 100 and the thermal Thiele modulus (1/Py) number is higher than 100 or 1400 [148, 150]. The conversion of the particle in the thermally thick regime is the most complex and cannot be simplified, so only a detailed description via the Eulerian-Lagrangian approach is valid (DDPM or DEM).

13.7.4 Appropriate Model for Different Kinds of Beds

For fixed bed reactors, the only limitation for the particle size are the reactor dimensions. Therefore relatively large biomass particles (e.g. logs or large chunks) can be processed in a fixed bed. For this reactor type, the movement of particles is negligible, and the mixing of solids is insignificant. Taking this into account, Wurzenberger et al. [151] proposed the Representative Particle Model (RPM), suitable for the description of the conversion of single particles in fixed bed reactors. The method assumes that parameters of biomass conversion can be treated as homogenous for the whole reactor, so all processed particles show the same behaviour. In consequence, it leads to the conclusion that for the RPM, the single-particle model needs to be solved for one representative particle only once for the applied boundary and initial conditions. Application of the RPM method for modelling fixed beds reduces the computational time significantly and shows moderately good agreement with experimental results [5, 152, 153].

An additional challenge is brought into the reactor model description for systems in which particles are in motion. Movement of particles can be driven by changes in pressure of a fluid (pneumatically driven) or by the physical forces transmitted to the particles via the reactor's rotary elements (mechanically driven). In the second scenario, the moving element also has an influence on the gas motion in the reactor, and this needs to be taken into account in the description of a model. The selection of the driver of the particle's movement imposes practical boundaries on the size of particles that can be processed in the reactor.

For fluidised beds, the particle size has to be significantly small to be able to be suspended and/or dragged by the fluidising gas. In general, the size of particles that can be applied in fluidised bed reactors does not exceed 2–3 mm. Application of such small particles in fluidised bed reactors opens possibilities for model simplification (Eulerian-Eulerian, DPM or DDPM). An implementation of the simplification leads to a significant reduction in complexity of the description and simultaneously, it lowers the computational burden.

In processing in rotary reactors (auger/screw or rotary kiln reactors), the size of particles is usually larger than in fluidised bed reactors. The maximum size of particles for rotary reactors is limited by the dimensions of the reactor and its moving parts (e.g. size of a screw and its pitch), the reactor's mechanical durability and the homogeneous distribution of solid material in the reactor. The particles processed in rotary reactors cannot be assigned to the kinetic thermal regime due to their large size. Therefore there is no possibility of applying the Euler-Euler approach for those systems. For rotary reactors, the Eulerian-Lagrangian approach of the two-phase flow has to be implemented (DPM, DDPM, or DEM). It has to be kept in mind that for rotary reactors, the influence of the movement of the reactor's elements as well as of the particles on the fluid phase has to be included in the model description. Models for rotary reactors are the most demanding, both for the modeller (interdependences between phases, number of correlations and parameters), as well as for the software and hardware used to conduct computation on such complex systems.

An extensive and comprehensive overview of the application of different approaches for describing specific reactor modelling cases can be found in the recent work of Xiong et al. [154]. The work contains numerous references to examples from literature, so the authors strongly recommend this review for readers interested in the subject. Among many others, the works of Subramaniam [155], Ku et al. [156, 157] and Xie et al. [158] are worth to mention, as they contain the mathematical description of the Eulerian-Lagrangian method, as well as the work of Funke et al. [159] in which, for the first time the heat transfer between particles in an auger type reactor was calculated using a combined fundamental heat transfer model with DEM simulation.

13.7.5 Reactor Model and Limitations

Increase in complexity of a comprehensive model for biomass conversion, in principle, is done to bring a model closer to reality and improve its prediction accuracy. Simultaneous application of detailed models causes the issue requiring a vast amount of data and correlations that need to be handled and computed by a solver. A vast increase in computational load requires simultaneously a higher need for hardware power to obtain adequate solving efficiency. A very complex model and limited computational resources result in elongated computational time, which does not allow for a rapid refinement of the model to the investigated scenario. Therefore, model complexity is a bottleneck for the investigation and the development of the reactor technology under study. For a modeller, it is crucial to select the level of complexity that simultaneously will fulfil a required, satisfactory accuracy of a prediction, will be technically possible in implementation and will be feasible in terms of time and cost.

The reactor submodel of a comprehensive biomass conversion model is the most difficult and the most complex part among all model parts, so a short elaboration on its problems is provided here. An increase of the computational demand needed to solve a reactor model, besides the increased complexity of solid phase movement description (e.g. via application of DEM), is caused by expanding the meshed geometry of a reactor domain as well as by an increase in number of particles that have to be considered. Besides high requirement of the hardware computational resources, an additional issue is connected to the application of the DEM method in the solid phase description. A detailed description of solid-phase interactions and mechanical changes of particles is not fully developed yet, so there is no certainty that already established solid-phase descriptions are accurate in their predictions. Another issue that hinders the effective use of the complex reactor models is the in-depth knowledge on how to use the computational resources in an economical and an effective manner (e.g. parallelisation of computation, adjustment of the procedure of a solver) [5]. From the side of practice, there also exists a problem with the insufficiently developed software, which can have problems with mesh adaptation in more complex scenarios which form a barrier in modelling, like in e.g. modelling a double-screw rotary reactor [160].

13.8 Conclusions

Numerical modelling is a very robust tool, which allows for cost-effective research and development of technologies within the field of biomass thermal processing. As it is indicated in this chapter, proper construction and use of a comprehensive model needs knowledge from different areas of science. Only through their combination in an efficient manner, the model will lead to reasonable and useful results.

Table 13.16 Descriptive summary of components of a comprehensive biomass conversion model

Submodel	Molecular	Single particle	Reactor
Used for	Investigation of biomass degradation chemistry depending on the initial feedstock composition	Investigation of particles' thermo-physical and structural changes and their influence on the pyrolysis product yields and composition	Investigation of the influence of large-scale production parameters on the product quality and process efficiency
Possible to predict	Pyrolysis product yields and composition	Yields and composition of pyrolysis products, mass loss, temperature distribution, pressure distribution, shape and porosity in single particles	Product streams and their composition, size distribution of solids, mass and heat transfer distribution in a reactor, production quality and efficiency
Particles size/ thermal regime	Only fine powders, which belong to the kinetic regime, for other thermal regimes the influence of structural and material thermo-physical factors will introduce bias	Theoretically applicable to every size of a given particle (and associated thermal regime), in practice it is not efficient to model kinetic regime	Applied simplification depends on the particle size (thermal regime): small particles, kinetic regime—Eulerian-Eulerian, medium size particles, thermally thin regime—Eulerian-Lagrangian (DPM, DDPM), large particles, thermally thick regime—Eulerian-Lagrangian (DEM)
Complexity	Simple, only needs thermodynamic data for the compounds in the kinetic scheme	Complex, besides the data for reaction kinetics, the model also requires material's thermo-physical and structural information and their changes with conversion	The most complex, requires data of molecular and single particle model as well as data of particle-wall, particle-particle and particle-reactor gas interactions
Computational burden	Low, numerical solver depends on the complexity of the applied kinetic scheme	Medium, numerical solver is needed, depends on the complexity of the molecular model, the structure-chemistry interconnections and thermo-physical changes description	High, robust numerical solver essential, depends on the single particle model complexity, number of modelled particles and applied simplifications
Knowledge-gap to fill urgently	Quantitative influence of the mineral matter and heating rate on the degradation mechanism	The link between pore size distribution, gas permeability and extent of conversion; a reliable model of solid thermal conductivity	A detailed description of the mechanical behaviour of particles and their interactions

In theory, there is no limitation to model every processing technology or to base the model on parameters for any range (feedstock or process-related). Nonetheless, from a practical point of view, the selected environment of conversion, as well as the applied processed material, imposes strong boundaries on the modelled system. Those boundaries impose limits on possibilities of the model's validation against experimental data, which is the only reliable method to assess model performance.

The level of complexity and the proper selection of components of the model has a significant influence on the model's accuracy and reliability. A general, descriptive summary of the submodels of a comprehensive biomass conversion model is presented in Table 13.16. In general, the balance between accuracy and computational efficiency as well as the technical feasibly have to be obtained. It is advised to apply the most detailed description when it is feasible, and always a check if the model cannot be simplified without loss in model accuracy. This balance has to be taken as one of the priorities in modelling practice.

Acknowledgements Work made within the **Greencarbon Project**, which has received funding from the European Union's Horizon 2020 research and innovation programme under the Marie Skłodowska-Curie grant agreement No 721991.

References

1. Antal MJ, Grønli M. The art, science, and technology of charcoal production. Ind Eng Chem Res. 2003;42(8):1619–40. https://doi.org/10.1021/ie0207919.
2. Dubois G. Modeling and simulation: challenges and best practices for industry. Boca Raton: CRC Press; 2018. https://doi.org/10.1201/9781351241137.
3. Balci O. Guidelines for successful simulation studies. In: Balci O, editor. 1990 winter simulation conference proceedings. San Diego: IEEE; 1990. p. 25–32. https://doi.org/10.1109/WSC.1990.129482.
4. Kennedy MC, O'Hagan A. Bayesian calibration of computer models. J R Stat Soc. 2001;63(3):425–64. https://doi.org/10.1111/1467-9868.00294.
5. Anca-Couce A. Reaction mechanisms and multi-scale modelling of lignocellulosic biomass pyrolysis. Prog Energy Combust Sci. 2016;53:41–79. https://doi.org/10.1016/j.pecs.2015.10.002.
6. Debiagi PEA, Pecchi C, Gentile G, Frassoldati A, Cuoci A, Faravelli T, Ranzi E. Extractives extend the applicability of multistep kinetic scheme of biomass pyrolysis. Energy Fuel. 2015;29(10):6544–55. https://doi.org/10.1021/acs.energyfuels.5b01753.
7. Vassilev SV, Baxter D, Andersen LK, Vassileva CG. An overview of the chemical composition of biomass. Fuel. 2010;89(5):913–33. https://doi.org/10.1016/j.fuel.2009.10.022.
8. Vassilev SV, Baxter D, Andersen LK, Vassileva CG, Morgan TJ. An overview of the organic and inorganic phase composition of biomass. Fuel. 2012;94:1–33. https://doi.org/10.1016/j.fuel.2011.09.030.
9. Vassilev SV, Baxter D, Andersen LK, Vassileva CG. An overview of the composition and application of biomass ash. Part 1. Phase–mineral and chemical composition and classification. Fuel. 2013;105:40–76. https://doi.org/10.1016/j.fuel.2012.09.041.
10. Vassilev SV, Baxter D, Andersen LK, Vassileva CG. An overview of the composition and application of biomass ash.: part 2. Potential utilisation, technological and ecological advantages and challenges. Fuel. 2013;105:19–39. https://doi.org/10.1016/j.fuel.2012.10.001.

11. Mamleev V, Bourbigot S, Le Bras M, Yvon J. The facts and hypotheses relating to the phenomenological model of cellulose pyrolysis: interdependence of the steps. J Anal Appl Pyrolysis. 2009;84(1):1–17. https://doi.org/10.1016/j.jaap.2008.10.014.
12. Thy P, Yu C, Jenkins BM, Lesher CE. Inorganic composition and environmental impact of biomass feedstock. Energy Fuel. 2013;27(7):3969–87. https://doi.org/10.1021/ef400660u.
13. Ronsse F, van Hecke S, Dickinson D, Prins W. Production and characterization of slow pyrolysis biochar: influence of feedstock type and pyrolysis conditions. GCB Bioenergy. 2012;5(2):104–15. https://doi.org/10.1111/gcbb.12018.
14. Ronsse F, Bai X, Prins W, Brown RC. Secondary reactions of levoglucosan and char in the fast pyrolysis of cellulose. Environ Prog Sustain Energy. 2012;31(2):256–60. https://doi.org/10.1002/ep.11633.
15. Patwardhan PR, Brown RC, Shanks BH. Product distribution from the fast pyrolysis of hemicellulose. ChemSusChem. 2011;4(5):636–43. https://doi.org/10.1002/cssc.201000425.
16. Faravelli T, Frassoldati A, Migliavacca G, Ranzi E. Detailed kinetic modeling of the thermal degradation of lignins. Biomass Bioenergy. 2010;34(3):290–301. https://doi.org/10.1016/j.biombioe.2009.10.018.
17. Nowakowska M, Herbinet O, Dufour A, Glaude P-A. Detailed kinetic study of anisole pyrolysis and oxidation to understand tar formation during biomass combustion and gasification. Combust Flame. 2014;161(6):1474–88. https://doi.org/10.1016/j.combustflame.2013.11.024.
18. Zheng M, Wang Z, Li X, Qiao X, Song W, Guo L. Initial reaction mechanisms of cellulose pyrolysis revealed by ReaxFF molecular dynamics. Fuel. 2016;177:130–41. https://doi.org/10.1016/j.fuel.2016.03.008.
19. Anca-Couce A, Mehrabian R, Scharler R, Obernberger I. Kinetic scheme of biomass pyrolysis considering secondary charring reactions. Energy Convers Manag. 2014;87:687–96. https://doi.org/10.1016/j.enconman.2014.07.061.
20. Yang H, Yan R, Chen H, Lee DH, Zheng C. Characteristics of hemicellulose, cellulose and lignin pyrolysis. Fuel. 2007;86(12–13):1781–8. https://doi.org/10.1016/j.fuel.2006.12.013.
21. Wu S, Shen D, Hu J, Zhang H, Xiao R. Cellulose-lignin interactions during fast pyrolysis with different temperatures and mixing methods. Biomass Bioenergy. 2016;90:209–17. https://doi.org/10.1016/j.biombioe.2016.04.012.
22. Nowakowski DJ, Bridgwater AV, Elliott DC, Meier D, de Wild P. Lignin fast pyrolysis: results from an international collaboration. J Anal Appl Pyrolysis. 2010;88(1):53–72. https://doi.org/10.1016/j.jaap.2010.02.009.
23. Vyazovkin S, Chrissafis K, Di Lorenzo ML, Koga N, Pijolat M, Roduit B, Sbirrazzuoli N, Suñol JJ. ICTAC Kinetics Committee recommendations for collecting experimental thermal analysis data for kinetic computations. Thermochim Acta. 2014;590:1–23. https://doi.org/10.1016/j.tca.2014.05.036.
24. Di Blasi C. Modeling chemical and physical processes of wood and biomass pyrolysis. Prog Energy Combust Sci. 2008;34(1):47–90. https://doi.org/10.1016/j.pecs.2006.12.001.
25. Liden AG, Berruti F, Scott DS. A kinetic model for the production of liquids from flash pyrolysis of biomass. Chem Eng Commun. 1988;65(1):207–21. https://doi.org/10.1080/00986448808940254.
26. Shafizadeh F, Chin PPS. Thermal deterioration of wood. In: Goldstein IS, editor. Wood technology: chemical aspects, ACS symposium series, vol. 43. Washington DC: American Chemical Society; 1977. p. 57–81. https://doi.org/10.1021/bk-1977-0043.ch005.
27. Anca-Couce A, Berger A, Zobel N. How to determine consistent biomass pyrolysis kinetics in a parallel reaction scheme. Fuel. 2014;123:230–40. https://doi.org/10.1016/j.fuel.2014.01.014.
28. Park WC, Atreya A, Baum HR. Experimental and theoretical investigation of heat and mass transfer processes during wood pyrolysis. Combust Flame. 2010;157(3):481–94. https://doi.org/10.1016/j.combustflame.2009.10.006.

29. Koufopanos CA, Lucchesi A, Maschio G. Kinetic modelling of the pyrolysis of biomass and biomass components. The. Can J Chem Eng. 1989;67(1):75–84. https://doi.org/10.1002/cjce.5450670111.
30. Ranzi E, Cuoci A, Faravelli T, Frassoldati A, Migliavacca G, Pierucci S, Sommariva S. Chemical kinetics of biomass pyrolysis. Energy Fuel. 2008;22(6):4292–300. https://doi.org/10.1021/ef800551t.
31. Cuoci A, Faravelli T, Frassoldati A, Granata S, Migliavacca G, Ranzi E. A general mathematical model of biomass devolatilization. Note 1. Lumped kinetic models of cellulose, hemicellulose and lignin. In: Scala F, editor. Proceedings of the 30th combustion meeting of the Italian Section of the Combustion Institute. Milan: ASICI; 2007. p. 2.1–6.
32. Cuoci A, Faravelli T, Frassoldati A, Granata S, Migliavacca G, Pierucci S. A general mathematical model of biomass devolatilization. Note 2. Detailed kinetics of volatile species. In: Scala F, editor. Proceedings of the 30th combustion meeting of the Italian Section of the Combustion Institute. Milan: ASICI; 2007. p. 3.1–6.
33. Corbetta M, Frassoldati A, Bennadji H, Smith K, Serapiglia MJ, Gauthier G, Melkior T, Ranzi E, Fisher EM. Pyrolysis of centimeter-scale woody biomass particles: kinetic modeling and experimental validation. Energy Fuel. 2014;28(6):3884–98. https://doi.org/10.1021/ef500525v.
34. Ranzi E, Debiagi PEA, Frassoldati A. Mathematical modeling of fast biomass pyrolysis and bio-oil formation. Note I: kinetic mechanism of biomass pyrolysis. ACS Sustain Chem Eng. 2017;5(4):2867–81. https://doi.org/10.1021/acssuschemeng.6b03096.
35. Ranzi E, Debiagi PEA, Frassoldati A. Mathematical modeling of fast biomass pyrolysis and bio-oil formation. Note II: secondary gas-phase reactions and bio-oil formation. ACS Sustain Chem Eng. 2017;5(4):2882–96. https://doi.org/10.1021/acssuschemeng.6b03098.
36. Debiagi P, Gentile G, Cuoci A, Frassoldati A, Ranzi E, Faravelli T. A predictive model of biochar formation and characterization. J Anal Appl Pyrolysis. 2018;134:326–35. https://doi.org/10.1016/j.jaap.2018.06.022.
37. Anca-Couce A, Scharler R. Modelling heat of reaction in biomass pyrolysis with detailed reaction schemes. Fuel. 2017;206:572–9. https://doi.org/10.1016/j.fuel.2017.06.011.
38. Anca-Couce A, Obernberger I. Application of a detailed biomass pyrolysis kinetic scheme to hardwood and softwood torrefaction. Fuel. 2016;167:158–67. https://doi.org/10.1016/j.fuel.2015.11.062.
39. Rath J, Wolfinger MG, Steiner G, Krammer G, Barontini F, Cozzani V. Heat of wood pyrolysis. Fuel. 2003;82(1):81–91. https://doi.org/10.1016/S0016-2361(02)00138-2.
40. Blondeau J, Jeanmart H. Biomass pyrolysis at high temperatures: prediction of gaseous species yields from an anisotropic particle. Biomass Bioenergy. 2012;41:107–21. https://doi.org/10.1016/j.biombioe.2012.02.016.
41. Mellin P, Kantarelis E, Yang W. Computational fluid dynamics modeling of biomass fast pyrolysis in a fluidized bed reactor, using a comprehensive chemistry scheme. Fuel. 2014;117:704–15. https://doi.org/10.1016/j.fuel.2013.09.009.
42. Anca-Couce A, Sommersacher P, Scharler R. Online experiments and modelling with a detailed reaction scheme of single particle biomass pyrolysis. J Anal Appl Pyrolysis. 2017;127:411–25. https://doi.org/10.1016/j.jaap.2017.07.008.
43. Norinaga K, Shoji T, Kudo S, Hayashi J-I. Detailed chemical kinetic modelling of vapour-phase cracking of multi-component molecular mixtures derived from the fast pyrolysis of cellulose. Fuel. 2013;103:141–50. https://doi.org/10.1016/j.fuel.2011.07.045.
44. Trendewicz A, Evans R, Dutta A, Sykes R, Carpenter D, Braun R. Evaluating the effect of potassium on cellulose pyrolysis reaction kinetics. Biomass Bioenergy. 2015;74:15–25. https://doi.org/10.1016/j.biombioe.2015.01.001.
45. Versteeg HK, Malalasekera W. An introduction to computational fluid dynamics: the finite volume method. 3rd ed. Harlow: Pearson Education Ltd.; 2011.
46. Tu J, Yeoh GH, Liu C. Computational fluid dynamics: a practical approach. 3rd ed. Oxford: Butterworth-Heinemann; 2018.

47. Childress S. An introduction to theoretical fluid mechanics. New York: AMS and the Courant Institute of Mathematical Sciences; 2009.
48. Batchelor GK. An introduction to fluid dynamics. Cambridge: Cambridge University Press; 2000. https://doi.org/10.1017/CBO9780511800955.
49. Whitaker S. Simultaneous heat, mass, and momentum transfer in porous media: a theory of drying. Adv Heat Tran. 1977;13:119–203. https://doi.org/10.1016/S0065-2717(08)70223-5.
50. Perré P, Turner IW. A 3-D version of TransPore: a comprehensive heat and mass transfer computational model for simulating the drying of porous media. Int J Heat Mass Transf. 1999;42(24):4501–21. https://doi.org/10.1016/S0017-9310(99)00098-8.
51. Nasrallah SB, Perre P. Detailed study of a model of heat and mass transfer during convective drying of porous media. Int J Heat Mass Transf. 1988;31(5):957–67. https://doi.org/10.1016/0017-9310(88)90084-1.
52. Grønli MG (1996) A theoretical and experimental study of thermal degradation of biomass (Ph.D. thesis). NTNU, Trondheim, Norway
53. Gentile G, Debiagi PEA, Cuoci A, Frassoldati A, Ranzi E, Faravelli T. A computational framework for the pyrolysis of anisotropic biomass particles. Chem Eng J. 2017;321:458–73. https://doi.org/10.1016/j.cej.2017.03.113.
54. Jalili M, Anca-Couce A, Zobel N. On the uncertainty of a mathematical model for drying of a wood particle. Energy Fuel. 2013;27(11):6705–17. https://doi.org/10.1021/ef401156s.
55. Melaaen MC. Numerical analysis of heat and mass transfer in drying and pyrolysis of porous media. Numer Heat Transf Pt A Appl. 1996;29(4):331–55. https://doi.org/10.1080/10407789608913796.
56. Anca-Couce A, Zobel N. Numerical analysis of a biomass pyrolysis particle model: solution method optimized for the coupling to reactor models. Fuel. 2012;97:80–8. https://doi.org/10.1016/j.fuel.2012.02.033.
57. Fatehi H, Bai XS. A comprehensive mathematical model for biomass combustion. Combust Sci Technol. 2014;186(4-5):574–93. https://doi.org/10.1080/00102202.2014.883255.
58. Shi X (2017) Computational fluid dynamics modelling of biomass slow pyrolysis in screw reactors for the production of biochar and charcoal (Ph.D. Thesis). Ghent University, Ghent
59. Chan W-CR, Kelbon M, Krieger BB. Modelling and experimental verification of physical and chemical processes during pyrolysis of a large biomass particle. Fuel. 1985;64(11):1505–13. https://doi.org/10.1016/0016-2361(85)90364-3.
60. Shrestha D, Cramer S, White R. Time-temperature profile across a lumber section exposed to pyrolytic temperatures. Fire Mater. 1994;18(4):211–20. https://doi.org/10.1002/fam.810180404.
61. Bryden KM, Hagge MJ. Modeling the combined impact of moisture and char shrinkage on the pyrolysis of a biomass particle. Fuel. 2003;82(13):1633–44. https://doi.org/10.1016/S0016-2361(03)00108-X.
62. Di Blasi CD, Branca C, Sparano S, La Mantia B. Drying characteristics of wood cylinders for conditions pertinent to fixed-bed countercurrent gasification. Biomass Bioenergy. 2003;25(1):45–58. https://doi.org/10.1016/S0961-9534(02)00180-0.
63. Peters B, Bruch C. Drying and pyrolysis of wood particles: experiments and simulation. J Anal Appl Pyrolysis. 2003;70(2):233–50. https://doi.org/10.1016/S0165-2370(02)00134-1.
64. Haberle I, Haugen NEL, Skreiberg Ø. Drying of thermally thick wood particles: a study of the numerical efficiency, accuracy, and stability of common drying models. Energy Fuel. 2017;31(12):13743–60. https://doi.org/10.1021/acs.energyfuels.7b02771.
65. Lu H, Robert W, Peirce G, Ripa B, Baxter LL. Comprehensive study of biomass particle combustion. Energy Fuel. 2008;22(4):2826–39. https://doi.org/10.1021/ef800006z.
66. Thunman H, Leckner B, Niklasson F, Johnsson F. Combustion of wood particles—a particle model for eulerian calculations. Combust Flame. 2002;129(1):30–46. https://doi.org/10.1016/S0010-2180(01)00371-6.
67. Raznjevic K. Handbook of thermodynamic tables and charts. Washington DC: Hemisphere Publishing; 1976.

68. Laboratory USFP. The encyclopedia of wood: wood as an engineering material. Agriculture handbook, vol. 72. London: Oak Tree Press; 1980.
69. Olek W, Perré P, Weres J. Inverse analysis of the transient bound water diffusion in wood. Holzforschung. 2005;59(1):38–45. https://doi.org/10.1515/HF.2005.007.
70. Moghtaderi B, Dlugogorski BZ, Kennedy EM, Fletcher DF. Effects of the structural properties of solid fuels on their re-ignition characteristics. Fire Mater. 1999;22(4):155–65. https://doi.org/10.1002/(SICI)1099-1018(1998070)22:4<155::AID-FAM651>3.0.CO;2-F.
71. Di Blasi C. Multi-phase moisture transfer in the high-temperature drying of wood particles. Chem Eng Sci. 1998;53(2):353–66. https://doi.org/10.1016/S0009-2509(97)00197-8.
72. Stanish MA, Schajer GS, Kayihan F. A mathematical model of drying for hygroscopic porous media. AIChE J. 1986;32(8):1301–11. https://doi.org/10.1002/aic.690320808.
73. Singh B. Biochar - a guide to analytical methods. 1st ed. Melbourne: CRC Press; 2017.
74. Plötze M, Niemz P. Porosity and pore size distribution of different wood types as determined by mercury intrusion porosimetry. Eur J Wood Wood Prod. 2011;69(4):649–57. https://doi.org/10.1007/s00107-010-0504-0.
75. Brewer CE, Chuang VJ, Masiello CA, Gonnermann H, Gao X, Dugan B, Driver LE, Panzacchi P, Zygourakis K, Davies CA. New approaches to measuring biochar density and porosity. Biomass Bioenergy. 2014;66:176–85. https://doi.org/10.1016/j.biombioe.2014.03.059.
76. Zauer M, Pfriem A, Wagenführ A. Toward improved understanding of the cell-wall density and porosity of wood determined by gas pycnometry. Wood Sci Technol. 2013;47(6):1197–211. https://doi.org/10.1007/s00226-013-0568-1.
77. Brown RA, Kercher AK, Nguyen TH, Nagle DC, Ball WP. Production and characterization of synthetic wood chars for use as surrogates for natural sorbents. Org Geochem. 2006;37(3):321–33. https://doi.org/10.1016/j.orggeochem.2005.10.008.
78. Emmett PH. Adsorption and pore-size measurements on charcoals and whetlerites. Chem Rev. 1948;43(1):69–148. https://doi.org/10.1021/cr60134a003.
79. Zelinka SL, Glass SV, Jakes JE, Stone DS. A solution thermodynamics definition of the fiber saturation point and the derivation of a wood–water phase (state) diagram. Wood Sci Technol. 2016;50(3):443–62. https://doi.org/10.1007/s00226-015-0788-7.
80. Gezici-Koç Ö, Erich SJF, Huinink HP, van der Ven LGJ, Adan OCG. Bound and free water distribution in wood during water uptake and drying as measured by 1D magnetic resonance imaging. Cellulose. 2017;24(2):535–53. https://doi.org/10.1007/s10570-016-1173-x.
81. Stamm AJ. Review of nine methods for determining the fiber saturation points of wood and wood products. Wood Sci. 1971;4:114–28.
82. Skaar C. Wood-water relations, vol IV. Springer series in wood science. 1st ed. Berlin: Springer; 1988.
83. Stamm AJ, Seborg RM. Absorption compression on cellulose and wood: I density measurements in benzene. J Phys Chem. 1935;39(1):133–42. https://doi.org/10.1021/j150361a010.
84. Siau JF. Transport processes in wood. Springer series in wood science. Berlin: Springer; 1984.
85. Spolek GA, Plumb OA. Capillary pressure in softwoods. Wood Sci Technol. 1981;15(3):189–99. https://doi.org/10.1007/BF00353471.
86. Perré P, Degiovanni A. Simulation par volumes finis des transferts couplés en milieux poreux anisotropes: séchage du bois à basse et à haute température. Int J Heat Mass Transf. 1990;33(11):2463–78. https://doi.org/10.1016/0017-9310(90)90004-E.
87. Lee CK, Chaiken RF, Singer JM. Charring pyrolysis of wood in fires by laser simulation. Symp Combust. 1977;16(1):1459–70. https://doi.org/10.1016/S0082-0784(77)80428-1.
88. Comstock GL. Directional permeability of softwoods. Wood Fiber Sci. 1970;4(1):283–9.
89. Perré P. Measurements of softwoods' permeability to air: Importance upon the drying model. Int Commun Heat Mass Transf. 1987;14(5):519–29. https://doi.org/10.1016/0735-1933(87)90016-9.
90. Turner IW, Perré P. Vacuum drying of wood with radiative heating: II. Comparison between theory and experiment. AIChE J. 2004;50(1):108–18. https://doi.org/10.1002/aic.10010.

91. Di Blasi C. Heat, momentum and mass transport through a shrinking biomass particle exposed to thermal radiation. Chem Eng Sci. 1996;51(7):1121–32. https://doi.org/10.1016/S0009-2509(96)80011-X.
92. Di Blasi C. Modelling the fast pyrolysis of cellulosic particles in fluid-bed reactors. Chem Eng Sci. 2000;55(24):5999–6013. https://doi.org/10.1016/S0009-2509(00)00406-1.
93. Larfeldt J, Lecher B, Melaaen MC. Modelling and measurements of drying and pyrolysis of large wood particles. In: Bridgwater AV, editor. Progress in thermochemical biomass conversion. Oxford: Blackwell Science Ltd; 2000. https://doi.org/10.1002/9780470694954.ch85.
94. Ronewicz K, Kluska J, Heda Ł, Kardaś D. Chemical and physical properties of pine wood during pyrolysis. Drvna Ind. 2017;68(1):29–36. https://doi.org/10.5552/drind.2017.1617.
95. Perré P, Turner IW. A dual-scale model for describing drier and porous medium interactions. AIChE J. 2006;52(9):3109–17. https://doi.org/10.1002/aic.10918.
96. Couture F, Jomaa W, Puiggali JR. Relative permeability relations: a key factor for a drying model. Transp Porous Media. 1996;23(3):303–35. https://doi.org/10.1007/BF00167101.
97. Perré P, Moser M, Martin M. Advances in transport phenomena during convective drying with superheated steam and moist air. Int J Heat Mass Transf. 1993;36(11):2725–46. https://doi.org/10.1016/0017-9310(93)90093-L.
98. Perré P, Turner I. Determination of the material property variations across the growth ring of softwood for use in a heterogeneous drying model. Part 2. Use of homogenisation to predict bound liquid diffusivity and thermal conductivity. Holzforschung. 2001;55:417–25. https://doi.org/10.1515/HF.2001.069.
99. Stamm AJ. Wood and cellulose science. New York: Ronald Press Co.; 1964.
100. Perré P, Turner I. Determination of the material property variations across the growth ring of softwood for use in a heterogeneous drying model part 1. Capillary pressure, tracheid model and absolute permeability. Holzforschung. 2001;55:318–23. https://doi.org/10.1515/HF.2001.052.
101. Bird RB, Stewart WE, Lightfoot EN. Transport phenomena. 2nd ed. New York: Wiley; 2006.
102. Poling BE, Prausnitz JM, O'Connell JP. The properties of gases and liquids. Boston: McGraw-Hill; 2007.
103. Atreya A, Olszewski P, Chen Y, Baum HR. The effect of size, shape and pyrolysis conditions on the thermal decomposition of wood particles and firebrands. Int J Heat Mass Transf. 2017;107:319–28. https://doi.org/10.1016/j.ijheatmasstransfer.2016.11.051.
104. Janse AMC, Westerhout RWJ, Prins W. Modelling of flash pyrolysis of a single wood particle. Chem Eng Process. 2000;39(3):239–52. https://doi.org/10.1016/S0255-2701(99)00092-6.
105. Bruggeman DAG. Berechnung verschiedener physikalischer Konstanten von heterogenen Substanzen. I. Dielektrizitätskonstanten und Leitfähigkeiten der Mischkörper aus isotropen Substanzen. Ann Phys. 1935;416(7):636–64. https://doi.org/10.1002/andp.19354160705.
106. Millington RJ, Quirk JP. Permeability of porous solids. Trans Faraday Soc. 1961;57:1200–7. https://doi.org/10.1039/TF9615701200.
107. Stanish MA. The roles of bound water chemical potential and gas phase diffusion in moisture transport through wood. Wood Sci Technol. 1986;20(1):53–70. https://doi.org/10.1007/BF00350694.
108. Bonneau P, Puiggali JR. Influence of heartwood-sapwood proportions on the drying kinetics of a board. Wood Sci Technol. 1993;28(1):67–85. https://doi.org/10.1007/BF00193878.
109. Fernandez ML, Howell JR. Convective drying model of southern pine. Dry Technol. 1997;15(10):2343–75. https://doi.org/10.1080/07373939708917365.
110. Dupont C, Chiriac R, Gauthier G, Toche F. Heat capacity measurements of various biomass types and pyrolysis residues. Fuel. 2014;115:644–51. https://doi.org/10.1016/j.fuel.2013.07.086.
111. Gorensek MB, Shukre R, Chen C-C. Development of a thermophysical properties model for flowsheet simulation of biomass pyrolysis processes. ACS Sustain Chem Eng. 2019;7(9):9017–27. https://doi.org/10.1021/acssuschemeng.9b01278.

112. Hunt JF, Gu H, Lebow P. Theoretical thermal conductivity equation for uniform density wood cells. Wood Fiber Sci. 2008;40(2):167–80.
113. WebBook NC (2018). https://webbook.nist.gov/chemistry/
114. McBride BJ, Gordon S. Computer program for calculation of complex chemical equilibrium compositions and applications II. User's manual and program description. Cleveland: NASA, Lewis Research Center; 1996.
115. Davidson TA (1993) A simple and accurate method for calculating viscosity of gaseous mixtures. Report of Investigations 9456 US Dept. of the Interior, Bureau of Mines, Pittsburgh
116. de Paiva Souza MEN, S. A. Heat and mass transfer model in wood chip drying. Wood Fiber Sci. 2000;32(2):153–63.
117. Williams CL, Westover TL, Petkovic LM, Matthews AC, Stevens DM, Nelson KR. Determining thermal transport properties for softwoods under pyrolysis conditions. ACS Sustain Chem Eng. 2017;5(1):1019–25. https://doi.org/10.1021/acssuschemeng.6b02326.
118. Rumble JR, Lide DR, Bruno TJ. CRC handbook of chemistry and physics: a ready-reference book of chemical and physical data. 99th ed. Boca Raton: CRC Press; 2018.
119. Maku T. Studies on the heat conduction in wood, Wood research bulletin, vol. 13. Kyoto: Japan Wood Research Institute, Kyoto University; 1954.
120. Suleiman BM, Larfeldt J, Leckner B, Gustavsson M. Thermal conductivity and diffusivity of wood. Wood Sci Technol. 1999;33(6):465–73. https://doi.org/10.1007/s002260050130.
121. Behazin E. Mechanical, chemical and physical properities of wood and perennial grass biochars for possible composite aplication. Bioresources. 2016;11(1):1334–48.
122. Yaws CL. Handbook of transport property data: viscosity, thermal conductivity, and diffusion coefficients of liquids and gases. Houston: Gulf Pub. Co.; 1995.
123. Vargaftik NB. Handbook of thermal conductivity of liquids and gases. Boca Raton: CRC Press; 1994.
124. Zaitseva LS, Yakush LV, Vanicheva NA. Thermal conductivities of benzene and toluene vapors. J Eng Phys. 1976;31(5):1292–5. https://doi.org/10.1007/BF00859307.
125. Panton RL, Rittmann JG. Pyrolysis of a slab of porous material. Symp Combust. 1971;13(1):881–91. https://doi.org/10.1016/S0082-0784(71)80089-9.
126. Olek W, Weres J, Guzenda R. Effects of thermal conductivity data on accuracy of modeling heat transfer in wood. Holzforschung. 2003;57:3. https://doi.org/10.1515/HF.2003.047.
127. Thunman H, Leckner B. Thermal conductivity of wood—models for different stages of combustion. Biomass Bioenergy. 2002;23(1):47–54. https://doi.org/10.1016/S0961-9534(02)00031-4.
128. Kollmann FFP, Côté WA. Principles of wood science and technology: I. Solid wood. Berlin: Springer; 1968. https://doi.org/10.1007/978-3-642-87928-9.
129. Caposciutti G, Almuina-Villar H, Dieguez-Alonso A, Gruber T, Kelz J, Desideri U, Hochenauer C, Scharler R, Anca-Couce A. Experimental investigation on biomass shrinking and swelling behaviour: particles pyrolysis and wood logs combustion. Biomass Bioenergy. 2019;123:1–13. https://doi.org/10.1016/j.biombioe.2019.01.044.
130. Bellais M (2007) Modelling of the pyrolysis of large wood particles (Ph.D. Thesis). KTH - Royal Institute of Technology, Stockholm, Sweden
131. Perré P (1999) How to get a relevant material model for wood drying simulation? In: First COST Action E15, 1st Workshop "State of the art for Kiln drying". Advances in the drying of wood (1999-2003), Edinburgh, Scotland
132. Pattanotai T, Watanabe H, Okazaki K. Gasification characteristic of large wood chars with anisotropic structure. Fuel. 2014;117:331–9. https://doi.org/10.1016/j.fuel.2013.09.030.
133. Davidsson KO, Pettersson JBC. Birch wood particle shrinkage during rapid pyrolysis. Fuel. 2002;81(3):263–70. https://doi.org/10.1016/S0016-2361(01)00169-7.
134. Çengel YA, Ghajar AJ, Kanoglu M. Heat and mass transfer: fundamentals and Applications. 5th ed. New York: McGraw-Hill Education; 2014.

135. Botterill JSM. Heat transfer to gas-fluidized beds. Powder Technol. 1970;4(1):19–26. https://doi.org/10.1016/0032-5910(70)80003-1.
136. Holman JP. Heat transfer (metric version). 10th ed. Boston: McGraw-Hill; 2009.
137. Di Natale F, Lancia A, Nigro R. Surface-to-bed heat transfer in fluidised beds: effect of surface shape. Powder Technol. 2007;174(3):75–81. https://doi.org/10.1016/j.powtec.2007.01.010.
138. Di Natale F, Lancia A, Nigro R. A single particle model for surface-to-bed heat transfer in fluidized beds. Powder Technol. 2008;187(1):68–78. https://doi.org/10.1016/j.powtec.2008.01.014.
139. Di Natale F, Nigro R. A critical comparison between local heat and mass transfer coefficients of horizontal cylinders immersed in bubbling fluidised beds. Int J Heat Mass Transf. 2012;55(25):8178–83. https://doi.org/10.1016/j.ijheatmasstransfer.2012.08.002.
140. Prins W (1987) Fluidized bed combustion of a single carbon particle (Ph.D. Thesis). University of Twente, Enschede
141. Babu BV, Chaurasia AS. Pyrolysis of biomass: improved models for simultaneous kinetics and transport of heat, mass and momentum. Energy Convers Manag. 2004;45(9):1297–327. https://doi.org/10.1016/j.enconman.2003.09.013.
142. Schröder E, Class A, Krebs L. Measurements of heat transfer between particles and gas in packed beds at low to medium Reynolds numbers. Exp Thermal Fluid Sci. 2006;30(6):545–58. https://doi.org/10.1016/j.expthermflusci.2005.11.002.
143. Kersten SRA, Wang X, Prins W, van Swaaij WPM. Biomass pyrolysis in a fluidized bed reactor. Part 1: literature review and model simulations. Ind Eng Chem Res. 2005;44(23):8773–85. https://doi.org/10.1021/ie0504856.
144. Sakai M. How should the discrete element method be applied in industrial systems? A review. KONA Powder Part J. 2016;33:169–78. https://doi.org/10.14356/kona.2016023.
145. Kloss C, Goniva C, Aichinger G, Pirker S (2009) Comprehensive DEM-DPM-CFD simulations-model synthesis, experimental validation and scalability. In: 7th International Conference on CFD in the Minerals and Process Industries, Melbourne, Australia. pp 1–7
146. Mettler MS, Mushrif SH, Paulsen AD, Javadekar AD, Vlachos DG, Dauenhauer PJ. Revealing pyrolysis chemistry for biofuels production: conversion of cellulose to furans and small oxygenates. Energy Environ Sci. 2012;5(1):5414–24. https://doi.org/10.1039/C1EE02743C.
147. Paulsen AD, Mettler MS, Dauenhauer PJ. The role of sample dimension and temperature in cellulose pyrolysis. Energy Fuel. 2013;27(4):2126–34. https://doi.org/10.1021/ef302117j.
148. Pyle DL, Zaror CA. Heat transfer and kinetics in the low temperature pyrolysis of solids. Chem Eng Sci. 1984;39(1):147–58. https://doi.org/10.1016/0009-2509(84)80140-2.
149. Mehrabian R, Zahirovic S, Scharler R, Obernberger I, Kleditzsch S, Wirtz S, Scherer V, Lu H, Baxter LL. A CFD model for thermal conversion of thermally thick biomass particles. Fuel Process Technol. 2012;95:96–108. https://doi.org/10.1016/j.fuproc.2011.11.021.
150. Villermaux J, Antoine B, Lede J, Soulignac F. A new model for thermal volatilization of solid particles undergoing fast pyrolysis. Chem Eng Sci. 1986;41(1):151–7. https://doi.org/10.1016/0009-2509(86)85208-3.
151. Wurzenberger JC, Wallner S, Raupenstrauch H, Khinast JG. Thermal conversion of biomass: comprehensive reactor and particle modeling. AIChE J. 2004;48(10):2398–411. https://doi.org/10.1002/aic.690481029.
152. Mahmoudi AH, Hoffmann F, Peters B. Detailed numerical modeling of pyrolysis in a heterogeneous packed bed using XDEM. J Anal Appl Pyrolysis. 2014;106:9–20. https://doi.org/10.1016/j.jaap.2013.12.001.
153. Anca-Couce A, Zobel N, Jakobsen HA. Multi-scale modeling of fixed-bed thermo-chemical processes of biomass with the representative particle model: Application to pyrolysis. Fuel. 2013;103:773–82. https://doi.org/10.1016/j.fuel.2012.05.063.
154. Xiong Q, Yang Y, Xu F, Pan Y, Zhang J, Hong K, Lorenzini G, Wang S. Overview of computational fluid dynamics simulation of reactor-scale biomass pyrolysis. ACS Sustain Chem Eng. 2017;5(4):2783–98. https://doi.org/10.1021/acssuschemeng.6b02634.

155. Subramaniam S. Lagrangian–Eulerian methods for multiphase flows. Prog Energy Combust Sci. 2013;39(2):215–45. https://doi.org/10.1016/j.pecs.2012.10.003.
156. Ku X, Li T, Løvås T. Influence of drag force correlations on periodic fluidization behavior in Eulerian–Lagrangian simulation of a bubbling fluidized bed. Chem Eng Sci. 2013;95:94–106. https://doi.org/10.1016/j.ces.2013.03.038.
157. Ku X, Li T, Løvås T. CFD–DEM simulation of biomass gasification with steam in a fluidized bed reactor. Chem Eng Sci. 2015;122:270–83. https://doi.org/10.1016/j.ces.2014.08.045.
158. Xie J, Zhong W, Jin B, Shao Y, Huang Y. Eulerian–Lagrangian method for three-dimensional simulation of fluidized bed coal gasification. Adv Powder Technol. 2013;24(1):382–92. https://doi.org/10.1016/j.apt.2012.09.001.
159. Funke A, Grandl R, Ernst M, Dahmen N. Modelling and improvement of heat transfer coefficient in auger type reactors for fast pyrolysis application. Chem Eng Process. 2018;130:67–75. https://doi.org/10.1016/j.cep.2018.05.023.
160. Jurtz N, Kraume M, Wehinger Gregor D. Advances in fixed-bed reactor modeling using particle-resolved computational fluid dynamics (CFD). Rev Chem Eng. 2018;35:139–90. https://doi.org/10.1515/revce-2017-0059.

Index

Z. Fang et al. (eds.), *Production of Biofuels and Chemicals with Pyrolysis*, Biofuels and Biorefineries 10, https://doi.org/10.1007/978-981-15-2732-6

D

K

L

M

N

O

T

The manufacturer's authorised representative in the EU is Springer Nature Customer Service Centre GmbH, Europaplatz 3, 69115 Heidelberg, Germany. If you have any concerns regarding our products, please contact ProductSafety@springernature.com

Printed and bound by CPI Group (UK) Ltd, Croydon, CR0 4YY

30/09/2026

02225330-0016